FOURTH EDITION

Principles of
Environmental
SCIENCE

Inquiry & Applications

William P. Cunningham
University of Minnesota

Mary Ann Cunningham
Vassar College

Higher Education

Boston   Burr Ridge, IL   Dubuque, IA   New York   San Francisco   St. Louis
Bangkok   Bogotá   Caracas   Kuala Lumpur   Lisbon   London   Madrid   Mexico City
Milan   Montreal   New Delhi   Santiago   Seoul   Singapore   Sydney   Taipei   Toronto

Higher Education

PRINCIPLES OF ENVIRONMENTAL SCIENCE: INQUIRY & APPLICATIONS, FOURTH EDITION

Published by McGraw-Hill, a business unit of The McGraw-Hill Companies, Inc., 1221 Avenue of the Americas, New York, NY 10020. Copyright © 2008 by The McGraw-Hill Companies, Inc. All rights reserved. No part of this publication may be reproduced or distributed in any form or by any means, or stored in a database or retrieval system, without the prior written consent of The McGraw-Hill Companies, Inc., including, but not limited to, in any network or other electronic storage or transmission, or broadcast for distance learning.

Some ancillaries, including electronic and print components, may not be available to customers outside the United States.

♻ This book is printed on recycled, acid-free paper containing 10% postconsumer waste.

1 2 3 4 5 6 7 8 9 0 QPD/QPD 0 9 8 7 6

ISBN 978–0–07–305089–8
MHID 0–07–305089–X

Publishers: *Thomas D. Timp/Margaret J. Kemp*
Senior Developmental Editor: *Joan M. Weber*
Outside Developmental Services: *Brian S. Loehr*
Senior Marketing Manager: *Tami Petsche*
Project Manager: *April R. Southwood*
Senior Production Supervisor: *Laura Fuller*
Lead Media Project Manager: *Judi David*
Media Producer: *Daniel M. Wallace*
Designer: *John Joran*
Cover/Interior Designer: *Jamie E. O'Neal*
(USE) Cover Image: *© David L. Hansen, University of Minnesota, Agricultural Experiment Station*
Senior Photo Research Coordinator: *Lori Hancock*
Photo Research: *LouAnn K. Wilson*
Compositor: *Techbooks*
Typeface: *10/12 Times Roman*
Printer: *Quebecor World Dubuque, IA*

The credits section for this book begins on page 393 and is considered an extension of the copyright page.

Library of Congress Cataloging-in-Publication Data

Cunningham, William P.
  Principles of environmental science : inquiry & applications / William Cunningham, Mary Ann Cunningham.—4th ed.
      p. cm.
  Includes index.
  ISBN 978–0–07–305089–8 — ISBN 0–07–305089–X (hard copy : alk. paper)
  1. Environmental sciences—Textbooks. I. Cunningham, Mary Ann. II. Title.

GE105.C865   2008b
363.7—dc22
                                            2006020498
                                            CIP

www.mhhe.com

WILLIAM P. CUNNINGHAM

William P. Cunningham is an emeritus professor at the University of Minnesota. In his 38-year career at the university, he taught a variety of biology courses, including Environmental Science, Conservation Biology, Environmental Health, Environmental Ethics, Plant Physiology, General Biology, and Cell Biology. He is a member of the Academy of Distinguished Teachers, the highest teaching award granted at the University of Minnesota. He was a member of a number of interdisciplinary programs for international students, teachers, and nontraditional students. He also carried out research or taught in Sweden, Norway, Brazil, New Zealand, China, and Indonesia.

Professor Cunningham has participated in a number of governmental and nongovernmental organizations over the past 40 years. He was chair of the Minnesota chapter of the Sierra Club, a member of the Sierra Club national committee on energy policy, vice president of the Friends of the Boundary Waters Canoe Area, chair of the Minnesota governor's task force on energy policy, and a citizen member of the Minnesota Legislative Commission on Energy.

In addition to environmental science textbooks, Cunningham edited three editions of an *Environmental Encyclopedia* published by Thompson-Gale Press. He has also authored or co-authored about 50 scientific articles, mostly in the fields of cell biology and conservation biology as well as several invited chapters or reports in the areas of energy policy and environmental health. His Ph.D. from the University of Texas was in botany.

His hobbies include backpacking, canoe and kayak building (and paddling), birding, hiking, gardening, and traveling. He lives in St. Paul, Minnesota with his wife, Mary. He has three children (one of whom is co-author of this book) and seven grandchildren.

MARY ANN CUNNINGHAM

Mary Ann Cunningham is an Assistant Professor of Geography at Vassar College, in New York's Hudson Valley. A biogeographer with interests in landscape ecology, geographic information systems (GIS), and remote sensing, she teaches environmental science, natural resource conservation, and land-use planning, as well as GIS and remote sensing. Field research methods, statistical methods, and scientific methods in data analysis are regular components of her teaching. As a scientist and educator, Mary Ann enjoys teaching and conducting research with both science students and non-science liberal arts students. As a geographer, she likes to engage students with the ways their physical surroundings and social context shape their world experience. In addition to teaching at a liberal arts college, she has taught at community colleges and research universities.

Mary Ann has been writing in environmental science for over a decade, and she has been co-author of this book since its first edition. She is also co-author of *Environmental Science* (now in its ninth edition), and an editor of the *Environmental Encyclopedia* (third edition, Thompson-Gale Press). She has published work on pedagogy in cartography, as well as instructional and testing materials in environmental science. With colleagues at Vassar, she has published a GIS lab manual, *Exploring Environmental Science with GIS,* designed to provide students with an easy, inexpensive introduction to spatial and environmental analysis with GIS.

In addition to environmental science, Mary Ann's primary research activities focus on land-cover change, habitat fragmentation, and distributions of bird populations. This work allows her to conduct field studies in the grasslands of the Great Plains as well as in the woodlands of the Hudson Valley. In her spare time she loves to travel, hike, and watch birds.

Mary Ann holds a bachelor's degree from Carleton College, a master's degree from the University of Oregon, and a Ph.D. from the University of Minnesota.

Brief Contents

Contents

CHAPTER 11 **Environmental Geology and Earth Resources** 263

CHAPTER 12 **Energy** 283

List of Case Studies

Preface

YOU CAN MAKE A DIFFERENCE

"Ask not what your country can do for you; ask what you can do for your country." This stirring challenge from President John F. Kennedy's 1961 inaugural address calls on all of us to contribute to our communities and to the needs of future generations. Increasingly, environmental quality is recognized as part of this caretaking task. The environment affects our health, our food and water, our politics, and our economy. What can you do to ensure that these effects are positive? That question is a central theme for this book.

Much has been written in recent years about the failure of environmental advocacy groups and environmental scientists to reach out to the public and convince people that threats to our natural resources are real and important. People claim that environmental protection is important, but few act on this conviction. One cause of this disconnect is that the public has become overwhelmed by gloom and doom stories. There's so much bad news that many people become overwhelmed and apathetic. We sympathize with the desire for some good news for a change. We've always emphasized positive stories and opportunities to take action, but in this edition we've increased the focus on positive actions and the ways environmental science can help us understand and resolve current problems.

As an illustration of how people are taking action to improve their environment, we've chosen a cover photo for the book that shows an outstanding example of how a small group of students have cleaned and restored a wetland on their campus. Besides helping to make the campus more beautiful, this revitalized wetland now provides wildlife habitat, recreational opportunities, and is useful as an outdoor laboratory for science classes. The knowledge gained in these classes provides a basis for restoration projects, but it also takes organization and plain hard work to make it happen. As Margaret Mead once wrote, "Never doubt that a small group of thoughtful, committed people can change the world; indeed it is the only thing that ever has."

Throughout *Principles of Environmental Science,* we examine issues honestly and openly, with a balanced, objective perspective that presents both sides of controversial issues. This book provides a solid foundation in scientific approaches to environmental problems and solutions. We integrate information from a wide range of disciplines from both the natural and social sciences. Although this book is suitable for non-science majors, you will find that it isn't simplistic or condescending. The presentation, while condensed, remains sophisticated and discerning.

WHAT SETS THIS BOOK APART FROM THE REST?

Concise and Affordable

In recent years, environmental science textbooks have gotten bigger, more encyclopedic, and increasingly expensive. Both instructors and students call for a concise and affordable text. This book was written to fill that need. Rather than the 25 to 30 chapters found in most environmental science textbooks, we've limited this book to 15 chapters, or about one chapter per week for a typical semester course. This more concise presentation focuses on key principles, scientific methods and ideas, and life-long learning skills for students. We have also included enough case studies and current events to provide the real-world context for the themes discussed here. Additional case studies and current issues are also available on the book's website to help enrich your course's content. At the same time, the moderate size and price of this book should allow you to add supplementary materials to meet your individual teaching/learning objectives.

The outline of this book follows a topic sequence widely used in many environmental science textbooks and courses, but we know that many instructors choose to organize their courses around their own outlines. We've written each chapter in a way that doesn't assume that students have already read other chapters in any particular order. If instructors prefer a different organization of course topics, chapters can be presented in any order that suits their course's needs.

Active Learning and Critical Thinking

Learning how scientists approach problems can help students develop habits of independent, orderly, and objective thought. But it takes active involvement to master these skills. Throughout this book, we encourage students to practice thinking for themselves. Data and interpretations aren't presented as immutable truths, but rather as evidence to be examined and tested. We try to give a balanced view of controversial topics. Orderly, critical assessment of complex problems is a key part of **scientific literacy,** which is essential for understanding current environmental science. In every chapter, students are invited to practice **critical thinking** and to apply new ideas. We also include case studies demonstrating how scientists have thought about important environmental questions.

Integration and Sustainability

Environmental problems and their solutions occur at the intersection of natural systems and the human systems that manipulate the natural world. In this book we present an **integrated approach** to physical sciences—biology, ecology, geology, air and water resources—and to human systems that affect nature—food and agriculture, population growth, urbanization, environmental health, resource economics, and policy. Although it is tempting to emphasize purely natural systems, we feel that students can never understand why coral reefs are threatened, or why tropical forests are being cut down if they don't know something about the cultural, economic, and political forces that shape our decisions.

This integrated approach is essential if we are to work toward sustainable solutions in our environment. Throughout the book, we present **sustainability** as an ultimate goal for both preserving nature and improving the lives of people everywhere. Sustainability implies that human well-being and environmental health need to be complementary, not contradictory, efforts. The goal of sustainability also requires a global view. We take a **global perspective** in this book because we believe that the most important and difficult environmental problems we now face involve worldwide resources and international institutions. To help students gain geographical literacy, we have included many maps and international case studies in this book, including a valuable set of world maps in the appendices.

Current and Accurate

Throughout this book, we present **up-to-date** art, tables, and graphs with the most current available data. We hope these data will give students an appreciation of the kinds of information available in environmental science. Among the sources we have called upon here are geographic information systems (GIS) data and maps, current census and population data, international news and data sources, and federal data collection agencies.

This text has had the benefit of input from more than 400 researchers, professionals, and instructors who have reviewed this book, or our larger text, *Environmental Science: A Global Concern.* These reviewers have helped us keep the text current and focused. We deeply appreciate their many helpful suggestions and comments.

NEW TO THIS EDITION

The fourth edition of *Principles of Environmental Science: Inquiry and Applications* is the result of extensive analysis of the text and evaluation of input from environmental science instructors who conscientiously reviewed chapters during the revision. We have utilized the constructive comments provided by these professionals in our continuing efforts to enhance the strengths of the text. The following is a list of global changes that have taken place,

along with a description of significantly revised chapter content. To see a more detailed list of chapter-by-chapter changes, please contact your McGraw-Hill sales representative.

Updated Case Studies

Every chapter opens with a case study (nearly all of them are new to this edition) that shows how the science presented in that chapter is of current interest. But even more, we've chosen positive examples for almost all those case studies in which people are taking action to repair and improve some aspect of their environment. In addition, each case study has a new **locator map** that highlights the area of interest being discussed.

Revised Art Program

Numerous new photos have been added throughout the text to depict real-life situations. Several illustrations, graphs, and charts are new or revised to present realistic and current information in a form that is easier to comprehend than if that same material were presented in text form.

Redesigned Beginning-of-Chapter Objectives

Each chapter opens with a **list of objectives** that will help students organize study priorities. Rather than being imperative requirements, these objectives have been changed to more active questions that lead rather than command.

New and Redesigned End-of-Chapter Study Tools

- ***New Data Analysis Exercises.*** The first step toward making change is acquiring the necessary skills and background. Because so many students need practice analyzing data, graphing data, and thinking analytically, we have redesigned the end-of-chapter exercises as data analysis projects. In each exercise, students are asked to create or analyze graphs, to map or inspect a data set, or take other steps to analyze data.

- ***New Critical Thinking and Discussion Questions.*** A more challenging, open-ended set of questions has been added to encourage students to think more deeply and independently about issues and principles presented in the chapter. These critical thinking and discussion questions make excellent starting points for discussion sections. They also could be used to practice for essay exams, or might even serve as an essay exam themselves.

- ***Additional Review Questions.*** We've **doubled** the number of review questions to provide more comprehensive coverage of important topics in each chapter. Answers to review questions will be posted to the text-specific website.

- ***Relocation of Chapter Summary and Key Terms.*** Both of these tools have been moved to the text-specific website to add more boldface terms and expand the list of review questions to make it a more thorough review of the chapter contents.

New Chapter Content

- Chapter 2 had a major revision of the section on matter, energy, chemistry, biomolecules, and cells. Additionally, we added a new section on ecological systems, and expanded the discussion on DNA.

- Chapter 3 was revised heavily to include a strengthened discussion of evolution, speciation, taxonomy, species interaction, keystone species, community complexity, and ecological disturbance.

- Chapter 5 also had a major revision on the discussion on biomes with a greatly expanded section on coastlines, reefs, mangroves, estuaries, salt marshes, and freshwater ecosystems.

- In chapter 6, we revised the section on world forests and grasslands; rewrote the section on subsidized logging on public lands; rewrote the section on habitat fragmentation and landscape ecology; and added several new boxes.

- In chapter 9, we added a new section on New Source Review and the 2005 Clean Air Interstate Rule; rewrote the section on evidence for climate change, including the 2005 hurricane season; and updated the section on controlling greenhouse gas emissions.

- Chapter 10 has a new section on dams, reservoirs, and water diversion schemes.

- In chapter 12, we added a discussion on rapid energy demand growth in China and India, and on LNG shipments. We expanded the discussion of energy conservation and renewable energy sources, including new sections on biofuels and wave power.

- Chapter 15 has undergone a major revision and reorganization including much more material on sustainability and what individuals can do, including environmental education, environmental literacy, citizen science, environmental careers, individual choices, green consumerism, working together, and student environmental groups. We rewrote the section on ecosystem management with a focus on Everglades restoration, and the chapter now ends with a new positive message about global sustainability and the key messages from the Millennium Assessment study.

LET US KNOW WHAT YOU THINK

We'd appreciate hearing from both students and instructors about where—and how—we could improve this text. You, the users, are the real test of whether we have accomplished our goal of presenting the principles of environmental science in an engaging and understandable way. Please let us know what you think. We value your comments and suggestions. Please send your recommendations to the Plants/Animals/Environment Division of McGraw-Hill at 2460 Kerper Blvd, Dubuque, Iowa, 52001.

ACKNOWLEDGMENTS

We thank Dr. Kim Chapman, who revised the ecology and population chapters and made many other contributions to this edition. In addition to being an excellent writer, Kim has 25 years of experience in ecological research, conservation planning, wetland remediation, prairie rehabilitation, and natural resource management. During this long career, he has worked as staff ecologist for governmental agencies, environmental groups, and environmental restoration companies. His knowledge and wealth of practical experience in resource management and restoration have been invaluable in helping us find a positive message about our environmental future.

We're also indebted to all the instructors who have reviewed the manuscript and made helpful suggestions, corrections, and recommendations for improving this book. Space does not permit inclusion of all the excellent ideas that were provided, but we will continue to do our best to incorporate the ideas that reviewers have given us. In addition, all of us owe a great debt to the many scholars whose work forms the basis of our understanding of environmental science. We stand on the shoulders of giants. If errors persist in spite of our best efforts to root them out, we accept responsibility.

We express our gratitude to the entire McGraw-Hill book team for their wonderful work in putting together this edition. Brian Loehr and Joan Weber oversaw the developmental stages and have made many creative contributions to this book. Cathy Conroy has done a superb job of copy editing, correcting errors, and improving our prose. April Southwood, production project manager, kept everything running smoothly. Lori Hancock and LouAnn Wilson found excellent photos for us. Laura Fuller, John Joran, Judi David, and Dan Wallace provided excellent production, design, and media assistance. Marge Kemp and Tami Petsche have supported this project with their enthusiasm and creative ideas.

The following individuals provided reviews for this book, and for the previous edition. We thank them for their suggestions.

Fourth Edition

Marilynn Bartels
Black Hawk College

Ruth E. Beattie
University of Kentucky

Geoffrey L. Buckley
Ohio University

Catherine W. Carter
Georgia Perimeter College

Richard Clements
Chattanooga State Technical Community College

Nancy A. Dalman
Cuesta College

Terese Dudek
Kishwaukee College

Robert Harrison
University of Washington

Barbara Hollar
University of Detroit–Mercy

Charles Ide
Western Michigan University

Walter A. Illman
University of Iowa

Robert G. Kremer
The Metropolitan State College of Denver

Ernesto Lasso de la Vega
Edison College

Tammy J. Liles
Bluegrass Community and Technical College

Jeanne Linsdell
San Jose State University

Timothy Lyon
Ball State University

Allan L. Markezich
Black Hawk College

Edward A. Martinez
California State University–Sacramento

Linda Mueller Fitzhugh
Gulf Coast Community College

Bruce Olszewski
San Jose State University

Thomas E. Pliske
Florida International University

Greg Pryor
Francis Marion University

Dawn Ranish
Broward Community College

Carlton L. Rockett
Bowling Green State University

Bruce A. Schulte
Georgia Southern University

Roy Sofield
Chattanooga State Technical Community College

Jana H. Svec
Moraine Valley Community College

Lynda Swander Ochs
Johnson County Community College

S. Kant Vajpayee
The University of Southern Mississippi

Jeff White
Lake Land College

Ray E. Williams
Rio Hondo College

J. Michael Wright
Truckee Meadows Community College

Third Edition

Shannon A. Bliss
Cerro Coso Community College

Robert Buschbacher
University of Florida

Catherine W. Carter
Georgia Perimeter College

Richard Clements
Chattanooga State Technical Community College

Michael L. Denniston
Georgia Perimeter College

Laurie S. Eberhardt
Valparaiso University

Brad C. Fiero
Pima Community College

Steven P. Frysinger
James Madison University–Harrisonburg

Zofia E. Gagnon
Marist College

Carey A. Gazis
Central Washington University

Marcia L. Gillette
Indiana University–Kokomo

Daniel F. Gleason
Georgia Southern University

Lawrence J. Gray
Utah Valley State College

Mark F. Hammer
Wayne State College

Charles Kaminski
Middlesex Community College

Paul Kimball
Northeast Iowa Community College

Ned J. Knight
Linfield College

Jeff Kushner
James Madison University

Matthew Laposata
Kennesaw State University

Kurt M. Leuschner
College of the Desert

Edward M. Lignowski
Holy Family College

Les M. Lynn
Bergen Community College

Timothy F. Lyon
Ball State University

Janet S. MacFall
Elon University

Nancy Jean Mann
Cuesta College

Ken R. Marion
University of Alabama–Birmingham

Thomas C. Moon
California University of Pennsylvania

Michele Morek
Brescia University

Michael J. Neilson
The University of Alabama–Birmingham

Pamela Pape-Lindstrom
Everett Community College

David R. Perault
Lynchburg College

Ervand M. Peterson
Sonoma State University

Julie Phillips
De Anza College

Mark D. Plunkett
Bellevue Community College

John M. Pratte
Kennesaw State University

Carlton Lee Rockett
Bowling Green State University

Robert J. Sager
Pierce College

Ronald L. Sass
Rice University

Jeffry A. Schneider
SUNY–Oswego

Bruce A. Schulte
Georgia Southern University

Julie A. Seiter
Oakland Community College

Jill Singer
Buffalo State College

Denise L. Stetson
Johnson and Wales University

L. Harold Stevenson
McNeese State University

Ronald C. Sundell
Northern Michigan University

John A. Tiedemann
Monmouth University

Michael Toscano
San Joaquin Delta College

Edward R. Wells
Wilson College

Arlene A. Westhoven
Ferris State University

Ray E. Williams
Rio Hondo College

Thomas B. Wilson
University of Arizona

Bruce C. Wyman
McNeese State University

William P. Cunningham
Mary Ann Cunningham

GUIDED TOUR

Science Inquiry and Applications

We have integrated a number of learning tools to help students gain an informed, thoughtful view of our environment.

An introductory **Case Study** at the beginning of each chapter illustrates an important current issue and shows how it relates to practical environmental concerns. These stories also help to demystify scientific investigation and help students understand how scientists study complex issues.

Each case study has a new **Locator Map** that highlights the area of interest being discussed.

Case Study   Cultural Choices and the Rate of Population Growth

The good news is that the human population on earth may never double again. The bad news is that it may take a century to get there. In between, the choices we make individually, and collectively through our political, economic, and cultural actions, will determine whether it will be a pleasant, or not so pleasant, hundred years. The choices are already being made in India, where two different states are taking two very different approaches to population growth.

In 1999, having added more than 180 million people in just a decade, India reached a population of 1 billion humans. If current growth rates persist, India will have 1.63 billion residents in 2050 and will surpass China as the world's most populous country. How will this country, where more than a quarter of its inhabitants live in abject poverty, feed, house, educate, and employ all those being added each year? A fierce debate is taking place about how to control India's population, with ramifications for the

Its annual growth rate of 1.7 percent suggests that India's population could double in 41 years, becoming the world's most populous country by 2050. Indian states have taken different approaches to slow population growth. Despite its poverty, Kerala has a total fertility rate lower than the United States.

first Indian state to do so. Although still one of the poorest places in the world economically, Kerala's fertility rate is comparable to that of many industrialized nations, including the United States. Both women and men have a nearly 100 percent literacy rate and share affordable and accessible health care, family planning, and educational opportunities; therefore, women have only the number of children they want, usually two. The Kerala experience suggests that increased wealth isn't a prerequisite for zero population growth.

Taking a far different path to birth reduction is the neighboring state of Andra Pradesh, which reached a stable growth rate in 2001. Boasting the most dramatic fertility decline of any large Indian state, Andra Pradesh has focused on targeted, strongly enforced sterilization programs. The poor are encouraged—some would say coerced—to be sterilized after having only one or two children. The incentives include cash payments. You might receive 500 rupees—equivalent to $11 (U.S.) or a month's wages for an illiterate farm worker—if you agree to have "the operation." In addition, participants are eligible for better housing, land, wells, and subsidized loans.

Case Study   The Next Pandemic?

Every year one or more new flu virus strains emerge somewhere in the world and quickly spread around the globe. For most people, in most years, flu merely means a few days of feeling terrible. Every couple of decades, however, an especially virulent strain appears. The worst flu **pandemic** (a widespread, deadly epidemic) in history occurred in 1918 when somewhere between 50 and 100 million people are thought to have died.

Now, health officials warn us, a highly contagious form of flu is moving through domestic and wild bird populations in Asia and Europe. First identified in 1997 in Hong Kong, the highly pathogenic H5N1 bird flu has spread to more than 50 countries in Asia, Europe, and Africa. More than 150 million domestic ducks and chickens were culled or died from the disease in its first two years. Of the 175 people known to have contracted this flu strain, at least 105 died, making it a highly deadly disease. As we'll discuss later in this chapter, wild geese and ducks are now spreading this dangerous pathogen between continents and are already infecting poultry.

Dr. David Nabarro, who coordinates the United Nations response to avian influenza, warned that if this virus starts a pandemic, it might kill 150 million people. There are now three times as many people in the world than in 1918, he pointed out; people travel farther and faster, and the H5N1 virus has killed nearly 60 percent of the people it has infected so far, compared with only 1 to 2 percent lethality in 1918. This grim estimate was quickly downplayed by the World Health Organization, which fears spreading panic, but many epidemiologists agree with Dr. Nabarro that the question is not *if* but *when* the next pandemic will occur.

A major breakthrough in understanding this disease occurred in 2005 when a team of U.S. scientists successfully sequenced the entire genome of the 1918 flu virus. They did this using tissue samples from people who died

officials did in 1918. We now have new antiviral drugs that can be mass produced and stockpiled to fight a pandemic. There is hope that having the genome sequence of H5N1 will help create an effective vaccine. World leaders are being more proactive in addressing the current epidemic. And China is being remarkably open and cooperative in dealing with the current bird flu. This is a marked contrast to the situation in 2003 when Severe Acute Respiratory Syndrome (SARS) emerged in China. In that outbreak, the Chinese government denied the existence and severity of the disease for months while it spread to 31 countries around the world and infected at least 8,500 people. Stung by international criticism of their handling of that epidemic, the Chinese are being much more forthcoming now. Instead of imposing an information blackout, the government has been quick to report bird flu developments.

In 2005, China announced plans to vaccinate 14 billion birds, and to ship 45 tons of vaccine to Vietnam. Some veterinary scientists doubt whether this plan can be effective. They point out that inoculation will prevent birds from dying, but won't prevent them from being infected and acting as a reservoir for the virus. Furthermore, most poultry in China is raised in small flocks in the backyards of rural homes. Reaching all those birds may be impossible, especially when poor farmers are suspicious of the government and would rather eat dead birds than report them to officials. Some experts call for a shutdown of all backyard poultry production and elimination of live-bird markets where animals are killed under very unsanitary conditions. This would be a severe economic and cultural blow to billions of people.

Domestic livestock represents a huge reservoir of pathogens that can spread to humans. Since 2004, China has killed 150 million ducks and chickens in an effort to contain a deadly new strain of bird flu. Will this be the next great pandemic?

DATA analysis   Extracting Data from a Graph

1. How can you extract data from a line plot? The process is simply the reverse of the way you just learned to create the graph. You draw lines from the axes to where they intersect on the graph at the point whose value you want to know. Let's look at figure 1, the line plot we examined in chapter 1.

• How many cells were there at the half point of the growth curve?

• How does that result compare to the cell population three hours earlier?

• When did the growth of this population start to slow?

A curve, such as this, that increases slowly at first but then rapidly accelerates is called a logistic curve. We'll discuss the mathematics that create this pattern in the chapter 3.

2. How do you extract data from a bar graph? Look again at the bar graph in figure 2. Using the same procedure described for a line plot, draw a horizontal line from the top of any bar straight to the Y-axis. Read the value for the frequency of that category on the Y-axis.

• Why don't you need to draw a vertical line with this type of graph?

• How many cities have 30 ug/m³?

• Why aren't there any bars between 50 and 65 ug/m³?

• Why are there so many more cities at 40 ug/m³?

• Why is one city at 70 ug/m³?

• How many more cities have 40 ug/m³ than 30 ug/m³?

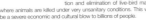

FIGURE 1 Growth of *Tetrahymena*

FIGURE 2 Urban Air Quality

Data Analysis Boxes

The end of each chapter has a new study tool that gives students further opportunities to apply skills and put into practice the knowledge they've gained. In each exercise, students are asked to create or analyze graphs, to map or inspect a data set, or take other steps to analyze data.

Science Inquiry and Applications

What Can You Do?

In keeping with the theme of taking personal responsibility for protecting and restoring our common environment, these boxes give practical suggestions for things individuals can do to help improve environmental quality and resource conservation

What Can You Do?

Reducing Waste

1. Buy foods that come with less packaging; shop at farmers' markets or co-ops, using your own containers.

2. Take your own washable, refillable beverage container to m or convenience stores.

3. When you have a choice at the grocery store among plastic, or metal containers for the same food, buy the reusable or ea recycle glass or metal.

4. Separate your cans, bottles, papers, and plastics for recyclin

5. Wash and reuse bottles, aluminum foil, plastic bags, and so your personal use.

6. Compost yard and garden wastes, leaves, and grass clipping

7. Help your school develop responsible systems for disposing tronics and other waste.

8. Write to your senators and representatives, and urge them to for container deposits, recycling, and safe incinerators or lar

Source: Data from Minnesota Pollution Control Agency.

Applications

These boxes invite students to stop for a moment and practice using the principles they've learned, showing the practical impact of science in our everyday lives.

APPLICATIONS

Comparing Biome Climates

Look back at the climate graphs for San Diego, California, an arid region, and Belém, Brazil, in the Amazon rainforest (fig. 5.4). How much colder is San Diego than Belém in January? In July? Which location has the greater range of temperature through the year? How much do the two locations differ in precipitation during their wettest months?

Compare the temperature and precipitation in these two places with those in the other biomes shown in the pages that follow. How wet are the wettest biomes? Which biomes have distinct dry seasons? How do rainfall and length of warm seasons explain vegetation conditions in these biomes?

Answers: San Diego is about 13°C colder in January, about 6°C colder in July; San Diego has the greater range of temperature; there is about 250 mm difference in precipitation in December–February.

Exploring Science

These essays give students some experience with tools and approaches scientists use to solve contemporary environmental problems. Each boxed essay shows how scientists work to solve complex environmental questions.

Exploring SCIENCE

Carbon-Enrichment Studies

Among the greenhouse gases, none has been studied more than carbon dioxide (CO_2). As you've learned already in this chapter, CO_2 has been analyzed in 400,000-year-old air bubbles taken from glacial ice, measured atop Hawaii's Mauna Loa volcano since 1957, and implicated in a whole host of current climate changes. No one disputes that CO_2 levels in the troposphere increased from 280 to 380 ppm between 1750 and 2005. Few argue that it may exceed 500 ppm—nearly twice its preindustrial concentration—before the twenty-first century ends, but we're not sure what the effects of these high CO_2 levels may be on ecological processes.

Plants need CO_2 to carry out photosynthesis. We know that many plant species flourish in elevated CO_2 levels. That's why greenhouses often pipe in extra CO_2 to accelerate growth. Some horticulturists predict that doubling CO_2 atmospheric concentrations will produce such lush vegetation that the earth will return to a verdant Eden. Others contend that how plants behave in the ideal conditions of a greenhouse or growth chamber is very different from what happens in the real world. How can we know what to expect if current trends continue?

In 1997, Peter Reich and his colleagues began experiments at the University of Minnesota's Cedar Creek Natural History Area that attempt to mimic natural conditions more faithfully than is possible in a growth chamber. At Cedar Creek, the researchers erected several large rings of towers from which CO_2 is released (fig. 1). Sensors track wind speeds and CO_2 concentrations within the ring, while a computer regulates gas flows to maintain 560 ppm throughout the experimental plot. Even when winds are swirling around the research site, the computer can adjust gas flows to keep CO_2 levels constant. Because the CO_2 isn't contained, these experiments are called Free Air Carbon Enrichment (FACE). Control plots have exactly the same design as the experimental ones except that the control towers blow plain air rather than CO_2-enriched air. The Cedar Creek Long-Term Ecological Research is an important center for the study of the relationship between biodiversity and prairie ecosystem stability, so the FACE research also included several different levels of native prairie

FIGURE 1 Student researchers collect samples for analysis at Cedar Creek's Free Air Carbon Enrichment study. The ring of striped poles blow CO_2 over the prairie plants growing in experimental plots.

species richness, as well as different fertilizer treatments, in experimental plots.

Dr. Reich and colleagues discovered that the response of plants to elevated CO_2 and nitrogen fertilization depended on how many species were in a test plot. In a paper published in 2001 in *Nature*, they showed that the amount of plant material (biomass) produced each year rose when CO_2 and nitrogen were added to the test plots. But test plots containing 9 or 16 plant species benefited much more than test plots with a single plant species (fig. 2). In a later paper in the *Proceedings of the National Academy of Sciences*, Dr. Reich and his team also demonstrated that, if test

FIGURE 2 The Cedar Creek BioCON study demonstrates strength in numbers. Single species plots did not respond as well to CO_2 increases as multi-species plots.
Data from Peter Reich.

plots contained ecologically different plants (e.g., a legume and a summer-blooming grass), more biomass developed than in test plots with a single one of those functional plant groups. In other words, vegetation that contained a single plant species responded positively to rising carbon dioxide, but several plant species with different ecological functions responded much better, even without fertilization. The different species and functional groups compensated for each other in the face of a varying climate and environment.

Similar research is being done on forest species and agricultural crops. Peter Curtis, an Ohio State University researcher, writing in the journal *Phytologist*, summarized the results of dozens of FACE experiments on a number of species. In grain crop plants the total seed weight—and therefore food yield—increased by up to 25 percent with a doubling of CO_2 concentration, but in all species except legumes the nutritional value of those plants fell. In one study of tall fescue (*Festuca alatior*), one of the most important pasture grasses in United States, the digestible protein that grazers could extract from the CO_2-stimulated grass actually fell below levels in existing forage, despite fertilization. So, while we may have luxuriant plant growth under elevated CO_2, it may be less valuable for certain uses than what we have today.

In the end, raising CO_2 through human activities represents a huge global experiment not unlike the FACE experiments, but with an important difference. If CO_2 reaches 500 ppm by 2100, we'll be stuck with those conditions—and their effects—for many years. What basic research tells us in the meantime about plant responses, including our important crop plants, will be important in understanding how to adapt to the challenges we face.

Dr. Reich says, "The idea that carbon dioxide will uniformly increase plant growth and therefore provide a large and predictable sink for carbon dioxide is naive at best. This type of model fails to consider the real complexity of the carbon dioxide fertilization effect. Species composition, community diversity, and the amount of nitrogen in the system interact to produce varying responses. In particular, reduced biodiversity and the widespread limited amount of nitrogen in the soil will reduce the CO_2 fertilization effect in terrestrial plant biomass."

Relevant Photos and Instructional Art

Numerous new photos have been added throughout the text to depict real-life situations. We present realistic illustrations that are easy to understand.

FIGURE 3.29 Biomass accumulates at different rates in the world's major ecosystems. Differences in net primary production among the ecosystems is chiefly due to factors limiting plant growth (temperature, rainfall, nutrients, altitude, soil organic matter) but also to interactions among species which boost productivity.

FIGURE 5.1 Biomes most likely to occur in the absence of human disturbance or other disruptions, according to average annual temperature and precipitation. *Note:* This diagram does not consider soil type, topography, wind speed, or other important environmental factors. Still, it is a useful general guideline for biome location.

This map shows the Yellowstone ecosystem complex, or biogeographical region, which extends far beyond the park's boundaries. Park managers and ecologists believe that it is necessary to manage the entire region if the park itself is to remain biologically viable.

Current and Global Coverage

Throughout this book, we present up-to-date art, tables, and graphs with the most current available data.

TABLE 4.3

Life Expectancy at Birth for Selected Countries in 1900 and 2005

Country	1900 Males	1900 Females	2005 Males	2005 Females
India	23	23	61	63
Russia	31	33	59	72
United States	46	48	75	80
Sweden	57	60	78	83
Japan	42	44	78	85

Source: Data from Population Reference Bureau, 2005.

Maps throughout the text serve as references for the whole book and should help students expand their geographical knowledge and global perspective. A large fold out world map is located inside the back cover for easy reference from any page of the text.

Teaching and Learning Supplements

McGraw-Hill offers various tools and technology products to support *Principles of Environmental Science*. Students can order supplemental study materials by contacting their local bookstore or by calling 800-262-4729. Instructors can obtain teaching aids by calling the Customer Service Department at 800-338-3987, visiting the McGraw-Hill website at www.mhhe.com, or by contacting their local McGraw-Hill sales representative.

Teaching Supplements for Instructors

 McGraw-Hill's ARIS—Assessment, Review, and Instruction System

(http://www.mhhe.com/cunningham4e) for *Principles of Environmental Science* is a complete, online tutorial, electronic homework, and course management system, designed for greater ease of use than any other system available. Instructors can create and share course materials and assignments with colleagues with a few clicks of the mouse. All PowerPoint lectures, assignments, quizzes, tutorials, and interactives are directly tied to text-specific materials in *Principles of Environmental Science,* but instructors can also edit questions, import their own content, and create announcements and due dates for assignments. ARIS has automatic grading and reporting of easy-to-assign homework, quizzing, and testing. All student activity within McGraw-Hill's ARIS website is automatically recorded and available to the instructor through a fully integrated grade book that can be downloaded to Excel.

ARIS Presentation Center (found at www.mhhe.com/cunningham4e)

Build instructional materials where-ever, when-ever, and how-ever you want!
ARIS Presentation Center is an online digital library containing assets such as photos, artwork, animations, PowerPoints, and other media types that can be used to create customized lectures, visually enhanced tests and quizzes, compelling course websites, or attractive printed support materials.

Access to your book, access to all books!
The Presentation Center library includes thousands of assets from many McGraw-Hill titles. This ever-growing resource gives instructors the power to utilize assets specific to an adopted textbook as well as content from all other books in the library.

Nothing could be easier!
Accessed from the instructor side of your textbook's ARIS website, Presentation Center's dynamic search engine allows you to explore by discipline, course, textbook chapter, asset type, or keyword. Simply browse, select, and download the files you need to build engaging course materials. All assets are copyright McGraw-Hill Higher Education but can be used by instructors for classroom purposes.

Instructors will find the following digital assets for *Principles of Environmental Science* at ARIS Presentation Center:

- **Color Art** Full-color digital files of ALL illustrations in the text can be readily incorporated into lecture presentations, exams, or custom-made classroom materials. These include all of the 3-D realistic art found in this edition, representing some of the most important concepts in environmental science.
- **Photos** Digital files of ALL photographs from the text can be reproduced for multiple classroom uses.

- **Additional Photos** 444 full-color bonus photographs are available in a separate file. These photos are searchable by content and will add interest and contextual support to your lectures.
- **Tables** Every table that appears in the text is provided in electronic format.
- **Videos** This special collection of 84 underwater video clips displays interesting habitats and behaviors for many animals in the ocean.
- **Animations** 94 full-color animations that illustrate many different concepts covered in the study of environmental science are available for use in creating classroom lectures, testing materials, or online course communication. The visual impact of motion will enhance classroom presentations and increase comprehension.
- **Active Art** These 24 special art pieces consist of key environmental science illustrations that are converted to a format that allows you to break down the art into core elements, and then group the various pieces to create customized images. This is especially helpful with difficult concepts because they can be explained to students step by step.
- **Global Base Maps** Twenty-two base maps for all world regions and major subregions are offered in four versions: black-and-white and full-color, both with labels and without labels. These choices allow instructors the flexibility to plan class activities, quizzing opportunities, study tools, and PowerPoint enhancements.
- **PowerPoint Lecture Outlines** Ready-made presentations that combine art and photos and lecture notes are provided for each of the 15 chapters of the text. These outlines can be used as they are or tailored to reflect your preferred lecture topics and sequences.
- **PowerPoint Slides** For instructors who prefer to create their lectures from scratch, all illustrations, photos, and tables are preinserted by chapter into blank PowerPoint slides for convenience.

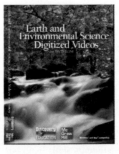 **Earth and Environmental Science DVD by Discovery Channel Education (ISBN: 978-0-07-352541-9; MHID: 0-07-352541-3)**

Begin your class with a quick peek at science in action. The exciting NEW DVD by Discovery Channel Education offers 50 short (3–5 minute) videos on topics ranging from conservation to volcanoes. Search by topic and download into your PowerPoint lecture.

McGraw-Hill's Biology Digitized Videos (ISBN: 978-0-07-312155-0; MHID: 0-07-312155-X)

Licensed from some of the highest quality life-science video producers in the world, these brief video clips on DVD range in length from 15 seconds to two minutes and cover all areas of general biology, from cells to ecosystems.

Engaging and informative, McGraw-Hill's digitized biology videos will help capture students' interest while illustrating key biological concepts, applications, and processes.

Instructor's Testing Resource CD-ROM

This CD-ROM contains a wealth of cross-platform (Windows and Macintosh) resources for the instructor. Supplements featured on this CDROM include a computerized test bank, which utilizes EZ Test software to quickly create customized exams. This flexible and user-friendly program allows instructors to search for questions by topic, format, or difficulty level, and edit existing questions or add new ones. Multiple versions of the test can be created, and any test can be exported for use with course management systems such as WebCT, Blackboard, or PageOut. Word files of the test bank are included for those instructors who prefer to work outside of the test-generator software. Other assets on the Instructor's Testing Resource CD-ROM are grouped within easy-to-use folders.

Transparencies (ISBN: 978-0-07-312794-1; MHID: 0-07-312794-9)

A set of 100 overhead transparencies includes key illustrations and tables from the text. The images are printed with better visibility and contrast than ever before, and labels are large and bold for clear projection.

eInstruction

This classroom performance system (CPS) utilizes wireless technology to bring interactivity into the classroom or lecture hall. Instructors and students receive immediate feedback through wireless response pads that are easy to use and engage students. eInstruction can assist instructors by:

- taking attendance
- administering quizzes and tests
- creating a lecture with intermittent questions
- using the CPS grade book to manage lectures and student comprehension
- integrating interactivity into PowerPoint presentations

Contact your local McGraw-Hill sales representative for more information.

Course Delivery Systems

With help from WebCT, Blackboard, and other course management systems, professors can take complete control of their course content. Course cartridges containing website content, online testing, and powerful student tracking features are readily available for use within these platforms.

Learning Supplements for Students

ARIS (http://www.mhhe.com/cunningham4e)

This site includes quizzes for each chapter, additional case studies, interactive base maps, and much more. Learn more about the exciting features provided for students through the *Principles of Environmental Science* website.

Field & Laboratory Exercises in Environmental Science, Seventh Edition by Enger and Smith (ISBN: 978-0-07-290913-5; MHID: 0-07-290913-7)

The major objectives of this manual are to provide students with hands-on experiences that are relevant, easy to understand, applicable to the student's life, and presented in an interesting, informative format. Ranging from field and lab experiments to conducting social and personal assessments of the environmental impact of human activities, the manual presents something for everyone, regardless of the budget or facilities of each class. These labs are grouped by categories that can be used in conjunction with any introductory environmental textbook.

Exploring Environmental Science with GIS by Stewart, Cunningham, Schneiderman, and Gold (ISBN: 978-0-07-297564-2; MHID: 0-07-297564-4)

This short book provides exercises for students and instructors who are new to GIS, but are familiar with the Windows operating system. The exercises focus on improving analytical skills, understanding spatial relationships, and understanding the nature and structure of environmental data. Because the software used is distributed free of charge, this text is appropriate for courses and schools that are not yet ready to commit to the expense and time involved in acquiring other GIS packages.

Annual Editions: Environment 06/07 by Allen (ISBN: 978-0-07-351542-7; MHID: 0-07-351542-6)

This twenty-fifth edition is a compilation of current articles from the best of the public press. The selections explore the global environment, the world's population, energy, the biosphere, natural resources, and pollutions.

Taking Sides: Clashing Views on Environmental Issues, Eleventh Edition Expanded by Easton (ISBN: 978-0-07-351441-3; MHID: 0-07-351441-1)

This book represents the arguments of leading environmentalists, scientists, and policymakers. The issues reflect a variety of viewpoints and are staged as "pro" and "con" debates. Issues are organized around four core areas: general philosophical and political issues, the environment and technology, disposing of wastes, and the environment and the future.

Global Studies: The World at a Glance, Second Edition by Tessema (ISBN: 978-0-07-340408-0; MHID: 0-07-340408-X)

This book features a compilation of up-to-date data and accurate information on some of the important facts about the world we live in. While it is close to impossible to be an expert in all areas, such as a nation's capital, type of government, currency, major languages, population, religions, political structure, climate, economics, etc., this book is intended to assist in understanding these essential facts in order to make useful applications.

Sources: Notable Selections in Environmental Studies, Second Edition by Goldfarb (ISBN: 978-0-07-303186-6; MHID: 0-07-303186-0)

This volume brings together primary source selections of enduring intellectual value—classic articles, book excerpts, and research studies—that have shaped environmental studies and our contemporary understanding of it. The book includes carefully edited selections from the works of the most distinguished environmental observers, past and present. Selections are organized topically around the following major areas of study: energy, environmental degradation, population issues and the environment, human health and the environment, and environment and society.

Student Atlas of Environmental Issues, by Allen (ISBN: 978-0-69-736520-0; MHID: 0-69-736520-4)

This atlas is an invaluable pedagogical tool for exploring the human impact on the air, waters, biosphere, and land in every major world region. This informative resource provides a unique combination of maps and data that help students understand the dimensions of the world's environmental problems and the geographical basis of these problems.

Traditional outrigger canoes and hand lines are still used by villagers on many islands in the southwestern Pacific, but these low-impact fishing methods are being threatened by trawlers, dynamite fishing, and other destructive techniques.

C H A P T E R

1

Understanding Our Environment

Today we are faced with a challenge that calls for a shift in our thinking, so that humanity stops threatening its life-support system.

–Wangari Maathai, winner of 2004 Nobel Peace Prize–

OBJECTIVES

Studying this chapter will help you answer the following questions:

- What is environmental science, and why is it important?
- Why does science support—but rarely prove—particular theories?
- How can scientists know if their research is reliable and important?
- How can critical thinking help us understand environmental issues?

- What does the history of conservation show us about our attitudes toward nature?
- Describe several of the most important environmental problems facing the world. Are there signs of hope for solving these problems?
- What do we mean by sustainability and sustainable development?

Case Study   Saving the Reefs of Apo Island

Coral reefs are among the most beautiful, species-rich, and productive biological communities on the planet. They serve as the nurseries for many open-water species. At least half the world's reefs are threatened by pollution, global climate change, destructive fishing methods, and other human activities, but they can be protected and restored if we care for them.

As their outrigger canoes glide gracefully onto Apo Island's beach after an early morning fishing expedition, villagers call to each other to ask how fishing was. "Tunay mabuti!" (very good!) is the cheerful reply. Nearly every canoe has a basketful of fish; enough to feed a family for several days with a surplus to send to the market. Life hasn't always been so good on the island. Thirty years ago, this island, like many others in the Philippines, suffered a catastrophic decline in the seafood that was the mainstay of their diet and livelihood. Rapid population growth coupled with destructive fishing methods such as dynamite or cyanide fishing, small mesh gill nets, deep-sea trawling, and *Muro-ami* (a technique in which fish are chased into nets by pounding on coral with weighted lines) had damaged the reef habitat and exhausted fish stocks.

In 1979, scientists from Silliman University on nearby Negros Island visited Apo to explain how establishing a marine sanctuary could help reverse this decline. The coral reef fringing the island acts as a food source and nursery for many of the marine species sought by fishermen. Protecting that breeding ground, they explained, is the key to preserving a healthy fishery. The scientists took villagers from Apo to the uninhabited Sumilon Island, where a no-take reserve was teeming with fish.

After much discussion, several families decided to establish a marine sanctuary along a short section of Apo Island shoreline. Initially, the area had high-quality coral but few fish. The participating families took turns watching to make sure that no one trespassed in the no-fishing zone. Within a few years, fish numbers and sizes in the sanctuary increased dramatically, and "spillover" of surplus fish led to higher catches in surrounding areas. In 1985, Apo villagers voted to establish a 500 m (0.3 mi) wide marine sanctuary around the entire island.

Fishing is now allowed in this reserve, but only by low-impact methods such as hand-held lines, bamboo traps, large mesh nets, spearfishing without SCUBA gear, and hand netting. Coral-destroying techniques, such as dynamite, cyanide, trawling, and *Muro-ami* fishing are prohibited. By protecting the reef, villagers are guarding the nursery that forms the base for their entire marine ecosystem. Young fish growing up in the shelter of the coral move out as adults to populate the neighboring waters and yield abundant harvests. Fishermen report that they spend much less time traveling to distant fishing areas now that fish around the island are so much more abundant.

Apo Island's sanctuary is so successful that it has become the inspiration for more than 400 marine preserves throughout the Philippines and many others around the world. Not all are functioning as well as they might, but many, like Apo, have made dramatic progress in restoring abundant fish populations in nearby waters.

The rich marine life and beautiful coral formations in Apo Island's crystal clear water now attract international tourists. Two small hotels and a dive shop provide jobs for island residents. Other villagers take in tourists as boarders or sell food and T-shirts to visitors. The island government collects a diving/snorkeling fee, which has been used to build schools, improve island water supplies, and provide electricity to most of the island's 145 households. Almost all the island men still fish as their main occupation, but the fact that they don't have to go so far or work so hard for the fish they need means they have time for other activities, such as guiding diving tours or helping with household chores.

Higher family incomes now allow most island children to attend high school on Negros. Many continue their education with college or technical programs. Some find jobs elsewhere in the Philippines, and the money they send back home is a big economic boost for Apo families. Others return to their home island as teachers or to start businesses such as restaurants or dive shops. Seeing that they can do something positive to improve their environment and living conditions has empowered villagers to take on self-improvement projects that they may not otherwise have attempted.

Finding ways to live sustainably within the limits of the resource base available to us and without damaging the life-support systems provided by our ecosystem is a preeminent challenge of environmental science. Our answers to these challenges must be ecologically sound, economically sustainable, and socially acceptable if they are to succeed in the long term. Sometimes, as this case study shows, actions based on ecological knowledge and local action can spread to have positive effects on a global scale. Economics, policy, planning, and social organization all play vital roles in finding answers to human/environment problems. We'll discuss those disciplines later in this book, but first we'll look at the basic principles of ecological science as a basis for understanding our global environment.

Understanding Our Environment

The example of preserving reefs and fish stocks around Apo Island illustrates some of the complexity and importance of contemporary environmental issues. Human actions are having widespread impacts on our world and the other organisms with which we share it. Science can help us understand both how things work and how we can make our environment safer, more comfortable, and more enduring. The knowledge being gained by scientists is fundamental to our ability to manage the earth's resources in a sustainable manner and to improve the quality of our lives and those of our children. Environmental scientists work on many problems that critically affect our well-being in many ways. Because of the significance of its findings, an understanding of environmental science is becoming increasingly necessary for any educated person. And there are practical ways that each of us can use scientific knowledge to protect and restore the quality of our shared environment.

We live on a marvelous planet

Before proceeding in our discussion of current dilemmas and how scientists are trying to understand them, we should pause for a moment to consider the extraordinary natural world that we inherited and that we hope to pass on to future generations in as good—or perhaps even better—condition than we found it.

Imagine that you are an astronaut returning to the earth after a long trip to the moon or Mars. What a relief it would be, after experiencing the hostile environment of outer space, to come back to this beautiful, bountiful planet (fig. 1.1). Although there are dangers and difficulties here, we live in a remarkably prolific and hospitable

FIGURE 1.2 Perhaps the most amazing feature of our planet is its rich diversity of life.

world that is, as far as we know, unique in the universe. Compared with the conditions on other planets in our solar system, temperatures on the earth are mild and relatively constant. Plentiful supplies of clean air, fresh water, and fertile soil are regenerated endlessly and spontaneously by biogeochemical cycles and biological communities (discussed in chapters 2 and 3).

Perhaps the most amazing feature of our planet is its rich diversity of life. Millions of beautiful and intriguing species populate the earth and help sustain a habitable environment (fig. 1.2). This vast multitude of life creates complex, interrelated communities where towering trees and huge animals live together with, and depend upon, such tiny life-forms as viruses, bacteria, and fungi. Together, all these organisms make up delightfully diverse, self-sustaining ecosystems, including dense, moist forests; vast, sunny savannas; and richly colorful coral reefs.

From time to time, we should pause to remember that, in spite of the challenges and complications of life on earth, we are incredibly lucky to be here. We should ask ourselves: what is our proper place in nature? What *ought* we do and what *can* we do to protect the irreplaceable habitat that produced and supports us? These are some of the central questions of environmental science.

What is environmental science?

We inhabit two worlds. One is the natural world of plants, animals, soils, air, and water that preceded us by billions of years and of which we are a part. The other is the world of social institutions and artifacts that we create for ourselves using science, technology, and political organization. Both worlds are essential to our lives, but integrating them successfully causes enduring tensions.

Environment (from the French *environner:* to encircle or surround) can be defined as (1) the circumstances and conditions that surround an organism or a group of organisms or (2) the social and cultural conditions that affect an individual or a community. Since humans inhabit the natural world as well as the "built" or technological, social, and cultural world, all constitute important parts of our environment (fig. 1.3).

Environmental science is the systematic study of our environment and our place in it. A relatively new field, environmental science is highly interdisciplinary. It integrates information from biology, chemistry, geography, agriculture, and many other fields. To apply this information to improve the ways we treat

FIGURE 1.1 The life-sustaining ecosystems on which we all depend are unique in the universe, as far as we know.

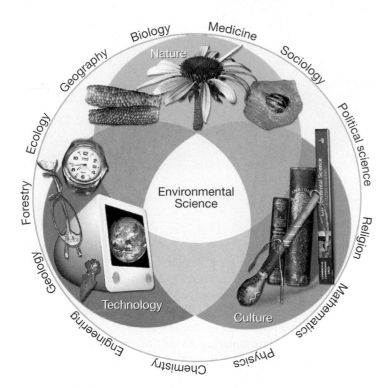

FIGURE 1.3 The intersections of the natural, cultural, and technological worlds outline the province of environmental science. Many disciplines contribute to our understanding and management of our environment.

SCIENCE AS A WAY OF KNOWING

Science is a process for producing empirical knowledge about the natural world through methodical and logical studies of nature. Derived from *scire*, "to know" in Latin, science depends on making precise observations of natural phenomena. We develop or test theories (proposed explanations of how a process works) using these observations. "Science" also refers to the cumulative body of knowledge produced by many scientists. Science is valuable because it helps us understand the world and meet practical needs, such as new medicines, new energy sources, or new foods. In this section, we'll investigate how and why science follows standard methods.

Science rests on the assumption that the world is knowable and that we can learn about it by careful observation and logical reasoning. (table 1.1). For early philosophers of science, this assumption was a radical departure from religious and philosophical approaches. In the Middle Ages, the ultimate sources of knowledge about matters, such as how crops grow, how diseases spread, or how the stars move, were religious authorities or cultural traditions. While these sources provided many useful insights, there was no way to test their explana-

our world, environmental scientists also incorporate knowledge of social organization, politics, and the humanities. In other words, environmental science is inclusive and holistic. Environmental science is also mission-oriented: it implies that we all have a responsibility to get involved and try to do something about the problems we have created.

As distinguished economist Barbara Ward pointed out, for an increasing number of environmental issues, the difficulty is not to identify remedies. Remedies are now well understood; the problem is to make them socially, economically, and politically acceptable. Foresters know how to plant trees, but not how to establish conditions under which villagers in developing countries can manage plantations for themselves. Engineers know how to control pollution, but not how to persuade factories to install the necessary equipment. City planners know how to design urban areas, but not how to make them affordable for the poorest members of society. The solutions to these problems increasingly involve human social systems as well as natural science.

Criteria for environmental literacy have been suggested by the National Environmental Education Advancement Project in Wisconsin. These criteria include awareness and appreciation of the natural and built environment, knowledge of natural systems and ecological concepts, understanding of current environmental issues, and ability to use analytical and problem-solving skills on environmental issues. These are good goals to keep in mind as you study this book.

TABLE 1.1

Basic Principles of Science

1. *Empiricism:* We can learn about the world by careful observation of empirical (real, observable) phenomena; we can expect to understand fundamental processes and natural laws by observation.

2. *Uniformitarianism:* Basic patterns and processes are uniform across time and space; the forces at work today are the same as those that shaped the world in the past, and they will continue to do so in the future.

3. *Parsimony:* When two plausible explanations are reasonable, the simpler (more parsimonious) one is preferable. This rule is also known as Ockham's razor, after the English philosopher who proposed it.

4. *Uncertainty:* Knowledge changes as new evidence appears, and explanations (theories) change with new evidence. Theories based on current evidence should be tested on additional evidence, with the understanding that new data may disprove the best theories.

5. *Repeatability:* Tests and experiments should be repeatable; if the same results cannot be reproduced, then the conclusions are probably incorrect.

6. *Proof is elusive:* We rarely expect science to provide absolute proof that a theory is correct, because new evidence may always undermine our current theories. Even evolution, the cornerstone of modern biology, ecology, and other sciences, remains a "theory" because we cannot absolutely prove its operation.

7. *Testable questions:* To find out whether a theory is correct, it must be tested; we formulate testable statements (hypotheses) to test theories.

tions independently and objectively. The benefit of scientific thinking is that it searches for testable evidence: If you suspect a disease spreads through contaminated water, you can close off access to the water source and see if the disease stops spreading.

Science depends on skepticism and accuracy

Ideally scientists are skeptical. They are cautious about accepting proposed explanations until there is substantial evidence to support them. Even then, every explanation is considered only provisionally true, because there is always a possibility that some additional evidence may appear to disprove it. Scientists also aim to be methodical and unbiased. Because bias and methodical errors are hard to avoid, scientific tests are subject to review by informed peers, who can evaluate results and conclusions (fig. 1.4). The peer review process is an essential part of ensuring that scientists maintain good standards in study design, data collection, and interpretation of results.

Scientists demand **reproducibility** because they are cautious about accepting conclusions. Making an observation or obtaining a result just once doesn't count for much. You have to produce the same result consistently to be sure that your first outcome wasn't a fluke. Even more important, you must be able to describe the conditions of your study so that someone else can reproduce your findings. Repeating studies or tests is known as **replication.**

Science also relies on accuracy and precision. Accuracy is correctness of measurements. Inaccurate data can produce sloppy

FIGURE 1.5 Making careful, accurate measurements and keeping good records are essential in scientific research.

and misleading conclusions (fig. 1.5). Precision means repeatability of results and level of detail. The classic analogy for repeatability is throwing darts at a dart board. You might throw ten darts and miss the center every time, but if all the darts hit nearly the same spot, they may not have been very accurate, but they were very precise.

Deductive and inductive reasoning are both useful

Ideally, scientists deduce conclusions from general laws that they know to be true. For example, if we know that massive objects attract each other (because of gravity), then it follows that an apple will fall to the ground when it releases from the tree. This logical reasoning from general to specific is known as **deductive reasoning.** Often, however, we do not know general laws that guide natural systems. We observe, for example, that birds appear and disappear as a year goes by. Through many repeated observations in different places, we can infer that the birds move from place to place. We can develop a general rule that birds migrate seasonally. Reasoning from many observations to produce a general rule is **inductive reasoning.** Although deductive reasoning is more logically sound than inductive reasoning, it only works when our general laws are correct. We often rely on inductive reasoning to understand the world because we have few immutable laws.

Sometimes it is insight, as much as reasoning, that leads us to an answer. Many people fail to recognize the role that insight, creativity, aesthetics, and luck play in research. Some of our most important discoveries were made not because of superior scientific method and objective detachment, but because the investigators were passionately interested in their topics and pursued hunches that appeared unreasonable to fellow scientists. A good example is Barbara McClintock, the geneticist who discovered that genes in corn can move and recombine spontaneously. Where other corn geneticists saw random patterns of color and kernel

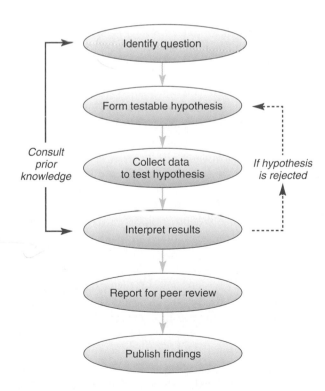

FIGURE 1.4 Ideally, scientific investigation follows a series of logical, orderly steps to formulate and test hypotheses.

size, McClintock's years of experience in corn breeding and an uncanny ability to recognize patterns, led her to guess that genes could recombine in ways that no one had yet imagined. Her intuitive understanding led to a theory that took other investigators years to accept.

Science studies problems methodically

You may already be using the scientific method without being aware of it. Suppose you have a flashlight that doesn't work. The flashlight has several components (switch, bulb, batteries) that could be faulty. If you change all the components at once, your flashlight might work, but a more methodical series of tests will tell you more about what was wrong with the system—knowledge that may be useful next time you have a faulty flashlight. So you decide to follow the standard scientific steps:

1. *Observe* that your flashlight doesn't light; also, there are three main components of the lighting system (batteries, bulb, and switch).

2. Propose a **hypothesis,** a testable explanation: "The flashlight doesn't work because the batteries are dead."

3. Develop a *test* of the hypothesis and *predict* the result that would indicate your hypothesis was correct: "I will replace the batteries; the light should then turn on."

4. Gather *data* from your test: After you replaced the batteries, did the light turn on?

5. *Interpret* your results: If the light works now, then your hypothesis was right; if not, then you should formulate a new hypothesis, perhaps that the bulb is faulty, and develop a new test for that hypothesis.

In systems more complex than a flashlight, it is almost always easier to prove a hypothesis wrong than to prove it unquestionably true. This is because we usually test our hypotheses with observations, but there is no way to make every possible observation. The philosopher Ludwig Wittgenstein illustrated this problem as follows: Suppose you saw hundreds of swans, and all were white. These observations might lead you to hypothesize that all swans were white. You could test your hypothesis by viewing thousands of swans, and each observation might support your hypothesis, but you could never be entirely sure that it was correct. On the other hand, if you saw just one black swan, you would know with certainty that your hypothesis was wrong.

As you'll read in later chapters, the elusiveness of absolute proof is a persistent problem in environmental policy and law. You can never absolutely prove that the toxic waste dump up the street is making you sick. The elusiveness of proof often decides environmental liability lawsuits.

When an explanation has been supported by a large number of tests, and when a majority of experts have reached a general consensus that it is a reliable description or explanation, we call it a **scientific theory.** Note that scientists' use of this term is very different from the way the public uses it. To many people, a theory is speculative and unsupported by facts. To a scientist, it

FIGURE 1.6 Data collection and reproducible tests support scientific theories. Here aquatic biologists examine and categorize samples they've collected.

means just the opposite: While all explanations are tentative and open to revision and correction, an explanation that counts as a scientific theory is supported by an overwhelming body of data and experience, and it is generally accepted by the scientific community, at least for the present (fig. 1.6).

Understanding probability helps reduce uncertainty

One strategy to improve confidence in the face of uncertainty is to focus on probability. **Probability** is a measure of how likely something is to occur. Usually, probability estimates are based on a set of previous observations or on standard statistical measures. Probability does not tell you what *will* happen, but it tells you what *is likely* to happen. If you hear on the news that you have a 20 percent chance of catching a cold this winter, that means that 20 of every 100 people are likely to catch a cold. This doesn't mean that you will catch one. In fact, it's more likely that you won't catch a cold than that you will. If you hear that 80 out of every 100 people will catch a cold, you still don't know whether you'll get sick, but there's a much higher chance that you will.

Science often involves probability, so it is important to be familiar with the idea. Sometimes probability has to do with random chance: If you flip a coin, you have a random chance of getting heads or tails. Every time you flip, you have the same 50 percent probability of getting heads. The chance of getting ten heads in a row is small (in fact, the chance is 1 in 2^{10}, or 1 in 1,024), but on any individual flip, you have exactly the same 50 percent chance, since this is a random test. Sometimes probability is weighted by circumstances: Suppose that about 10 percent of the students in your class earn an A each semester. Your likelihood of being in that 10 percent depends a great deal on how much time you spend studying, how many questions you ask in class, and other factors. Sometimes there is a combination of chance

Principles of Environmental Science   http://www.mhhe.com/cunningham4e

Calculating Probability

An understanding of statistical analysis and probability is fundamental in most areas of modern science. Working with these concepts is critical to your ability to comprehend scientific information.

Every time you flip a coin, the chance that heads will end up on top is 1 in 2 (50 percent, assuming you have a normal coin). The odds of getting heads two times in a row is $1/2 \times 1/2$, or $1/4$.

1. What are the odds of getting heads five times in a row?

2. As you start the fifth flip, what are the odds of getting heads?

3. If there are 100 students in your class and everybody flips a coin five times, how many people are likely to get five heads in a row?

Answers: 1. (1 in 2 to the 5th) = 1/32; 2. 1 in 2; 3. 100 students × 1/32 = about 3.

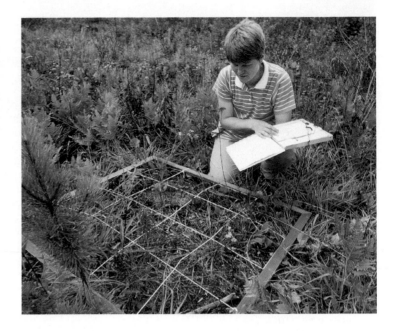

FIGURE 1.7 What is an appropriate sample size for an investigation? How many individuals would you need to examine to have a representative sample of a population? These are statistical questions of variability, probability, and accuracy.

and circumstances: The probability that you will catch a cold this winter depends partly on whether you encounter someone who is sick (largely random chance) and whether you take steps to stay healthy (get enough rest, wash your hands frequently, eat a healthy diet, and so on).

Statistics can calculate the probability that your results were random

Statistics can help in experimental design as well as in interpreting data (see Exploring Science, p. 8). Many statistical tests focus on calculating the probability that observed results could have occurred by chance. Often, the degree of confidence we can assign to results depends on sample size as well as the amount of variability between groups (fig. 1.7).

Ecological tests are generally considered significant if there is less than 5 percent probability that the results were achieved by random chance. A probability of less than 1 percent gives still greater confidence in the results.

As you read this book, you will encounter many statistics, including many measures of probability. When you seen these numbers, stop and think: Is the probability high enough to worry about? How high is it compared to other risks or chances you've read about? What are the conditions that make probability higher or lower? Science involves many other aspects of statistics.

Experimental design can reduce bias

The study of colds and sleep deprivation is an example of an observational experiment, one in which you observe natural events and interpret a causal relationship between the variables. This kind of study is also called a **natural experiment,** one that involves observation of events that have already happened. Many scientists depend on natural experiments: A geologist, for instance, might want to study mountain building, or an ecologist might want to learn about how species coevolve, but neither scientist can spend millions of years watching the process happen. Similarly, a toxicologist cannot give people a disease just to see how lethal it is.

Other scientists can use **manipulative experiments,** in which conditions are deliberately altered, and all other variables are held constant. Most manipulative experiments are done in the laboratory, where conditions can be carefully controlled. Suppose you were interested in studying whether lawn chemicals contributed to deformities in tadpoles. You might keep two groups of tadpoles in fish tanks, and expose one to chemicals. In the lab, you could ensure that both tanks had identical temperatures, light, food, and oxygen. By comparing a treatment (exposed) group and a control (unexposed) group, you have also made this a **controlled study.**

Often, there is a risk of experimenter bias. Suppose the researcher sees a tadpole with a small nub that looks like it might become an extra leg. Whether she calls this nub a deformity might depend on whether she knows that the tadpole is in the treatment group or the control group. To avoid this bias, **blind experiments** are often used, in which the researcher doesn't know which group is treated until after the data have been analyzed. In health studies, such as tests of new drugs, **double-blind experiments** are used, in which neither the subject (who receives a drug or a placebo) nor the researcher knows who is in the treatment group and who is in the control group.

Statistics are numbers that let you evaluate and compare things. "Statistics" is also a field of study that has developed meaningful methods of comparing those numbers. By both definitions, statistics are widely used in environmental sciences, partly because they can give us a useful way to assess patterns in a large population, and partly because the numbers can give us a measure of confidence in our research or observations. Understanding the details of statistical tests can take years of study, but a few basic ideas will give you a good start toward interpreting statistics.

1. *Descriptive statistics help you assess the general state of a group.* In many towns and cities, the air contains a dust, or particulate matter, as well as other pollutants. From personal experience you might know your air isn't as clean as you'd like, but you may not know how clean or dirty it is. You could start by collecting daily particulate measurements to find average levels. An averaged value is more useful than a single day's values, because daily values may vary a great deal, but general, long-term conditions affect your general health. Collect a sample every day for a year; then divide the sum by the number of days, to get a **mean** (average) dust level. Suppose you found a mean particulate level of 30 micrograms per cubic meter ($\mu g/m^3$) of air. Is this level high or low? In 1997 the EPA set a standard of 50 $\mu g/m^3$ as a limit for allowable levels of coarse particulates (2.5–10 micrometers in diameter). Higher levels tend to be associated with elevated rates of asthma and other respiratory diseases. Now you know that your town, with an annual average of 30 $\mu g/m^3$, has relatively clean air, after all.

2. *Statistical samples.* Although your town is clean by EPA standards, how does it compare with the rest of the cities in the country? Testing the air in *every* city is probably not possible. You could compare your town's air quality with a **sample,** or subset of cities, however. A large, random sample of cities should represent the general "population" of cities reasonably well. Taking a large sample reduces the effects of outliers (unusually high or low values) that might be included. A random sample minimizes the chance that you're getting only the worst sites, or only a collection of sites that are close together, which might all have similar conditions. Suppose you get average annual particulate levels from a sample of 50 randomly selected cities. You can draw a frequency distribution, or histogram, to display your results (fig. 1). The mean value of this group is 36.8 $\mu g/m^3$, so by comparison your town (at 30 $\mu g/m^3$) is relatively clean.

Many statistical tests assume that the sample has a normal, or Gaussian, frequency distribution, often described as a bell-shaped curve (fig. 2). In this distri-

FIGURE 1 Average annual airborne dust levels for 50 cities in 2001.
Source: Data from U.S. Environmental Protection Agency.

bution, the mean is near the center of the range of values, and most values are fairly close to the mean. Large and random samples are more likely to fit this shape than are small and nonrandom samples.

3. *Confidence.* How do you know that the 50 cities you sampled really represent all the cities in the country? You can't ever be completely certain, but you can use estimates, such as confidence limits, to express the reliability of your mean statistic. Depending on the size of your sample (not 10, not 100, but 50) and the amount of variability in the sample data, you can calculate a confidence interval that the mean represents the whole population (all cities). Confidence levels, or confidence intervals, represent the likelihood that

In each of these studies there is one **dependent variable** and one, or perhaps more, **independent variables.** The dependent variable, also known as a response variable, is affected by the independent variables. In a graph, the dependent variable is on the vertical (Y) axis, by convention. Independent variables are rarely really independent (they are affected by the same environmental conditions as the dependent variable, for example). Many people prefer to call them explanatory variables, because we hope they will explain differences in the dependent variable.

Science is a cumulative process

The scientific method outlined in figure 1.4 is the process used to carry out individual studies. Larger-scale accumulation of scientific knowledge involves cooperation and contributions from countless people. Good science is rarely carried out by a single individual working in isolation. Instead, a community of scientists collaborates in a cumulative, self-correcting process. You often hear about big breakthroughs and dramatic discoveries that change our understanding overnight, but in reality these changes are usually the culmination of the labor of many people, each working on different aspects of a common problem, each adding small insights to solve a problem. Ideas and information are exchanged, debated, tested, and re-tested to arrive at **scientific consensus,** or general agreement among informed scholars.

The idea of consensus is important. For those not deeply involved in a subject, the multitude of contradictory results can be bewildering: Are coral reefs declining, and does it matter? Is climate changing, and how much? Among those who have performed and read many studies, there tends to emerge a general agreement about the state of a problem. Scientific consensus now holds that many coral reefs are in danger, though opinions vary

FIGURE 2 A normal distribution.

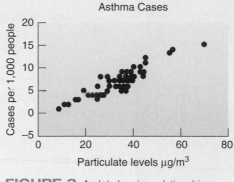

FIGURE 3 A plot showing relationships between variables.

your statistics represent the entire population correctly. For the mean of your sample, a confidence interval tells you the probability that your sample is similar to other random samples of the population. A common convention is to compare values with a 95 percent confidence level, or a probability of 5 percent or less that your conclusions are misleading. Using statistical software, we can calculate that, for our 50 cities, the mean is 36.8 $\mu g/m^3$, and the confidence interval is 35.0 to 38.6. This suggests that, if you take 1,000 samples from the entire population of cities, 95 percent of those samples ought to be within 2 $\mu g/m^3$ of your mean. This indicates that your mean is reliable and representative.

4. *Is your group unusual?* Once you have described your group of cities, you can compare it with other groups. For example, you might believe that Canadian cities have cleaner air than U.S. cities. You can compare mean air quality levels for the two groups. Then you can calculate confidence intervals for the difference between the means, to see if the difference is meaningful.

5. *Evaluating relationships between variables.* Are respiratory diseases correlated with air pollution? For each city in your sample, you could graph pollution and asthma rates (fig. 3). If the graph looks like a loose cloud of dots, there is no clear relationship. A tight, linear pattern of dots trending upward to the right indicates a strong and positive relationship. You can also use a statistical package to calculate an equation to describe the relationship and, again, confidence intervals for the equation. This is known as a regression equation.

6. *Lies, damned lies, and statistics.* Can you trust a number to represent a complex or large phenomenon? One of the devilish details of representing the world with numbers is that those numbers can be tabulated in many ways. If we want to assess the greatest change in air quality statistics, do we report rates of change or the total amount of change? Do we look at change over five years? Twenty-five years? Do we accept numbers selected by the EPA, by the cities themselves, by industries, or by environmental groups? Do we trust that all the data were collected with a level of accuracy and precision that we would accept if we knew the hidden details in the data-gathering process? Like all information, statistics need to be interpreted in terms of who produced them, when, and why. Awareness of some of the standard assumptions behind statistics, such as sampling, confidence, and probability, will help you interpret statistics that you see and hear.

For more information about creating and interpreting graphs, see the Data Analysis exercise at the end of this chapter.

on how severe the problem is. Consensus is that global climates are changing, though models differ somewhat on how rapidly they will change under different policy scenarios.

Sometimes new ideas emerge that cause major shifts in scientific consensus. Two centuries ago, geologists explained many earth features in terms of Noah's flood. The best scientists held that the flood created beaches well above modern sea level, scattered boulders erratically across the landscape, and gouged enormous valleys where there is no water now (fig. 1.8). Then the Swiss glaciologist Louis Agassiz and others suggested that the earth had once been much colder and that glaciers had covered large areas. Periodic ice ages better explained changing sea levels, boulders transported far from their source rock, and the great, gouged valleys. This new idea completely altered the way geologists explained their subject. Similarly, the idea of tectonic plate movement, in which continents shift slowly around the earth's surface, revolutionized the ways geologists, biogeographers, ecologists, and others explained the development of the earth and its life-forms.

These great changes in explanatory frameworks were termed **paradigm shifts** by Thomas Kuhn (1967), who studied revolutions in scientific thought. According to Kuhn, paradigm shifts occur when a majority of scientists accept that the old explanation no longer explains new observations very well. The shift is often contentious and political, because whole careers and worldviews, based on one sort of research and explanation, can be undermined by a new model. Sometimes a revolution happens rather quickly. Quantum mechanics and Einstein's theory of relativity, for example, overturned classical physics in only about 30 years. Sometimes a whole generation of scholars has to retire before new paradigms can be accepted.

As you study this book, try to identify some of the paradigms that guide our investigations, explanations, and actions today.

FIGURE 1.8 Paradigm shifts change the ways we explain our world. Geologists now attribute Yosemite's valleys to glaciers where once they believed Noah's flood carved its walls.

This is one of the skills involved in critical thinking, which we will discuss in further detail later in the chapter.

What is sound science?

Ideally, **sound science** follows the principles outlined in this chapter for precision, reliability, reproducibility, and skepticism. It should meet the test of a large body of peer-reviewed data and consistent results that are accepted by a majority of the scientific community as the best available explanation of phenomena.

In every controversial topic, however, a few contrarian scientists champion a view diametrically opposed to the majority position of the scientific community. There are several reasons to adopt a contradictory view. One might be sincerely convinced, for instance, that the majority opinion is wrong. Often the most brilliant and original thinkers are regarded as hopelessly off track by conventional scientists. Others may be motivated simply by a desire to consider all possibilities in case unexpected evidence comes up.

Some people also may be tempted to take an opposing viewpoint because they know it will generate publicity and even, perhaps, generous payments from those who benefit from countertestimony. In many court cases, for example, opposing sides are able to produce paid "experts" who are willing to present a partisan agenda rather than objective science. A common tactic is to use scientific uncertainty as an excuse to postpone or reverse an action or a policy that a vast majority of the scientific community believes to be prudent. Opponents will claim that the evidence doesn't constitute conclusive proof that the action needs to be taken, or that the policy is based only on a scientific "theory" and therefore has no validity.

Harvard's Edward O. Wilson writes, "We will always have contrarians whose sallies are characterized by willful ignorance, selective quotations, disregard for communications with genuine experts, and distructive campaigns to attract the attention of the media rather than scientists. They are the parasite load on scholars who earn success through the slow process of peer review and approval."

Some politicians and idealogues demand "sound science," by which they mean data and conclusions that support their own agenda. They denounce everything that contradicts their position as "junk science." How can we identify "junk science" that presents bogus analysis dressed up in quasi-scientific jargon but that really represents a perversion of objective inquiry? For an example of a current controversy concerning "sound science," see "A Skeptical Environmentalist" at www.mhhe.com/cunningham4e. The astronomer Carl Sagan proposed a "Baloney Detection Kit" containing the questions in table 1.2.

Is environmental science the same as environmentalism?

Environmental science is the use of scientific methods to study processes and systems in the environment in which we live. Environmentalism is working to influence attitudes and policies that affect our environment. The two are often separate goals. Petroleum geologists, for example, are environmental scientists who study geology in order to find oil resources. They work to increase access to resources, to benefit their company, perhaps to benefit our standard of living, but usually not to protect the environment in which they work and live. Other environmental scientists are also environmentalists. E.O. Wilson is an ecologist who actively works to

---

TABLE 1.2

Questions for Baloney Detection

1. How reliable are the sources of this claim? Is there reason to believe that they might have an agenda to pursue in this case?

2. Have the claims been verified by other sources? What data are presented in support of this opinion?

3. What position does the majority of the scientific community hold in this issue?

4. How does this claim fit with what we know about how the world works? Is this a reasonable assertion or does it contradict established theories?

5. Are the arguments balanced and logical? Have proponents of a particular position considered alternate points of view or only selected supportive evidence for their particular beliefs?

6. What do you know about the sources of funding for a particular position? Are they financed by groups with partisan goals?

7. Where was evidence for competing theories published? Has it undergone impartial peer review or is it only in proprietary publication?

defend the ecosystems he studies. Many environmental scientists work in the public interest, to serve public health, for example, without necessarily being interested in nature or other species.

Whether we use science to pursue public health, economic success, environmental quality, or other goals depends on issues outside of science. Many of these issues have to do with world-views and ethics, which are discussed below.

THINKING ABOUT THINKING

Perhaps the most valuable skill you can learn in any of your classes is the ability to think clearly, creatively, and purposefully. Much of the most important information in environmental science is highly contested. As previous sections of this chapter show, facts can vary, depending on when and by whom they were gathered. How can you make sense out of a welter of conflicting information? Developing your ability to learn new skills, examine new facts, evaluate new theories, and formulate your own interpretations is essential for keeping up in a changing world.

A flood of information and misinformation inundates us every day. Competing explanations and contradictory ideas battle for our attention. The rapidly growing complexity of our world and our lives intensifies the difficulties of knowing what to believe or how to act. Consider how the communications revolution has brought us computers, email, cell phones, mobile faxes, pagers, the World Wide Web, hundreds of channels of satellite TV, and direct mail or electronic marketing that overwhelm us with conflicting information. We have more choices than we can possibly manage, and we know more about the world around us than ever before but, perhaps, understand less.

An additional complication is that distinguished authorities vehemently disagree about many important topics. Some cynics say that, for any expert, there is always an equal and opposite expert. How can you decide what is true and meaningful? Is it simply a matter of what feels good at the moment or what supports our preconceived notions? Or can we use logical, orderly, creative thinking procedures to reach decisions?

Critical thinking is one approach to knowledge and meaning

A number of skills, attitudes, and approaches can help us evaluate information and make decisions. **Analytical thinking** asks, "How can I break this problem down into its constituent parts?" **Creative thinking** asks, "How might I approach this problem in new and inventive ways?" **Logical thinking** asks, "How can orderly, deductive reasoning help me think clearly?" **Critical thinking** asks, "What am I trying to accomplish here, and how will I know when I've succeeded?" **Reflective thinking** asks, "What does it all mean?"

While much of science is based on analytical and logical thinking, critical thinking adds elements of contextual sensitivity and empathy that can be very helpful in your study of environmental science. It challenges us to examine theories, facts, and options in a systematic, purposeful, and responsible manner. It shares many

TABLE 1.3

Steps in Critical Thinking

1. What is the purpose of my thinking?
2. What precise question am I trying to answer?
3. Within what point of view am I thinking?
4. What information am I using?
5. How am I interpreting that information?
6. What concepts or ideas are central to my thinking?
7. What conclusions am I aiming toward?
8. What am I taking for granted; what assumptions am I making?
9. If I accept the conclusions, what are the implications?
10. What would the consequences be if I put my thoughts into action?

methods and approaches with other methods of reasoning but adds some important skills, attitudes, and dispositions. Furthermore, it challenges us to plan methodically and to assess the process of thinking as well as the implications of our decisions. Thinking critically can help us discover hidden ideas and meanings, develop strategies for evaluating reasons and conclusions in arguments, recognize the differences between facts and values, and avoid jumping to conclusions. Richard Paul, chair of the National Council for Critical Thinking, identifies ten steps in critical thinking (table 1.3).

Notice that many critical thinking processes are self-reflective and self-correcting. This form of thinking is sometimes called "thinking about thinking." It is an attempt to plan rationally how to analyze a problem, to monitor your progress while you are doing it, and to evaluate how your strategy worked and what you have learned when you are finished. It is not critical in the sense of finding fault, but it makes a conscious, active, disciplined effort to be aware of hidden motives and assumptions; to uncover bias; and to recognize the reliability or unreliability of sources.

While critical thinking shares many of the orderly, systematic approaches of formal logic, it also invokes traits such as empathy, sensitivity, courage, and humility. Formulating intelligent opinions about some of the complex issues you will encounter in environmental science requires more than simple logic. Developing these attitudes and skills is not easy or simple. It takes practice. You have to develop your mental faculties just as you need to train for a sport. Intellectual integrity, modesty, fairness, compassion, and fortitude are not traits you use only occasionally. They must be cultivated until they become a part of your normal way of thinking.

What do you need to think critically?

We all use critical or reflective thinking at times. Suppose a television commercial tells you that a new breakfast cereal is tasty and good for you. You may be suspicious and ask yourself a few questions: What do they mean by *good?* Good for whom or what? Does *tasty* simply mean more sugar and salt? Might the sources of this information have other motives besides your health and happiness? Although you may not have been aware of it, you

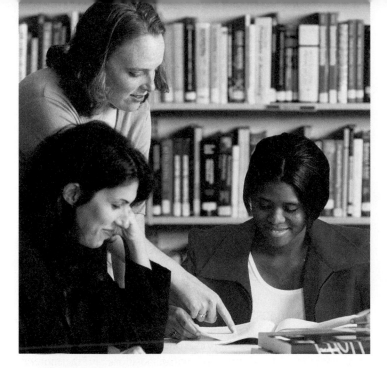

FIGURE 1.9 Explaining ideas to others, debating points, or simply verbalizing your understanding of something is an excellent way to clarify your thoughts.

already have used some of the techniques of critical analysis. Working to expand these skills helps you recognize how information and analysis can be distorted, misleading, prejudiced, superficial, unfair, or otherwise defective. Often, discussing complex ideas with others is a good way to clarify your own thoughts (fig. 1.9).

Here are some steps in critical thinking:

1. *Identify and evaluate premises and conclusions in an argument.* What is the basis for the claims made here? What evidence is presented to support these claims, and what conclusions are drawn from this evidence? If the premises and evidence are correct, does it follow that the conclusions are necessarily true?

2. *Acknowledge and clarify uncertainties, vagueness, equivocation, and contradictions.* Do the terms used have more than one meaning? If so, are all participants in the argument using the same meanings? Is ambiguity or equivocation deliberate? Can all the claims be true simultaneously?

3. *Distinguish between facts and values.* Can claims be tested? (If so, these are statements of fact and should be verifiable by gathering evidence.) Are claims made about the worth or lack of worth of something? (If so, these are value statements or opinions and probably cannot be verified objectively.) For example, claims of what we ought to do to be moral or righteous or to respect nature are generally value statements.

4. *Recognize and assess assumptions.* Given the backgrounds and views of the protagonists, what underlying reasons might there be for the premises, evidence, or conclusions presented? Does anyone have an "axe to grind" or a personal agenda in this issue? What does s/he think you know, need, want, or believe? Is there a subtext based on race, gender, ethnicity, economics, or some belief system that distorts this discussion?

5. *Distinguish source reliability or unreliability.* What qualifies the experts on this issue? What special knowledge or information do they have? What evidence do they present? How can we determine whether the information offered is accurate, true, or even plausible?

6. *Recognize and understand conceptual frameworks.* What are the basic beliefs, attitudes, and values that this person, group, or society holds? What dominating philosophy or ethics control their outlooks and actions? How do these beliefs and values affect the way people view themselves and the world around them? If there are conflicting or contradictory beliefs and values, how can these differences be resolved?

"Unpacking" an argument helps you understand it

A logical argument is made up of one or more introductory statements (called premises) and a conclusion that supposedly follows logically from the premises. In ordinary conversation, different kinds of statements often are mixed together, so distinguishing among them or deciphering hidden or implied meanings is difficult. Social theorists call the process of separating and analyzing textual components "unpacking." Applying this type of analysis to an argument can be useful.

An argument's premises are usually claimed to be based on facts; conclusions are usually opinions and values drawn from, or used to interpret, those facts. Words that often introduce a premise include *as, because, assume that, given that, since, whereas,* and *we all know that.* Words that generally indicate a conclusion or statement of opinion or values include *and, so, thus, therefore, it follows that, consequently, the evidence shows,* and *we can conclude that.*

For instance, in the example we used earlier, the television cereal ad might have said, "Since we all need vitamins, and since this cereal contains vitamins, the cereal must be good for you." Which are the premises, and which is the conclusion? Does one necessarily follow from the other? Remember that, even if the facts in a premise are correct, the conclusions drawn from those facts may not be. Information may be withheld from the argument, such as the fact that the cereal is also loaded with unhealthy amounts of sugar.

Critical thinking helps you learn environmental science

In this book, you will have many opportunities to practice critical thinking skills. Every chapter includes many facts, figures, opinions, and theories. Are all of them true? No, probably not. They were the best information available when this text was written, but much in environmental science is in a state of flux. Data change constantly, as does our interpretation of data. Do the ideas presented here give a complete picture of the state of our environment? Unfortunately, they probably don't. No matter how comprehensive our discussion is of this complex, diverse subject, it can never capture everything worth knowing, nor can it reveal all possible points of view.

When reading this text, try to distinguish between statements of fact and opinion. Ask yourself if the premises support the conclusions drawn from them. Although we have tried to be fair and even-handed in presenting controversies, our personal biases and values—some of which we may not even recognize—affect how we see issues and present arguments. Watch for cases in which you need to think for yourself, and use your critical and reflective thinking skills to uncover the truth.

You'll find more on critical thinking, as well as some useful tips on how to study effectively, on our web page at www.mhhe.com/cunningham4e.

A BRIEF HISTORY OF CONSERVATION AND ENVIRONMENTALISM

Although some early societies had negative impacts on their surroundings, others lived in relative harmony with nature. In modern times, however, growing human populations and the power of our technology have heightened concern about what we are doing to our environment. We can divide conservation history and environmental activism into at least four distinct stages: (1) pragmatic resource conservation, (2) moral and aesthetic nature preservation, (3) a growing concern about health and ecological damage caused by pollution, and (4) global environmental citizenship. These stages aren't necessarily mutually exclusive, however; parts of each persist today in the environmental movement, and one person may embrace them all simultaneously.

Nature protection has historic roots

Recognizing human misuse of nature is not unique to modern times. Plato complained in the fourth century B.C. that Greece once was blessed with fertile soil and clothed with abundant forests of fine trees. After the trees were cut to build houses and ships, however, heavy rains washed the soil into the sea, leaving only a rocky "skeleton of a body wasted by disease." Springs and rivers dried up, while farming became all but impossible. Many classical authors regarded the earth as a living being, vulnerable to aging, illness, and even mortality. Periodic threats about the impending death of nature from human misuse have persisted to our own times. As Mostafa K. Tolba, former director of the United Nations Environment Programme said, "The problems that overwhelm us today are precisely those we failed to solve decades ago."

Some of the earliest scientific studies of environmental damage were carried out in the eighteenth century by French or British colonial administrators, many of whom were trained scientists and who considered responsible environmental stewardship as an aesthetic and moral priority, as well as an economic necessity. These early conservationists observed and understood the connections between deforestation, soil erosion, and local climate change. The pioneering British plant physiologist Stephen Hales, for instance, suggested that conserving green plants preserves rainfall. His ideas were put into practice in 1764 on the Caribbean island of Tobago, where about 20 percent of the land was marked as "reserved in wood for rains."

Pierre Poivre, an early French governor of Mauritius, an island in the Indian Ocean, was appalled at the environmental and social devastation caused by destruction of wildlife (such as the flightless dodo) and the felling of ebony forests on the island by early European settlers. In 1769 Poivre ordered that one-quarter of the island be preserved in forests, particularly on steep mountain slopes and along waterways. Mauritius remains a model for balancing nature and human needs. Its forest reserves shelter a larger percentage of its original flora and fauna than most other human-occupied islands.

Resource waste triggered pragmatic resource conservation

Many historians consider the publication of *Man and Nature* in 1864 by geographer George Perkins Marsh as the wellspring of environmental protection in North America. Marsh, who also was a lawyer, politician, and diplomat, traveled widely around the Mediterranean as part of his diplomatic duties in Turkey and Italy. He read widely in the classics (including Plato) and personally observed the damage caused by excessive grazing by goats and sheep and by the deforestation of steep hillsides. Alarmed by the wanton destruction and profligate waste of resources still occurring on the American frontier in his lifetime, he warned of its ecological consequences. Largely because of his book, national forest reserves were established in the United States in 1873 to protect dwindling timber supplies and endangered watersheds.

Among those influenced by Marsh's warnings were U.S. President Theodore Roosevelt and his chief conservation adviser, Gifford Pinchot (fig. 1.10 a and b). In 1905 Roosevelt, who was

(a) President Teddy Roosevelt

(b) Gifford Pinchot

(c) John Muir

(d) Aldo Leopold

FIGURE 1.10 Some early pioneers of the American conservation movement. President Teddy Roosevelt (a) and his main advisor Gifford Pinchot (b) emphasized pragmatic resource conservation, while John Muir (c) and Aldo Leopold (d) focused on ethical and aesthetic relationships.

the leader of the populist, progressive movement, moved forest management out of the corruption-filled Interior Department into the Department of Agriculture. Pinchot, who was the first American-born professional forester, became the first chief of the new Forest Service. He put resource management on an honest, rational, and scientific basis for the first time in American history. Together with naturalists and activists such as John Muir, William Brewster, and George Bird Grinnell, Roosevelt and Pinchot established the framework of the national forest, park, and wildlife refuge system. They passed game protection laws and tried to stop some of the most flagrant abuses of the public domain. In 1908 Pinchot organized and chaired the White House Conference on Natural Resources, perhaps the most prestigious and influential environmental meeting ever held in the United States. Pinchot also was governor of Pennsylvania and founding head of the Tennessee Valley Authority, which provided inexpensive power to the southeastern United States.

The basis of Roosevelt's and Pinchot's policies was pragmatic **utilitarian conservation.** They argued that the forests should be saved "not because they are beautiful or because they shelter wild creatures of the wilderness, but only to provide homes and jobs for people." Resources should be used "for the greatest good, for the greatest number, for the longest time." "There has been a fundamental misconception," Pinchot wrote, "that conservation means nothing but husbanding of resources for future generations. Nothing could be further from the truth. The first principle of conservation is development and use of the natural resources now existing on this continent for the benefit of the people who live here now. There may be just as much waste in neglecting the development and use of certain natural resources as there is in their destruction." This pragmatic approach still can be seen in the multiple-use policies of the U.S. Forest Service.

Ethical and aesthetic concerns inspired the preservation movement

John Muir (fig. 1.10c), amateur geologist, popular author, and first president of the Sierra Club, strenuously opposed Pinchot's utilitarian policies. Muir argued that nature deserves to exist for its own sake, regardless of its usefulness to us. Aesthetic and spiritual values formed the core of his philosophy of nature protection. This outlook has been called **biocentric preservation** because it emphasizes the fundamental right of other organisms—and nature as a whole—to exist and to pursue their own interests. Muir wrote, "The world, we are told, was made for man. A presumption that is totally unsupported by the facts. . . . Nature's object in making animals and plants might possibly be first of all the happiness of each one of them. . . . Why ought man to value himself as more than an infinitely small unit of the one great unit of creation?"

Muir, who was an early explorer and interpreter of California's Sierra Nevada range, fought long and hard for establishment of Yosemite and Kings Canyon National Parks. The National Park Service, established in 1916, was first headed by Muir's disciple, Stephen Mather, and has always been oriented toward preservation of nature in its purest state. It has often been at odds with

Pinchot's utilitarian Forest Service. One of Muir and Pinchot's biggest battles was over the damming of Hetch Hetchy Valley in Yosemite. Muir regarded flooding the valley a sacrilege against nature. Pinchot, who championed publicly owned utilities, viewed the dam as a way to free San Francisco residents from the clutches of greedy water and power monopolies.

In 1935, pioneering wildlife ecologist Aldo Leopold (fig. 1.10d) bought a small, worn-out farm in central Wisconsin. A dilapidated chicken shack, the only remaining building, was remodeled into a rustic cabin. Working together with his children, Leopold planted thousands of trees in a practical experiment in restoring the health and beauty of the land. "Conservation," he wrote, "is the positive exercise of skill and insight, not merely a negative exercise of abstinence or caution." The shack became a writing refuge and became the main focus of *A Sand County Almanac,* a much beloved collection of essays about our relation with nature. In it, Leopold wrote, "We abuse land because we regard it as a commodity belonging to us. When we see land as a community to which we belong, we may begin to use it with love and respect." Together with Bob Marshall and two others, Leopold was a founder of the Wilderness Society.

Rising pollution levels led to the modern environmental movement

The undesirable effects of pollution probably have been recognized as long as people have been building smoky fires. In 1273 King Edward I of England threatened to hang anyone burning coal in London because of the acrid smoke it produced. In 1661 the English diarist John Evelyn complained about the noxious air pollution caused by coal fires and factories and suggested that sweet-smelling trees be planted to purify city air. Increasingly dangerous smog attacks in Britain led, in 1880, to formation of a national Fog and Smoke Committee to combat this problem.

The tremendous expansion of chemical industries during and after World War II added a new set of concerns to the environmental agenda. *Silent Spring,* written by Rachel Carson (fig. 1.11a) and published in 1962, awakened the public to the threats of pollution and toxic chemicals to humans as well as other species. The movement she engendered might be called **modern environmentalism** because its concerns extended to include both natural resources and environmental pollution. Among some other pioneers of this movement were activist David Brower and scientist Barry Commoner (fig. 1.11b and c). Brower, as executive director of the Sierra Club, Friends of the Earth, and Earth Island Institute, introduced many of the techniques of environmental lobbying and activism, including litigation, intervention in regulatory hearings, book and calendar publishing, and use of mass media for publicity campaigns. Commoner, who was trained as a molecular biologist, has been a leader in analyzing the links between science, technology, and society. Both activism and research remain hallmarks of the modern environmental movement.

Under the leadership of a number of other brilliant and dedicated activists and scientists, the environmental agenda was expanded in the 1970s, to most of the issues addressed in this

Principles of Environmental Science

(a) Rachel Carson

(b) David Brower

(c) Barry Commoner

(d) Wangari Maathai

FIGURE 1.11 Among many distinguished environmental leaders in modern times. Rachel Carson (a), David Brower (b), Barry Commoner (c), and Wangari Maathai (d) stand out for their dedication, innovation, and bravery.

textbook, such as human population growth, atomic weapons testing and atomic power, fossil fuel extraction and use, recycling, air and water pollution, and wilderness protection. Environmentalism has become well established in the public agenda since the first national Earth Day in 1970. A majority of Americans now consider themselves environmentalists, although there is considerable variation in what that term means.

Global interconnections have expanded environmentalism

Increased opportunities to travel and expanded international communications now enable us to know about daily events in places unknown to our parents or grandparents. We have become, as Marshal McLuhan famously announced in the 1960s, a global village. As in a village, we are all interconnected in various ways. Events that occur on the other side of the globe have profound and immediate effects on our lives.

Photographs of the earth from space (see fig. 1.1) provide a powerful icon for the fourth wave of ecological concern, which might be called **global environmentalism.** Such photos remind us how small, fragile, beautiful, and rare our home planet is. We all share an environment at this global scale. As Ambassador Adlai Stevenson noted, in his 1965 farewell address to the United Nations, we now need to worry about the life-support systems of the planet as a whole. He went on to say in this speech, "We

cannot maintain it half fortunate, half miserable, half confident, half despairing, half slave to the ancient enemies of mankind and half free in a liberation of resources undreamed of until this day. No craft, no crew, can travel with such vast contradictions. On their resolution depends the security of us all."

In 1977, Professor Wangari Maathai (fig. 1.11d) founded the Green Belt Movement in her native Kenya as a way of both organizing poor rural women and restoring their environment. Beginning at a small, local scale, this organization has grown to more than 600 grassroots networks across Kenya. They have planted more than 30 million trees while mobilizing communities for self-determination, justice, equity, poverty reduction, and environmental conservation. Dr. Maathai was elected to the Kenyan Parliament and served as Assistant Minister for Environment and Natural Resources. Her leadership has helped bring democracy and good government to her country. In 2004, she received the Nobel Peace Prize for her work, the first time a Nobel has been awarded for environmental action. In her acceptance speech, she said, "Working together, we have proven that sustainable development is possible; that reforestation of degraded land is possible; and that exemplary governance is possible when ordinary citizens are informed, sensitized, mobilized and involved in direct action for their environment."

Some other leaders of the worldwide movement combining environmental protection with social justice have been British economist Barbara Ward, French/American scientist René Dubois, Norwegian prime minister Gro Harlem Brundtland, and Canadian diplomat Maurice Strong. All have been central in major international environmental conventions, such as the 1972 UN Conference on the Human Environment in Stockholm or the 1992 UN Earth Summit in Rio de Janeiro. Once again, new issues have become part of the agenda as our field of vision widens. We have begun to appreciate the links between poverty, injustice, oppression, and exploitation of humans and our environment. A concern about the human dimensions of environmental science is an important focus of this textbook.

We face many environmental problems

Every year the World Resources Institute, the United Nations Environment Program, and the World Bank issue Earth Trends, a comprehensive assessment of current world environmental conditions. The most recent version describes many serious environmental and social challenges. With more than 6.5 billion humans currently, we're adding about 85 million more to the world every year. While demographers report a transition to slower growth rates in most countries, present trends project a population between 8 and 10 billion by 2050. The impacts of that many people on our natural resources and ecological systems is a serious concern.

Water may well be the most critical resource in the twenty-first century. Already at least 1.1 billion people lack access to safe drinking water, and twice that many don't have adequate sanitation. Polluted water contributes to the death of more than 15 million people every year, most of them children under age 5. About 40 percent of the world population lives in countries where water

Making Sense of Data

You will see many numbers in this book. Instead of simply glossing over them, think about what they mean relative to some benchmark numbers.

1. For example, you just read that we are adding 85 million people to the world's population each year. How does that compare with the number of people in the United States? In Canada? In Mexico? (If you don't know, where can you find those numbers?)

2. What proportion of the current global population of about 6.5 billion is the 85 million people added each year?

Answers: 1. The United States has about 295 million people, Canada has about 32 million, and Mexico has about 106 million. **2.** About 1.3 percent.

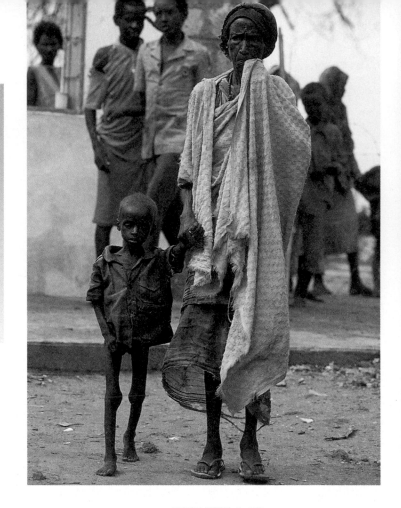

FIGURE 1.12 More than 850 million people are chronically undernourished, and at least 60 million face life-threatening food shortages.

demands now exceed supplies, and the UN projects that by 2025 as many as three-fourths of us could live under similar conditions.

Over the past century, global food production has more than kept pace with human population growth, but there are worries about whether we will be able to maintain this pace. Soil scientists report that about two-thirds of all agricultural lands show signs of degradation. The biotechnology and intensive farming techniques responsible for much of our recent production gains are too expensive for many poor farmers. Can we find ways to produce the food we need without further environmental degradation? And will that food be distributed equitably? In a world of food surpluses, currently more than 850 million people are chronically undernourished, and at least 60 million face acute food shortages due to bad weather or politics (fig. 1.12).

How we obtain and use energy is likely to play a crucial role in our environmental future. Fossil fuels (oil, coal, and natural gas) presently provide around 80 percent of the energy used in industrialized countries. Supplies of these fuels are diminishing, however, and problems associated with their acquisition and use—air and water pollution, mining damage, shipping accidents, and geopolitics—may limit what we do with remaining reserves. Cleaner, renewable energy resources—solar, wind, geothermal, and biomass power—together with conservation could give us cleaner, less destructive options if we invest in appropriate technology.

Burning fossil fuels, making cement, cultivating rice paddies, clearing forests, and other human activities release carbon dioxide

Low   High

PERSISTANCE OF GREENING

FIGURE 1.13 Satellite images and surface temperature data show that polar regions, especially in Eurasia, are becoming green earlier and staying green longer than ever in recorded history. This appears to be evidence of a changing global climate.
Source: NASA, 2002.

and other so-called greenhouse gases, which trap heat in the atmosphere. Over the past 200 years, atmospheric CO_2 concentrations have increased about 30 percent. Climatologists warn that by 2100, if current trends continue, mean global temperatures will probably warm between 1.5° and 6°C (2.7° and 11°F). Global climate change already is affecting a wide variety of biological species (fig. 1.13). Further warming is likely to cause increasingly severe weather events, including droughts in some areas and floods in others. Melting alpine glaciers and snowfields could threaten water supplies on which millions of people depend. Rising sea levels already are flooding low-lying islands and coastal regions, while habitat losses and climate changes are affecting many biological species. Canadian Environment Minister David Anderson said that global climate change is a greater threat than terrorism because it could force hundreds of millions of people from their homes and trigger an economic and social catastrophe.

Air quality has worsened dramatically in many areas. Over southern Asia, for example, satellite images recently revealed a 3-km (2-mile)-thick toxic haze of ash, acids, aerosols, dust, and photochemical products, which regularly covers the entire Indian subcontinent for much of the year. Nobel laureate Paul Crutzen estimates that at least 3 million people die each year from diseases triggered by air pollution. Worldwide, the United Nations estimates, more than 2 billion metric tons of air pollutants (not including carbon dioxide or wind-blown soil) are released each year. Air pollution no longer is merely a local problem. Mercury, polychlorinated biphenyls (PCB), DDT, and other long-lasting pollutants accumulate in arctic ecosystems and native people after being transported by air currents from industrial regions thousands of kilometers to the south. And during certain days, as much as 75 percent of the smog and particulate pollution recorded on the west coast of North America can be traced to Asia.

Biologists report that habitat destruction, overexploitation, pollution, and introduction of exotic organisms are eliminating species at a rate comparable to the great extinction that marked the end of the age of dinosaurs. The UN Environment Programme reports that, over the past century, more than 800 species have disappeared and at least 10,000 species are now considered threatened. This includes about half of all primates and freshwater fish, together with around 10 percent of all plant species. Top predators, including nearly all the big cats in the world, are particularly rare and endangered. A nationwide survey of the United Kingdom in 2004 found that most bird and butterfly populations had declined between 50 and 75 percent over the previous 20 years. At least half of the forests existing before the introduction of agriculture have been cleared, and much of the diverse "old growth" on which many species depend for habitat is rapidly being cut and replaced by secondary growth or monoculture.

As the opening case study for this chapter shows, the ocean is an irreplaceable food resource for many people. More than a billion people in developing countries depend on seafood for their main source of animal protein, but most commercial fisheries around the world are in steep decline (fig. 1.14). According to the World Resources Institute, more than three-quarters of the 441 fish stocks for which information is available are severely depleted or in urgent need of better management. Canadian researchers estimate that 90 percent of all the large predators, including bluefin tuna, marlin, swordfish, sharks, cod, and halibut have been removed from the ocean.

There are also signs of hope

The dismal litany of problems facing us seems overwhelming, doesn't it? Is there hope that we can find solutions to these dilemmas? We think so. The example of Apo Island shows that ordinary people can make important progress both in protecting nature and in improving their lives. Around the world, people who depend on seafood for their livelihood and sustenance are finding that setting aside marine reserves can restore fish populations as well as

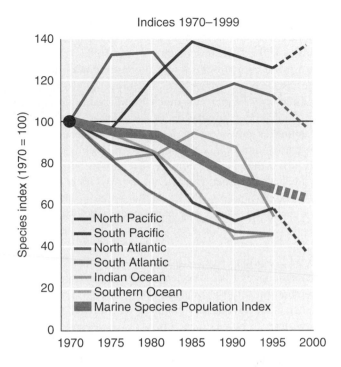

FIGURE 1.14 Changes in abundance of 217 species of marine fish, birds, mammals, and reptiles, from 1970 to 1999, by continent. **Source:** World Wide Fund for Nature (WWF), 2000.

FIGURE 1.15 Fishing has improved dramatically on Apo Island since the reefs have been protected and destructive fishing techniques have been outlawed.

promote human development (fig. 1.15). Showing that these projects can be ecologically sound, economically sustainable, and socially acceptable on the local scale can lead to wider applications. In 2006, Britain's Royal Commission on Environmental

Pollution proposed that one-third of the world's oceans be declared marine preserves similar to those around Apo Island.

As you will see in subsequent chapters in this book, progress has been made in many areas in reducing pollution and curbing wasteful resource use. Many cities in Europe and North America, for example, are cleaner and much more livable now than they were a century ago. Population has stabilized in most industrialized countries and even in some very poor countries where social security and democracy have been established. Over the past 25 years, the average number of children born per woman worldwide has decreased from 6.1 to 2.6. By 2050, the UN Population Division predicts, all developed countries and 75 percent of the developing world will experience a below-replacement fertility rate of 2.1 children per woman. This suggests that the world population will stabilize at about 8.9 billion, rather than the 9.3 billion previously estimated.

The incidence of life-threatening infectious diseases has been reduced sharply in most countries during the past century, while life expectancies have nearly doubled, on average. Smallpox has been completely eradicated and polio has been vanquished except in a few countries. Since 1990 more than 800 million people have gained access to improved water supplies and modern sanitation. In spite of population growth that added nearly a billion people to the world during the 1990s, the number facing food insecurity and chronic hunger during this period actually declined by about 40 million.

A recent study by the University of British Columbia Human Security Centre showed that, since the end of the Cold War in 1992, the number and size of interstate armed conflicts has declined steadily, and the number of battle deaths has fallen even more sharply. The authors conclude that this dramatic improvement in human security is due largely to conflict resolution and United Nations peace building activities in the Cold War aftermath.

Deforestation has slowed in Asia, from more than 8 percent during the 1980s to less than 1 percent in the 1990s. Nature preserves and protected areas have increased dramatically over the past few decades. In 2006, according to UNESCO, there were more than 100,000 parks and nature preserves in the world, representing more than 20 million km^2 (about 7.7 million mi^2), or about 13.5 percent of the world's land area. Ecoregion and habitat protection remains uneven, however, and some areas are protected in name only. Still, this is dramatic progress in biodiversity protection.

Encouraging progress is being made in a transition to renewable energy sources. The European Union has announced a goal of obtaining 22 percent of its electricity and 12 percent of all energy from renewable sources by 2010. British Prime Minister Tony Blair has laid out ambitious plans to fight global warming by cutting carbon dioxide emissions in Britain by 60 percent through energy conservation and a switch to renewables. If nonpolluting, sustainable energy technology is made available to the world's poorer countries, it may be possible to enhance human development while simultaneously reducing environmental damage.

The increased speed at which information and technology now flow around the world holds promise that we can continue to find solutions to our environmental dilemmas. Although we continue to face many challenges, collectively we may be able to implement sustainable development that raises living standards for everyone while also reducing our negative environmental impacts.

HUMAN DIMENSIONS OF ENVIRONMENTAL SCIENCE

Because we live in both the natural and social worlds, and because we and our technology have become such dominant forces on the planet, environmental science must take human institutions and the human condition into account. We live in a world of haves and havenots; a few of us live in increasing luxury, while many others lack the basic necessities for a decent, healthy, productive life. The World Bank estimates that more than 1.4 billion people—about one-fifth of the world's population—live in acute poverty with an income of less than $1 (U.S.) per day. These poorest of the poor generally lack access to an adequate diet, decent housing, basic sanitation, clean water, education, medical care, and other essentials for a humane existence. Seventy percent of those people are women and children. In fact, four out of five people in the world live in what would be considered poverty in the richer countries (fig. 1.16).

Policymakers are becoming aware that eliminating poverty and protecting our common environment are inextricably interlinked because the world's poorest people are both the victims and the agents of environmental degradation. The poorest people

FIGURE 1.16 While many of us live in luxury, more than 1.4 billion people lack access to food, housing, clean water, sanitation, education, medical care, and other essentials for a healthy, productive life. Helping them meet their needs is not only humane, it is essential to protect our mutual environment.

are often forced to meet short-term survival needs at the cost of long-term sustainability. Desperate for croplands to feed themselves and their families, many move into virgin forests or cultivate steep, erosion-prone hillsides, where soil nutrients are exhausted after only a few years. Others migrate to the grimy, crowded slums and ramshackle shantytowns that now surround most major cities in the developing world. With no way to dispose of wastes, the residents often foul their environment further and contaminate the air they breathe and the water on which they depend for washing and drinking.

The cycle of poverty, illness, and limited opportunities can become a self-sustaining process that passes from one generation to another. People who are malnourished and ill can't work productively to obtain food, shelter, or medicine for themselves or their children, who also are malnourished and ill. About 250 million children—mostly in Asia and Africa and some as young as 4 years old—are forced to work under appalling conditions weaving carpets, making ceramics and jewelry, or working in the sex trade. Growing up in these conditions leads to educational, psychological, and developmental deficits that condemn these children to perpetuate this cycle.

Faced with immediate survival needs and few options, these unfortunate people often have no choice but to overharvest resources; in doing so, however, they diminish not only their own options but also those of future generations. And in an increasingly interconnected world, the environments and resource bases damaged by poverty and ignorance are directly linked to those on which we depend.

Where do the rich and poor live?

About one-fifth of the world's population lives in the 20 richest countries, where the average per capita income is above $25,000 (U.S.) per year. Most of these countries are in North America or Western Europe, but Japan, Singapore, Australia, New Zealand, the United Arab Emirates, and Israel also fall into this group. Almost every country, however, even the richest, such as the United States and Canada, has poor people. No doubt everyone reading this book knows about homeless people or other individuals who lack resources for a safe, productive life. According to Bread for the World, 35 million Americans—one-third of them children—live in households without sufficient food.

The other four-fifths of the world's population lives in middle- or low-income countries, where nearly everyone is poor by North American standards. More than 3 billion people live in the poorest nations, where the average per capita income is below $620 (U.S.) per year. China and India are the largest of these countries, with a combined population of about 2.3 billion people. Among the 41 other nations in this category, 33 are in sub-Saharan Africa. All the other lowest-income nations, except Haiti, are in Asia. Although poverty levels in countries such as China and Indonesia have fallen in recent years, most countries in sub-Saharan Africa and much of Latin America have made little progress. The destabilizing and impoverishing effects of earlier colonialism continue to play important roles in the

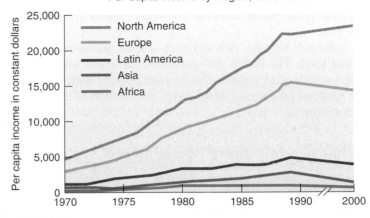

Growing Disparities in Incomes among Regions
Per Capita Income by Region, 1970–2000

FIGURE 1.17 Although the percentage of the world's population living in poverty has decreased slightly over the past 30 years, the relative gap between the rich and poor nations has increased sharply.
Source: United Nations, 2001.

ongoing problems of these unfortunate countries. Meanwhile, the relative gap between rich and poor has increased dramatically (fig. 1.17).

As table 1.4 shows, the gulf between the richest and poorest nations affects many quality-of-life indicators. The average individual in the highest income countries has an annual income more than 100 times that of those in the lowest-income nations. Because of high infant mortality rates, the average family in the poorest countries has more than four times as many children as

TABLE 1.4

Average Indicators of Quality of Life for the 20 Richest and Poorest Countries

Indicator	Poor Countries	Rich Countries
GDP/capita[1]	$230	$27,460
Total fertility[2]	6.3	1.5
Life expectancy	49.3 years	77.8 years
Infant mortality[3]	100	5
Safe drinking water	44%	99%
Adult literacy	38%	99%
Annual population growth	2.6%	0.3%

[1]Annual gross domestic product, in US dollars
[2]Average number of children per woman.
[3]Per 1,000 live births.
Source: UNDP Human Development Indicators, 2005.

those in richer countries. If it were not for immigration, the population in most of the richer countries now would be declining; that in the poorer countries continues to grow at 2.6 percent per year.

The gulf between rich and poor is even greater at the individual level. The richest 200 people in the world have a combined wealth of $1 trillion. This is more than the total owned by the 3 billion people who make up the poorest half of the world's population.

What is a fair share of resources?

The affluent lifestyle that many of us in the richer countries enjoy consumes an inordinate share of the world's natural resources and produces a shockingly high proportion of pollutants and wastes. The United States, for instance, with less than 5 percent of the total population, consumes about one-quarter of most commercially traded commodities, such as oil, and produces a quarter to half of most industrial wastes, such as greenhouse gases, pesticides, and other persistent pollutants.

To get an average American through the day takes about 450 kg (nearly 1,000 lbs) of raw materials, including 18 kg (40 lbs) of fossil fuels, 13 kg (29 lbs) of other minerals, 12 kg (26 lbs) of farm products, 10 kg (22 lbs) of wood and paper, and 450 liters (119 gal) of water. Every year, Americans throw away some 160 million tons of garbage, including 50 million tons of paper, 67 billion cans and bottles, 25 billion styrofoam cups, 18 billion disposable diapers, and 2 billion disposable razors (fig. 1.18).

This profligate resource consumption and waste disposal strains the planet's life-support systems. If everyone in the world tried to live at consumption levels approaching ours, the results would be disastrous. Unless we find ways to curb our desires and produce the things we truly need in less destructive ways, the sustainability of human life on our planet is questionable.

Sustainability is a central theme

Sustainability is a search for ecological stability and human progress that can last over the long term. Of course, neither ecological systems nor human institutions can continue forever. We can work, however, to protect the best aspects of both realms, and to encourage resiliency and adaptability in both of them. World Health Organization Director Gro Harlem Brundtland has defined **sustainable development** as "meeting the needs of the present without compromising the ability of future generations to meet their own needs." In these terms, development means bettering people's lives. Sustainable development, then, means progress in human well-being that we can extend or prolong over many generations, rather than just a few years. To be truly enduring, the benefits of sustainable development must be available to all humans and not just to the members of a privileged group. We will discuss this topic further in chapter 14.

Indigenous peoples are guardians of much of the world's biodiversity

In both rich and poor countries, native, or **indigenous people** are generally the least powerful, most neglected groups in the world. Typically descendants of the original inhabitants of an area taken over by more powerful outsiders, they are distinct from their country's dominant language, culture, religion, and racial communities. Of the world's nearly 6,000 recognized cultures, 5,000 are indigenous ones that account for only about 10 percent of the total world population. In many countries, traditional caste systems, discriminatory laws, economics, or prejudice repress indigenous people. Their unique cultures are disappearing, along with biological diversity as natural habitats are destroyed to satisfy industrialized world appetites for resources. Traditional ways of life are disrupted further by dominant Western culture sweeping around the globe.

At least half of the world's 6,000 distinct languages are dying because they are no longer taught to children. When the last few elders who still speak the language die, so will the culture that was its origin. Lost with those cultures will be a rich repertoire of knowledge about nature and a keen understanding of a particular environment and way of life (fig. 1.19).

Nonetheless, in many places, the 500 million indigenous people who remain in traditional homelands still possess valuable ecological wisdom and remain the guardians of little-disturbed habitats that are refuges for rare and endangered species and undamaged ecosystems. In his book *The Future of Life*, the eminent ecologist E. O. Wilson argues that the cheapest and most effective way to preserve species is to protect the natural ecosystems in which they now live. Interestingly, just 12 countries account for 60 percent of all human languages (fig. 1.20). Seven of these are also among the "megadiversity" countries that contain more than half of all unique plant and

FIGURE 1.18 "And may we continue to be worthy of consuming a disproportionate share of this planet's resources."

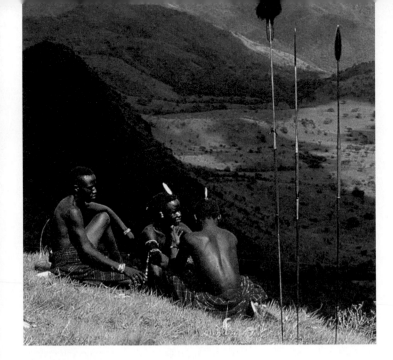

FIGURE 1.19 Do indigenous people have unique knowledge about nature and inalienable rights to traditional territories?

	Highest cultural diversity	Highest biological diversity
Nigeria	Indonesia	Madagascar
Cameroon	New Guinea	South Africa
Australia	Mexico	Malaysia
Congo	China	Cuba
Sudan	Brazil	Peru
Chad	United States	Ecuador
Nepal	Philippines	New Zealand

FIGURE 1.20 Cultural diversity and biodiversity often go hand in hand. Seven of the countries with the highest cultural diversity in the world are also on the list of "megadiversity" countries with the highest number of unique biological organisms (listed in decreasing order of importance).
Source: Norman Myers, Conservation International and Cultural Survival Inc., 2002.

animal species. Conditions that support evolution of many unique species seem to favor development of equally diverse human cultures as well.

Recognizing native land rights and promoting political pluralism can be among the best ways to safeguard ecological processes and endangered species. As the Kuna Indians of Panama say, "Where there are forests, there are native people, and where there are native people, there are forests." A few countries, such as Papua New Guinea, Fiji, Ecuador, Canada, and Australia, acknowledge indigenous title to extensive land areas.

Other countries, unfortunately, ignore the rights of native people. Indonesia, for instance, claims ownership of nearly three-quarters of its forestland and all waters and offshore fishing rights, ignoring the interests of indigenous inhabitants. Similarly, the Philippine government claims possession of all uncultivated land in its territory, while Cameroon and Tanzania recognize no rights at all for forest-dwelling pygmies who represent one of the world's oldest cultures.

REVIEW QUESTIONS

1. Describe how fishing has changed at Apo Island, and the direct and indirect effects on people's lives.
2. Define the terms *environment* and *environmental science*.
3. What are some basic assumptions of science?
4. Distinguish between a hypothesis and a theory.
5. Describe the steps in the scientific method.
6. What is probability? Give an example.
7. Define *sound science*. Give some examples of how the term might be misused.
8. What is shown in figure 3 on p. 9? Describe the meaning of this graph in your own words.
9. How do analytical, creative, critical, logical, and reflective thinking differ?
10. What's the first step in critical thinking according to table 1.3?
11. Distinguish between utilitarian conservation and biocentric preservation. Name two environmental leaders associated with each of these philosophies.
12. What is global environmentalism, and why is it the fourth wave in environmental history? Name a few of its leaders.
13. Why do some experts regard water as the most critical natural resource for the twenty-first century?
14. How many people in the world face famine and chronic hunger?
15. Where in figure 1.13 do the largest areas of persistence of greening occur? What is persistence of greening?
16. Describe some signs of hope in overcoming global environmental problems.
17. What is the link between poverty and environmental quality?
18. Which of the indicators of environmental quality in table 1.4 show the greatest differences between rich and poor countries?
19. Define *sustainability* and *sustainable development*.
20. Why is cultural survival a concern for environmental science?

CRITICAL THINKING AND DISCUSSION

Apply the principles you have learned in this chapter to discuss these questions with other students.

1. How do you think the example of Apo Island's marine sanctuary meets the criteria of being scientifically sound, economically sustainable, and socially acceptable? If you were studying this situation, what information would you look for to support or refute this conclusion?

2. Do you believe the world is ultimately knowable? What values or evidence shape your views on this question? Why might some people have an attitude different from your own?

3. Many social theorists argue that there is no such thing as objective truth or an impartial observer. If you were researching a controversial topic—perhaps whether marine reserves improve fishing and lead to economic development—what steps would you take to try to maintain objectivity and impartiality?

4. Does the world have enough resources for 8 or 10 billion people to live decent, secure, happy, fulfilling lives? What do those terms mean to you? Try to imagine what they mean to others in our global village.

5. Suppose you wanted to study the environmental impacts of a rich versus a poor country. What factors would you examine, and how would you compare them?

SUGGESTED READINGS

Blair, Tony. 2003. Meeting the sustainable development challenge. *Environment* 45(4):20–28.

Diamond, Jared. 2005. *Collapse*. Viking.

Flavin, Christopher, et al. 2005. *State of the World 2005: Redefining Global Security*. Worldwatch Institute.

Marten, Gerald, et al. 2005. Environmental tipping points. *World Watch* 18(6):10–14.

Maypa, A. P., et al. 2002. Long-term trends in yield and catch rates of the coral reef fishery at Apo Island, central Philippines. *Marine and Freshwater Research* 53(2)207–13.

Mumby, Peter J., et al. 2006. Fishing, trophic cascades, and the process of grazing on coral reefs. *Science* 311(5757): 98–101.

Plenderleith, Kristina, ed. 2004. *Indigenous Knowledge and Ethics. A Darrel Posey Reader*. Routledge Press.

Prugh, Thomas, and Erik Assadourian. 2003. What is sustainability, anyway? *World Watch* 16(5):10–21.

Sachs, Jeffery. 2005. *The End of Poverty: Economic Possibilities for Our Time*. Penguin Press.

Speath, J. G. 2004. *Red Sky at Morning*. Yale Univ. Press.

United Nations Environment Programme. 2004. *Geo Yearbook 2004/05: An Overview of Our Changing Environment*. UNEP.

Watkins, K., et al. 2005. *Human Development Report 2005. International Cooperation at a Crossroads: Aid, Trade and Security in an Unequal World*. United Nations Development Programme.

World Resources Institute. 2005. *The Wealth of the Poor: Managing Ecosystems to Fight Poverty*. WRI.

DATA analysis   Working with Graphs

Graphs are visual presentations of data that help us identify trends and understand relationships. A table of numbers is more precise, but most of us have difficulty visualizing patterns in a field of numbers. We can look at the patterns in a graphic representation and make comparisons much more easily than we could with raw data. The ability to read graphs and plot data are essential skills for students of environmental science. You'll see graphs throughout this book. In this exercise, we'll explore some basic principles and applications of the four most common types of graphs.

Line Plots

A line plot represents a data set that involves a sequence of some sort, or a change over time. Table 1.5 shows growth of a population of a single-cell protozoan called *Tetrahymena* in a laboratory culture. To start the experiment, 10 cells were placed in a growth medium (this is time 0 in the table). Each hour for the next 12 hours, the cell number was counted and recorded. How can you convert data such as these into a graph? First, you lay out the axes. By convention, the horizontal (X) axis shows the **independent variable** (one whose values presumably don't depend on the other factor). The vertical axis (Y) records the **dependent variable** (which responds to changes in the other factor). Choose a range for each axis that approximately fits the range of values: this way your data will fill most of the graph, and the patterns should be big enough to see. Now plot the data: for hour 0, find the number of cells (10). Draw a dot *above the 0 on the X-axis* and *level with the 10 on the Y-axis*. Repeat these steps for hour 1, hour 2, and so on. After you've added all your dots, you can connect them with a line. The line makes it easier to see trends in the data. Your graph should look something like figure 1.

TABLE 1.5 Growth of *Tetrahymena*

Hours	Cells
0	10
1	15
2	20
3	40
4	80
5	160
6	320
7	450
8	550
9	600
10	620
11	630
12	640

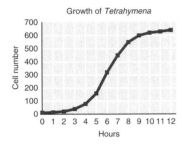

FIGURE 1 Line plot.

Can you extract data from a line plot? Approximately. The process is simply the reverse of the way you just learned to create the graph. You draw a line from any point on the curve to the spot on the axes directly under and parallel to that point. For instance, the point on the curve for hour 6 corresponds to slightly more than 300 cells on the Y-axis.

To make comparisons between two different sets of dependent variables, it's often useful to plot them on the same graph. Look, for example, at figures 1.14 and 1.17 on pp. 17 and 19. You can see they have as many as seven categories of independent variables in a single graph. To make the graphs simpler, the individual data points have been removed and each curve is a simple continuous line. You could still find the data for a specific point on either of these graphs with the process outlined above.

FIGURE 2 Bar graphs.

Bar Graphs and Pie Charts

Bar graphs help you compare *values* of different *categories*, such as the number of apples and the number of oranges. Figure 2, for example (which was fig. 1 on p. 8), shows levels of urban air pollution. The particulate quantities are classified by size rather than a sequence of dates. Thus you can read from the graph that 40 $\mu g/m^3$ occurred about 25 times during the year studied, although you can't tell when that happened. As we discussed in the box on p. 8, this type of graph allows you to see quickly which category is most abundant, where the approximate mean is, and to recognize *outliers* (unusually high or low values, such as 70 $\mu g/m^3$). Notice that the frequency distribution in this graph creates a roughly bell-shaped curve. A bell-shaped curve is often called a *normal distribution,* or a Gaussian distribution, because most large, randomly selected groups have approximately that distribution. More measurements would probably make this histogram closer to a normal, or Gaussian, distribution.

Pie charts, like bar graphs, compare categories. The difference is that categories in a pie chart add up to 100 percent of something. Figure 3, for example, shows major energy sources for the United States. To make comparison between these energy sources easier,

FIGURE 3 Pie chart.

each is expressed as a percentage of the total energy supply. They are grouped together in a circle, the circumference of which is made to be 100 percent. In effect, the outside of the circle is the X-axis, and the size of each sector shows its value on the Y (dependent variable) axis. Because neither of these axes have scales, it's customary to label the sectors with the name of the category and its value.

FIGURE 4 Scatter plot.

Scatter Plots

A scatter plot (fig. 4) shows the distribution of many observations that have no particular sequence. Trends in the scatter plot show a relationship between the two variables. A field of dots trending upward indicates that the dependent variable (Y-axis) increases as the independent variable (X-axis) increases. A tight field of dots shows that Y responds closely to X. A loose cloud shows that Y responds only generally to X. Often scatter plots show no trend, and no relationship between X and Y. Figure 4, for example, shows the number of asthma cases per 1,000 people in an urban area at different particulate levels. Although the data points don't fall on a single line, you can see a trend in which higher pollution levels are associated with higher numbers of asthma cases. To test whether this trend is reliable or merely a chance occurrence, you could use statistical procedures such as regression analysis. These are discussed in the box on statistics on p. 8–9.

Think about these principles as you examine graphs throughout this book. In subsequent chapters, we'll have more discussion of data analysis techniques.

Arcata, California, built an artificial marsh as a low-cost, ecologically based treatment system for sewage effluent.

Environmental Systems

Connections, Cycles, Flows, and Feedback Loops

Most institutions demand unqualified faith; but the institution of science makes skepticism a virtue.

–Robert King Merton–

OBJECTIVES

Studying this chapter will help you answer the following questions:

- What are systems, and how do feedback loops regulate them?
- Chemical bonds hold atoms and molecules together. What would our world be like if there were no chemical bonds, or if they were so strong they never broke apart?
- Ecologists say there is no "away" to throw things, and that everything in the universe tends to slow down and fall apart. What do they mean?

- All living things—except for some unusual organisms living in extreme conditions—depend on the sun as their ultimate energy source. Explain.
- What qualities make water so unique and essential for life as we know it?
- Why are big, fierce animals so rare?
- How and why do elements such as carbon, nitrogen, phosphate, and sulfur cycle through ecosystems?

Case Study   A Natural System for Wastewater Treatment

Arcata, a small town in northwestern California's redwood country, sits at the north end of Humbolt Bay, about 450 km (280 mi) north of San Francisco. Logging, fishing, and farming, long the economic engines of the area, are now giving way to tourism, education, and diversified industries. Many of the 17,000 residents of Arcata choose to live there because of the outstanding environmental quality and recreational opportunities of the area. Until the early 1970s, however, Arcata's sewage system discharged unchlorinated sewage effluent directly into the nearly enclosed bay. Algal blooms discolored the water, while fishing and swimming appeal declined.

In 1974 California enacted a policy prohibiting discharge of untreated wastewater into bays and estuaries. State planners proposed a large and expensive regional sewage treatment plant for the entire Humbolt Bay area. The plan was for large interceptor sewers to encircle the bay with a major underwater pipeline crossing the main navigation channel. Effluent from the proposed plant was to be released offshore into an area of shifting sea bottom in heavy winter storms. The cost and disruption required by this project prompted city officials to look for alternatives.

Ecologists from Humbolt State University pointed out that natural processes could be harnessed to solve Arcata's wastewater problems at a fraction of the cost and disturbance of a conventional treatment plant. After several years of study and experimentation, the city received approval to build a constructed wetland for wastewater treatment. This solved two problems at the same time. Arcata's waterfront was blighted by an abandoned lumbermill pond, channelized sloughs, marginal pasture land, and an abandoned sanitary landfill. Building a new wetland on this degraded area would beautify it while also solving the sewage treatment problem.

Today, Arcata's waterfront has been transformed into 40 ha (about 100 acres) of freshwater and saltwater marshes, brackish ponds, tidal sloughs, and estuaries. This diverse habitat supports a wide variety plants and animals, and is now a wildlife sanctuary and a major tourist attraction. As a home or rest stop for over 200 bird species, the wetland is regarded as one of the best birding sites along the Pacific North Coast. In 1987, the city was awarded a prize by the Ford Foundation for innovative local government projects that included $100,000 to build an interpretive center in the marsh. The wetland is now an outstanding place for outdoor education, scientific study, and recreation.

Arcata's constructed wetlands remove pathogens, pollutants, and nutrients from municipal wastewater using native biological communities and ecological processes in a setting that also serves as a wildlife sanctuary and tourist attraction.

Source: Environmental Protection Agency.

Arcata's constructed wetland is also an unqualified success in wastewater treatment. After primary clarification to remove grit and sediment, wastewater passes through several oxidation ponds that remove about half the organic material (measured by biological oxygen demand) and suspended solids. Sludge captured by the clarifiers goes to digesters that generate methane gas, which is burned to provide heat that speeds the digestion process. Effluent from the oxidation ponds passes through treatment marshes, where sunlight and oxygen kill pathogens while aquatic plants and animals remove remaining organic material along with nitrogen, phosphorus, and other plant nutrients. The effluent is treated with chlorine gas to kill any residual pathogens. Finally, the treated water trickles through several large enhancement marshes that allow chlorine to evaporate and complete pollutant removal.

After 20 years of successful operation, Arcata's constructed wetland has inspired many other communities to find ecological solutions to their problems. Arcata Bay now produces more than half the oysters grown in California. The city also operates a wastewater aquaculture project where salmon, trout, and other fish species are raised in a mixture of wastewater effluent and seawater. Economically, the project is a success, providing excellent water quality treatment far below the cost of conventional systems. And the innovative approach of working with nature rather than against it has made Arcata an important ecotourist destination. Around the world, hundreds of communities have followed Arcata's model and now use constructed wetlands to eliminate up to 98 percent of the pollutants from their wastewater.

We'll discuss water quality and pollution control further in chapter 10. First, however, we'll look, in this chapter, at some of the ecological relationships that determine energy flows and nutrient recycling in biological communities, such as those in Arcata's marshes and wetlands. We'll study how those communities function as systems, as well as the feedback loops that regulate those systems. Understanding these ecological principles will help us find natural solutions to other environmental problems we face.

SYSTEMS

Environmental science seeks to understand **systems,** or networks of interactions among interdependent components, and processes, or compartments and flows. Finding solutions to environmental problems often involves understanding the factors that shape and regulate natural and social systems. Before we look at environmental systems in particular, we should examine some of the properties of systems in general.

Systems can be physical, as in the biological and chemical cycles that move nutrients around the world, or they can be abstract, as in a computer program or a mathematical model. Some systems are **closed,** or self-contained and isolated from outside influences. Others are **open,** in the sense that material, energy, or information is exchanged between parts of the system and external sources or sinks. The constructed wetland in the opening Case Study for this chapter, for example, is an open system. Solar energy warms the wetland and radiant energy (heat) leaves it. Water enters as rain or snow and leaves via evaporation or runoff. Nutrients come from the sewage effluent and surface runoff, and are deposited in bottom in sediment, carried away by departing organisms, or released into the ocean. A nuclear submarine traveling under the polar ice cap, on the other hand, is a mostly closed system. For all practical purposes, nothing enters or leaves the system (except, perhaps, a small amount of heat) while it is underway. Eventually, however, the sub must surface and replenish its supplies. Few systems can be completely self-contained indefinitely.

Some of the most important characteristics of a system are the interconnections between parts of the system and the regulatory mechanisms that control the characteristics of the system. Some of these controlling mechanisms are external to the system. Water level in a pond, for example, is determined mainly by the temperature and rainfall. The pond itself has very little influence on its own volume. Some more interesting regulatory controls are internal. The number of fish in a pond is regulated primarily by food supplies, predation, mortality, and reproductive rates.

Sometimes a system's output can serve as an input to that same system, a circular process known as a **feedback loop.** A **positive feedback loop** uses the output from a process to augment that process. You've probably experienced this in a sound system, where a microphone picks up sound from a speaker and amplifies it in a cycle that gets louder and louder. The output from the speaker becomes an input for the microphone, and the interaction between them drives the system ever further in one direction (loudness). If unchecked, positive feedback loops can lead a system to become increasingly unbalanced and unstable. If someone doesn't turn off the microphone, for example, it can easily destroy the speaker.

Negative feedback, on the other hand, reverses the direction in which a system is moving. A thermostat, for example, regulates room temperature via a pair of negative feedback loops with a furnace. When the room becomes too hot, the thermostat turns the furnace off; if it becomes too cold, it turns the furnace on. When regulatory processes balance each other, as is the case with a well-functioning thermostat, we say that the system is in dynamic equilibrium or **homeostasis,** the tendency of a system to maintain constant or stable internal conditions. Your body is generally in a state of homeostasis. Counterbalancing effects of hormones, physiological processes, and conscious actions maintain most bodily functions within reasonable limits.

Feedback mechanisms in nature also can exacerbate or moderate interactions at both large and small scales. In the Arcata marsh, for example, some of the interactions act as positive feedbacks, while others are negative. The metabolic wastes (urea, CO_2) produced by fish and other herbivores act as a positive feedback to stimulate plant growth (fig. 2.1). Excess plant biomass, however, acts as a negative feedback by blocking sunlight and reducing further plant growth. If too many herbivores reduce plant biomass below a critical level, there won't be enough food to support the fish biomass, again acting as a negative feedback. Altogether, these regulatory mechanisms generally maintain a balance between the pond organisms and their environment.

Disturbances, periodic, destructive events such as fire or floods, are a normal part of many natural systems. They may even depend on these disturbances. Floodplains often need periodic floods to replenish soil nutrients and maintain vegetative diversity; prairies depend on fires to recycle nutrients, slow tree growth, and reduce built-up, undecayed plant litter; ponderosa pine forests need regular fires to clear understory vegetation. When a combination of positive and negative feedbacks allow a system to recover from some external disturbance, we say the system has **resilience.**

Emergent properties are characteristics of a whole system that are greater than the sum of the system's parts. For example, you are a system consisting of tissues, organs, and metabolic reactions, but you are also much more than the physical materials and processes from which you are made. You can sing, dance, plan, produce ideas and art, and share feelings with those around you. All of these properties emerge because you function well as a system. Similarly, a tree is more than a mass of stored carbon. It provides structure to a forest, habitat for other organisms, shades and cools the ground, and holds soil on a hillside with its roots. Often, we describe systems in terms of their flows, compartments, and feedbacks, but it is the emergent properties that make them interesting to study (fig. 2.2).

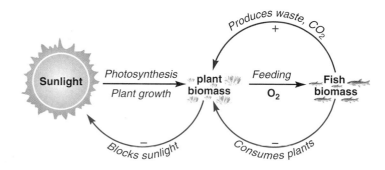

FIGURE 2.1 Systems consist of compartments (also known as state variables) such as fish and plants, and flows of resources, such as nutrients or O_2. Feedback loops enhance or inhibit flows and the growth of compartments.

Principles of Environmental Science

FIGURE 2.2 Emergent properties of a systems, including beautiful sights and sounds, make them exciting to study.

ELEMENTS OF LIFE

How are nutrients and energy exchanged between plants and animals in Arcata's constructed wetlands? And what energy source keeps this whole system running? These questions are at the core of **ecology**, the scientific study of relationships between organisms and their environment. In this chapter we'll introduce a number of concepts that are essential to understanding how living things function in their environment.

As an introduction to principles of ecology, this chapter first reviews the nature of matter and energy, then explores the ways organisms acquire and use energy and chemical elements. Then we'll investigate feeding relationships among organisms—the ways that energy and nutrients are passed from one living thing to another—forming the basis of ecosystems. Finally, we'll review some of the key substances that cycle through organisms, ecosytems, and our environment.

In chapters 3 and 5 we'll continue investigating concepts of ecology: general organization of ecosystems by environmental types and landscapes, and principles of population growth.

In a sense, every organism is a chemical factory that captures matter and energy from its environment and transforms them into structures and processes that make life possible. To understand how these processes work, we will begin with some of the fundamental properties of matter and energy.

Matter is made of atoms, molecules, and compounds

Everything that takes up space and has mass is **matter.** Matter exists in three distinct states—solid, liquid, and gas—due to differences in the arrangement of its constitutive particles. Water, for example, can exist as ice (solid), as liquid water, or as water vapor (gas).

Under ordinary circumstances, matter is neither created nor destroyed but rather is recycled over and over again. Some of the molecules that make up your body probably contain atoms that once made up the body of a dinosaur and most certainly were part of many smaller prehistoric organisms, as chemical elements are used and reused by living organisms. Matter is transformed and combined in different ways, but it doesn't disappear; everything goes somewhere. These statements paraphrase the physical principle of **conservation of matter.**

How does this principle apply to human relationships with the biosphere? Particularly in affluent societies, we use natural resources to produce an incredible amount of "disposable" consumer goods. If everything goes somewhere, where do the things we dispose of go after the garbage truck leaves? As the sheer amount of "disposed-of stuff" increases, we are having greater problems finding places to put it. Ultimately, there is no "away" where we can throw things we don't want any more.

Matter consists of **elements,** which are substances that cannot be broken down into simpler forms by ordinary chemical reactions. Each of the 115 known elements (92 natural, plus 23 created under special conditions) has distinct chemical characteristics. Just four elements—oxygen, carbon, hydrogen, and nitrogen—are responsible for more than 96 percent of the mass of most living organisms.

All elements are composed of **atoms,** which are the smallest particles that exhibit the characteristics of the element. Atoms are composed of positively charged protons, negatively charged electrons, and electrically neutral neutrons. Protons and neutrons, which have approximately the same mass, are clustered in the nucleus in the center of the atom (fig. 2.3). Electrons, which are tiny in comparison to the other particles, orbit the nucleus at the speed of light.

Each element has a characteristic number of protons per atom, called its **atomic number.** The number of neutrons in different atoms of the same element can vary slightly. Thus, the atomic mass, which is the sum of the protons and neutrons in each nucleus, also can vary. We call forms of a single element that differ in atomic mass **isotopes.** For example, hydrogen, the lightest element, normally has only one proton (and no neutrons) in its nucleus. A small percentage of hydrogen atoms have one proton and one neutron. We call this isotope deuterium (^{2}H). An even smaller percentage of natural hydrogen called tritium (^{3}H) has one proton plus two neutrons. The heavy form of nitrogen (^{15}N) has one more neutron in its nucleus than does the more common ^{14}N. Both these nitrogen isotopes are

FIGURE 2.3 As difficult as it may be to imagine when you look at a solid object, all matter is composed of tiny, moving particles, separated by space and held together by energy. This model represents a carbon-12 atom, with a nucleus containing six protons and six neutrons; the six electrons are represented as a fuzzy cloud of potential locations, rather than as individual particles.

- 6 protons
- 6 neutrons
- 6 electrons

stable but some isotopes are unstable—that is, they may spontaneously emit electromagnetic energy, or subatomic particles, or both. Radioactive waste, and nuclear energy, result from unstable isotopes of elements such as uranium and plutonium.

Chemical bonds hold molecules together

Atoms often join to form **compounds,** or substances composed of different kinds of atoms (fig. 2.4). A pair or group of atoms that can exist as a single unit is known as a **molecule.** Some elements commonly occur as molecules, such as molecular oxygen (O_2) or molecular nitrogen (N_2), and some compounds can exist as molecules, such as glucose ($C_6H_{12}O_6$). In contrast to these molecules, sodium chloride (NaCl, table salt) is a compound that cannot exist as a single pair of atoms. Instead it occurs in a large mass of Na and Cl atoms or as two ions, Na^+ and Cl^-, in solution. Most molecules consist of only a few atoms. Others, such as proteins and nucleic acids, can include millions or even billions of atoms.

When ions with opposite charges form a compound, the electrical attraction holding them together is an *ionic bond.* Sometimes atoms form bonds by *sharing* electrons. For example, two hydrogen atoms can bond by sharing a single electron—it orbits the two hydrogen nuclei equally and holds the atoms together. Such electron-sharing bonds are known as *covalent bonds.* Carbon (C) can form covalent bonds simultaneously with four other atoms, so carbon can create complex structures such as sugars

and proteins. Atoms in covalent bonds do not always share electrons evenly. An important example in environmental science is the covalent bonds in water (H_2O). The oxygen atom attracts the shared electrons more strongly than do the two hydrogen atoms. Consequently, the hydrogen portion of the molecule has a slight positive charge, while the oxygen has a slight negative charge. These charges create a mild attraction between water molecules, so that water tends to be somewhat cohesive. This fact helps explain some of the remarkable properties of water (see Exploring Science, p. 29).

When an atom gives up one or more electrons, we say it is *oxidized* (because it is very often oxygen that takes the electron, as bonds are formed with this very common and highly reactive element). When an atom gains electrons, we say it is *reduced.* Chemical reactions necessary for life involve oxidation and reduction: Oxidation of sugar and starch molecules, for example, is an important part of how you gain energy from food.

Breaking bonds requires energy, while forming bonds generally releases energy. Burning wood in your fireplace breaks up large molecules, such as cellulose, and forms many smaller ones, such as carbon dioxide and water. The net result is a release of energy (heat). Generally, some energy input (activation energy) is needed to initiate these reactions. In your fireplace, a match might provide the needed activation energy. In your car, a spark from the battery provides activation energy to initiate the oxidation (burning) of gasoline.

FIGURE 2.4 These common molecules, with atoms held together by covalent bonds, are important components of the atmosphere or important pollutants.

Electrical charge is an important chemical characteristic

Atoms frequently gain or lose electrons, acquiring a negative or positive electrical charge. Charged atoms (or combinations of atoms) are called **ions.** Negatively charged ions (with one or more extra electrons) are *anions.* Positively charged ions are *cations.* A hydrogen (H) atom, for example, can give up its sole electron to become a hydrogen ion (H^+). Chlorine (Cl) readily gains electrons, forming chlorine ions (Cl^-).

Substances that readily give up hydrogen ions in water are known as **acids.** Hydrochloric acid, for example, dissociates in water to form H^+ and Cl^- ions. In later chapters, you may read about acid rain (which has an abundance of H^+ ions), acid mine drainage, and many other environmental problems involving acids. In general, acids cause environmental damage because the H^+ ions react readily with living tissues (such as your skin or tissues of fish larvae) and with nonliving substances (such as the limestone on buildings, which erodes under acid rain).

Substances that readily bond with H^+ ions are called **bases** or alkaline substances. Sodium hydroxide (NaOH), for example, releases hydroxide ions (OH^-) that bond with H^+ ions in water. Bases can be highly reactive, so they also cause significant environmental problems. Acids and bases can also be essential to living things: The acids in your stomach dissolve food, for example, and acids in soil help make nutrients available to growing plants.

If travelers from another solar system were to visit our lovely, cool, blue planet, they might call it Aqua rather than Terra because of its outstanding feature: the abundance of streams, rivers, lakes, and oceans of liquid water. Our planet is the only place we know where water exists as a liquid in any appreciable quantity. Water covers nearly three-fourths of the earth's surface and moves around constantly through evaporation, precipitation, and runoff that distribute nutrients, replenish freshwater supplies, and shape the land. Water makes up 60 to 70 percent of the weight of most living organisms. It fills cells, giving form and support to tissues. Among water's unique, almost magical qualities, are the following:

1. Water molecules are polar, that is, they have a slight positive charge on one side and a slight negative charge on the other side. Therefore, water readily dissolves polar or ionic substances, including sugars and nutrients, and carries materials to and from cells.
2. Water is the only inorganic liquid that exists under normal conditions at temperatures suitable for life. Most substances exist as either a solid or a gas, with only a very narrow liquid temperature range. Organisms synthesize organic compounds such as oils and alcohols that remain liquid at ambient temperatures and are therefore extremely valuable to life, but the original and predominant liquid in nature is water.
3. Water molecules are cohesive, tending to stick together tenaciously. You have

experienced this property if you have ever done a belly flop off a diving board. Water has the highest surface tension of any common, natural liquid. Water also adheres to surfaces. As a result, water is subject to *capillary action:* it can be drawn into small channels. Without capillary action, movement of water and nutrients into groundwater reservoirs and through living organisms might not be possible.

4. Water is unique in that it expands when it crystallizes. Most substances shrink as they change from liquid to solid. Ice floats because it is less dense than liquid water. When temperatures fall below freezing, the surface layers of lakes, rivers, and oceans cool faster and freeze before deeper water. Floating ice then insulates underlying layers, keeping most water bodies liquid (and aquatic organisms alive) throughout the winter in most places. Without this feature, many aquatic systems would freeze solid in winter.
5. Water has a high heat of vaporization, using a great deal of heat to convert from liquid to vapor. Consequently, evaporating water is an effective way for organisms to shed excess heat. Many animals pant or sweat to moisten evaporative cooling surfaces. Why do you feel less comfortable on a hot, humid day than on a hot, dry day? Because the water vapor–laden air inhibits the rate of evaporation from

Surface tension is demonstrated by the resistance of a water surface to penetration, as when it is walked upon by a water strider.

your skin, thereby impairing your ability to shed heat.

6. Water also has a high specific heat; that is, a great deal of heat is absorbed before it changes temperature. The slow response of water to temperature change helps moderate global temperatures, keeping the environment warm in winter and cool in summer. This effect is especially noticeable near the ocean, but it is important globally.

All these properties make water a unique and vitally important component of the ecological cycles that move materials and energy and make life on earth possible.

We describe the strength of an acid and base by its **pH,** the negative logarithm of its concentration of H^+ ions (fig. 2.5). Acids have a pH below 7; bases have a pH greater than 7. A solution of exactly pH 7 is "neutral." Because the pH scale is logarithmic, pH 6 represents *ten times* more hydrogen ions in solution than pH 7.

A solution can be neutralized by adding buffers, or substances that accept or release hydrogen ions. In the environment, for example, alkaline rock can buffer acidic precipitation, decreasing its acidity. Lakes with acidic bedrock, such as granite, are especially vulnerable to acid rain because they have little buffering capacity.

Organic compounds have a carbon backbone

Organisms use some elements in abundance, others in trace amounts, and others not at all. Certain vital substances are

concentrated within cells, while others are actively excluded. Carbon is a particularly important element because chains and rings of carbon atoms form the skeletons of **organic compounds,** the material of which biomolecules, and therefore living organisms, are made.

The four major categories of organic compounds in living things ("bio-organic compounds") are lipids, carbohydrates, proteins, and nucleic acids. Lipids (including fats and oils) store energy for cells, and they provide the core of cell membranes and other structures. Many hormones are also lipids. Lipids do not readily dissolve in water, and their basic structure is a chain of carbon atoms with attached hydrogen atoms. This structure makes them part of the family of hydrocarbons (fig. 2.6a). Carbohydrates (including sugars, starches, and cellulose) also store energy and provide structure to cells. Like lipids, carbohydrates

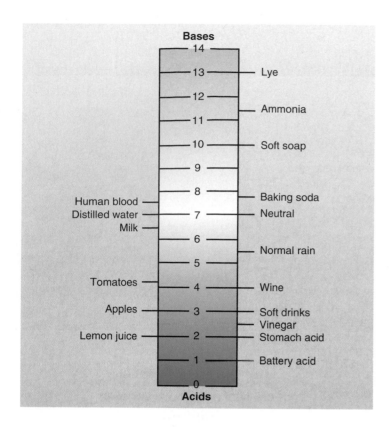

FIGURE 2.5 The pH scale. The numbers represent the negative logarithm of the hydrogen ion concentration in water. Alkaline (basic) solutions have a pH greater than 7. Acids (pH less than 7) have high concentrations of reactive H⁺ ions.

have a basic structure of carbon atoms, but hydroxide (OH) groups replace half the hydrogen atoms in their basic structure, and they usually consist of long chains of nucleic acids. Glucose (fig. 2.6b) is an example of a very simple sugar.

Proteins are composed of chains of subunits called amino acids (fig. 2.6c). Folded into complex three-dimensional shapes, proteins provide structure to cells and are used for countless cell functions. Enzymes, such as those that release energy from lipids and carbohydrates, are proteins. Proteins also help identify disease-causing microbes, make muscles move, transport oxygen to cells, and regulate cell activity.

Nucleotides are complex molecules made of a five-carbon sugar (ribose or deoxyribose), one or more phosphate groups, and an organic nitrogen-containing base called either a purine or pyrimidine (fig. 2.6d). Nucleotides are extremely important as signaling molecules (they carry information between cells, tissues, and organs) and as sources of intracellular energy. They also form long chains called *ribonucleic acid* (RNA) or **deoxyribonucleic acid (DNA)** that are essential for storing and expressing genetic information. Only four kinds of nucleotides (adenine, guanine, cytosine, and uracil) occur in DNA, but there can be billions of these molecules lined up in a very specific sequence. Groups of three nucleotides (called codons) act as the letters in a message that directs the amino acid sequences in proteins. The thousands of different kinds of proteins, for which DNA provides

FIGURE 2.6 The four major groups of organic molecules are based on repeating subunits of these carbon-based structures. Basic structures are shown for (a) butyric acid (a building block of lipids) and a hydrocarbon, (b) a simple carbohydrate, (c) a protein, and (d) a nucleic acid.

the code (or instructions), make up most of the structure and machinery of your body.

Long chains of DNA bind together to form a stable double helix (fig. 2.7). These chains separate for replication in preparation for cell division or to express their genetic information during protein synthesis. Molecular biologists have developed techniques for extracting DNA from cells and reading its nucleotide sequence. These techniques are proving to be tremendously powerful in medical genetics and agriculture. They also are extremely useful in fields such as forensics and taxonomy. Because every individual has a unique set of DNA molecules, sequencing their nucleotide content can provide a distinctive individual identification. See the related story "Bar-Coding Life" at www.mhhe.com/cunningham4e.

Cells are the fundamental units of life

All living organisms are composed of **cells,** minute compartments within which the processes of life are carried out (fig. 2.8). Microscopic organisms such as bacteria, some algae, and protozoa are composed of single cells. Most higher organisms are multicellular, usually with many different cell varieties. Your body, for instance, is composed of several trillion cells of about two hundred distinct types. Every cell is surrounded by a thin but dynamic membrane of lipid and protein that receives information about the exterior world and regulates the flow of materials between the cell and its environment. Inside, cells are subdivided into tiny organelles and subcellular particles that provide the machinery for life. Some of these organelles store and release energy. Others manage and distribute information. Still others create the internal structure that gives the cell its shape and allows it to fulfill its role.

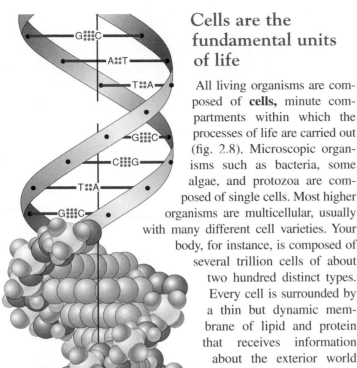

FIGURE 2.7 A composite molecular model of DNA. The lower part shows individual atoms, while the upper part has been simplified to show the strands of the double helix held together by hydrogen bonds (small dots) between matching nucleotides (A, T, G, and C). A complete DNA molecule contains millions of nucleotides and carries genetic information for many specific, inheritable traits.

All of the chemical reactions required to create these various structures, provide them with energy and materials to carry out their functions, dispose of wastes, and perform other functions of life at the cellular level are carried out by a special class of proteins called **enzymes.** Enzymes are molecular catalysts, that is, they regulate chemical reactions without being used up or inactivated in the process. Think of them as tools: Like hammers or wrenches, they do their jobs without being consumed or damaged as they work. There are generally thousands of different kinds of enzymes in every cell, all necessary to carry out the many processes on which life depends. Altogether, the multitude of enzymatic reactions performed by an organism is called its **metabolism.**

✹ ENERGY

If matter is the material of which things are made, energy provides the force to hold structures together, tear them apart, and move them from one place to another. In this section we will look at some fundamental characteristics of these components of our world.

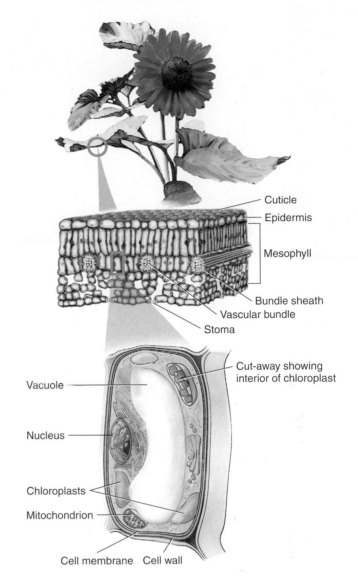

FIGURE 2.8 Plant tissues and a single cell's interior. Cell components include a cellulose cell wall; a nucleus; a large, empty vacuole; and several chloroplasts, which carry out photosynthesis.

Energy occurs in different types and qualities

Energy is the ability to do work such as moving matter over a distance or causing a heat transfer between two objects at different temperatures. Energy can take many different forms. Heat, light, electricity, and chemical energy are examples that we all experience. The energy contained in moving objects is called **kinetic energy.** A rock rolling down a hill, the wind blowing through the trees, water flowing over a dam (fig. 2.9), or electrons speeding around the nucleus of an atom are all examples of kinetic energy.

Potential energy is stored energy that is latent but available for use. A rock poised at the top of a hill and water stored behind a dam are examples of potential energy. **Chemical energy** stored in the food that you eat and the gasoline that you put into your car are also examples of potential energy that can be released to do useful work. Energy is often measured in units of heat (calories)

FIGURE 2.9 Water stored behind this dam represents potential energy. Water flowing over the dam has kinetic energy, some of which is converted to heat.

or work (joules). One joule (J) is the work done when one kilogram is accelerated at one meter per second per second. One calorie is the amount of energy needed to heat one gram of pure water one degree Celsius. A calorie can also be measured as 4.184 J.

Heat describes the energy that can be transferred between objects of different temperature. When a substance absorbs heat, the kinetic energy of its molecules increases, or it may change state: A solid may become a liquid, or a liquid may become a gas. We sense change in heat content as change in temperature (unless the substance changes state).

An object can have a high heat content but a low temperature, such as a lake that freezes slowly in the fall. Other objects, like a burning match, have a high temperature but little heat content. Heat storage in lakes and oceans is essential to moderating climates and maintaining biological communities. Heat absorbed in changing states is also critical. As you will read in chapter 10, evaporation and condensation of water in the atmosphere helps distribute heat around the globe.

Energy that is diffused, dispersed, and low in temperature is considered low-quality energy because it is difficult to gather and use for productive purposes. The heat stored in the oceans, for instance, is immense but hard to capture and use, so it is low quality. Conversely, energy that is intense, concentrated, and high in temperature is high-quality energy because of its usefulness in carrying out work. The intense flames of a very hot fire or high-voltage electrical energy are examples of high-quality forms that are valuable to humans. Many of our alternative energy sources (such as wind) are diffuse compared to the higher-quality, more concentrated chemical energy in oil, coal, or gas.

Thermodynamics regulates energy transfers

Atoms and molecules cycle endlessly through organisms and their environment, but energy flows in a one-way path. A constant supply of energy—nearly all of it from the sun—is needed to keep biological processes running. Energy can be used repeatedly as it flows through the system, and it can be stored temporarily in the chemical bonds of organic molecules, but eventually it is released and dissipated.

The study of thermodynamics deals with how energy is transferred in natural processes. More specifically, it deals with the rates of flow and the transformation of energy from one form or quality to another. Thermodynamics is a complex, quantitative discipline, but you don't need a great deal of math to understand some of the broad principles that shape our world and our lives. The **first law of thermodynamics** states that energy is *conserved;* that is, it is neither created nor destroyed under normal conditions. Energy may be transformed, for example, from the energy in a chemical bond to heat energy, but the total amount does not change.

The **second law of thermodynamics** states that, with each successive energy transfer or transformation in a system, less energy is available to do work. That is, energy is degraded to lower-quality forms, or it dissipates and is lost, as it is used. When you drive a car, for example, the chemical energy of the gas is degraded to kinetic energy and heat, which dissipates, eventually, to space. The second law recognizes that disorder, or **entropy,** tends to increase in all natural systems. Consequently, there is always less *useful* energy available when you finish a process than there was before you started. Because of this loss, everything in the universe tends to fall apart, slow down, and get more disorganized.

How does the second law of thermodynamics apply to organisms and biological systems? Organisms are highly organized, both structurally and metabolically. Constant care and maintenance is required to keep up this organization, and a continual supply of energy is required to maintain these processes. Every time some energy is used by a cell to do work, some of that energy is dissipated or lost as heat. If cellular energy supplies are interrupted or depleted, the result—sooner or later—is death.

ENERGY FOR LIFE

Where does the energy needed by living organisms come from? How is it captured and used to do work? For nearly all plants and animals living on the earth's surface, the sun is the ultimate energy source, but for organisms living deep in the earth's crust or at the bottom of the oceans, where sunlight is unavailable, chemicals derived from rocks provide alternate energy sources. We'll consider this alternate energy pathway first because it seems to be more ancient. Before green plants existed, we believe that ancient bacteria-like cells probably lived by processing chemicals in hot springs.

Extremophiles live in severe conditions

Until recently, the deep ocean floor was believed to be a biological desert. Cold, dark, subject to crushing pressures, and without any known energy supply, it was a place where scientists thought nothing could survive. Undersea explorations in the 1970s, however, revealed dense colonies of animals—blind shrimp, giant tube worms, strange crabs, and bizarre clams—clustered around vents called black smokers, where boiling hot, mineral-laden water bubbles up through cracks in the earth's crust. But how were these organisms getting energy? The answer is **chemosynthesis,** the process in which inorganic chemicals, such as hydrogen sulfide (H_2S) or hydrogen gas (H_2), serve as an energy source for synthesis of organic molecules.

Discovering organisms living under the severe conditions of deep-sea hydrothermal vents led to an interest in other sites that seem exceptionally harsh to us. A variety of interesting organisms have been discovered in hot springs in thermal areas such as Yellowstone National Park, in intensely salty lakes, and even in deep rock formations (up to 1,500 m, or nearly a mile deep) in Columbia River basalts, for example. Some species are amazingly hardy. An organism called *Pyrolobus fumarii* can withstand temperatures up to 113°C (235°F). Most of these extremophiles are archaea, single-celled organisms intermediate between bacteria or eukaryotic organisms (those with their genetic material enclosed in a nucleus (see fig. 2.8). Archaea are thought to be the most primitive of all living organisms, and the conditions under which they live are thought to be similar to those in which life first evolved.

Deep-sea explorations of areas without thermal vents also have found abundant life (fig. 2.10). We now know that archaea live in oceanic sediments in astonishing numbers. The deepest of these species (they can be 800 m or more below the ocean floor) make methane from gaseous hydrogen (H_2) and carbon dioxide (CO_2), derived from rocks. Other species oxidize methane using sulfur to create hydrogen sulfide (H_2S), which is consumed by bacteria that serve as a food source for more complex organisms such as tubeworms. But why should we care about this exotic community? It's estimated that the total mass of microbes (microscopic organisms) living beneath the seafloor represents nearly one-third of all the biomass (organic material) on the planet. Furthermore, the vast supply of methane generated by this community could be either a great resource or a terrible threat to us.

The total amount of methane made by these microbes is probably greater than all the known reserves of coal, gas, and oil. If we could safely extract the huge supplies of methane hydrate in ocean sediments, it could supply our energy needs for hundreds of years. Of greater immediate importance is that if methane-eating microbes weren't intercepting the methane produced by their neighbors, more than 300 million tons per year of this potent greenhouse gas would probably be bubbling to the surface, and we'd have run-away global warming. Some geologists believe that sudden "burps" of methane from melting hydrate deposits may have been responsible for catastrophic landslides and tsunamis, sudden climatic shifts, and mass extinctions in the past. We'll look further at issues of global climate change in chapter 9.

Green plants get energy from the sun

Our sun is a star, a fiery ball of exploding hydrogen gas. Its thermonuclear reactions emit powerful forms of radiation, including potentially deadly ultraviolet and nuclear radiation (fig. 2.11), yet life here is nurtured by, and dependent upon, this searing, energy source. Solar energy is essential to life for two main reasons.

First, the sun provides warmth. Most organisms survive within a relatively narrow temperature range. In fact, each species has its own range of temperatures within which it can function normally. At high temperatures (above 40°C), biomolecules begin to break down or become distorted and nonfunctional. At low temperatures (near 0°C), some chemical reactions of metabolism occur too slowly to enable organisms to grow and reproduce. Other planets in our solar system are either too hot or too cold to support life as we know it. The earth's water and atmosphere help to moderate, maintain, and distribute the sun's heat.

Second, nearly all organisms on the earth's surface organisms depend on solar radiation for life-sustaining energy, which is captured by green plants, algae, and some bacteria in a process called **photosynthesis.** Photosynthesis converts radiant energy into useful, high-quality chemical energy in the bonds that hold together organic molecules.

How much of the available solar energy is actually used by organisms? The amount of incoming solar radiation is enormous, about 1,372 watts/m^2 at the top of the atmosphere (1 watt = 1 J per second). However, more than half of the incoming sunlight may be reflected or absorbed by atmospheric clouds, dust, and gases. In particular, harmful, short wavelengths are filtered out by gases (such as ozone) in the upper atmosphere; thus, the atmosphere is a valuable shield, protecting life-forms from harmful doses of ultraviolet and other forms of radiation. Even with these energy reductions, however, the sun provides much more

FIGURE 2.10 A colony of tube worms and mussels clusters over a cool, deep-sea methane seep in the Gulf of Mexico.

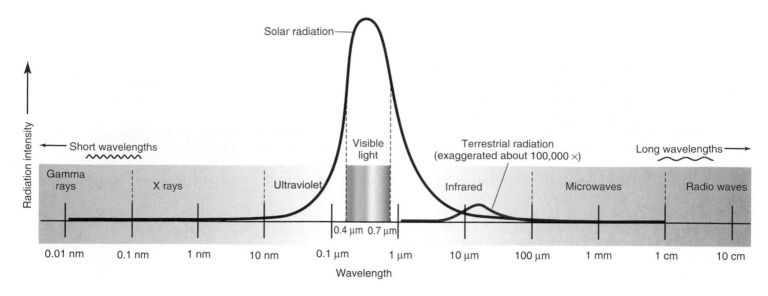

FIGURE 2.11 The electromagnetic spectrum. Our eyes are sensitive to light wavelengths, which make up nearly half the energy that reaches the earth's surface (represented by the area under the "solar radiation" curve). Photosynthesizing plants use the most abundant solar wavelengths (light and infrared). The earth reemits lower-energy, longer wavelenghts (shown by the "terrestrial radiation" curve), mainly the infrared part of the spectrum.

energy than biological systems can harness, and more than enough for all our energy needs if technology could enable us to tap it efficiently.

Of the solar radiation that does reach the earth's surface, about 10 percent is ultraviolet, 45 percent is visible, and 45 percent is infrared. Most of that energy is absorbed by land or water or is reflected into space by water, snow, and land surfaces. (Seen from outer space, Earth shines about as brightly as Venus.)

Of the energy that reaches the earth's surface, photosynthesis uses only certain wavelengths, mainly red and blue light. (Most plants reflect green wavelengths, so that is the color they appear to us.) Half of the energy plants absorb is used in evaporating water. In the end, only 1 to 2 percent of the sunlight falling on plants is available for photosynthesis. This small percentage represents the energy base for virtually all life in the biosphere!

How does photosynthesis capture energy?

Photosynthesis occurs in tiny membranous organelles called chloroplasts that reside within plant cells (see fig. 2.8). The most important key to this process is chlorophyll, a unique green molecule that can absorb light energy and use it to create high-energy chemical bonds in compounds that serve as the fuel for all subsequent cellular metabolism. Chlorophyll doesn't do this important job all alone, however. It is assisted by a large group of other lipid, sugar, protein, and nucleotide molecules. Together these components carry out two interconnected cyclic sets of reactions (fig. 2.12).

Photosynthesis begins with a series of steps called light-dependent reactions: These occur only while the chloroplast is receiving light. Enzymes split water molecules and release

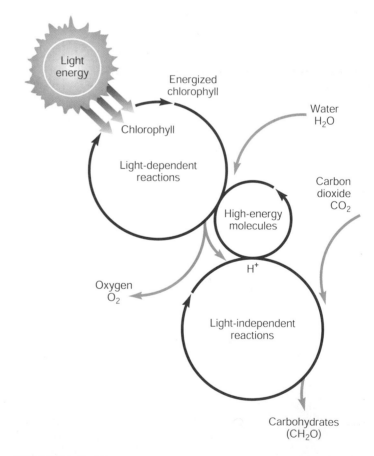

FIGURE 2.12 Photosynthesis involves a series of reactions in which chlorophyll captures light energy and forms high-energy molecules, ATP and NADPH. Light-independent reactions then use energy from ATP and NADPH to fix carbon (from air) in organic molecules.

Principles of Environmental Science

molecular oxygen (O_2). This is the source of all the oxygen in the atmosphere on which all animals, including you, depend for life. The light-dependent reactions also create mobile, high-energy molecules (adenosine triphosphate, or ATP, and nicotinamide adenine dinucleotide phosphate, or NADPH), which provide energy for the next set of processes, the light-independent reactions. As their name implies, these reactions do not use light directly. Here, enzymes extract energy from ATP and NADPH to add carbon atoms (from carbon dioxide) to simple sugar molecules, such as glucose. These molecules provide the building blocks for larger, more complex organic molecules.

In most temperate-zone plants, photosynthesis can be summarized in the following equation:

$$6H_2O + 6CO_2 + \text{solar energy} \xrightarrow[\text{chlorophyll}]{} C_6H_{12}O_6 \text{ (sugar)} + 6O_2$$

We read this equation as "water plus carbon dioxide plus energy produces sugar plus oxygen." The reason the equation uses six water and six carbon dioxide molecules is that it takes six carbon atoms to make the sugar product. If you look closely, you will see that all the atoms in the reactants balance with those in the products. This is an example of conservation of matter.

You might wonder how making a simple sugar benefits the plant. The answer is that glucose is an energy-rich compound that serves as the central, primary fuel for all metabolic processes of cells. The energy in its chemical bonds—the ones created by photosynthesis—can be released by other enzymes and used to make other molecules (lipids, proteins, nucleic acids, or other carbohydrates), or it can drive kinetic processes such as movement of ions across membranes, transmission of messages, changes in cellular shape or structure, or movement of the cell itself in some cases. This process of releasing chemical energy, called **cellular respiration,** involves splitting carbon and hydrogen atoms from the sugar molecule and recombining them with oxygen to re-create carbon dioxide and water. The net chemical reaction, then, is the reverse of photosynthesis:

$$C_6H_{12}O_6 + 6O_2 \xrightarrow{\hspace{1cm}} 6H_2O + 6CO_2 + \text{released energy}$$

Note that in photosynthesis, energy is *captured,* while in respiration, energy is *released.* Similarly, photosynthesis *consumes* water and carbon dioxide to *produce* sugar and oxygen, while respiration does just the opposite. In both sets of reactions, energy is stored temporarily in chemical bonds, which constitute a kind of energy currency for the cell. Plants carry out both photosynthesis and respiration, but during the day, if light, water, and CO_2 are available, they have a net production of O_2 and carbohydrates.

We animals don't have chlorophyll and can't carry out photosynthetic food production. We do have the components for cellular respiration, however. In fact, this is how we get all our energy for life. We eat plants—or other animals that have eaten plants—and break down the organic molecules in our food through cellular respiration to obtain energy (fig. 2.13). In the process, we also consume oxygen and release carbon

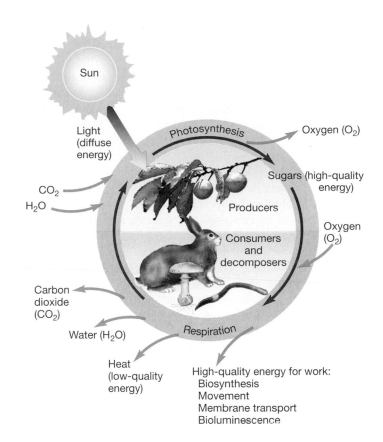

FIGURE 2.13 Energy exchange in ecosystems. Plants use sunlight, water, and carbon dioxide to produce sugars and other organic milecules. Consumers use oxygen and break down sugars during cellular respiration. Plants also carry out respiration, but during the day, if light, water, and CO_2 are available, they have a net production of O_2 and carbohydrates.

dioxide, thus completing the cycle of photosynthesis and respiration. Later in this chapter we will see how these feeding relationships work.

FROM SPECIES TO ECOSYSTEMS

While cellular and molecular biologists study life processes at the microscopic level, ecologists study interactions at the species, population, biotic community, or ecosystem level. In Latin, *species* literally means *kind.* In biology, **species** refers to all organisms of the same kind that are genetically similar enough to breed in nature and produce live, fertile offspring. There are several qualifications and some important exceptions to this definition of species (especially among bacteria and plants), but for our purposes this is a useful working definition.

Organisms occur in populations, communities, and ecosystems

A **population** consists of all the members of a species living in a given area at the same time. Chapter 4 deals further with population growth and dynamics. All of the populations of organisms living and interacting in a particular area make up a **biological community.** What populations make up the biological community of which you are a part? The population sign marking your city limits announces only the number of humans who live there, disregarding the other populations of animals, plants, fungi, and microorganisms that are part of the biological community within the city's boundaries. Characteristics of biological communities are discussed in more detail in chapter 3.

An ecological system, or **ecosystem,** is composed of a biological community and its physical environment. Arcata's constructed wetlands, described in the opening case study for this chapter, is a good example of an ecosystem. Its environment includes abiotic factors (nonliving components), such as climate, water, minerals, and sunlight, as well as biotic factors, such as organisms, their products (secretions, wastes, and remains), and effects in a given area. It is useful to think about the biological community and its environment together, because energy and matter flow through both. Understanding how those flows work is a major theme in ecology.

Food chains, food webs, and trophic levels link species

Photosynthesis (and rarely chemosynthesis) is the base of all ecosystems. Organisms that photosynthesize, mainly green plants and algae, are therefore known as **producers.** One of the most important properties of an ecosystem is its **productivity,** the amount of **biomass** (biological material) produced in a given area during a given period of time. Photosynthesis is described as *primary productivity* because it is the basis for almost all other growth in an ecosystem. Manufacture of biomass by organisms that eat plants is termed *secondary productivity.* A given ecosystem may have very high total productivity, but if decomposers decompose organic material as rapidly as it is formed, the *net primary productivity* will be low. Remote sensing from aircraft or satellites allows us to measure productivity in large systems (see Exploring Science, p. 37).

In ecosystems, some consumers feed on a single species, but most consumers have multiple food sources. Similarly, some species are prey to a single kind of predator, but many species in an ecosystem are beset by several types of predators and parasites. In this way, individual food chains become interconnected to form a **food web.** Figure 2.14 shows feeding relationships among some of the larger organisms in a woodland and lake community. If we were to add all the insects, worms, and microscopic organisms that belong in this picture, however, we would have overwhelming complexity. Perhaps you can imagine the challenge ecologists face in trying to quantify and interpret the precise matter and energy transfers that occur in a natural ecosystem!

An organism's feeding status in an ecosystem can be expressed as its **trophic level** (from the Greek *trophe,* food). In our first example, the corn plant is at the producer level; it trans-

FIGURE 2.14 Each time an organism feeds, it becomes a link in a food chain. In an ecosystem, food chains become interconnected when predators feed on more than one kind of prey, thus forming a food web. The arrows in this diagram indicate the direction in which matter and energy are transferred through feeding relationships.

Principles of Environmental Science http://www.mhhe.com/cunningham4e

Measuring primary productivity is important for understanding individual plants and local environments. Understanding the rates of primary productivity is also key to understanding global processes, such as material cycling, and biological activity:

- In global carbon cycles, how much carbon is stored by plants, how quickly is it stored, and how does carbon storage compare in contrasting environments, such as the Arctic and the tropics?
- How does this carbon storage affect global climates (chapter 9)?
- In global nutrient cycles, how much nitrogen and phosphorus wash offshore, and where?

How can environmental scientists measure primary production (photosynthesis) at a global scale? In a small, relatively closed ecosystem, such as a pond, ecologists can collect and analyze samples of all trophic levels. But that method is impossible for large ecosystems, especially for oceans, which cover 70 percent of the earth's surface. One of the newest methods of quantifying biological productivity involves remote sensing, or using data collected from satellite sensors that observe the energy reflected from the earth's surface.

As you have read in this chapter, chlorophyll in green plants *absorbs* red and blue wavelengths of light and *reflects* green wavelengths. Your eye receives, or senses, these green wavelengths. A white-sand beach, on the other hand, reflects approximately equal amounts of all light wavelengths that reach it from the sun, so it looks white (and bright!) to your eye. In a similar way, different surfaces of the earth reflect characteristic wavelengths. Snow-covered surfaces reflect light wavelengths; dark-green forests with abundant chlorophyll-rich leaves—and ocean surfaces rich in photosynthetic algae and plants—reflect greens and near-infrared wavelengths.

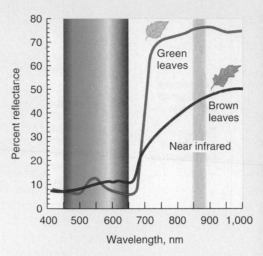

FIGURE 1 Energy wavelengths reflected by green and brown leaves.

Dry, brown forests with little active chlorophyll reflect more red and less infrared energy than do dark-green forests (fig. 1).

To detect land cover patterns on the earth's surface, we can put a sensor on a satellite that orbits the earth. As the satellite travels, the sensor receives and transmits to earth a series of "snapshots." One of the best known earth-imaging satellites, *Landsat 7,* produces images that cover an area 185 km (115 mi) wide, and each pixel represents an area of just 30 × 30 m on the ground. *Landsat* orbits approximately from pole to pole, so as the earth spins below the satellite, it captures images of the entire surface every 16 days. Another satellite, *SeaWIFS,* was designed mainly for monitoring biological activity in oceans (fig. 2). *SeaWIFS* follows a path similar to *Landsat's* but it revisits each point on the earth every day and produces images with a pixel resolution of just over 1 km.

Since satellites detect a much greater range of wavelengths than our eyes can, they are able to monitor and map chlorophyll

FIGURE 2 *SeaWiFS* image showing chlorophyll abundance in oceans and plant growth on land (normalized difference vegetation index).

abundance. In oceans, this is a useful measure of ecosystem health, as well as carbon dioxide uptake. By quantifying and mapping primary production in oceans, climatologists are working to estimate the role of ocean ecosystems in moderating climate change: for example, they can estimate the extent of biomass production in the cold, oxygen-rich waters of the North Atlantic (fig. 2). Oceanographers can also detect near-shore areas where nutrients washing off the land surface fertilize marine ecosystems and stimulate high productivity, such as near the mouth of the Amazon or Mississippi River. Monitoring and mapping these patterns helps us estimate human impacts on nutrient flows from land to sea.

forms solar energy into chemical energy, producing food molecules. Other organisms in the ecosystem are **consumers** of the chemical energy harnessed by the producers. An organism that eats producers is a primary consumer. An organism that eats primary consumers is a secondary consumer, which may, in turn, be eaten by a tertiary consumer, and so on. Most terrestrial food chains are relatively short (seeds → mouse → owl), but aquatic food chains may be quite long (microscopic algae → copepod → minnow → crayfish → bass → osprey). The length of a food chain also may reflect the physical characteristics of a particular ecosystem. A harsh arctic landscape generally has a much shorter food chain than a temperate or tropical one.

Organisms can be identified both by the trophic level at which they feed and by the *kinds* of food they eat (fig. 2.15).

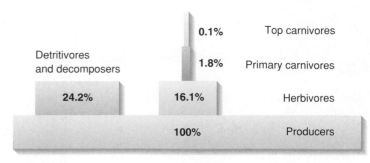

FIGURE 2.15 Organisms in an ecosystem can be identified by how they obtain food for their life processes (producer, herbivore, cornivore, omnivore, scavenger, decomposer, reducer) or by consumer level (producer; primary, secondary, or tertiary consumer) or by trophic level (first, second, third, fourth).

FIGURE 2.16 A classic example of an energy pyramid from Silver Springs, Florida. The numbers in each bar show the percentage of the energy captured at the primary producer level that is incorporated in the biomass at each succeeding level. Detritivores and decomposers feed at every level but are shown attached to the producer bar because this level provides most of their energy.

Source: Data from Howard T. Odum, "Trophic Structure and Productivity of Silver Springs, Florida" in *Ecological Monographs*, 27:55–112, 1957, Ecological Society of America.

Herbivores are plant eaters, **carnivores** are flesh eaters, and **omnivores** eat both plant and animal matter. What are humans? We are natural omnivores, by history and by habit.

One of the most important trophic levels is occupied by the many kinds of organisms that remove and recycle the dead bodies and waste products of others. **Scavengers** such as crows, jackals, and vultures clean up dead carcasses of larger animals. **Detritivores** such as ants and beetles consume litter, debris, and dung, while **decomposer** organisms such as fungi and bacteria complete the final breakdown and recycling of organic materials. It could be argued that these microorganisms are second in importance only to producers, because without their activity nutrients would remain locked up in the organic compounds of dead organisms and discarded body wastes, rather than being made available to successive generations of organisms.

Ecological pyramids describe trophic levels

If we arrange the organisms in a food chain according to trophic levels, they often form a pyramid with a broad base representing primary producers and only a few individuals in the highest trophic levels. This pyramid arrangement is especially true if we look at the energy content of an ecosystem (fig. 2.16). True to the second principle of thermodynamics, less food energy is available to the top trophic level than is available to preceding levels. For example, it takes a huge number of plants to support a modest colony of grazers such as prairie dogs. Several colonies of prairie dogs, in turn, might be

required to feed a single coyote. And a very large top carnivore like a tiger may need a home range of hundreds of square kilometers to survive.

Why is there so much less energy in each successive level in figure 2.16? In the first place, some of the food that organisms eat is undigested and doesn't provide usable energy. Much of the energy that is absorbed is used in the daily processes of living or lost as heat when it is transformed from one form to another and thus isn't stored as biomass that can be eaten.

Furthermore, predators don't operate at 100 percent efficiency. If there were enough foxes to catch all the rabbits available in the summer when the supply is abundant, there would be too many foxes in the middle of the winter when rabbits are scarce. A general rule of thumb is that only about 10 percent of the energy in one consumer level is represented in the next higher level (fig. 2.17). The amount of energy available is often expressed in biomass. For example, it generally takes about 100 kg of clover to make 10 kg of rabbit and 10 kg of rabbit to make 1 kg of fox.

The total number of organisms and the total amount of biomass in each successive trophic level of an ecosystem also may form pyramids (fig. 2.18) similar to those describing energy content. The renowned ecologist Charles Elton once started a research article with the Chinese proverb, "One hill cannot shelter two tigers." Top predators, such as tigers, grizzly bears, and great white sharks, have always been rare because they require huge territories. It takes many individuals at lower trophic levels to support one of these large carnivores. Furthermore, the characteristics that make them superb predators also make them a threat to us. We have eliminated many of the large fierce animals in the world either directly or by removing their prey and habitat.

The relationship between biomass and numbers is not as dependable as energy, however. The biomass pyramid, for instance, can be inverted by periodic fluctuations in producer populations (for example, low plant and algal biomass present

FIGURE 2.17 The energy pyramid is understood more clearly if it is related to a biomass pyramid, which represents the amount of biomass at each trophic level in a food chain. This figure illustrates how nutrients and energy become increasingly less available to successive consumers.

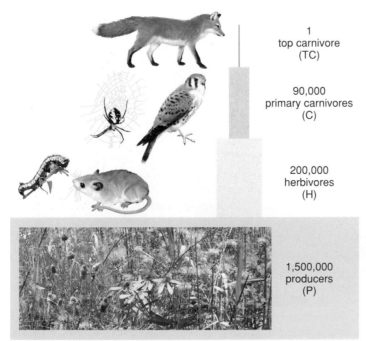

Grassland in summer

FIGURE 2.18 Usually, smaller organisms are eaten by larger organisms, and it takes numerous small organisms to feed one large organism. The classic study represented in this pyramid shows numbers of individuals at each trophic level per 1,000 m^2 of grassland and reads like this: to support 1 individual at the top carnivore level, there were 90,000 primary carnivores feeding upon 200,000 herbivores, which in turn fed upon 1,500,000 producers.

during winter in temperate aquatic ecosystems). The numbers pyramid also can be inverted. One coyote can support numerous tapeworms, for example, or one large tree can support thousands of caterpillars.

APPLICATIONS

Food Webs

To what food webs do you belong? Think about what you have eaten today, and trace the energy it contained back to its photosynthetic source. Are you at the same trophic level in all the food webs in which you participate? Are there ways that you could change your ecological role? Might that make more food available for other people? Why, or why not?

BIOGEOCHEMICAL CYCLES AND LIFE PROCESSES

The elements and compounds that sustain us are cycled endlessly through living things and through the environment. As the great naturalist John Muir said, "When one tugs at a single thing in nature, he finds it attached to the rest of the world." On a global scale, this movement is referred to as biogeochemical cycling. Substances can move quickly or slowly: you might store carbon for hours or days, while carbon is stored in the earth for millions of years. When human activity alters flow rates or storage times in these natural cycles, overwhelming the environment's ability to process them, these materials can become pollutants. Sulfur, nitrogen, carbon dioxide, and phosphorus are some of the more serious examples of these. Global-scale movement of energy is discussed in detail in chapter 9. Here we will explore some of the paths involved in cycling several important elements: water, carbon, nitrogen, sulfur, and phosphorus.

The hydrologic cycle

The path of water through our environment is perhaps the most familiar material cycle, and it is discussed in greater detail in chapter 10 (fig. 2.19). Most of the earth's water is stored in the oceans, but solar energy continually evaporates this water, and winds distribute water vapor around the globe. Water that condenses over land surfaces, in the form of rain, snow, or fog, supports all terrestrial (land-based) ecosystems. Living organisms emit the moisture they have consumed through respiration and perspiration. Eventually this moisture reenters the atmosphere or enters lakes and streams, from which it ultimately returns to the ocean again.

As it moves through living things and through the atmosphere, water is responsible for metabolic processes within cells, for maintaining the flows of key nutrients through ecosystems, and for global-scale distribution of heat and energy (chapter 9). Water performs countless services because of its unusual properties (see Exploring Science, p. 37). Water is so important that, when astronomers look for signs of life on distant planets, traces of water are the key evidence they seek.

The carbon cycle

Carbon serves a dual purpose for organisms: (1) it is a structural component of organic molecules, and (2) chemical bonds in carbon compounds provide metabolic energy. The **carbon cycle** begins with photosynthetic organisms taking up carbon dioxide (CO_2) (fig. 2.20). This is called carbon-fixation because carbon is changed from gaseous CO_2 to less mobile organic molecules. Once a carbon atom is incorporated into organic compounds, its path to recycling may be very quick or extremely slow. Imagine for a moment what happens to a simple sugar molecule you swallow in a glass of fruit juice.

FIGURE 2.19 The hydrologic cycle. Most exchange occurs with evaporation from oceans and precipitation back to oceans. About one-tenth of water evaporated from oceans falls over land, is recycled through terrestrial systems, and eventually drains back to oceans in rivers.

The sugar molecule is absorbed into your bloodstream, where it is made available to your cells for cellular respiration or the production of more complex biomolecules. If it is used in respiration, you may exhale the same carbon atom as CO_2 in an hour or less, and a plant could take up that exhaled CO_2 the same afternoon.

Alternatively, your body may use that sugar molecule to make larger organic molecules that become part of your cellular structure. The carbon atoms in the sugar molecule could remain a part of your body until it decays after death. Similarly, carbon in the wood of a thousand-year-old tree will be released only when fungi and bacteria digest the wood and release carbon dioxide as a by-product of their respiration.

Sometimes recycling takes a very long time. Coal and oil are the compressed, chemically altered remains of plants and microorganisms that lived millions of years ago. Their carbon atoms (and hydrogen, oxygen, nitrogen, sulfur, etc.) are not released until the coal and oil are burned. Enormous amounts of carbon also are locked up as calcium carbonate ($CaCO_3$), used to build shells and skeletons of marine organisms from tiny protozoans to corals. The world's extensive surface limestone deposits are biologically formed calcium carbonate from ancient oceans, exposed by geological events. The carbon in limestone has been locked away for millennia, which is probably the fate of carbon currently being deposited in ocean sediments. Eventually, even the deep ocean deposits are recycled as they are drawn into deep molten layers and released via volcanic activity. Geologists estimate that every carbon atom on the earth has made about 30 such round trips over the past 4 billion years.

Materials that store carbon, including geologic formations and standing forests, are known as carbon sinks. When carbon is released from these sinks, as when we burn fossil fuels and inject CO_2 into the atmosphere, or when we clear extensive forests, natural recycling systems may not be able to keep up. This is the root of the global warming problem, discussed in chapter 9. Alternatively, extra atmospheric CO_2 could support faster plant growth, speeding some of the recycling processes.

The nitrogen cycle

Organisms cannot exist without amino acids, peptides, and proteins, all of which are organic molecules that contain nitrogen. Nitrogen is therefore an extremely important nutrient for living things. (Nitrogen is a primary component of many household and agricultural fertilizers.) Even though nitrogen makes up about 78 percent of the air around us, plants cannot use N_2, the stable diatomic (two-atom) molecule in the air.

FIGURE 2.20   The carbon cycle. Numbers indicate approximate exchange of carbon in gigatons (Gt) per year. Natural exchanges are balanced, but human sources produce a net increase of CO_2 in the atmosphere.

Lightning and volcanoes (10 Tg)
Fossil fuel burning, commercial and agricultural nitrogen fixation (140 Tg)

Atmospheric Nitrogen (N₂)

Runoff
(36Tg)

Inorganic Nitrogen Fixation

Biological Nitrogen Fixation

Ammonification

Decomposition
of organic compounds
(40 Tg)

Nitrates, ammonia, or ammonium
absorbed by plants are used to make
organic nitrogen compounds

Assimilation

Denitrification

Nitrifying bacteria
convert ammonia
into nitrates

Ammonia (NH₃)
Ammonium (NH₄⁺)   **Nitrification**   Nitrates (NO₃⁻)

Denitrifying bacteria
convert nitrates to
molecular nitrogen

Nitrogen-fixing bacteria
some live in root nodules of certain plants;
others live in soil, freshwater and saltwater (80 Tg)

FIGURE 2.21   The nitrogen cycle. Human sources of nitrogen fixation (conversion of molecular nitrogen to ammonia or ammonium) are now about 50 percent greater than natural sources. Bacteria convert ammonia to nitrates, which plants use to create organic nitrogen. Eventually, nitrogen is stored in sediments or converted back to molecular nitrogen (1 Tg = 10^{12} g).

Plants acquire nitrogen through an extremely complex **nitrogen cycle** (fig. 2.21). The key to this cycle is nitrogen-fixing bacteria (including some blue-green algae or cyanobacteria). These organisms have a highly specialized ability to "fix" nitrogen, or combine gaseous N_2 with hydrogen to make ammonia (NH_3).

Other bacteria then combine ammonia with oxygen to form nitrites (NO_2^-). Another group of bacteria converts nitrites to nitrates (NO_3^-), which green plants can absorb and use. After plant cells absorb nitrates, the nitrates are reduced to ammonium (NH_4^+), which cells use to build amino acids that become the building blocks for peptides and proteins.

Members of the bean family (legumes) and a few other kinds of plants are especially useful in agriculture because nitrogen-fixing bacteria actually live in their root tissues (fig. 2.22). Legumes and their associated bacteria add nitrogen to the soil, so interplanting and rotating legumes with crops, such as corn, that use but cannot replace soil nitrates are beneficial farming practices that take practical advantage of this relationship.

Nitrogen reenters the environment in several ways. The most obvious path is through the death of organisms. Fungi and bacteria decompose dead organisms, releasing ammonia and ammonium ions, which then are available for nitrate formation.

FIGURE 2.22   Bumps called nodules cover the roots of this adzuki bean plant. Each nodule is a mass of root tissue containing many bacteria that help convert nitrogen in the soil to a form that the bean plant can assimilate and use to manufacture amino acids.

Principles of Environmental Science   http://www.mhhe.com/cunningham4e

Organisms don't have to die to donate proteins to the environment, however. Plants shed their leaves, needles, flowers, fruits, and cones; animals shed hair, feathers, skin, exoskeletons, pupal cases, and silk. Animals also produce excrement and urinary wastes that contain nitrogenous compounds. Urine is especially high in nitrogen because it contains the detoxified wastes of protein metabolism. All of these by-products of living organisms decompose, replenishing soil fertility (see related Case Study, "Why Trees Need Salmon," at www.mhhe.com/cunningham4e).

How does nitrogen reenter the atmosphere, completing the cycle? Denitrifying bacteria break down nitrates into N_2 and nitrous oxide (N_2O), gases that return to the atmosphere; thus, denitrifying bacteria compete with plant roots for available nitrates. However, denitrification occurs mainly in waterlogged soils that have low oxygen availability and a large amount of decomposable organic matter. These are suitable growing conditions for many wild plant species in swamps and marshes, but not for most cultivated crop species, except for rice, a domesticated wetland grass.

In recent years, humans have profoundly altered the nitrogen cycle. By using synthetic fertilizers, cultivating nitrogen-fixing crops, and burning fossil fuels, we now convert more nitrogen to ammonia and nitrates than all natural land processes combined. This excess nitrogen input causes algal blooms and excess plant growth in water bodies, called eutrophication, which we will discuss in more detail in chapter 10. Excess nitrogen also causes serious loss of soil nutrients such as calcium and potassium; acidification of rivers and lakes; and rising atmospheric concentrations of nitrous oxide, a greenhouse gas. It also encourages the spread of weeds into areas such as prairies, where native plants are adapted to nitrogen-poor environments.

The phosphorus cycle

Minerals become available to organisms after they are released from rocks. Two mineral cycles of particular significance to organisms are phosphorus and sulfur. Why do you suppose phosphorus is a primary ingredient in fertilizers? At the cellular level, energy-rich phosphorus-containing compounds are primary participants in energy-transfer reactions, as discussed earlier.

The amount of available phosphorus in an environment, therefore, can dramatically affect productivity. Abundant phosphorus stimulates lush plant and algal growth, making phosphorus a major contributor to water pollution.

The phosphorus cycle begins when phosphorus compounds leach from rocks and minerals over long periods of time (fig. 2.23). Because phosphorous has no atmospheric form, it is usually transported in water. Producer organisms take in inorganic phosphorus, incorporate it into organic molecules, and then pass it on to consumers. Phosphorus returns to the environment by decomposition. An important aspect of the phosphorus cycle is the very long time

FIGURE 2.23 The phosphorus cycle. Natural movement of phosphorus is slight, involving recycling within ecosystems and some erosion and sedimentation of phosphorus-bearing rock. Use of phosphate (PO_4^{-3}) fertilizers and cleaning agents increases phosphorus in aquatic systems, causing eutrophication. Units are teragrams (Tg) phosphorus per year.

it takes for phosphorus atoms to pass through it. Deep ocean sediments are significant phosphorus sinks of extreme longevity. Phosphate ores that now are mined to make detergents and inorganic fertilizers represent exposed ocean sediments that are millennia old. You could think of our present use of phosphates, which are washed out into the river systems and eventually the oceans, as an accelerated mobilization of phosphorus from source to sink. Aquatic ecosystems often are dramatically affected in the process because excess phosphates can stimulate explosive growth of algae and photosynthetic bacteria populations (algae blooms), upsetting ecosystem stability (see related Case Study, "The Environmental Chemistry of Phosphorus," at www.mhhe.com/cunningham4e). Can you think of ways we could reduce the amount of phosphorus we put into our environment?

The sulfur cycle

Sulfur plays a vital role in organisms, especially as a minor but essential component of proteins. Sulfur compounds are important determinants of the acidity of rainfall, surface water, and soil. In addition, sulfur in particles and tiny airborne droplets may act as critical regulators of global climate. Most of the earth's sulfur is tied up underground in rocks and minerals, such as iron disulfide

(pyrite) and calcium sulfate (gypsum). Weathering, emissions from deep seafloor vents, and volcanic eruptions release this inorganic sulfur into the air and water (fig. 2.24).

The sulfur cycle is complicated by the large number of oxidation states the element can assume, producing hydrogen sulfide (H_2S), sulfur dioxide (SO_2), sulfate ion (SO_4^{2-}), and others. Inorganic processes are responsible for many of these transformations, but living organisms, especially bacteria, also sequester sulfur in biogenic deposits or release it into the environment. Which of the several kinds of sulfur bacteria prevails in any given situation depends on oxygen concentrations, pH level, and light level.

Human activities also release large quantities of sulfur, primarily through burning fossil fuels. Total yearly anthropogenic sulfur emissions rival those of natural processes, and acid rain (caused by sulfuric acid produced as a result of fossil fuel use) is a serious problem in many areas (see chapter 9). Sulfur dioxide and sulfate aerosols cause human health problems, damage buildings and vegetation, and reduce visibility. They also absorb ultraviolet (UV) radiation and create cloud cover that cools cities and may be offsetting greenhouse effects of rising CO_2 concentrations.

Interestingly, the biogenic sulfur emissions of oceanic phytoplankton may play a role in global climate regulation. When ocean water is warm, tiny, single-celled organisms release

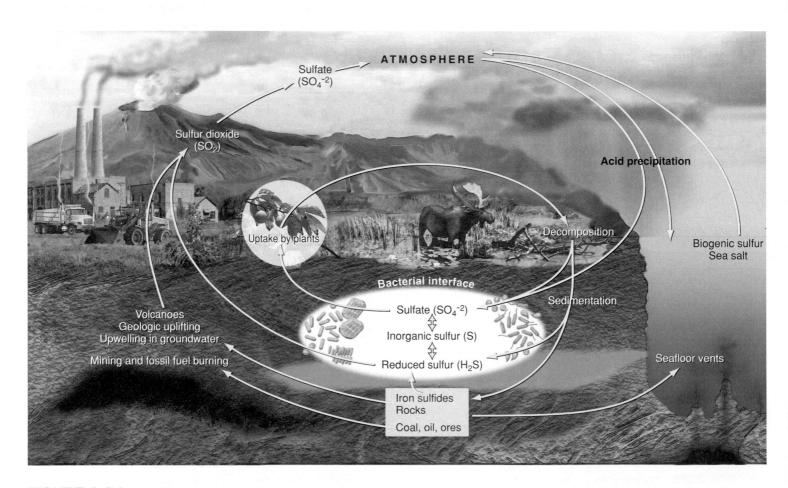

FIGURE 2.24 The sulfur cycle. Sulfur is present mainly in rocks, soil, and water. It cycles through ecosystems when it is taken in by organisms. Combustion of fossil fuels causes increased levels of atmospheric sulfur compounds, which create problems related to acid precipitation.

dimethylsulfide (DMS), which is oxidized to SO_2 and then SO_4^{2-} in the atmosphere. Acting as cloud droplet condensation nuclei, these sulfate aerosols increase the earth's albedo (reflectivity) and cool the earth. As ocean temperatures drop because less sunlight gets through, phytoplankton activity decreases, DMS production falls, and clouds disappear. Thus, DMS, which may account for half of all biogenic sulfur emissions, could be a feedback mechanism that keeps temperature within a suitable range for all life.

REVIEW QUESTIONS

1. What two problems did Arcata, California, solve with its constructed wetland?

2. What are systems and how do feedback loops regulate them?

3. Your body contains vast numbers of carbon atoms. How is it possible that some of these carbons may have been part of the body of a prehistoric creature?

4. List six unique properties of water. Describe, briefly, how each of these properties makes water essential to life as we know it.

5. What is DNA, and why is it important?

6. The oceans store a vast amount of heat, but this huge reservoir of energy is of little use to humans. Explain the difference between high-quality and low-quality energy.

7. In the biosphere, matter follows circular pathways, while energy flows in a linear fashion. Explain.

8. Which wavelengths do our eyes respond to, and why? (Refer to fig. 2.11.) What is the ratio of short ultraviolet wavelengths to microwave lengths?

9. Where do extremophiles live? How do they get the energy they need for survival?

10. Ecosystems require energy to function. From where does this energy come? Where does it go?

11. How do green plants capture energy, and what do they do with it?

12. Define the terms *species, population,* and *biological community*.

13. Why are big fierce animals so rare?

14. Most ecosystems can be visualized as a pyramid with many organisms in the lowest trophic levels and only a few individuals at the top. Give an example of an inverted numbers pramid.

15. What is the ratio of human-caused carbon releases into the atmosphere shown in figure 2.20 compared to the amount released by terrestrial respiration?

CRITICAL THINKING AND DISCUSSION

Apply the principles you have learned in this chapter to discuss these questions with other students.

1. Ecosystems are often defined as a matter of convenience because we can't study everything at once. How would you describe the characteristics and boundaries of the ecosystem in which you live. In what respects is your ecosystem an open one?

2. Think of some practical examples of increasing entropy in everyday life. Is a messy room really evidence of thermodynamics at work, or merely personal preference?

3. Some chemical bonds are weak and have a very short half-life (fractions of a second, in some cases); others are strong and stable, lasting for years or even centuries. What would our world be like if all chemical bonds were either very weak or extremely strong?

4. If you had to design a research project to evaluate the relative biomass of producers and consumers in an ecosystem, what would you measure? (*Note:* This could be a natural system or a human-made one.)

5. Understanding storage compartments is essential to understanding material cycles, such as the carbon cycle. If you look around your backyard, how many carbon storage compartments are there? Which ones are the biggest? Which ones are the longest lasting?

SUGGESTED READINGS

Carpenter, Steve R., et al. 2001. Trophic cascades, nutrients, and lake productivity: Whole-lake experiments. *Ecological Monographs* 71:163–86.

Chapelle, Francis H., et al. 2002. A hydrogen-based subsurface microbial community dominated by methanogens. *Nature* 415:312–15.

Falkowski, Paul G. 2002. The ocean's invisible forest. *Scientific American* 287(2):54–61.

Govindarajulu, M., et al. 2005. Nitrogen transfer in the arbuscular mycorrhizal symbiosis. *Nature* 435:819–23.

Imhoff, Marc L., et al. 2004. Global patterns in human consumption of net primary production. *Nature* 429(6994):870–73.

Knight, T. M., et al. 2005. Trophic cascades across ecosystems. *Nature* 437:880–83.

Millennium Ecosystem Assessment. 2003. *Ecosystems and Human Well-Being.* Island Press.

Pace, Michael L., et al. 2004. Whole-lake carbon-13 additions reveal terrestrial support of aquatic food webs. *Nature* 427:240–43.

Primack, Richard B. 2004. *Essentials of Conservation Biology.* 3rd ed. Sinauer Associates.

Reich, Peter B., and Jacek Oleksyn. 2004. Global patterns of plant leaf N and P in relation to temperature and latitude. *Proceedings of the National Academy of Sciences (U.S.)* 101(30):11001–06.

DATA analysis   Extracting Data from a Graph

1. How can you extract data from a line plot? The process is simply the reverse of the way you just learned to create the graph. You draw lines from the axes to where they intersect on the graph at the point whose value you want to know. Let's look at figure 1, the line plot we examined in chapter 1.

- How many cells were there at the half point of the growth curve?

- How does that result compare to the cell population three hours earlier?

- When did the growth of this population start to slow?

A curve, such as this, that increases slowly at first but then rapidly accelerates is called a logistic curve. We'll discuss the mathematics that create this pattern in the chapter 3.

2. How do you extract data from a bar graph? Look again at the bar graph in figure 2. Using the same procedure described for a line plot, draw a horizontal line from the top of any bar straight to the Y-axis. Read the value for the frequency of that category on the Y-axis.

- Why don't you need to draw a vertical line with this type of graph?

- How many cities have 30 ug/m^3?

- Why aren't there any bars between 50 and 65 ug/m^3?

- Why are there so many more cities at 40 ug/m^3?

- Why is one city at 70 ug/m^3?

- How many more cities have 40 ug/m^3 than 30 ug/m^3?

FIGURE 1   Growth of *Tetrahymena*

FIGURE 2   Urban Air Quality

For Additional Help in Studying This Chapter, please visit our website at www.mhhe.com/cunningham4e. You will find practice quizzes, a chapter summary, key terms, answers to review questions, additional case studies, regional examples, and an extensive reading list, all of which will help you understand the material in this chapter.

Students labor for science and future generations, weeding experimental grassland plots at Cedar Creek Natural History Area in Minnesota.

CHAPTER

Species Populations, Interactions, and Communities

When I view all beings not as special creations, but as lineal descendents of some few beings which have lived long before the first bed of the Cambrian system was deposited, they seem to me to become ennobled.

–Charles Darwin–

OBJECTIVES

Studying this chapter will help you answer the following questions:

- How does species diversity arise?
- Why do species live in different locations?
- How do interactions among species affect their fates and that of communities?
- If a species has unlimited growth potential, why doesn't it fill the earth?

- What special properties does a community of species have, and why are they important?
- What is the relationship between species diversity and community stability?
- What is disturbance, and how does it affect communities?

Case Study Species Diversity Promotes Community Recovery

Does diversity matter? Ecologists long believed that a diversity of species in a community is not just beautiful and fascinating, but also functionally important. If disturbed, do communities recover quickly when they contain many species? Are they more stable, in other words? This question, raised by ecologists in the 1950s, suggests a strong reason to preserve the earth's biodiversity, but not until 1994 did someone present data that helped answer it. In a long-term study of prairie plant communities, ecologist Dave Tilman and his colleagues planted dozens of experimental plots at Cedar Creek Natural History Area, each plot containing different numbers of species. In 1988, central Minnesota experienced its hottest, driest summer in 50 years. As Tilman's team watched their plots recover from the drought, they measured growth in the plots by carefully clipping, counting, and weighing every plant. By 1992, it was clear that plots with five or fewer species were recovering slowly, while higher-diversity plots had reached or exceeded their pre-drought productivity.

Because the question of diversity and recovery was controversial, the strength of Tilman's results depended on duplicated plots (replicates) and a long period of study. If he had used only one plot for each level of species diversity, chance alone might have depressed productivity. Another question in any experiment is how well experimental results represent other situations and other samples. Tilman's team approximated an answer to this question by reporting a range of values around the average recovery rates, in this case the "standard error." This standard statistical measure and its related value, confidence interval, show the range within which nearly all means should fall if someone else did the same experiment. In this way we describe how confident we are that the results can be applied to other situations.

Later experiments at Cedar Creek showed that species-rich plots recovered more fully after drought because some species were

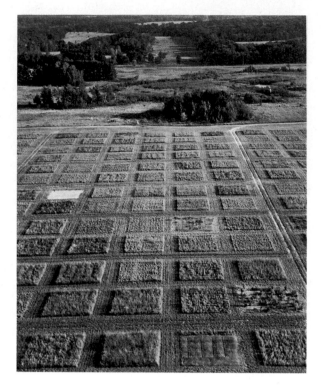

Experimental grassland plots at Cedar Creek.

By 1992, plots with five or fewer species were still below pre-drought productivity, while higher-diversity plots had recovered. (Bars show standard error, or variability, around the means.)
Source: Tilman & Dowling (1994) *Nature* 367:363–365.

severely harmed by drought, others went dormant but didn't die, and still others continued growing but at a slow rate. When rain returned, surviving plants revived filling the plots with green leaves and stems. The chance that a species-poor plot contained quickly recovering species was lower than for a species-rich plot. Consequently, low-diversity plots were more likely to attain their former productivity slowly, if at all. This explanation was bolstered by another experimental discovery that nitrogen—a limiting factor in the sterile, sandy soils of the plots—was used up more completely in species-rich plots than in species-poor plots.

The experimental data from Cedar Creek prairies stimulated tests in other ecosystems. Is resilience after disturbance always evident in species-rich plant communities, such as rainforests of Central and South America? (Preliminary results show that it depends on how big and severe the disturbance is.) Do species-poor plant communities, such as dune grasslands and salt marshes, recover only slowly following disturbance? Actually, some simple plant communities appear quite stable where disturbances like wind, waves, and storms are frequent.

What are the implications of Tilman's research for agricultural fields populated by a single crop plant. One important application may be in biofuels production. Ethanol (a biomass fuel) made from starch or cellulose may be an attractive alternative to petroleum-based fuels (chapter 12). Tilman's study suggests that, given variable weather over several years, a diverse biomass crop can achieve an economical net energy yield while providing greater environmental benefits than does a single-species crop.

The general lesson is that keeping a diversity of species as a backdrop to human existence may serve us best in the long run; and if species diversity is maintained, nature may be more resilient than we imagine.

HOW SPECIES DIVERSITY ARISES

Why do some species live in one place but not another? A more important question to environmental scientists is, what are the mechanisms that promote the great variety of species on earth and that determine which species will survive in one environment but not another? In this section you will come to understand (1) concepts behind the theory of speciation by means of natural selection and adaptation (evolution); (2) the characteristics of species that make some of them weedy and others endangered; and (3) the limitations species face in their environments and implications for their survival. First we'll start with the basics: How do species arise?

Natural selection and adaptation modify species

How does a polar bear stand the long, sunless, super-cold arctic winter? How does the saguaro cactus survive blistering temperatures and extreme dryness of the desert? We commonly say that each species is *adapted* to the environment where it lives, but what does that mean? **Adaptation,** the acquisition of traits that allow a species to survive in its environment, is one of the most important concepts in biology.

We use the term *adapt* in two ways. An individual organism can respond immediately to a changing environment in a process called acclimation. If you keep a houseplant indoors all winter and then put it out in full sunlight in the spring, the leaves become damaged. If the damage isn't severe, your plant may grow new leaves with thicker cuticles and denser pigments that block the sun's rays. However, the change isn't permanent. After another winter inside, it will still get sun-scald in the following spring. The leaf changes are not permanent and cannot be passed on to offspring, or even carried over from the previous year. Although the capacity to acclimate is inherited, houseplants in each generation must develop their own protective leaf epidermis.

Another type of adaptation affects populations consisting of many individuals. Genetic traits are passed from generation to generation and allow a species to live more successfully in its environment. This process of adaptation to environment is explained by the theory of evolution. It was simultaneously proposed by Charles Darwin (1809–1882) based on his studies in the Galápagos Islands (fig. 3.1) and elsewhere, and by Alfred Russel Wallace (1823–1913), who investigated the species of the Malay Archipelago. The basic idea of evolution is that species change over generations because individuals compete for scarce resources. Better competitors in a population survive—they have greater reproductive potential or fitness—and their offspring inherit the beneficial traits. Over generations, those traits become common in a population. The process of better-selected individuals passing their traits to the next generation is called **natural selection.** The traits are encoded in a species' DNA, but from where does the original DNA coding come, which then gives some individuals greater fitness? Every organism has a dizzying array of genetic diversity in its DNA. It has been demonstrated in experiments and by observing natural populations that changes to the DNA coding sequence of individuals occurs, and that the changed sequences are inherited by offspring. Exposure to ionizing radiation and toxic materials, and random recombination and mistakes in replication of DNA strands during reproduction are the main causes of genetic mutations. Sometimes a single mutation has a large effect, but evolutionary change is mostly brought about by many mutations accumulating over time. Only mutations in reproductive cells (gametes) matter; body cell changes—cancers, for example—are not inherited. Most mutations have no effect on fitness, and many actually have a negative effect. During the course of a species' life span—a million or more years—some mutations are thought to have given those individuals an advantage under the **selection pressures** of their environment at that time. The result is a species population that differs from those of numerous preceding generations.

FIGURE 3.1 The Galápagos Islands. It was in these relatively young, barren, isolated islands that the relatively young Charles Darwin made his observations which changed the way people viewed life's progression through vast eons of time.

All species live within limits

Environmental factors exert selection pressure and influence the fitness of individuals and their offspring. For this reason, species are limited in where they can live. Limitations include the following: (1) physiological stress due to inappropriate levels of some critical environmental factor, such as moisture, light, temperature, pH, or specific nutrients; (2) competition with other species; (3) predation, including parasitism and disease; and (4) luck. In some cases, the individuals of a population that survive environmental catastrophes or find their way to a new habitat, where they start a new population, may simply be lucky rather than more fit than their contemporaries.

An organism's physiology and behavior allow it to survive only in certain environments. Temperature, moisture level, nutrient supply, soil and water chemistry, living space, and other

FIGURE 3.2 Saguaro cacti inhabit the Sonoran desert, and their northern limit is partly controlled by a critical environmental factor—freezing temperatures. Local topography allows some individuals to push that boundary, such that temperature alone doesn't explain the exact edge of the cactus's northern range.

environmental factors must be at appropriate levels for organisms to persist. In 1840, the chemist Justus von Liebig proposed that the single factor in shortest supply relative to demand is the **critical factor** determining where a species lives. The giant saguaro cactus (*Carnegiea gigantea*), which grows in the dry, hot Sonoran desert of southern Arizona and northern Mexico, offers an example (fig. 3.2). Saguaros are extremely sensitive to freezing temperatures. A single winter night with temperatures below freezing for 12 or more hours kills growing tips on the branches, preventing further development. Thus the northern edge of the saguaro's range corresponds to a zone where freezing temperatures last less than half a day at any time.

Ecologist Victor Shelford (1877–1968) later expanded Liebig's principle by stating that each environmental factor has both minimum and maximum levels, called **tolerance limits,** beyond which a particular species cannot survive or is unable to reproduce (fig. 3.3). The single factor closest to these survival limits, Shelford postulated, is the critical factor that limits where a particular organism can live. At one time, ecologists tried to identify unique factors limiting the growth of every plant and animal population. We now know that several factors working together, even in a clear-cut case like the saguaro, usually determine a species' distribution. If you have ever explored the rocky coasts of New England or the Pacific Northwest, you have probably noticed that mussels and barnacles grow thickly in the intertidal zone, the place between high and low tides. No one factor decides this pattern. Instead, the distribution of these animals is determined by a combination of temperature extremes, drying time between tides, salt concentrations, competitors, and food availability.

In some species, tolerance limits affect the distribution of young differently than adults. The desert pupfish, for instance, lives in small, isolated populations in warm springs in the northern Sonoran desert. Adult pupfish can survive temperatures between 0° and 42°C (a remarkably high temperature for a fish) and tolerate an equally wide range of salt concentrations. Eggs and juvenile fish, however, can survive only between 20° and 36°C and are killed by high salt levels. Reproduction, therefore, is limited to a small part of the range of the adult fish.

Sometimes the requirements and tolerances of species are useful **indicators** of specific environmental characteristics. The presence or absence of such species indicates something about the community and the ecosystem as a whole. Lichens and eastern white pine, for example, are indicators of air pollution because they are extremely sensitive to sulfur dioxide and ozone, respectively. Bull thistle and many other plant weeds grow on disturbed soil but are not eaten by cattle; therefore, a vigorous population of bull thistle or certain other plants in a pasture indicates it is being overgrazed. Similarly, anglers know that trout

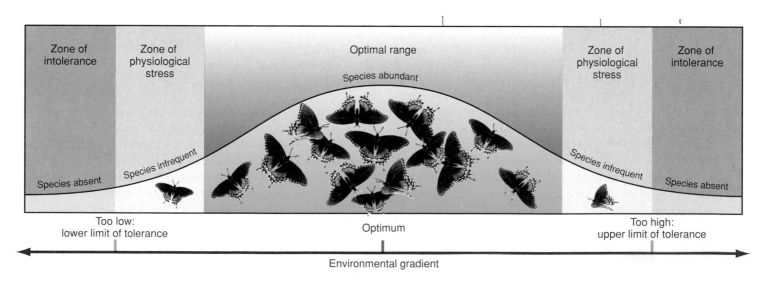

FIGURE 3.3 Tolerance limits affect species populations. For every environmental factor, there is an optimal range where a species is most abundant. As an organism moves away from locations with optimal conditions, the harder life becomes. At some point along the environmental gradient for that factor, individuals become physiologically stressed and few, if any, survive those conditions.

species require cool, clean, well-oxygenated water; the presence or absence of trout is used as an indicator of good water quality.

The ecological niche is a species' role and environment

Habitat describes the place or set of environmental conditions in which a particular organism lives. A more functional term, **ecological niche,** describes either the role played by a species in a biological community or the total set of environmental factors that determine a species distribution. The concept of niche was first defined in 1927 by the British ecologist Charles Elton (1900–1991). To Elton, each species had a role in a community of species, and the niche defined its way of obtaining food, the relationships it had with other species, and the services it provided to its community. Thirty years later, the American limnologist G. E. Hutchinson (1903–1991) proposed a more biophysical definition of niche. Every species, he pointed out, exists within a range of physical and chemical conditions (temperature, light levels, acidity, humidity, salinity, etc.) and also biological interactions (predators and prey present, defenses, nutritional resources available, etc.). The niche is more complex than the idea of a critical factor (fig. 3.4). A graph of a species niche would be multidimensional, with many factors being simultaneously displayed, almost like an electron cloud.

For a generalist, like the brown rat, the ecological niche is broad. In other words, a generalist has a wide range of tolerance for many environmental factors. For others, such as the giant panda (*Ailuropoda melanoleuca*), only a narrow ecological niche exists (fig. 3.5). Bamboo is low in nutrients, but provides 95 percent of a panda's diet, requiring it to spend as much as 16 hours a day eating. There are virtually no competitors for bamboo, except other pandas, yet the species is endangered, primarily due to shrinking habitat. Giant pandas, like many species

FIGURE 3.4   The bears, birds, and fish of the McNeil River, Alaska. Each occupies an ecological niche.

FIGURE 3.5   The giant panda eats almost nothing but bamboo. It has a digestive system like a carnivore, but it is a vegetarian. Periodic die-offs of bamboo, combined with destruction of habitat, cause pandas to starve, with the last large die-off in the 1970s.

on earth, are habitat specialists. Specialists have more exacting habitat requirements, tend to have lower reproductive rates, and care for their young longer. They may be less resilient in response to environmental change. Plants can also be habitat specialists—for instance, **endemic** plant species exist on serpentine and other unusual types of rock outcrops, and nowhere else.

Over time, niches change as species develop new strategies to exploit resources. Species of greater intelligence or complex social structures, such as elephants, chimpanzees, and dolphins, learn from their social group how to behave and can invent new ways of doing things when presented with novel opportunities or challenges. In effect, they alter their ecological niche by passing on cultural behavior from one generation to the next. Most organisms, however, are restricted to their niche by their genetically determined bodies and instinctive behaviors. When two such species compete for limited resources, one eventually gains the larger share, while the other finds different habitat, dies out, or experiences a change in its behavior or physiology so that competition is minimized. The idea that "complete competitors cannot coexist" was proposed by the Russian biologist G. F. Gause (1910–1986) to explain why mathematical models of species competition always ended with one species disappearing. The **competitive exclusion principle,** as it is called, states that no two species can occupy the same ecological niche for long. The one that is more efficient in using available resources will exclude the other. We call this process of niche evolution **resource partitioning** (fig. 3.6). Partitioning can allow several species to utilize different parts of the same

The invincible Norway rat in lab coat white.

Why Are Generalists and Invasive Species So Successful?

Vilified for its supposedly dirty habits, diseases, and incursions in human habitations—the brown or Norway rat (*Rattus norvegicus*) is probably the most successful wild mammal species on earth. Possibly originating in China, it traveled the globe primarily by ship. It appeared in England by 1720, in the Western Hemisphere by 1770, and today lives on every continent except Antarctica. Besides being a scourge in cities and towns, the brown rat plagues farmers by consuming grain and eating the eggs and young of poultry. In natural settings they also devour the eggs and young of birds, and eat other small vertebrates. Ironically the brown rat (in white fur) is one of the most studied of laboratory animals. The white lab rat not only helps in discovering cures for disease and providing insight into animal behavior, it also delights pet owners with its intelligence and affection.

What is the secret to the brown rat's success? It is a **habitat generalist** of the first order—a species that seems to do well just about anywhere because of its biological characteristics. Other generalists are well known: the German cockroach and house sparrow of cities, the raccoon and European rabbit of urban and rural settings. Thousands of these generalists were introduced by accident or intentionally to new environments where their populations have grown rapidly. Also called weedy or opportunistic species, they share common biological traits. First, they reproduce early and often. The brown rat, for example, is sexually mature by 12 weeks old, bears young within 3–4 weeks of mating—several times a year if food supplies are adequate—and gives birth to 6–9 young at a single go. By the end of its brief life (2–3 years), a single female may have borne nearly 200 offspring! Second, the offspring of generalists are ready to take on the world after only a brief period of maternal care. Brown rats breed and live in large colonies where any mother may suckle any newborn; by the tender age of 3–4 weeks, a brown ratling leaves its nest. Third, generalists have a broad ecological niche, the combination of environmental factors they tolerate. The brown rat consumes almost anything, including electrical cords, tolerates a wide temperature range, and

inhabits any structure that people build. They are intelligent foragers and recall where food was found, and where they stored it.

When a generalist species is introduced to a new environment and becomes highly successful, we call it an **invasive species** (chapter 5). After being introduced to islands in the Pacific Ocean, the brown rat (together with the Polynesian rat *R. exulans,* and tree rat *R. rattus*) apparently contributed to the extinction of dozens of endemic bird species by eating their eggs and young. Rat invasions altered entire communities of birds, which affected plants and other species with which the birds interacted. Many ecologists consider invasive species the most pressing hazard for biological communities in the twenty-first century (chapter 5).

Often a species is deliberately introduced because it is successful in its home territory. Kudzu (*Pueraria montana* var. *lobata*), a vine from Japan, was widely planted in the southeastern United States beginning in the 1930s to control soil erosion (chapter 5). An excellent **pioneer species,** it grows rapidly on bare soil and soon covers it with lush vegetation—while outcompeting native plants for sunlight and water.

Generalist species, especially those that are invasives, raise many questions: Should we introduce them to new habitats? Or should we not think badly of them just because they are common and successful? What about another wildly successful mammal—human beings? We are generalists too. Are we also an invasive or weedy species? Or do humans behave sometimes as generalists, sometimes as specialists, depending on the choices we make?

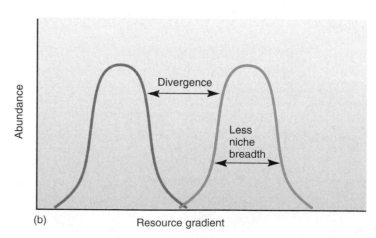

FIGURE 3.6 Competition causes resource partitioning and niche specialization. (a) Where niches of two species overlap along a resource gradient, competition occurs (shaded area). Individuals in this part of the niche have less success producing young. (b) Over time the traits of the populations diverge, leading to specialization, narrower niche breadth, and less competition between species.

Cape May
warbler

Blackburnian
warbler

Black-throated
green warbler

Bay-breasted
warbler

Yellow-rumped
warbler

60 ft
50 ft
40 ft
30 ft
20 ft
10 ft
Ground

FIGURE 3.7 Several species of insect-eating wood warblers occupy the same forests in eastern North America. The competitive exclusion principle predicts that the warblers should partition the resource—insect food—in order to reduce competition. And in fact, the warblers feed in different parts of the forest.
Source: Original observations by R. H. MacArthur (1958) *Ecology* 39:599–619.

resource and coexist within a single habitat (fig. 3.7). Species can specialize in time, too. Swallows and insectivorous bats both catch insects, but some insect species are active during the day and others at night, providing noncompetitive feeding opportunities for day-active swallows and night-active bats. The competitive exclusion principle does not explain all situations, however. For example, many similar plant species coexisted in the Cedar Creek experimental plots. Do they avoid competition in ways we cannot observe, or are resources so plentiful that no competition need occur?

Speciation maintains species diversity

As an interbreeding species population becomes better adapted to its ecological niche, its genetic heritage (including mutations passed from parents to offspring) gives it the potential to change further as circumstances, dictate. In the case of Galápagos finches, evidence from body shape, behavior, and genetics leads to the idea that modern Galápagos finches look, behave, and bear DNA related to an original seed-eating finch species (fig. 3.8). It is proposed to have blown to the islands from the mainland where a similar species still exists. Today there are 13 distinct species on the islands that differ markedly in appearance, food preferences, and habitat. Fruit eaters have thick, parrot-like bills; seed eaters have heavy, crushing bills; insect eaters have thin, probing beaks to catch their prey. One of the most unusual species is the woodpecker finch, which pecks at tree bark for hidden insects. Lacking the woodpecker's long tongue, the finch uses a cactus spine as a tool to extract bugs.

The development of a new species is called **speciation.** No one has observed a new species springing into being, but in some organisms, especially plants, it is inferred to occur somewhat frequently. Nevertheless, given the evidence, speciation is a reasonable proposal for how species arise. Speciation may be relatively rapid on millennial timescales (punctuated equilibrium).

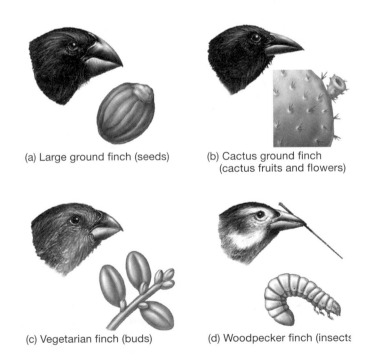

(a) Large ground finch (seeds)

(b) Cactus ground finch (cactus fruits and flowers)

(c) Vegetarian finch (buds)

(d) Woodpecker finch (insects

FIGURE 3.8 Four of the 13 species of finches in the Galápagos with their preferred foods. Distinctive beaks and behaviors arose from the ancestral, pointy-beaked warbler finch (*Certhidia olivacea,* not shown) in response to different foods available on the islands. Recent research on the finches' DNA suggests that beak shape changed first, followed by divergence in beak size.

For example, after a long period of stability, a new species may arise from parents following the appearance of a new food source, predator, or competitor, or a change in climate. One mechanism of speciation is **geographic isolation.** This is termed **allopatric speciation**—species arise in non-overlapping geographic locations. The original Galápagos finches were separated from the rest of the population on the mainland, could no longer share genetic material, and became reproductively isolated.

The barriers that divide subpopulations are not always physical. For example, two virtually identical tree frogs (*Hyla versicolor, H. chrysoscelis*) live in similar habitats of eastern North America but have different mating calls. This is an example of behavioral isolation. It also happens that one species has twice the chromosomes of the other. This example of **sympatric speciation** takes place in the same location as the ancestor species. Fern species and other plants seem prone to sympatric speciation by doubling or quadrupling the chromosome number of their ancestors.

Once isolation is imposed, the two populations begin to diverge in genetics and physical characteristics. Genetic drift ensures that DNA of two formerly joined populations eventually diverges; in several generations, traits are lost from a population during the natural course of reproduction. Under more extreme circumstances, a die-off of most members of an isolated population strips much of the variation in traits from the survivors. The cheetah experienced a genetic bottleneck about 10,000 years ago and exists today as virtually identical individuals.

In isolation, selection pressures shape physical, behavioral, and genetic characteristics of individuals, causing population traits to shift over time (fig. 3.9). From an original range of characteristics, the shift can be toward an extreme of the trait (directional selection), it can narrow the range of a trait (stabilizing selection), or it can cause traits to diverge to the extremes (disruptive selection). Directional selection is implied by increased pesticide resistance in German cockroaches (*Blattella germanica*). Apparently some individuals can make an enzyme that detoxifies pesticides. Individual cockroaches that lack this characteristic are dying out, and as a result, populations of cockroaches with pesticide resistance are developing.

A small population in a new location—island, mountaintop, unique habitat—encounters new environmental conditions that favor some individuals over others (fig. 3.10). The physical and behavioral traits these individuals have are passed to the next generation, and the frequency of the trait shifts in the population. Where a species may have existed but has died out, others arise and contribute to the incredible variety of life-forms seen in nature. The fossil record is one of ever-increasing species diversity, despite several catastrophes, which were recorded in different

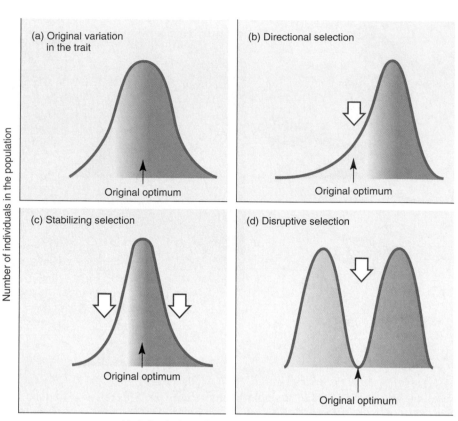

FIGURE 3.9 A species trait, such as beak shape, changes in response to selection pressure. (a) The original variation is acted on by selection pressure (arrows) that (b) shifts the characteristics of that trait in one direction, (c) or to an intermediate condition. (d) Disruptive selection moves characteristics to the extremes of the trait. Which selection type plausibly resulted in two distinct beak shapes among Galápagos finches—narrow in tree finches versus stout in ground finches?

geological strata and which wiped out a large proportion of the earth's species each time.

Taxonomy describes relationships among species

Taxonomy is the study of types of organisms and their relationships. With it you can trace how organisms have descended from common ancestors. Taxonomic relationships among species are displayed like a family tree. Botanists, ecologists, and other scientists often use the most specific levels of the tree, genus and species, to compose **binomials.** Also called scientific or Latin names, they identify and describe species using Latin, or Latinized nouns and adjectives, or names of people or places. Scientists communicate about species using these scientific names instead of common names (e.g., lion, dandelion, or ant lion), to avoid confusion. A common name can refer to any number of species in different places, and a single species might have many common names. The bionomial *Pinus resinosa*, on the other hand, always is the same tree, whether you call it a red pine, Norway pine, or just pine.

Principles of Environmental Science

1. Single population

2. Geographically isolated populations

FIGURE 3.10 Geographic isolation is a mechanism in allopatric speciation. In cool, moist glacial periods, Arizona was forested and red squirrels traveled and interbred freely. As the climate warmed and dried, desert replaced forests on the plains, while cooler mountaintops remained wooded and the only place in Arizona that red squirrels could survive. When mountaintop red squirrel populations became reproductively isolated from one another, they began to develop different characteristics.

Taxonomy also helps organize specimens and subjects in museum collections and research. You are *Homo sapiens* (human) and eat chips made of *Zea mays* (corn or maize). Both are members of two well-known kingdoms (table 3.1). Scientists, however, recognize six kingdoms (fig. 3.11): animals, plants, fungi (molds and mushrooms), protists (algae, protozoans, slime molds), bacteria (or eubacteria), and archaebacteria (ancient, single-celled organisms that live in harsh environments, such as hot springs). Within these kingdoms are millions of different species, which you will learn more about in chapter 5.

SPECIES INTERACTIONS SHAPE COMMUNITIES OF SPECIES

We have learned that adaptation to one's environment, determination of ecological niche, and even speciation is affected not just by bodily limits and behavior, but also by competition and predation. Don't despair. Not all biological interactions are antag-

TABLE 3.1

Taxonomy of Two Common Species

Taxonomic Level	Humans	Corn
Kingdom	Animalia	Plantae
Phylum	Chordata	Anthophyta
Class	Mammalia	Monocotyledons
Order	Primates	Commenales
Family	Hominidae	Poaceae
Genus	*Homo*	*Zea*
Species	*Homo sapiens*	*Zea mays*
Subspecies	*H. sapiens sapiens*	*Zea mays mays*

onistic, and many, in fact, involve cooperation or at least benign interactions and tolerance. In some cases, different organisms depend on each other to acquire resources. Now we will look at the interactions within and between species that affect their success and shape biological communities.

Competition leads to resource allocation

Competition is a type of antagonistic relationship within a biological community. Organisms compete for resources that are in limited supply: energy and matter in usable forms, living space, and specific sites to carry out life's activities. Plants compete for growing space to develop root and shoot systems so that they can

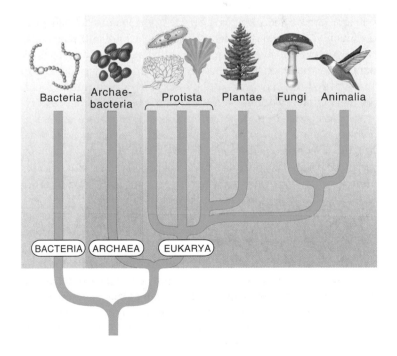

FIGURE 3.11 The six great kingdoms representing all life on earth. The Kingdoms are grouped in domains indicating common origins.

FIGURE 3.12 In this tangled Indonesian rainforest, space and light are at a premium. Plants growing beneath the forest canopy have adaptations that help them secure these limited resources. The ferns and bromeliads seen here are epiphytes; they find space and get closer to the sun by perching on limbs and tree trunks. Strangler figs start out as epiphytes, but send roots down to the forest floor and, once contact is made, put on a growth spurt that kills the supporting tree. These are just some of the adaptations to life in the dark jungle.

absorb and process sunlight, water, and nutrients (fig. 3.12). Animals compete for living, nesting, and feeding sites, and also for mates. Competition among members of the same species is called **intraspecific competition,** whereas competition between members of different species is called **interspecific competition.** Recall the competitive exclusion principle as it applies to interspecific competition. Competition shapes a species population and biological community by causing individuals and species to shift their focus from one segment of a resource type to another. Thus, warblers all competing with each other for insect food in New England tend to specialize on different areas of the forest's trees, reducing or avoiding competition. Since the 1950s there have been hundreds of interspecific competition studies in natural populations. In general, scientists assume it does occur, but not always, and in some groups—carnivores and plants—it has little effect.

In intraspecific competition, members of the same species compete directly with each other for resources. Several avenues exist to reduce competition in a species population. First, the young of the year disperse. Even plants practice dispersal; seeds are carried by wind, water, and passing animals to less crowded conditions away from the parent plants. Second, by exhibiting strong territoriality, many animals force their offspring or trespassing adults out of their vicinity. In this way territorial species, which include bears, songbirds, ungulates, and even fish, minimize competition between individuals and generations. A third way to reduce intraspecific competition is resource partitioning between generations. The adults and juveniles of these species occupy different ecological niches. For instance, monarch caterpillars munch on milkweed leaves, while metamorphosed

butterflies lap nectar. Crabs begin as floating larvae and do not compete with bottom-dwelling adult crabs.

We think of competition among animals as a battle for resources—"nature red in tooth and claw" is the phrase. In fact, many animals avoid fighting if possible, or confront one another with noise and predictable movements. Bighorn sheep and many other ungulates, for example, engage in ritualized combat, with the weaker animal knowing instinctively when to back off. It's worse to be injured than to lose. Instead, competition often is simply about getting to food or habitat first, or being able to use it more efficiently. As we discussed, each species has tolerance limits for nonbiological (abiotic) factors. Studies often show that, when two species compete, the one living in the center of its tolerance limits for a range of resources has an advantage and, more often than not, prevails in competition with another species living outside its optimal environmental conditions.

Predation affects species relationships

All organisms need food to live. Producers make their own food, while consumers eat organic matter created by other organisms. As we saw in chapter 2, photosynthetic plants and algae are the producers in most communities. Consumers include herbivores, carnivores, omnivores, scavengers, detritivores, and decomposers. You may think only carnivores are predators, but ecologically a predator is any organism that feeds directly on another living organism, whether or not this kills the prey (fig. 3.13). Herbivores, carnivores, and omnivores, which feed on live prey, are predators, but scavengers, detritivores, and decomposers, which feed on dead things, are not. In this sense, parasites (organisms that feed on a host organism or steal resources from it without necessarily killing it) and even pathogens (disease-causing organisms) can be considered predator organisms. Herbivory is the type of predation practiced by grazing and browsing animals on plants.

Predation is a powerful but complex influence on species populations in communities. It affects (1) all stages in the life cycles

FIGURE 3.13 Consider the innocuous caterpillar, a predatory animal feeding on helpless plant life. Actually, insect herbivores consume the vast majority of biomass in the world, while top carnivores consume comparatively little.

FIGURE 3.14 A large part of the world's biomass is taken up by microscopic plants and animals living in the world's oceans. Plankton forms the base of the Antarctic Ocean and other aquatic food chains. It is often made up of larvae that occupy different habitats and have different feeding techniques than their adult forms.

of predator and prey species; (2) many specialized food-obtaining mechanisms; and (3) the evolutionary adjustments in behavior and body characteristics that help prey escape being eaten, and predators more efficiently catch their prey. Predation also interacts with competition. In **predator-mediated competition,** a superior competitor in a habitat builds up a larger population than its competing species; predators take note and increase their hunting pressure on the superior species, reducing its abundance and allowing the weaker competitor to increase its numbers. To test this idea, scientists remove predators from communities of competing species. Often the superior competitors eliminated other species from the habitat. In a classic example, the ochre starfish (*Pisaster ochraceus*) was removed from Pacific tidal zones and its main prey, the common mussel (*Mytilus californicus*), exploded in numbers and crowded out other intertidal species.

Knowing how predators affect prey populations has direct application to human needs, such as pest control in cropland. The cyclamen mite (*Phytonemus pallidus*), for example, is a pest of California strawberry crops. Its damage to strawberry leaves is reduced by predatory mites (*Typhlodromus* and *Neoseiulus*), which arrive naturally or are introduced into fields. Pesticide spraying to control the cyclamen mite can actually increase the infestation because it also kills the beneficial predatory mites.

Predatory relationships may change as the life stage of an organism changes. In marine ecosystems, crustaceans, mollusks, and worms release eggs directly into the water where they and hatchling larvae join the floating plankton community (fig. 3.14). Planktonic animals eat each other and are food for larger carnivores, including fish. As prey species mature, their predators change. Barnacle larvae are planktonic and are eaten by small

fish, but as adults their hard shells protect them from fish, but not starfish and predatory snails. Predators often switch prey in the course of their lives. Carnivorous adult frogs usually begin their lives as herbivorous tadpoles. Predators also switch prey when it becomes rare, or something else becomes abundant. Many predators have morphologies and behaviors that make them highly adaptable to a changing prey base, but some, like the polar bear are highly specialized in their prey preferences.

Some adaptations help avoid predation

Predator-prey relationships exert selection pressures that favor evolutionary adaptation. In this world, predators become more efficient at searching and feeding, and prey become more effective at escape and avoidance. Toxic chemicals, body armor, extraordinary speed, and the ability to hide are a few strategies organisms use to protect themselves. Plants have thick bark, spines, thorns, or distasteful and even harmful chemicals in tissues—poison ivy and stinging nettle are examples. Arthropods, amphibians, snakes, and some mammals produce noxious odors or poisonous secretions that cause other species to leave them alone. Animal prey are adept at hiding, fleeing, or fighting back. On the Serengeti Plain of East Africa, the swift Thomson's gazelle and even swifter cheetah are engaged in an arms race of speed, endurance, and quick reactions. The gazelle escapes often because the cheetah lacks stamina, but the cheetah accelerates from 0 to 72 kph in 2 seconds, giving it the edge in a surprise attack. The response of predator to prey and vice versa, over tens of thousands of years, produces physical and behavioral changes in a process known as **coevolution.** Coevolution can be mutually beneficial: many plants and pollinators have forms and behaviors that benefit each other. A classic case is that of fruit bats, which pollinate and disperse seeds of fruit-bearing tropical plants.

Often species with chemical defenses display distinct coloration and patterns to warn away enemies (fig. 3.15). In a neat evolutionary twist, certain species that are harmless resemble poisonous or distasteful ones, gaining protection against predators who

FIGURE 3.15 Poison arrow frogs of the family Dendrobatidae display striking patterns and brilliant colors that remind potential predators about the extremely toxic secretions on their skin. Indigenous people in Latin America use the toxin to arm blowgun darts.

(a)

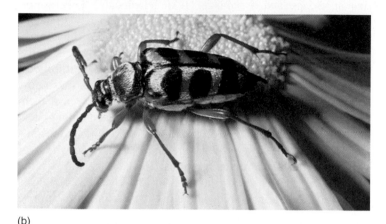

(b)

FIGURE 3.16 In a clear example of Batesian mimicry, (a) a dangerous wasp carries bold yellow and black bands that (b) are imitated by the much rarer but harmless longhorn beetle (family Cerambycidae). The beetle even behaves like a wasp, tricking predators which have experienced wasps into avoiding the beetle.

remember a bad experience with the actual toxic organism. This is called **Batesian mimicry,** after the English naturalist H. W. Bates (1825–1892), a traveling companion of Alfred Wallace. Many wasps, for example, have bold patterns of black and yellow stripes to warn off potential predators (fig. 3.16a). The much rarer longhorn beetle has no stinger but looks and acts much like a wasp, tricking predators into avoiding it (fig. 3.16b). The distasteful monarch and benign viceroy butterflies are a classic case of Batesian mimicry. Another form of mimicry, **Müllerian mimicry** (after the biologist Fritz Müller) involves two unpalatable or dangerous species who look alike. When predators learn to avoid either species, both benefit. Species also display forms, colors, and patterns that help avoid being discovered. Insects that look exactly like dead leaves or twigs are among the most remarkable examples. Unfortunately for prey, predators also use camouflage to conceal themselves as they lie in wait for their next meal (fig. 3.17).

Symbiosis: Intimate relations among species

In contrast to predation and competition, some interactions between organisms can be nonantagonistic, even beneficial (table 3.2). In

FIGURE 3.17 This is a scorpion fish whose perfect camouflage conceals it from its prey until it is too late. Natural selection has favored this adaptation because it increases the success of this predator.

such relationships, called **symbiosis,** two or more species live intimately together, with their fates linked. Symbiotic relationships often enhance the survival of one or both partners. In lichens, a fungus and a photosynthetic partner (either an alga or a cyanobacterium) combine tissues to mutual benefit (fig. 3.18a). This association is called **mutualism.** Some ecologists believe that cooperative, mutualistic relationships may be more important in evolution than commonly thought (fig. 13.18b). Survival of the fittest may also mean survival of organisms that can live together.

Symbiotic relationships often entail some degree of coevolution of the partners, shaping—at least in part—their structural and behavioral characteristics. This mutualistic coadaptation is evident between swollen thorn acacias (*Acacia collinsii*) and the ants (*Pseudomyrmex ferruginea*) that tend them in Central and South America. Acacia ant colonies live inside the swollen thorns on the acacia tree branches. Ants feed on nectar that is produced in glands at the leaf bases and also eat special protein-rich structures that are produced on leaflet tips. The acacias thus provide

TABLE 3.2

Types of Species Interactions
(+beneficial; −harmful;
0 neutral; ±varies)

Interaction between Two Species	Effect on First Species	Effect on Second Species
Mutualism	+	+
Commensalism	+	0
Parasitism	+	−
Predation	+	−
Competition	±	±

(a) Symbiosis

(b) Mutualism

(c) Commensalism

FIGURE 3.18 Symbiotic relationships. (a) Lichens represent an obligatory mutualism between a fungus and alga or cyanobacterium. (b) Mutualism between a parasite-eating red-billed oxpecker and parasite-infested impala. (c) Commensalism between a tropical tree and free-loading bromeliad.

shelter and food for the ants. Although they spend energy to provide these services, the trees are not harmed by the ants. What do the acacias get in return? Ants aggressively defend their territories, driving away herbivorous insects that want to feed on the acacias. Ants also trim away vegetation that grows around the tree, reducing competition by other plants for water and nutrients. You can see how mutualism is structuring the biological community in the vicinity of acacias harboring ants, just as competition or predation shapes communities.

Mutualistic relationships can develop quickly. In 2005 the Harvard entomologist E. O. Wilson pieced together evidence to explain a 500-year-old agricultural mystery in the oldest Spanish settlement in the New World, Hispaniola. Using historical accounts and modern research, Dr. Wilson reasoned that mutualism developed between the tropical fire ant (*Solenopsis geminata*), native to the Americas, and a sap-sucking insect that was probably introduced from the Canary Islands in 1516 on a shipment of plantains. The plantains were planted, the sap-suckers were distributed across Hispaniola, and in 1518 a great die-off of crops occurred. Apparently the native fire ants discovered the foreign sap-sucking insects, consumed their excretions of sugar and protein, and protected them from predators, thus allowing the introduced insect population to explode. The Spanish assumed the fire ants caused the agricultural blight, but a little ecological knowledge would have led them to the real culprit.

Commensalism is a type of symbiosis in which one member clearly benefits and the other apparently is neither benefited nor harmed. Many mosses, bromeliads, and other plants growing on trees in the moist tropics are considered commensals (fig. 3.18c). These epiphytes are watered by rain and obtain nutrients from leaf litter and falling dust, and often they neither help nor hurt the trees on which they grow. Robins and sparrows that inhabit suburban yards are commensals with humans. **Parasitism,** a form of predation, may also be considered symbiosis because of the dependency of the parasite on its host.

Keystone species: Influence all out of proportion

A **keystone species** plays a critical role in a biological community that is out of proportion to its abundance. Originally, keystone species were thought to be top predators—lions, wolves, tigers—which limited herbivore abundance and reduced the herbivory of plants. Scientists now recognize that less-conspicuous species also play keystone roles. Tropical figs, for example, bear fruit year-round at a low but steady rate. If figs are removed from a forest, many fruit-eating animals (frugivores) would starve in the dry season when fruit of other species is scarce. In turn, the disappearance of frugivores would affect plants that depend on them for pollination and seed dispersal. It is clear that the effect of a keystone species on communities often ripples across trophic levels.

Keystone functions have been documented for vegetation-clearing elephants, the predatory ochre sea star, and frog-eating salamanders in coastal North Carolina. Even microorganisms can play keystone roles. In many temperate forest ecosystems, groups of fungi that are associated with tree roots (mycorrhizae) facilitate the uptake of essential minerals. When fungi are absent, trees grow poorly or not at all. Overall, keystone species seem to be more common in aquatic habitats than in terrestrial ones.

The role of keystone species can be difficult to untangle from other species interactions. Off the northern Pacific coast, a giant brown alga (*Macrocystis pyrifera*) forms dense "kelp forests," which shelter fish and shellfish species from predators, allowing them to become established in the community (fig. 3.19). It turns out, however, that sea otters eat sea urchins living in the kelp forests; when sea otters are absent, the urchins graze on and eliminate kelp forests. To complicate things, around 1990, killer whales began preying on otters because of the dwindling stocks of seals and sea lions, thereby creating a cascade of effects. Is the kelp, otter, or orca the keystone here? Whatever

FIGURE 3.19 Giant kelp is a massive alga that forms dense "forests" off the Pacific coast. It is a keystone species in that it provides food, shelter, and structure essential for a whole community. Removing sea otters allows sea urchin populations to explode. When the urchins destroy the kelp, many other species suffer as well, so are sea otters also keystones here?

the case, keystone species exert their influence by changing competitive relationships. In some communities, perhaps we should call it a "keystone set" of organisms. The functional redundancy seen in the Cedar Creek experimental plots, for instance, suggests that a diversity of species—no single keystone—ensures that essential ecological functions will continue without much change.

THE GROWTH OF SPECIES POPULATIONS

Many biological organisms can produce unbelievable numbers of offspring if environmental conditions are right. Consider the common housefly (*Musca domestica*). Each female fly lays 120 eggs (assume half female) in a generation. In 56 days those eggs become mature adults, able to reproduce. In one year, with seven generations of flies being born and reproducing, that original fly would be the proud parent of 5.6 trillion offspring. If this rate of reproduction continued for ten years, the entire earth would be covered in several meters of housefly bodies. Luckily housefly reproduction, as for most organisms, is constrained in a variety of ways—scarcity of resources, competition, predation, disease, accident. The housefly merely demonstrates the remarkable amplification—the **biotic potential**—of unrestrained biological

reproduction. Population dynamics describes these changes in the number of organisms in a population over time.

Growth without limits is exponential

As you learned in chapter 2, a population consists of all the members of a single species living in a specific area at the same time. The growth of the housefly population just described is **exponential,** having no limit and possessing a distinctive shape when graphed over time. An exponential growth rate (increase in numbers per unit of time) is expressed as a constant fraction, or exponent, which is used as a multiplier of the existing population. The mathematical formula for exponential growth is:

$$\frac{dN}{dt} = rN$$

That is, the change in numbers of individuals (dN) per change in time (dt) equals the rate of growth (r) times the number of individuals in the population (N). The r term (intrinsic capacity for increase) is a fraction representing the average individual contribution to population growth. If r is positive, the population is increasing. If r is negative, the population is shrinking. If r is zero, there is no change, and $dN/dt = 0$.

A graph of exponential population growth is described as a **J curve** (fig. 3.20) because of its shape. As you can see, the number of individuals added to a population at the beginning of an exponential growth curve can be rather small. But within a very short time, the numbers begin to increase quickly because a fixed percentage leads to a much larger increase as the population size grows.

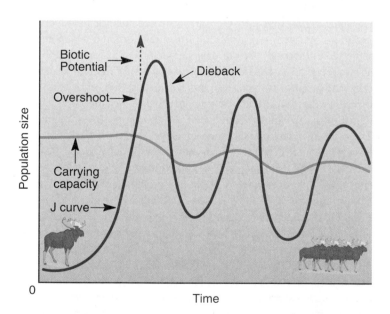

FIGURE 3.20 J curve, or exponential growth curve, with overshoot of carrying capacity. Exponential growth in an unrestrained population (*left side of curve*) leads to a population crash and oscillations below former levels. After the overshoot, carrying capacity may be reduced because of damage to the resources of the habitat. Moose on Isle Royale in Lake Superior may be exhibiting this growth pattern in response to their changing environment.

The exponential growth equation is a very simple model; it is an idealized, simple description of the real world. The same equation is used to calculate growth in your bank account due to compounded interest rates; achieving the potential of your savings depends on you never making a withdrawal. Just as species populations lose individuals and experience reduced biotic potential, not all of your dollars will survive to a ripe old age and contribute fully to your future cash position.

Carrying capacity relates growth to its limits

In the real world there are limits to growth. Around 1970, ecologists developed the concept of **carrying capacity** to mean the number or biomass of animals that can be supported (without harvest) in a certain area of habitat. The concept is now used more generally to suggest a limit of sustainability that an environment has in relation to the size of a species population. Carrying capacity is helpful in understanding the population dynamics of some species, perhaps even humans.

When a population overshoots or exceeds the carrying capacity of its environment, resources become limited and death rates rise. If deaths exceed births, the growth rate becomes negative and the population may suddenly decrease, a change called a population crash or dieback (fig 3.20). Populations may oscillate from high to low levels around the habitat's carrying capacity, which may be lowered if the habitat is damaged. Moose and other browsers or grazers sometimes overgraze their food plants such that future populations of herbivores in the same habitat find less preferred food to sustain them, at least until the habitat recovers. Some species go through predictable cycles if simple factors are involved, such as the seasonal light- and temperature-dependent bloom of algae in a lake. Cycles can be irregular if complex environmental and biotic relationships exist. Irregular cycles include migratory locust outbreaks in the Sahara, or tent caterpillars in temperate forests—these represent irruptive population growth. Population dynamics are also affected by the emigration of organisms from an overcrowded habitat, or immigration of individuals into new habitat, as occurred in 2005 when owls suddenly invaded the northern United States due to a food shortage in their Canadian habitat.

Sometimes predator and prey populations oscillate in synchrony with each other (fig. 3.21). This classic study employed the 200-year record of furs sold at Hudson Bay Company trading posts in Canada (the figure shows a portion of that record). Charles Elton, responsible for the diversity-stability hypothesis, was the ecological detective who showed that numbers of Canada lynx (*Lynx canadensis*) fluctuate on about a ten-year cycle which mirrors, slightly out of phase, the population peaks of snowshoe hares (*Lepus americanus*). When the hare population is high and food is plentiful, life for the lynx is very good, it reproduces well, and populations grow. Eventually, declining food supplies shrink hare populations. For a while the lynx benefits because starving hares are easier to catch than healthy ones. As hares become scarce, however, so do lynx. When hares are at their lowest levels, their food supply recovers and the whole cycle starts over again.

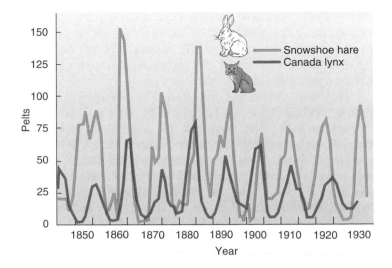

FIGURE 3.21 Ten-year oscillations in the population of snowshoe hare and lynx in Canada suggest a close linkage of predator and prey, but may not tell the whole story. These data are based on the number of pelts received by the Hudson Bay Company each year, meaning fur-traders were unwitting accomplices in later scientific research.
Source: Data from D. A. MacLulich. Fluctuations in the Numbers of the Varying Hare (*Lepus americus*). Toronto: University of Toronto Press, 1937, reprinted 1974.

This predator-prey oscillation is described mathematically in the Lotka-Volterra model, after the scientists who developed it.

Feedback produces logistic growth

Not all biological populations cycle through exponential overshoot and catastrophic dieback. Many species are regulated by both internal and external factors and come into equilibrium with their environmental resources while maintaining relatively stable population sizes. When resources are unlimited, they may even grow exponentially, but this growth slows as the carrying capacity of the environment is approached. This population dynamic is called **logistic growth** because of its constantly changing rate.

Mathematically, this growth pattern is described by the following equation, which adds a feedback term for carrying capacity (K) to the exponential growth equation:

$$\frac{dN}{dt} = rN\left(1 - \frac{N}{K}\right)$$

The logistic growth equation says that the change in numbers over time (dN/dt) equals the exponential growth rate (rN) times the portion of the carrying capacity (K) not already taken by the current population size (N). The term ($1 - N/K$) establishes the relationship between population size at any given time step and the number of individuals the environment can support. Depending on whether N is less than or greater than K, the rate of growth will be positive or negative (see Applications).

The logistic growth curve has a different shape than the exponential growth curve. It is a sigmoidal-shaped, or **S curve** (fig. 3.22). It describes a population that decreases if its numbers exceed the carrying capacity of the environment. In the Data Analysis exercise at the end of this chapter, you will learn how the terms in the exponential and logistic equations influence the size of a population at any time.

Population growth rates are affected by external and internal factors. External factors are habitat quality, food availability, and interactions with other organisms. As populations grow, food becomes scarcer and competition for resources more intense. With a larger population, there is an increased risk that disease or parasites will spread, or that predators will be attracted to the area. Some organisms become physiologically stressed when conditions are crowded, but other internal factors of maturity, body size, and hormonal status may cause them to reduce their reproductive output. Overcrowded house mice ($>1,600/m^3$), for instance, average 5.1 baby mice per litter, while uncrowded house mice ($<34/m^3$) produce 6.2 babies per litter. All these factors are **density-dependent,** meaning as population size increases, the effect intensifies. With density-independent factors, a population is affected no matter what its size. Drought, an early killing frost, flooding, landslide, or habitat destruction by people—all increase mortality rates regardless of the population size. Density-independent limits to population are often nonbiological, capricious acts of nature.

Species respond to limits differently: *r-* and *K*-selected species

The story of the race between the hare and the tortoise has parallels to the way that species deal with limiting factors in their environment. Some organisms, such as dandelions and barnacles,

depend on a high rate of reproduction and growth (*rN*) to secure a place in the environment. These organisms are called ***r*-selected species** because they employ a high reproductive rate (*r*) to overcome the high mortality of virtually ignored offspring. These species may even overshoot carrying capacity and experience population crashes, but as long as vast quantities of young are produced, a few will survive. Other organisms that reproduce more conservatively—longer generation times, late sexual maturity, fewer young—are referred to as ***K*-selected species,** because their growth slows as the carrying capacity (*K*) of their environment is approached.

Many species blend exponential (*r*-selected) and logistic (*K*-selected) growth characteristics. Still, it's useful to contrast the advantages and disadvantages of organisms at the extremes of the continuum. It also helps if we view differences in terms of "strategies" of adaptation and the "logic" of different reproductive modes (table 3.3).

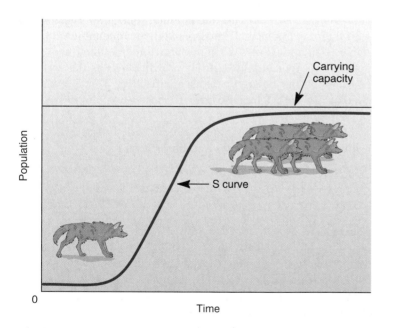

FIGURE 3.22 S curve, or logistic growth curve, describes a population's changing numbers over time in response to feedback from the environment or its own population density. Over the long run, a conservative and predictable population dynamic may win the race over an exponential population dynamic. Species with this growth pattern tend to be *K*-selected.

TABLE 3.3
Reproductive Strategies

r-Selected Species	*K*-Selected Species
1. Short life	1. Long life
2. Rapid growth	2. Slower growth
3. Early maturity	3. Late maturity
4. Many, small offspring	4. Few, large offspring
5. Little parental care and protection	5. High parental care or protection
6. Little investment in individual offspring	6. High investment in individual offspring
7. Adapted to unstable environment	7. Adapted to stable environment
8. Pioneers, colonizers	8. Later stages of succession
9. Niche generalists	9. Niche specialists
10. Prey	10. Predators
11. Regulated mainly by intrinsic factors	11. Regulated mainly by extrinsic factors
12. Low trophic level	12. High trophic level

Principles of Environmental Science

FIGURE 3.23 Exhibiting an *r*-selected reproductive strategy, this dandelion and other generalist species spread quickly in a variety of environments.

Organisms with *r*-selected, or exponential, growth patterns tend to occupy low trophic levels in their ecosystems (chapter 2) or they are successional pioneers. Consider the brown rat discussed earlier. This and other generalists occupy disturbed or new environments, grow rapidly, mature early, and produce many offspring with excellent dispersal abilities (fig. 3.23). As individual parents, they do little to care for their offspring or protect them from predation. They invest their energy in producing huge numbers of young and count on some surviving to adulthood.

A female clam, for example, can release up to 1 million eggs in her lifetime. The vast majority of young clams die before reaching maturity, but if even a few survive, the species will continue. Many marine invertebrates, parasites, insects, rodents, and annual plants follow this reproductive strategy. Also included in this group are most invasive and pioneer organisms, weeds, pests, and nuisance species.

So-called *K*-selected organisms are usually larger, live long lives, mature slowly, produce few offspring in each generation, and have few natural predators. Elephants, for example, are not reproductively mature until they are 18 to 20 years old. In youth and adolescence, a young elephant belongs to an extended family that cares for it, protects it, and teaches it how to behave. A female elephant normally conceives only once every 4 or 5 years. The gestation period is about 18 months; thus, an elephant herd doesn't produce many babies in any year. Since elephants have few enemies and live a long life (60 or 70 years), this low reproductive rate produces enough elephants to keep the population stable, given good environmental conditions and no poachers.

When you consider the species you recognize from around the world, can you pigeonhole them into categories of *r*- or *K*-selected species (fig. 3.24)? What strategies seem to be operating for ants, bald eagles, cheetahs, clams, giraffes, pandas, or sharks? An important question in the back of your mind might be: Where

(a) Survive to old age

(b) Die randomly

(c) High infant mortality

(d) Long adult life span

FIGURE 3.24 Which of these appears to be *r*- and which *K*-selected in their reproductive strategy? (a) Most elephants live long lives, but reproduce relatively late. (b) Seagulls die at all ages, mostly from accidents. (c) Antelope have high infant mortality, but survive well after that. (d) Seedlings of redwood trees die off in huge numbers, but the oldest redwood is at least 2,200 years old.

do people fit? Are we like wolves and elephants or more similar to clams and rabbits in our population growth strategy?

PROPERTIES OF COMMUNITIES DEPEND ON SPECIES DIVERSITY

No species is an island. It always lives with other species in a biological community in a particular environment. You've seen how interactions among species affect biological communities. In this section, you will learn how species interact to create the fundamental properties of biological communities and ecosystems: (1) diversity and abundance; (2) community structure and patchiness; and (3) complexity, resilience, productivity, and stability.

Diversity and abundance

In the opening Case Study, diversity was a driver of ecological processes, but what is diversity exactly? **Diversity** is the number of different species per unit area. The number of herbaceous plant species per square meter of African savanna, the number of bird species in Costa Rica, the insect species on earth (a big number!)—all describe diversity. Diversity is important because it indicates the variety of ecological niches and genetic variation in a community. **Abundance** refers to the number of individuals of a species in an area. It is expressed by density of individuals of either a single species or multiple species: the number of diatoms floating in 1 m^3 of sea water, the number of sparrows per hectare of urban habitat. Diversity and abundance are often related. Communities with high diversity often have few individuals of any one species. Most communities contain a few common species and many rarer ones.

As a general rule, diversity is greatest at the equator and drops toward the poles. For many species, the abundance of individuals tends to increase along the same gradient. Individual mosquitoes are abundant in the Arctic, but there are few insect species overall. In the tropics, hundreds of thousands of insect species have developed bizarre forms and behaviors, but at any one location only a few individuals of a species are found. The same pattern is seen in trees of high-latitude boreal forests versus tropical rainforests, and in bird populations. In Greenland there reside 56 species of breeding birds, while in Colombia, with one-fifth the area, there are 1,395. Why are there so many more species in Colombia? Climate and history play important roles. Greenland has a harsh climate and a short, cool growing season that restricts the biological activity that can take place. Limited energy overwhelms all other factors and prevents specialization and the development of ecological niches—hence, fewer species. Furthermore, because glaciers covered Greenland until about 10,000 years ago, new species have had little time to develop.

Many areas in the tropics, by contrast, have abundant rainfall and warm temperatures year-round, so that ecosystems there are highly productive. The year-round availability of food, moisture, and warmth supports an exuberance of life and allows a high degree of specialization in physical shape and behavior. Many niches exist in small areas, with associated high species diversity. Coral reefs are similarly stable, productive, and conducive to proliferation of diverse and exotic life-forms. An enormous abundance of brightly colored and fantastically shaped fishes, corals, sponges, and arthropods live in the reef community. Glaciers never covered tropical forests or coral reefs, although the past climate of the tropics was different from today's.

Species patterns create community structure

Just like jig-saw puzzle pieces, boundaries of species populations and communities form patterns that fit together: (1) individuals and species are spaced throughout communities in different ways; (2) the communities themselves are arranged over a large geographic area or landscape; and (3) communities have relatively uniform interiors ("cores") and also "edges" that meet. **Community** (or ecological) **structure** refers to these patterns of spatial distribution of individuals, species, and communities.

Individuals in communities are distributed in various ways

Even in a relatively uniform environment, individuals of a species population are distributed randomly, arranged in uniform patterns, or clustered together. In randomly distributed populations, individuals live wherever resources are available and chance events allowed (fig. 3.25a). Uniform patterns arise from the physical environment also, but more often are caused by competition. For example, penguins or seabirds compete fiercely for nesting sites in their colonies. Each nest tends to be just out of reach of neighbors sitting on their own nests. Constant squabbling produces a highly regular pattern (fig. 3.25b). Plants also compete, producing a uniform pattern. Sagebrush releases toxins from roots and fallen leaves, which inhibit the growth of competitors and create a circle of bare ground around each bush. Neighbors grow up to the limit of this chemical barrier, and regular spacing results.

Other species cluster together for protection, mutual assistance, reproduction, or access to an environmental resource. Ocean and freshwater fish form dense schools, increasing their chances of detecting and escaping predators (fig. 3.25c). Meanwhile many predators—whether wolves or humans—hunt in packs. When blackbirds flock in a cornfield, or baboons troop across the African savanna, their bands help them evade predators and find food efficiently. Plants also cluster for protection in harsh environments. You often see groves of wind-sheared evergreens at mountain treelines or behind foredunes at seashores. These treelines protect the plants from wind damage, and incidentally shelter other animals and plants.

Individuals can also be distributed vertically in a community. Forests, for instance, have many layers, each with different environmental conditions and combinations of species. Distinct communities of plants, animals, and microbes live in the treetops,

(a) Random

(b) Uniform

(c) Clustered

FIGURE 3.25 Distribution of members of a population in a given space can be (a) random, (b) uniform, or (c) clustered. These patterns in the distribution of individuals give structure to communities.

Principles of Environmental Science

at mid-canopy level, and near the ground. This vertical layering is best developed in tropical rainforests (fig 3.26). Aquatic communities also are often stratified into layers formed by species responding to varying light levels, temperature, salinity, nutrient supply, and pressure.

Communities are distributed in patterns across a landscape

If you have flown in an airplane, you may have noticed that the land consists of patches of different colors and shapes. Some appear long and narrow (hedgerows or rivers), while others are rectangular (pastures and cropfields), or green and lumpy in summer (forests). Each patch represents a biological community with its own set of species and environmental conditions. The name for this pattern of community distribution across a landscape is patchiness or habitat patchiness. The largest patches contain **core habitat,** a mostly uniform environment big enough to support nearly all the plants and animals that are typically found in that community. For example, the largest patches of old coniferous forest in the Pacific Northwest contain the most persistent populations of northern spotted owl, a threatened species. In smaller patches the owl fares poorly, possibly due to competition with the barred owl. A single pair of northern spotted owls may require over 1,000 hectares of core habitat to survive.

Outside habitat cores, species encounter different habitat in what ecologists call the **ecotone,** or border between two communities (fig. 3.27). Ecotones are often rich in species because individuals from both environments occupy the boundary area. Species that prefer ecotones and use the resources of both environments are called edge species. The white-tailed deer is one, browsing in fields but hiding in forests. Sometimes a community edge seems sharp and distinct, but other times one habitat type grades gradually into another.

Where communities meet, the environmental conditions blend and the species and microclimate of one community can penetrate the other. Called **edge effects,** the penetrating influences may extend hundreds of meters into an adjacent community. Grassland makes an adjacent forest edge sunnier, drier, hotter, and more susceptible to storm damage than the center of the forest. Generalist grassland species, including weeds and predators, move into the forest and negatively affect forest species. Depending on how far the edge effects penetrate, the shape of a community patch makes a big difference (fig. 3.28). In a narrow, irregularly shaped patch, far-reaching edge effects would leave no core habitat. In a similar-sized square patch, however, interior species would still find core habitat.

Many game animals (deer, pheasant, etc.) are most plentiful in ecotones and adapt well to human disturbance. With this knowledge, in the 1930s, North American game managers began to boost game populations by cutting openings in forests and planting trees and shrubs in grasslands. Since 1980, however, wildlife biologists recognized that creating edge effects everywhere is harmful to a region's biota, and that core habitat can benefit edge species by serving as refuges. In this view of habitat management,

FIGURE 3.26 Vertical layering of plants and animals in a tropical rainforest is a type of community structure.

large community patches are preserved, connected to each other, and linked to smaller patches (see fig. 6.22). You can learn about bird life, core habitat, and edge effects by reading "Where Have All the Songbirds Gone?" at www.mhhe.com/cunningham4e.

Community properties emerge from diversity and structure

You now know that more-diverse communities may recover faster from disturbance. But there's more to resilience than diversity and structure, and that is community complexity. **Complexity** refers to the number of trophic levels in a community and to the number of species at each of those trophic levels. A diverse community may not be complex if all of its species cluster in a few trophic levels and form a simple food chain. Take one key level or species away, and the community may unravel.

A complex, interconnected community might have many trophic levels and groups of species performing the same functions

FIGURE 3.27 Ecological edges are called ecotones. Ecotones exist between lake and wetland, and wetland and forest. The strip of forest in the center of the photo is all edge—sunny, hotter conditions on both sides penetrate to its center. The forest is a travel route, or corridor, for forest animals crossing the wetland.

(table 3.4). In the Antarctic, the sun powers the system, but detritus that consists of dead organisms and animal excrement is a key energy source. Whales and leopard seals are top predators, while krill (tiny crustaceans) play a key energy-transfer role. In tropical rainforests and many other communities, herbivores form guilds based on the specialized ways they feed on plants. There may be fruit-eaters, leaf-nibblers, root-borers, seed-gnawers, and sap-suckers—each guild

Total area: 47 ha   Core area: 0 ha

Total area: 47 ha   Core area: 20 ha

FIGURE 3.28 Shape can be as important as size in small preserves. While these areas are similar in size, no place in the top figure is far enough from the edge to have characteristics of core habitat, while the bottom patch has a significant core.

TABLE 3.4

Community Complexity in the Antarctic Ocean

Type of Function	Members of Functional Group
Top ocean predator	Sperm and killer whales, leopard and elephant seals
Aerial predator	Albatross, skuas
Other ocean predator	Weddell and Ross seals, king penguin, pelagic fish
Krill/plankton-feeder	Minke, humpback, fin, blue, and sei whales
Ocean herbivore	Krill, zooplankton (many species)
Ocean-bottom predator	Many species of octopods and bottom-feeding fish
Ocean-bottom herbivore	Many species of echinoderms, crustaceans, mollusks
Photosynthesizer	Many species of phytoplankton and algae

is composed of species of different sizes, shapes, and even biological kingdoms, but they feed in similar ways.

In 1955, Robert MacArthur (1930–1972), then a graduate student at Yale, proposed that the more complexity a community possesses, the more resilient it is when disturbance strikes. He reasoned that, if many different species occupy each trophic level, some can fill in if others are stressed or eliminated by external forces. The whole community has **resilience** and either resists or recovers quickly from disturbance, as witnessed in the Cedar Creek experiment.

In complex ecosystems, like kelp forests or intertidal zones, keystone species confuse the picture. For example, cutting all the fig trees from a tropical forest may destroy pollinators and fruit dispersers. We might replant the figs, but could we bring back the entire web of relationships that previously existed? In such cases, complexity seems to make a community less resilient, rather than more. While the concept that diversity and complexity lead to resilience is not perfect, it helps explain many observations about communities and urges us to preserve communities in as diverse and complex a condition as we can.

When we say something is productive, we mean it produces a lot of something. Communities produce biomass by converting solar energy into chemical energy that is stored in living (or once-living) organisms. **Primary productivity,** a community's annual output of biomass or energy, is expressed as units of biomass or energy per unit area per year. Since cellular respiration in organisms uses much of that energy, a more useful term is **net primary productivity,** or the amount of biomass stored after respiration. Productivity depends on light levels, temperature, moisture, and nutrient availability. Major ecosystems differ in how fast they produce biomass (fig. 3.29). Tropical forests, coral reefs, and the bays and drowned valleys where rivers meet the ocean (estuaries) have high productivity because of abundant resource supply. In deserts, a lack of water limits photosynthesis, and productivity is low. On the arctic tundra or on high mountains, low temperatures inhibit plant growth and productivity. In the open ocean, a lack of nutrients reduces the ability of algae to make use of plentiful sunshine and water.

Even the most photosynthetically active ecosystems capture only a small percentage of the available sunlight and use it to make energy-rich compounds. In a temperate-climate oak forest, leaves absorb only about half the available light on a midsummer day. Of this absorbed energy, 99 percent is used in respiration and in evaporating water to cool leaves. A large oak tree can transpire (evaporate) several thousand liters of water on a warm, dry, sunny day, while making only a few kilograms of sugars and other energy-rich organic compounds.

Stability is a tricky concept. When we say a community or ecosystem is stable, we mean it resists changes despite disturbance, springs back resiliently after disturbance, and supports the same species in about the same numbers as before the disturbance. Yet sometimes only one of these happens. When the Great Plains experienced devastating droughts in the 1930s and early 1940s, overall productivity declined, some species populations practically vanished, and others held their own but at lower abundance. When the rains returned, the plant communities were different, especially because before, during, and after the drought millions of cattle overgrazed the range. Today's Great Plains grasslands are very different from the grasslands George Armstrong Custer viewed from his horse in the late 1800s. Yet it remains grassland, produces forage, and feeds cattle. Are the Great Plains grasslands stable because they are still relatively productive, or unstable because species diversity and abundance changed since the 1800s? If the range was grazed according to ecological principles, would that restore the original species diversity and abundance, and raise productivity? Thinking about stability in this way allows us to figure out how best to use resources, rather than to simply say that diversity equals stability—which may be true or not, depending on your definition of stability.

FIGURE 3.29 Biomass accumulates at different rates in the world's major ecosystems. Differences in net primary production among the ecosystems is chiefly due to factors limiting plant growth (temperature, rainfall, nutrients, altitude, soil organic matter) but also to interactions among species which boost productivity.

1,000 kcal/m²/year

COMMUNITIES ARE DYNAMIC AND CHANGE OVER TIME

If fire sweeps through a biological community, it's destroyed, right? Not so fast. Fire may be good for that community. Up until now, we've focused on the day-to-day interactions of organisms with their environments, set in a context of adaptation and selection. In this section, we'll step back and look at more dynamic aspects of communities and how they change over time.

The nature of communities is debated

For several decades starting in the early 1900s, ecologists in North America and Europe argued about the basic nature of communities. It doesn't make interesting party conversation, but those discussions affected how we study and understand communities, view the changes taking place within them, and ultimately use them. Both J. E. B. Warming (1841–1924) in Denmark and Henry Chandler Cowles (1869–1939) in the United States came up with the idea that communities develop in a sequence of stages, starting either from bare rock or after a severe disturbance. They worked in sand dunes and watched the changes as plants first took root in bare sand and, with further development, created forest. This example represents constant change, not stability. In sand dunes, the community that developed last and lasted the longest was called the **climax community.**

The idea of climax community was first championed by the biogeographer F. E. Clements (1874–1945). He viewed the process as a relay—species replace each other in predictable groups and in a fixed, regular order. He argued that every landscape has a characteristic climax community, determined mainly by climate. If left undisturbed, this community would mature to a characteristic set of organisms, each performing its optimal functions. A climax community to Clements represented the maximum complexity and stability that was possible. He and others made the analogy that the development of a climax community resembled the maturation of an organism. Both communities and organisms, they argued, began simply and primitively, maturing until a highly integrated, complex community developed.

This organismal theory of community was opposed by Clements' contemporary, H. A. Gleason (1882–1975), who saw community history as an unpredictable process. He argued that species are individualistic, each establishing in an environment according to its ability to colonize, tolerate the environmental conditions, and reproduce there. This idea allows for myriad temporary associations of plants and animals to form, fall apart, and reconstitute in slightly different forms, depending on environmental conditions and the species in the neighborhood. Imagine a time-lapse movie of a busy airport terminal. Passengers come and go; groups form and dissipate. Patterns and assemblages that seem significant may not mean much a year later. Gleason suggested that we think ecosystems are uniform and stable only because our lifetimes are too short and our geographic scope too limited to understand their actual dynamic nature.

Ecological succession describes a history of community development

In any landscape, you can read the history of biological communities. That history is revealed by the process of ecological succession. During succession, organisms occupy a site and change the environmental conditions. In **primary succession** land that is bare of soil—a sandbar, mudslide, rock face, volcanic flow—is colonized by living organisms where none lived before (fig. 3.30). When an existing community is disturbed, a new one develops from the biological legacy of the old in a process called **secondary succession.** In both kinds of succession, organisms change the environment by modifying soil, light levels, food supplies, and microclimate. This change permits new species to colonize and eventually replace the previous species, a process known as ecological development or facilitation.

In primary succession on land, the first colonists are hardy **pioneer species,** often microbes, mosses, and lichens that can withstand a harsh environment with few resources. When they die, the bodies of pioneer species create patches of organic matter. Organics and other debris accumulate in pockets and crevices, creating soil where seeds lodge and grow. As succession proceeds, the community becomes more diverse and interspecies competition arises. Pioneers disappear as the environment favors new colonizers that have competitive abilities more suited to the new environment.

You can see secondary succession all around you, in abandoned farm fields, in clear-cut forests, and in disturbed suburbs and lots. Soil and possibly plant roots and seeds are present. Because soil lacks vegetation, plants that live one or two years (annuals and biennials) do well. Their light seeds travel far on the wind, and

FIGURE 3.30 Primary succession in boreal forest involves five stages (*left to right*). Bare rock is colonized by lichens and mosses which trap moisture and build soil for grasses, shrubs, and eventually trees. Natural, periodic outbreaks of insects kill or weaken trees. Wildfires often follow, initiating secondary succession. Aspens spring back from roots, jack pine cones open and spread seeds when heated by wildfires, and a new forest begins.

Principles of Environmental Science

http://www.mhhe.com/cunningham4e

their seedlings tolerate full sun and extreme heat. When they die, they lay down organic material that improves the soil's fertility and shelters other seedlings. Soon long-lived and deep-rooted perennial grasses, herbs, shrubs, and trees take hold, building up the soil's organic matter and increasing its ability to store moisture. Forest species that cannot survive bare, dry, sunny ground eventually find ample food, a diverse community structure, and shelter from drying winds and low humidity.

Generalists figure prominently in early succession. Over thousands of years, however, competition should decrease as niches proliferate and specialists arise. In theory, long periods of community development lead to greater community complexity, high nutrient conservation and recycling, stable productivity, and great resistance to disturbance—an ideal state to be in when the slings and arrows of misfortune arrive.

FIGURE 3.31 The barrens community at Kingston Plains in Michigan's eastern Upper Peninsula was created when clear-cutting of a dense white pine forest coincided with several devastating fires set by people. Note the stumps from trees cut over 100 years ago.

Appropriate disturbances can benefit communities

Disturbances are plentiful on earth: landslides, mudslides, hailstorms, earthquakes, hurricanes, tornadoes, tidal waves, wildfires, and volcanoes, to name just the obvious. A **disturbance** is any force that disrupts the established patterns of species diversity and abundance, community structure, or community properties. Animals can cause disturbance. African elephants rip out small trees, trample shrubs, and tear down tree limbs as they forage and move about, opening up forest communities and creating savannas. People also cause disturbances with agriculture, forestry, new roads and cities, and construction projects for dams and pipelines. It is customary in ecology to distinguish between natural disturbances and human-caused (or anthropogenic) disturbances, but a subtle point of clarification is needed. Aboriginal populations have disturbed and continue to disturb communities around the world, setting fire to grasslands and savannas, practicing slash-and-burn agriculture in forests, and so on. Because their populations often are or were relatively small, the disturbances are patchy and limited in scale in forests, or restricted to quickly passing wildfires in grasslands, savannas, or woodland, which are comprised of species already adapted to fire. In other words, the human-caused disturbances by technologically advanced and numerous people are very different from the disturbances caused by small groups of aborigines. In the Kingston Plains, clear-cut logging followed by repeated human-set wildfires from 1880 to 1900 caused a change in basic ecological conditions such that the white pine forest has not regenerated (fig. 3.31). Given the right combination of disturbances by modern people, or by nature, it may take hundreds of years for a community to return to its predisturbance state.

Ecologists generally find that disturbance benefits most species, much as predation does, because it sets back supreme competitors and allows less-competitive species to persist. In northern temperate forests, maples (especially sugar maple) are more prolific seeders and more shade tolerant at different stages of growth than nearly any other tree species. Given decades of succession, maples outcompete other trees for a place in the forest canopy. Most species of oak, hickory, and other light-requiring trees diminish in abundance, as do species of forest herbs. The dense shade of maples basically starves other species for light. When windstorms, tornadoes, wildfires, or ice storms hit a maple forest, trees are toppled, branches broken, and light again reaches the forest floor and stimulates seedlings of oaks and hickories, as well as forest herbs. Breaking the grip of a supercompetitor is the helpful role disturbances often play. The 1988 Yellowstone fires knocked back lodgepole pine (*Pinus contorta*) that had expanded its acreage in the park (fig. 3.32). In a couple of years a greater variety of plant species grew where previously there was mostly pine. Wildlife responded vigorously to the greater diversity.

Some landscapes never reach a stable climax in a traditional sense because they are characterized by periodic disturbance and are made up of **disturbance-adapted species** that survive fires underground, or resist the flames, and then reseed quickly after fires. Grasslands, the chaparral scrubland of California and the Mediterranean region, savannas, and some kinds of coniferous forests are shaped and maintained by periodic fires that have long been a part of their history. In fact, many of the dominant plant species in these communities need fire to suppress competitors, to prepare the ground for seeds to germinate, or to pop open cones or split thick seed coats and release seeds. Without fire, community structure would be quite different.

People taking an organismal view of such communities believe that disturbance is harmful. In the early 1900s this view merged with the desire to protect timber supplies from ubiquitous wildfires, and to store water behind dams while also controlling floods. Fire suppression and flood control became the central policies in American natural resource management (along with predator control) for most of the twentieth century. Recently, new concepts about natural disturbances are entering land management discussions and bringing change to land management policies. Grasslands and some forests are now considered "fire-adapted" and fires are allowed to burn in them if weather conditions are appropriate. Floods also are seen as crucial for

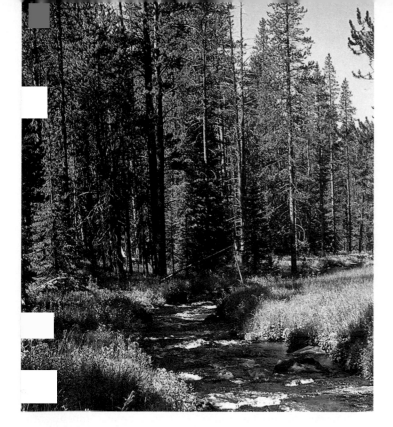

FIGURE 3.32 In 1988 (the driest year on record) lightning-ignited wildfires burned 1.4 million acres in the Greater Yellowstone area. Before the fires, 80 percent of Yellowstone National Park was dominated by lodgepole pine forest like this one. After the fires, some forests became savannas, while others grew into dense thickets. Overall, fires increased variety in the region and promoted species that need more light than is available in forests.
Source: National Park Service http://www.nps.gov/yell/publications/pdfs/fire/htmls/cover.htm.

maintaining floodplain and river health. Policymakers and managers increasingly consider ecological information when deciding on new dams and levee construction projects. You can read more about this in chapter 6 and in the related stories "Prescribed Fires in New Mexico" and "Fires in the Boundary Waters Canoe Area Wilderness" at www.mhhe.com/cunningham4e.

From another view, disturbance resets the successional clock that always operates in every community. Even though all seems chaotic after a disturbance, it may be that preserving species diversity by allowing in natural disturbances (or judiciously applied human disturbances) actually ensures stability over the long run, just as diverse prairies managed with fire recover after drought.

Knowing Where You Live

You think you live at 4275 Locust Lane, Anytown, USA. True, but you also live in an ecological setting defined by a certain climate, underlaid by specific types of rocks and soils, and filled with hundreds of species in many communities and habitats. You interact with these species and change their environment (even if you don't know it) just as those species interact and change their environment. Since you have an excellent memory, you are in a better position than the other species to know the landscape, ecosystems, communities, and species where you live. As someone studying environmental science, you might want to learn more about how your local environment is put together, what it needs to survive, and the effects of human activity on it. If you do, try learning about some of the items below.

- What ecosystems and biological communities existed in your area before European settlement?
- What impact, if any, did aboriginal people have on the flora and fauna of your area?
- What are the dominant species (besides humans) in your neighborhood? Where did they originate?
- How much rain falls in your region each year? Is there a dry or wet season? Is water a limiting factor for your region's biological communities?
- What are the high and low temperatures where you live? When is it coldest or hottest? How do native plants and animals adapt to temperature?
- Is there a keystone species or group of species especially important in your local ecosystems? Are keystone species missing? If yes, do you want them back?
- Do you use a nearby park or wildlife refuge for recreation, fishing, or hunting? Is it home to rare, threatened, or endangered species? Why are they rare, threatened, or endangered?
- Do you want to do something to improve the environment of the place you live? What do you want to do? Are you going to do it?

In time, community structure and productivity get back to normal, species diversity is preserved, and nature seems to reach its dynamic balance.

REVIEW QUESTIONS

1. Look at figure 3.10. The Mount Graham red squirrel (*Tamiasciurus hudsonicus grahamensis*) is an endangered subspecies related to red squirrels that live in northern forests, but it lives on an isolated mountaintop in Arizona. When a cooler climate gave way to a warmer climate 10,000–15,000 years ago, they are believed to have become isolated from the main population of northern red squirrels. Describe the process of adaptation and natural selection that may have occurred, leading to speciation.

2. Explain how tolerance limits to environmental factors determine the distribution of a highly specialized species, such as the desert pupfish. Compare this with the distribution of a generalist species, such as common carp (*Cyprinus carpio*) or European starling (*Sturna vulgaris*).

3. What dangers to community diversity, structure, and function are posed by invasive species. Browse the Web to learn about how introduced arctic foxes (*Alopex lagopus*)

affected the Aleutian islands. Even though arctic foxes are present on the Alaskan mainland, would you consider it invasive in the Aleutians? What ecological impacts, or characteristics of the fox itself, lead you to consider it invasive or not invasive?

4. To the best of your ability, describe the ecological niche occupied by a bird of prey, such as a hawk or an owl. Explain how many hawks and owls can use resources in the same community and yet not eliminate each other from the habitat?

5. Competition for limited resources occurs among species in all communities and ecosystems. Explain several ways that competition might be reduced within a single species population, or between several species using a similar resource.

6. Predators can be carnivorous or herbivorous. How might a carnivore affect its prey in terms of adaptation and natural selection? Since a carnivore's prey usually dies, but an herbivore's prey typically bounces back, would the adaptation and natural selection be different for an herbivore's prey? Describe the differences, if any.

7. Conduct Web research to find a symbiotic relationship not explained in this book. Is the relationship mutualistic or commensalistic?

8. Define *keystone species* and explain its importance in shaping a community's structure and functions.

9. Would a generalist species likely be *r*-selected or *K*-selected? What about a specialist species? Describe in terms of numbers how the feedback term in the logistic growth equation limits the exponential growth of a population.

10. What are some factors that prevent exponential growth in a population? For each one, say whether they are density-dependent or density-independent. Explain how carrying capacity could affect a generalist species. Is the reproductive strategy of a generalist species always a good strategy to avoid paying the consequences of exceeding carrying capacity?

11. Describe the differences between core habitat and edge habitat, and list some species that might occur in each. Would core habitat most likely have greater species diversity than edge habitat? Explain why or why not.

12. Explain the relationship between species diversity and community complexity, and between resilience and productivity. Explain the exceptions to the "diversity equals stability" concept.

13. All communities on the planet exhibit diversity, structure, complexity, resilience, and productivity. Describe how these characteristics apply to the ecosystem in which you live.

14. Why is the organismal concept of communities not as useful as the individualistic concept when thinking about disturbance? Give specific examples of communities when explaining your answer.

15. Describe secondary succession on disturbed land for your locality. About how many years does it take for a pioneer community in your neighborhood to become a climax community?

16. Why do mature communities often exhibit a higher level of stability than is found in earlier successional stages?

17. Would you call a community of disturbance-adapted species a climax community? Is there a better name?

CRITICAL THINKING AND DISCUSSION

Apply the principles you have learned in this chapter to discuss these questions with other students.

1. What is the difference between saying that a duck has webbed feet because it needs them to swim and saying that a duck is able to swim because it has webbed feet?

2. The concepts of natural selection and evolution are central to how most biologists understand and interpret the world, yet the theory of evolution appears to question the beliefs of many religious groups. Why do you think this theory is so widely accepted in science? Does it necessarily question religious beliefs?

3. Given the information in this chapter, do you suspect scientists of having an ulterior motive in coming up with the idea of speciation through natural selection and adaptation? How would you communicate these scientific ideas to a nonscientist?

4. The concept of keystone species is controversial among ecologists because most organisms are highly interdependent. If each trophic level in a community depends on all the others, how can we say that one is most important? Choose a familiar ecosystem and decide whether it has a keystone species or keystone set.

5. Some scientists look at the boundary between two biological communities and see a sharp dividing line. Others looking at the same boundary see a gradual transition with much intermixing of species and many interactions between communities. Is there a difference in education or personality that might cause one scientist to see clarity, and another complexity in the same landscape?

6. Ecologists debate whether biological communities have self-sustaining, self-regulating characteristics or are highly variable, accidental assemblages of individually acting species. What outlook or worldview might lead scientists to favor one or the other of these theories?

SUGGESTED READINGS

Botkin, Daniel B. 1989. *Discordant Harmonies: A New Ecology for the Twenty-First Century.* Oxford University Press.

Darwin, Charles. 1837. *The Voyage of the Beagle.* (http://www.literature.org/authors/darwin-charles/the-voyage-of-the-beagle).

Ehrlich, Paul R., and Peter H. Raven. 1967. Butterflies and plants: A study in coevolution. *Evolution* 18:586–608.

Eldredge, Nils, and Stephen J. Gould. 1972. Punctuated equilibria: An alternative to phyletic gradualism. In: Schopf, T. J. M. (ed.) *Models in Paleobiology.* Freeman, Cooper and Co.

Elton, Charles. 1958. *The Ecology of Invasions by Animals and Plants.* Methuen.

Gleason, Henry A. 1926. The individualistic concept of the plant association. *Bulletin of the Torrey Botanical Club* 53:7–26.

MacArthur, Robert H. 1955. Fluctuations of animal populations and a measure of community stability. *Ecology* 35:533–36.

Mayr, Ernst. 1942. *Systematics and the Origin of Species.* Columbia University Press.

Paracer, S., and V. Ahmadjian. 2000. *Symbiosis: An Introduction to Biological Associations.* Oxford University Press.

Simberloff, Daniel. 1997. Flagships, umbrellas, and keystones: Is single-species management passé in the landscape era? *Biological Conservation* 83:247–57.

Tilman, David, et al. 1997. The influence of functional diversity and composition on ecosystem processes. *Science* 277(5330):1300–02.

DATA analysis   Comparing Exponential to Logistic Population Growth

Exponential growth occurs in a series of time steps—days, months, years. Imagine cockroaches in a room multiplying (or some other species, if you must). Picture a population of ten cockroaches that together produce enough young to increase at a rate of 150 percent per month. What is r for this population?

first month), you begin with ten roaches, and end (N_e) with a larger number that depends on r, the intrinsic rate of growth. The beginning of the second time step (1) starts with that the number at the end of step 0. Round N to the nearest whole number. When you are done, graph the results. At the end of 7 months, how large did this population become? What is the shape of the growth curve?

Time Step (t)	Begin Step (N_b)	Intrinsic Growth Rate (r)	End Step (N_e)
0	10		15
1			
2			
3			
4			
5			
6			
7			

To find out what how this population grows, fill out the table shown. (*Hint: r* remains constant.) Remember, for time step 0 (the

For Additional Help in Studying This Chapter, please visit our website at www.mhhe.com/cunningham4e. You will find practice quizzes, a chapter summary, key terms, answers to review questions, additional case studies, regional examples, and an extensive reading list, all of which will help you understand the material in this chapter.

More than 1.1 billion people live in India.

C H A P T E R

Human Populations

4

*For every complex problem there is an answer that is
clear, simple, and wrong.*

–H. L. Mencken–

OBJECTIVES

Studying this chapter will help you answer the following questions:

- How fast is the human population growing?
- Will the world's population double again as it did between 1965 and 2000?
- What is the relationship between population growth and impact on the environment?

- Why has the human population grown so rapidly since 1800?
- How is human population growth changing in different parts of the world?
- How does population growth change as a society develops?
- What factors slow down or speed up human population growth?

Case Study Cultural Choices and the Rate of Population Growth

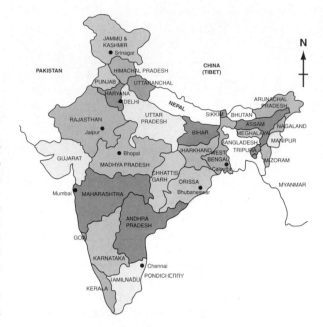

Its annual growth rate of 1.7 percent suggests that India's population could double in 41 years, becoming the world's most populous country by 2050. Indian states have taken different approaches to slow population growth. Despite its poverty, Kerala has a total fertility rate lower than the United States.

The good news is that the human population on earth may never double again. The bad news is that it may take a century to get there. In between, the choices we make individually, and collectively through our political, economic, and cultural actions, will determine whether it will be a pleasant, or not so pleasant, hundred years. The choices are already being made in India, where two different states are taking two very different approaches to population growth.

In 1999, having added more than 180 million people in just a decade, India reached a population of 1 billion humans. If current growth rates persist, India will have 1.63 billion residents in 2050 and will surpass China as the world's most populous country. How will this country, where more than a quarter of its inhabitants live in abject poverty, feed, house, educate, and employ all those being added each year? A fierce debate is taking place about how to control India's population, with ramifications for the rest of the world as well.

On one side of this issue are those who believe that the best way to reduce the number of children born is poverty eradication and progress for women. Drawing on social justice principles established at the 1994 UN Conference on Population and Development in Cairo, some argue that responsible economic development, a broad-based social welfare system, education and empowerment of women, and high-quality health care—including family planning services—are essential components of population control. Without progress in these areas, they believe, efforts to provide contraceptives or encourage sterilization are futile.

On the other side of this debate are those who contend that, while social progress is an admirable goal, India doesn't have the time or resources to wait for an indirect approach to population control. The government must push aggressively, they argue, to reduce births now or the population will be so huge and its use of resources so great that only mass starvation, class war, crime, and disease will be able to bring it down to a manageable size.

Unable to reach a consensus on population policy, the Indian government decided in 2000 to let each state approach the problem in its own way. Some states have chosen to focus on social justice, while others have adopted more direct, interventionist policies.

The model for the social justice approach is the southern state of Kerala, which achieved population stabilization in the mid-1980s, the first Indian state to do so. Although still one of the poorest places in the world economically, Kerala's fertility rate is comparable to that of many industrialized nations, including the United States. Both women and men have a nearly 100 percent literacy rate and share affordable and accessible health care, family planning, and educational opportunities; therefore, women have only the number of children they want, usually two. The Kerala experience suggests that increased wealth isn't a prerequisite for zero population growth.

Taking a far different path to birth reduction is the neighboring state of Andra Pradesh, which reached a stable growth rate in 2001. Boasting the most dramatic fertility decline of any large Indian state, Andra Pradesh has focused on targeted, strongly enforced sterilization programs. The poor are encouraged—some would say coerced—to be sterilized after having only one or two children. The incentives include cash payments. You might receive 500 rupees—equivalent to $11 (U.S.) or a month's wages for an illiterate farm worker—if you agree to have "the operation." In addition, participants are eligible for better housing, land, wells, and subsidized loans.

The pressure to be sterilized is overwhelmingly directed at women, for whom the procedure is major abdominal surgery. Sterilizations often are done by animal husbandry staff and carried out in government sterilization camps. This practice raises troubling memories of the 1970s for many people, when then-Prime Minister Indira Gandhi suspended democracy and instituted a program of forced sterilization of poor people. There were reports at the time of people being rounded up like livestock and castrated or neutered against their will.

While many feminists and academics regard Andra Pradesh's policies as appallingly intrusive and coercive to women and the poor, the state has successfully reduced population growth. By contrast, the hugely populous northern states of Uttar Pradesh and Bihar have increased their growth rates slightly over the past two decades to a current rate above 2.5 percent per year. How will they slow this exponential growth, and what might be the social and environmental costs of not doing so?

India's population problems introduce several of the important themes of this chapter. What are the trends in human populations around the world, what do those trends mean for our resources and environment, and what is the best way to approach population planning? Keep in mind, as you read this chapter, that resource limits aren't simply a matter of total number of people; they also depend on consumption levels and the types of technology used to produce the things we use and consume.

PAST AND CURRENT POPULATION GROWTH ARE VERY DIFFERENT

Every second, on average, four or five children are born, somewhere on the earth. In that same second, two other people die. This difference between births and deaths means a net gain of roughly 2.3 more humans per second in the world's population. The U.S. Census Bureau estimates that the total world population is over 6.5 billion people and growing at 1.14 percent per year. This means we are adding 74.6 million more people per year. Humans are now one of the most numerous vertebrate species on the earth. We also are more widely distributed and manifestly have a greater global environmental impact than any other species. For the families to whom these children are born, this may well be a joyous and long-awaited event (fig. 4.1). But is a continuing increase in humans good for the planet in the long run?

Many people worry that overpopulation will cause—or perhaps already is causing—resource depletion and environmental degradation that threaten the ecological life-support systems on which we all depend. These fears often lead to demands for immediate, worldwide birth-control programs to reduce fertility rates and to eventually stabilize or even shrink the total number of humans.

Others believe that human ingenuity, technology, and enterprise can expand the world's carrying capacity and allow us to overcome any problems we encounter. From this perspective, more people may be beneficial, rather than disastrous. A larger population means a larger workforce, more geniuses, more ideas about what to do. Along with every new mouth comes a pair of hands. Proponents of this worldview argue that continued economic and technological growth can both feed the world's billions and enrich everyone enough to end the population explosion voluntarily.

Still another perspective on this subject derives from social justice concerns. According to this worldview, resources are sufficient for everyone. Current shortages are only signs of greed, waste, and oppression. The root cause of environmental degradation, in this view, is inequitable distribution of wealth and power rather than merely population size. Fostering democracy, empowering women and minorities, and improving the standard of living of the world's poorest people are what is really needed. A narrow focus on population growth only fosters racism and an attitude that blames the poor for their problems, while ignoring the deeper social and economic forces at work.

Whether human populations will continue to grow at present rates and what that growth would imply for environmental quality and human life are among the most central and pressing questions in environmental science. In this chapter, we will look at some causes of population growth, as well as at how populations are measured and described. Family planning and birth control are essential for stabilizing populations. The number of children a couple decides to have and the methods they use to regulate fertility, however, are strongly influenced by culture, religion, politics, and economics, as well as basic biological and medical considerations. We will examine how some of these factors influence human demographics.

FIGURE 4.1 A Mayan family in Guatemala with four of their six living children. Decisions on how many children to have are influenced by many factors, including culture, religion, a need by parents for security in old age, family finances, household labor needs, child survival rates, and power relationships in the family.

Human populations grew slowly until recently

For most of our history, humans were not very numerous, compared with other species. Studies of hunting and gathering societies suggest that the total world population was probably only a few million people before the invention of agriculture and the domestication of animals around 10,000 years ago. The agricultural revolution produced a larger and more secure food supply and allowed the human population to grow, reaching perhaps 50 million people by 5000 B.C. For thousands of years, the number of humans increased very slowly. Archaeological evidence and historical descriptions suggest that only about 300 million people were living at the time of Christ (table 4.1).

TABLE 4.1

World Population Growth and Doubling Times

Date	Population	Doubling Time
5000 B.C.	50 million	?
800 B.C.	100 million	4,200 years
200 B.C.	200 million	600 years
A.D. 1200	400 million	1,400 years
A.D. 1700	800 million	500 years
A.D. 1900	1,600 million	200 years
A.D. 1965	3,200 million	65 years
A.D. 2000	6,100 million	51 years
A.D. 2050 (estimate)	8,920 million	215 years

Source: United Nations Population Division.

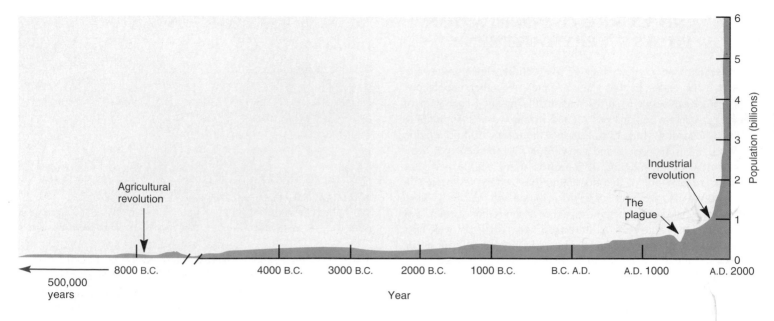

FIGURE 4.2 Human population levels throughout history. It is clear from the J-shaped growth curve that human population is growing exponentially. When will the growth curve assume an S shape and population growth level off? Choices made in the next 50 years will decide.

As you can see in figure 4.2, human populations began to increase rapidly after about A.D. 1600. Many factors contributed to this rapid growth. Increased sailing and navigating skills stimulated commerce and communication among nations. Agricultural developments, better sources of power, and better health care and hygiene also played a role. We are now in an exponential, or J curve, pattern of growth, described in chapter 3.

It took all of human history to reach 1 billion people in 1804 but little more than 150 years to reach 3 billion in 1960. It took us only 12 years—from 1987 to 1999—to add the sixth billion. Another way to look at population growth is that the number of humans tripled during the twentieth century. Will it do so again in the twenty-first century? If it does, will we overshoot our environment's carrying capacity and experience a catastrophic dieback similar to those described in chapter 3? As you will see later in this chapter, there is some evidence that population growth already is slowing, but whether we will reach equilibrium soon enough and at a size that can be sustained over the long term remains a difficult but important question.

PERSPECTIVES ON POPULATION GROWTH

As with many topics in environmental science, people have widely differing opinions about population and resources. Some believe that population growth is the ultimate cause of poverty and environmental degradation. Others argue that poverty, environmental degradation, and overpopulation are all merely symptoms of deeper social and political factors. The worldview we choose to believe will profoundly affect our approach to population issues. In this section, we will examine some of the major figures and their arguments in this debate.

Does environment or culture control human population growth?

Since the time of the Industrial Revolution, when the world population began growing rapidly, individuals have argued about the causes and consequences of population growth. In 1798 Thomas Malthus (1766–1834) wrote *An Essay on the Principle of Population,* changing the way European leaders thought about population growth. Malthus marshaled evidence to show that populations tended to increase at an exponential, or compound, rate while food production either remained stable or increased only slowly. Eventually human populations would outstrip their food supply and collapse into starvation, crime, and misery. He converted most economists of the day from believing that high fertility increased gross domestic output to believing that per capita output actually fell with rapidly rising population.

In Malthusian terms, growing human populations stop growing when disease or famine kills many, or when constraining social conditions compel others to reduce their birth rates—late marriage, insufficient resources, celibacy, and "moral restraint." Several decades later, the economist Karl Marx (1818–1883) presented an opposing view, that population growth resulted from poverty, resource depletion, pollution, and other social ills. Slowing population growth, said Marx, required that people be treated justly, and that exploitation and oppression be eliminated from social arrangements.

Both Marx and Malthus developed their theories about human population growth when understanding of the world, technology, and society were much different than they are today. Still, as the difference between Kerala and Andra Pradesh shows, the two perspectives still play out. On the one hand, some believe

Principles of Environmental Science http://www.mhhe.com/cunningham4e

that we are approaching, or may have surpassed, the earth's carrying capacity. Joel Cohen, a mathematical biologist at Rockefeller University, reviewed published estimates of the maximum human population size the planet can sustain. The estimates, spanning 300 years of thinking, converged on a median value of 10–12 billion (see What Do You Think? p. 78). We are more than 6.5 billion strong today, and growing, an alarming prospect for some. Cornell University entomologist David Pimental, for example, has said: "By 2100, if current trends continue, twelve billion miserable humans will suffer a difficult life on Earth." In this view, birth control should be our top priority. The political leaders of Andra Pradesh seemed to echo this sentiment in the programs they established to reduce birth rates. On the other hand, Kerala adopted policies that improved social conditions and educational levels, achieving the same level of birth reduction. In this other perspective, the earth is bountiful in its resource base, but poverty and high birth rates result from oppressive social relationships that unevenly distribute wealth and resources. Consequently, this position believes, technological development, education, and just social conditions are the means of achieving population control. Mohandas Gandhi stated it succinctly: "There is enough for everyone's need, but not enough for anyone's greed."

A more complex perspective is that of Jared Diamond in his 2005 book *Collapse: How Societies Choose to Fail or Succeed.* Starting with the case of societal collapse on Easter Island in the remote South Pacific, Diamond shows that choices that leaders and individuals in a society make will determine their future, as much as environmental damage, climate change, the loss of friendly trading partners, and incursions by enemies. No factor exclusively causes any other, but each influences and, in turn, is influenced by the others.

APPLICATIONS

Population Doubling Time

If the world population is growing at 1.14 percent per year and continues at that rate, how long before it doubles? The "rule of 70" is a useful way to calculate the approximate doubling time in years for anything growing exponentially. For example, a savings account (or biological population) growing at a compound interest rate of 1 percent per year will double in about 70 years. Using the formula below, calculate doubling time for the world (growth rate = 1.14 percent/year), Uganda (3.2 percent), Nicaragua (2.7 percent), India (1.7 percent), United States (0.6 percent), Japan (0.1 percent), and Russia (−0.6 percent).

Example: $\dfrac{70 \text{ years}}{(\text{growth percent})}$ = doubling time in years

Answers: World = 61 years; Uganda = 22 years; Nicaragua = 26 years; India = 41 years; United States = 117 years; Japan = 700 years; Russia = never

Technology increases carrying capacity for humans

Optimists argue that Malthus was wrong in his predictions of famine and disaster 200 years ago because he failed to account for scientific and technical progress. In fact, food supplies have increased faster than population growth since Malthus' time. For example, according to the UN FAO Statistics Division, each person on the planet averaged 2,435 calories of food per day in 1970, while in 2000 the caloric intake reached 2,807 calories. Even poorer, developing countries saw a rise, from an average of 2,135 calories per day in 1970 to 2,679 in 2000. In that same period the world population went from 3.7 to more than 6 billion people. Certainly terrible famines have stricken different locations in the past 200 years, but they were caused more by politics and economics than by lack of resources or population size. Whether the world can continue to feed its growing population remains to be seen, but technological advances have vastly increased human carrying capacity so far (chapter 7).

The burst of world population growth that began 200 years ago was stimulated by scientific and industrial revolutions. Progress in agricultural productivity, engineering, information technology, commerce, medicine, sanitation, and other achievements of modern life have made it possible to support approximately 1,000 times as many people per unit area as was possible 10,000 years ago. Economist Stephen Moore of the Cato Institute in Washington, D.C., regards this achievement as "a real tribute to human ingenuity and our ability to innovate." There is no reason, he argues, to think that our ability to find technological solutions to our problems will diminish in the future.

Much of our agricultural growth and rising standard of living in the past 200 years, however, has been based on easily acquired natural resources, especially cheap, abundant fossil fuels. Whether rising prices of fossil fuels will constrain that production and result in a crisis in food production and distribution, or in some other critical factor in human society, concerns many people.

Population growth could bring benefits

Think of the gigantic economic engine that China is becoming as it continues to industrialize and its population becomes more affluent. More people mean larger markets, more workers, and efficiencies of scale in mass production of goods. Moreover, adding people boosts human ingenuity and intelligence that will create new resources by finding new materials and discovering new ways of doing things. Economist Julian Simon (1932–1998), a champion of this rosy view of human history, believed that people are the "ultimate resource" and that no evidence suggests that pollution, crime, unemployment, crowding, the loss of species, or any other resource limitations will worsen with population growth. In a famous bet in 1980, Simon challenged Paul Ehrlich, author of *The Population Bomb,* to pick five commodities that

What Do You Think?

Humanity's Ecological Footprint

To you the world may seem infinitely large, with plenty of room for people. At a distance, seen from space, the picture is different. Can the earth sustain our current lifestyles? Will there be adequate natural resources for future generations? It appears that the world's population will stabilize around 40 percent above current levels some 100 years from now. But what about rising affluence, not just of that additional 40 percent, but of the majority of people on earth who use a fraction of the resources that people in many westernized countries consume? With 9.8 percent economic growth in 2005, the 1.3 billion Chinese now alive are sending a strong signal that they are striving for greater material wealth. Can the planet provide nearly 10 billion people with a standard of living equivalent to that of the average Englishman?

These questions are among the most important in environmental science today. We depend on nature for food, water, energy, oxygen, waste disposal, and other life-support systems. Sustainability implies that we cannot turn our resources into waste faster than nature can recycle that waste and replenish the resources on which we depend. Although we may be able to overspend nature's budget temporarily, future generations will have to pay the debts we leave them. In the words of the Brundtland Commission, living sustainably means "meeting our own vital needs without compromising the ability of future generations to meet their own needs."

On the island of Java, Indonesia, 160 million people, most of them farm workers, live and work in an area the size of Iowa, and half the land is occupied by volcanoes. Three crops of rice per year are harvested from

(a) 8,000 B.C.

(b) A.D. 5

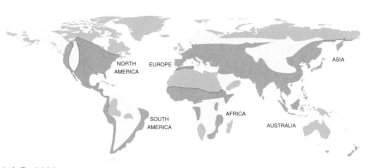

(c) A.D. 2005

The ecological footprint of humanity has expanded over several thousand years. Pink areas have high population density and intense resource use, yellow areas are a mixture of development levels, and green areas experience little human influence. (a) In 8,000 B.C. most people lived in small groups of hunter-gatherers, while a few were growing the first crops in small settlements. (b) In 5 A.D. the Roman Empire held sway over much of Europe and the Near East, while other civilizations with large urban centers existed in Central and North America, India, China, and Southeast Asia. (c) Today the combined effect of population size and resource use has led to global human dominance.

Source: Detailed data for 2005 is at the Human Footprint Project, Center for International Earth Science Information Network, Columbia University (http://www.ciesin.org).

fields that average less than one-tenth of a hectare (ha), the size of a city lot. Many factors reduce the average Javanese's global environmental impacts to a fraction of a typical Canadian's. They don't heat their homes, few drive cars or fly in airplanes, and their food is locally grown. If they adopted a European lifestyle, would Java's inhabitants need another island or two?

The impacts of resource use and population size on planetary resources were central questions in Mathis Wackernagel's doctoral research with his advisor, William Rees. They hit on the idea of the **human ecological footprint,** or the amount of productive land and water surface required to support all the needs of a person. You, personally, have an ecological footprint. Go to http://www.myfootprint.org, a site run by the nonprofit Redefining Progress, and take the footprint quiz. It computes the biological demands you place on nature by your level of affluence and the technological sophistication of your society. Each country has an ecological footprint. For example, the ecological footprint of Canada is 8.8 ha/person (average amount of land needed to feed, clothe, house, dispose of wastes, etc.) $\times$ 32.2 million people (2005 population) = 283.4 million ha of productive land. By contrast, Indonesia has a footprint of 1.1 ha/person $\times$ 221.9 million people = 244.1 million ha. Thus Indonesia, with a population nearly seven times larger than Canada, has a slightly smaller ecological footprint. This concept, the **I=PAT formula,** incorporates the idea that a country's impact (I) is not just from its population (P), but also its affluence (A) and the technology (T) it uses to provide the goods and services it consumes. The larger and more affluent a population is, as well as how inefficient or polluting its technology is, the greater its impact on global resources.

Like any model, an ecological footprint is useful, but has limits. It assumes that (1) resource consumption and waste flows can be measured and translated into a required "biologically productive area"; (2) all land and resource uses can be standardized into equivalent units of area; and (3) all the world's

Principles of Environmental Science http://www.mhhe.com/cunningham4e

biologically productive land is divisible into nonoverlapping uses. The ecological footprint also ignores technological improvements, such as agricultural practices that increase yields, or renewable energy sources that eliminate fossil fuels. These improvements increase the world's carrying capacity for people. Energy consumption by itself comprises half the calculated footprint, and reducing this single factor makes a huge difference in the footprint size.

What do you think? Do people in wealthier nations have a responsibility to reduce their ecological footprint, or should people in China and India restrain themselves from using more

Intensive cultivation, rich volcanic soil, a tropical climate, and plenty of moisture help support a high population density on verdant Mt. Lawu, Central Java. Despite a higher population density than many industrialized countries, Indonesia has a smaller ecological footprint.

resources? Is it your responsibility, or your grandchildren's, to solve the potential problems that come with rising affluence and population size? Or do you think it is far-fetched that we could outstrip the earth's capacity to absorb our waste materials, and to provide us with food, clothing, shelter, recreation, and other necessities of life?

would become more expensive by the end of the decade. Ehrlich chose metals that actually became cheaper, and he lost the bet. Leaders of many developing countries share this outlook and insist that, instead of being obsessed with population growth, we should focus on the inordinate consumption of the world's resources by people in richer countries (see What Do You Think? p. 78). For his part, Ehrlich did not really want to bet about metals, but about renewable resources or certain critical measures of environmental health. He and Simon were negotiating a second bet just before Simon's death.

MANY FACTORS DETERMINE POPULATION GROWTH

Demography (derived from the Greek words *demos* [people] and *graphein* [to write or to measure]) encompasses vital statistics about people, such as births, deaths, and where they live, as well as total population size. In this section, we will investigate ways to measure and describe human populations and discuss demographic factors that contribute to population growth.

How many of us are there?

The U.S. Census Bureau estimate of over 6.5 billion people in the world in mid-2006 is only an educated guess. Even in this age of information technology and communication, counting the number of people in the world is like shooting at a moving target. People continue to be born and die. Furthermore, some countries have never even taken a census, and those that have been done may not be accurate. Governments may overstate or understate their populations to make their countries appear larger and more important or smaller and more stable than they really are. Individuals, especially if they are homeless, refugees, or illegal aliens, may not want to be counted or identified.

We really live in two very different demographic worlds. One of these worlds is poor, young, and growing rapidly, while the other is rich, old, and shrinking in population size (fig. 4.3). The poorer world is occupied by the vast majority of people who live in the less-developed countries of Africa, Asia, and Latin America. These countries represent 80 percent of the world population but will contribute more than 90 percent of all projected future growth (fig. 4.4). The richer world is made up of North America, Western Europe, Japan, Australia, and New Zealand. The average age in richer countries is 40, and life expectancy of their residents will exceed 90 by 2050. With many couples choosing to have either one or no children, the populations of these countries are expected to decline significantly over the next century (see Data Analysis, p. 93).

The highest population growth rates occur in a few "hot spots" in the developing world, such as sub-Saharan Africa and the Middle East, where economics, politics, religion, and civil unrest keep birth rates high and contraceptive use low. In Chad and the Democratic Republic of Congo, for example, annual population growth is above 3.2 percent. Less than 10 percent of all couples use any form of birth control, women average more than seven children each, and nearly half the population is less than 15 years old. Even faster growth rates occur in Oman and Palestine, where the population doubling time is only 18 years.

Some countries in the developing world are growing so fast that they will reach immense population sizes by the middle of the twenty-first century (table 4.2). While China was the most populous country throughout the twentieth century, India is expected to pass China in the twenty-first century. Nigeria, which had only 33 million residents in 1950, is forecast to have 258 million in 2050. Ethiopia, with about 18 million people 50 years ago, is likely to grow almost tenfold over a century. In many of these countries, rapid population growth poses huge challenges to food supplies and stability. Bangladesh, about the size of Iowa, is already overcrowded at 144 million people. Another 87 million people by 2050 will only add to current problems.

(a)

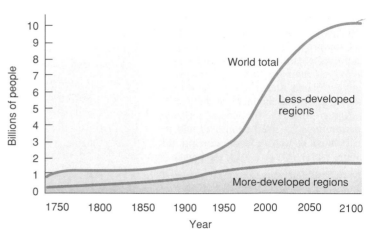

FIGURE 4.4 Since about 1950, less-developed regions of the planet have experienced exponential growth while growth in more-developed regions has slowed. More than 90 percent of future population growth will occur in less-developed countries.

(b)

FIGURE 4.3 Many people, two different worlds. (a) A typical Indian man living in Varanasi on the Ganges River earns less than $2 a day, has a life expectancy of 61 years, and lives in a country whose population will be 48 percent larger in 2050. Thirty-six percent of his countrymen are 15 years or younger, and women on average have 3 children each. (b) A typical male of Tokyo, Japan, lives where the gross national income per capita is $30,040, has a life expectancy of 78 years, and will see his country's population fall 21 percent by 2050. Only 14 percent of the people around him are younger than 15, and women on average bear 1.3 children each.
Source: 2005 World Population Data Sheet, Population Reference Bureau (http://www.prb.org).

It isn't only wealthy countries that have declining populations. Russia, for instance, is now declining by nearly 1 million people per year as death rates have soared and birth rates have plummeted. A collapsing economy, hyperinflation, crime, corruption, and despair have demoralized the population. Horrific pollution levels left from the Soviet era, coupled with poor nutrition and health care, have resulted in high levels of genetic abnormalities, infertility, and infant mortality. Abortions are twice as common as live births. Death rates, especially among adult men, have risen dramatically. According to some demographers, the life expectancy of Russian males fell by 10 years after 1990; in 2005 it began creeping back up, to 59 years. By 2050 Russia is

On the other hand, shrinking populations characterize some richer countries. Japan, which has 128 million residents now, is expected to shrink to about 100 million by 2050. Europe, which now makes up about 12 percent of the world population, will constitute less than 7 percent in 50 years, if current trends continue. Even the United States and Canada would have nearly stable populations if immigration were stopped. For the moment, the U.S. population continues to grow. In October 2006, the U.S. population reached 300 million.

TABLE 4.2

The World's Largest Countries

2005		2050	
Country	Population (millions)	Country	Population (millions)
China	1,304	India	1,628
India	1,104	China	1,437
United States	297	United States	420
Indonesia	222	Indonesia	308
Brazil	184	Pakistan	295
Pakistan	162	Brazil	260
Bangladesh	144	Nigeria	258
Russia	143	Bangladesh	231
Nigeria	132	Congo, Dem. Rep. of	183
Japan	128	Ethiopia	170

Source: Data from Population Reference Bureau, 2005.

Principles of Environmental Science

expected to have a smaller population than Vietnam, the Philippines, and the Democratic Republic of Congo. Bulgaria will lose a third of its population, simply due to declining birth rates.

The situation is complicated in many African countries, where AIDS and other communicable diseases are killing people at a terrible rate. In Zimbabwe, Botswana, Swaziland, and South Africa, for example, up to 30 percent or more of the population age 15–49 have AIDS or are HIV-positive. Health officials predict that more than two-thirds of the 15-year-olds now living in Botswana will die of AIDS before age 50. Some of these countries may experience declining populations. Overall, however, Africa is expected to grow by 1.5 billion over the next century.

The world population density map in appendix 2 on p. 381 shows human population distribution around the world. Notice the high densities supported by fertile river valleys of the Nile, Ganges, Yellow, Yangtze, and Rhine Rivers and the well-watered coastal plains of India, China, and Europe. Historic factors, such as technology diffusion and geopolitical power, also play a role in geographic distribution.

Fertility varies among cultures and at different times

Fecundity is the physical ability to reproduce, while fertility is the actual production of offspring. Those without children may be fecund but not fertile. The most accessible demographic statistic of fertility is usually the **crude birth rate,** the number of births in a year per thousand persons. It is statistically "crude" in the sense that it is not adjusted for population characteristics, such as the number of women of reproductive age.

The **total fertility rate** is the number of children born to an average woman in a population during her entire reproductive life. Upper-class women in seventeenth- and eighteenth-century Europe, whose babies were given to wet nurses immediately after birth, often had 25 or 30 pregnancies. The highest recorded total fertility rates for working-class people are among some Anabaptist agricultural groups in North America, who have averaged up to 12 children per woman. In most tribal or traditional societies, food shortages, health problems, and cultural practices limit total fertility to about six or seven children per woman, even without modern methods of birth control.

Fertility is usually calculated as births per woman because, in many cases, establishing paternity is difficult. Nevertheless, a few demographers argue that we should pay more attention to birth rates per male because in some cultures men have far more children, on average, than do women. In Cameroon, for instance, due to multiple marriages, extramarital affairs, and a high rate of female mortality, men are estimated to have 8.1 children in their lifetime, while women average only 4.8.

Zero population growth (ZPG) occurs when births plus immigration in a population just equal deaths plus emigration. It takes several generations of replacement-level fertility (in which people just replace themselves) to reach ZPG. Where infant

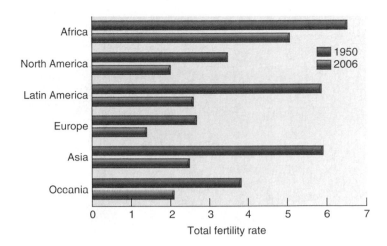

FIGURE 4.5 In all regions of the world, fertility rates declined from 1950 to 2006.
Source: Data from Population Reference Bureau, 2006.

mortality rates are high, the replacement level may be 5 or more children per couple. In the more highly developed countries, however, this rate is usually about 2.1 children per couple because some people are infertile, have children who do not survive, or choose not to have children.

Fertility rates have declined dramatically in every region of the world except Africa over the past 50 years (fig. 4.5). In the 1960s, total fertility rates above 6 were common in many countries. The average family in Mexico in 1975, for instance, had 7 children. By 2005, however, the average Mexican woman had only 2.6 children. According to the World Health Organization, 61 of the world's 190 countries are now at or below a **replacement rate** of 2.1 children per couple. The greatest fertility reduction has been in Southeast Asia, where rates have fallen by more than half, primarily in the past few decades. Contrary to what many demographers expected, some of the poorest countries in the world have been remarkably successful in lowering growth rates. Bangladesh, for instance, reduced its fertility rate from 6.9 in 1980 to only 3.0 children per woman in 2005.

China's one-child-per-family policy decreased the fertility rate from 6 in 1970 to 1.8 in 1990 (fig. 4.6). This policy, however, has sometimes resulted in abortions, forced sterilizations, and even infanticide.

Although the number of boys and girls normally should be fairly balanced in a population, 20 years of female infanticides and sex-based abortions in China have created ratios as high as 140 boys to 100 girls in some regions. This shortage of women has created a flourishing trade in abducted brides. A crackdown in 2002 released 110,000 women who had been kidnapped and sold into marriage, but it's thought that the total number abducted is much higher.

While the world as a whole still has an average fertility rate of 2.7, growth rates are now lower than at any time since World War II. If fertility declines like those in Bangladesh and China were to occur everywhere in the world, our total population could begin to decline by the end of the twenty-first century.

FIGURE 4.6 China's one-child-per-family policy has been remarkably successful in reducing birth rates. It had unintended social consequences, however, including infanticide and bride kidnapping.

Mortality offsets births

A traveler to a foreign country once asked a local resident, "What's the death rate around here?" "Oh, the same as anywhere," was the reply, "about one per person." In demographics, however, **crude death rates** (or crude mortality rates) are expressed in terms of the number of deaths per thousand persons in any given year. Countries in Africa where health care and sanitation are limited may have mortality rates of 20 or more per 1,000 people. Wealthier countries generally have mortality rates around 10 per 1,000. The number of deaths in a population is sensitive to the population's age structure. Rapidly growing, developing countries, such as Belize or Costa Rica, have lower crude death rates (4 per 1,000) than do the more-developed, slowly growing countries, such as Denmark (12 per 1,000). This is because a rapidly growing country has proportionately more youths and fewer elderly than a more slowly growing country. Declining mortality, not rising fertility, was the primary cause of most population growth in the past 300 years. Crude death rates began falling in Western Europe during the late 1700s.

Life expectancy is rising worldwide

Life span is the oldest age to which a species is known to survive. Although there are many claims in ancient literature of kings living for a thousand years or more, the oldest age that can be certified by written records was that of Jeanne Louise Calment of Arles, France, who was 122 years old at her death in 1997. While modern medicine has made it possible for many of us to survive much longer than our ancestors, it doesn't appear that the maximum life span has increased much at all. Apparently, cells in our bodies have a limited ability to repair damage and produce new components. Sooner or later they simply wear out, and we fall victim to disease, degeneration, accidents, or senility.

Life expectancy is the average age that a newborn infant can expect to attain in any given society. It is another way of expressing the average age at death. For most of human history, life expectancy in most societies probably has been 35 to 40 years. This doesn't mean that no one lived past age 40 but, rather, that many people died at earlier ages (mostly early childhood), which balanced out those who managed to live longer.

The twentieth century saw a global transformation in human health unmatched in history. This revolution can be seen in the dramatic increases in life expectancy in most places (table 4.3). Worldwide, the average life expectancy rose from about 40 to 65.5 years over the past century. The greatest progress was in developing countries. For example, in 1900, the average Indian man could expect to live less than 23 years, while the average woman would reach just over 23 years. By 2005, although India had an annual per capita income of less than $440 (U.S.), the average life expectancy for both men and women had nearly tripled and was very close to that of countries with ten times its income level. Longer lives were due primarily to better nutrition, improved sanitation, clean water, and education, rather than to miracle drugs or high-tech medicine. While the gains were not as great for the already industrialized countries, residents of the United States, Sweden, and Japan, for example, now live about half-again as long as they did at the beginning of the twentieth century, and they can expect to enjoy much of that life in relatively good health. The Disability Adjusted Life Years (DALYs, a measure of disease burden that combines premature death with loss of healthy life resulting from illness or disability) that someone living in Japan can expect is now 74.5 years, compared with only 64.5 DALYs two decades ago.

As figure 4.7 shows, annual income and life expectancy are strongly correlated up to about $4,000 (U.S.) per person. Beyond that level—which is generally enough for adequate food, shelter, and sanitation for most people—life expectancies level out at about 75 years for men and 85 for women.

TABLE 4.3

Life Expectancy at Birth for Selected Countries in 1900 and 2005

Country	1900		2005	
	Males	Females	Males	Females
India	23	23	61	63
Russia	31	33	59	72
United States	46	48	75	80
Sweden	57	60	78	83
Japan	42	44	78	85

Source: Data from Population Reference Bureau, 2005.

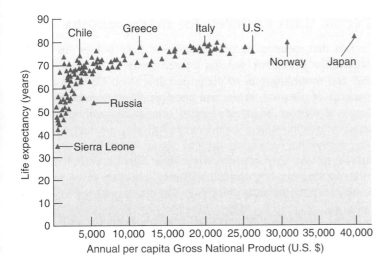

FIGURE 4.7 Up to about $4,000 (U.S.), rising income increases life expectancy. However, in some poorer countries people live as long as in rich countries. In Chile, for example, people live 76 years on average, while in the United States with over four times the per capita gross national income, people live 78 years.

Sources: United Nations (UN) Population Division, World Population Prospects, 1950–2050 (The 1996 Revision). The United Nations, New York, 1996. Population Reference Bureau, 2005, World Population Data Sheet.

Large discrepancies in how the benefits of modernization and social investment are distributed within countries are revealed in differential longevities of various groups. The greatest life expectancy reported anywhere in the United States is for women in Stearns County, Minnesota, who live to an average age of 86. By contrast, Native American men on the Pine Ridge Indian Reservation in neighboring South Dakota live, on average, only to age 45. Only a few countries in Africa have a lower life expectancy. The Pine Ridge Reservation is the poorest area in America, with an unemployment rate near 75 percent and high rates of poverty, alcoholism, drug use, and alienation. Similarly, African-American men in Washington, D.C., live, on average, only 57.9 years, which is less than the life expectancy in Lesotho or Swaziland.

Living longer has profound social implications

A population growing rapidly by natural increase has more young people than does a stationary population. One way to show these differences is to graph age classes in a histogram, as figure 4.8 shows. In Niger, which is growing at a rate of 3.5 percent per year, 47.8 percent of the population is in the prereproductive category (below age 15). Even if total fertility rates fell abruptly, the total number of births, and the population size, would continue to grow for some years as these young people entered reproductive age. This phenomenon is called population momentum.

By contrast, a country with a relatively stable population will have nearly the same number in most cohorts. Notice that females outnumber males in Sweden's oldest group because of differences in longevity between sexes. A rapidly declining population, such as Singapore's, can have a pronounced bulge in

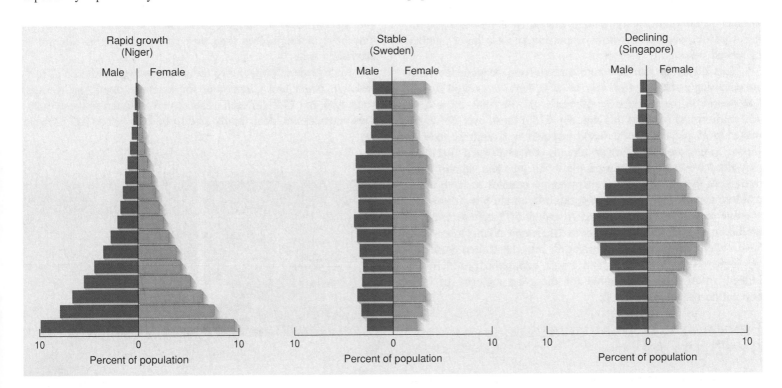

FIGURE 4.8 The shape of each age class histogram is distinctive for a rapidly growing (Niger), stable (Sweden), and declining population (Singapore). Horizontal bars represent the percentage of the country's population in consecutive age classes (0–5 yrs, 6–10 yrs, etc.)
Source: U.S. Census Bureau, 2003.

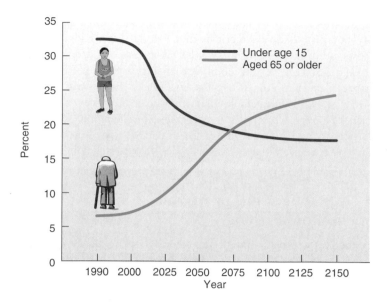

FIGURE 4.9 By the mid-twenty-first century, children under age 15 will make up a smaller percentage of world population, while people over age 65 will contribute a larger and larger share of the population.

middle-age cohorts as fewer children are born than in their parents' generation.

Both rapidly growing countries and slowly growing countries can have a problem with their **dependency ratio,** or the number of nonworking compared with working individuals in a population. In Niger, for example, each working person supports a high number of children. In the United States, by contrast, a declining working population is now supporting an ever larger number of retired persons.

This changing age structure and shifting dependency ratio are occurring worldwide (fig. 4.9). In 1950, there were only 130 million people in the world over 65 years old. In 2000, more than 420 million had reached this age. By 2150, those over 65 might make up 25 percent of the world population. Countries such as Japan, Singapore, and Taiwan already are concerned that they may not have enough young people to fill jobs and support their retirement system. They are encouraging couples to have more children and are recruiting immigrants who might bring down the average age of the population. A group of business leaders, including former Federal Reserve Chairman Alan Greenspan, have called for increased immigration into the United States to fill jobs, prevent inflation, and boost economic productivity. Indeed, immigration accounted for the great majority of U.S. population growth after 2000.

FERTILITY IS INFLUENCED BY CULTURE

A number of social and economic pressures affect decisions about family size, which in turn affects the population at large. In this section, we will examine both positive and negative pressures on reproduction.

People want children for many reasons

Factors that increase people's desires to have babies are called **pronatalist pressures.** Raising a family may be the most enjoyable and rewarding part of many people's lives. Children can be a source of pleasure, pride, and comfort. They may be the only source of support for elderly parents in countries without a social security system. Where infant mortality rates are high, couples may need to have many children to ensure that at least a few will survive to take care of them when they are old. Where there is little opportunity for upward mobility, children give status in society, express parental creativity, and provide a sense of continuity and accomplishment otherwise missing from life. Often children are valuable to the family not only for future income but even more as a source of current income and help with household chores. In much of the developing world, small children tend domestic animals and younger siblings, fetch water, gather firewood, help grow crops, or sell things in the marketplace (fig. 4.10). Parental desire for children rather than an unmet need for contraceptives may be the most important factor in population growth in many cases.

Society also has a need to replace members who die or become incapacitated. This need often is codified in cultural or religious values that encourage bearing and raising children. Some societies look upon families with few or no children with pity or contempt. The idea of deliberately controlling fertility may be shocking, even taboo. Women who are pregnant or have small children have special status and protection. Boys frequently are more valued than girls because they carry on the family name and are expected to support their parents in old age. Couples may have more children than they really want in an attempt to produce a son.

Male pride often is linked to having as many children as possible. In Niger and Cameroon, for example, men, on average, want 12.6 and 11.2 children, respectively. Women in these countries consider the ideal family size to be only about half of what

(a)

(b)

FIGURE 4.10 In rural areas with little mechanized agriculture (a) children are needed to tend livestock, care for younger children, and help parents with household chores. Where agriculture is mechanized (b) rural families view children just as urban families do—helpful, but not critical to survival. This affects the decision about how many children to have.

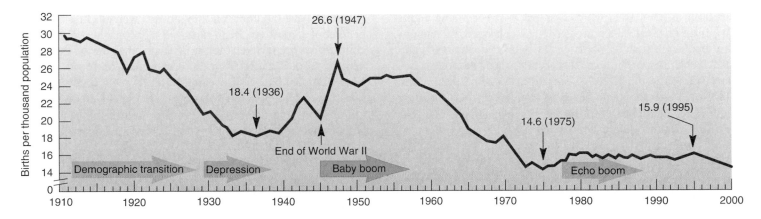

FIGURE 4.11 Birth rates in the United States, 1910–2000. The falling birth rate from 1910 to 1929 represents a demographic transition from an agricultural to an industrial society. The baby boom following World War II lasted from 1945 to 1965. A much smaller "echo boom" occurred around 1980 when the baby boomers started to reproduce.
Source: Data from Population Reference Bureau and U.S. Bureau of the Census.

their husbands desire. Even though a woman might desire fewer children, however, she may have few choices and little control over her own fertility. In many societies, a woman has no status outside of her role as wife and mother. Yet without children, she may have no source of support in her old age.

Education and income affect the desire for children

In more highly developed countries, many pressures tend to reduce fertility. Higher education and personal freedom for women often result in decisions to limit childbearing. A desire to spend time and money on other goods and activities offsets the desire to have children. When women have opportunities to earn a salary, they are less likely to stay home and have many children. Not only do many women find the challenge and variety of a career attractive, but the money that they earn outside the home becomes an important part of the family budget. Thus, education and socioeconomic status are usually inversely related to fertility in richer countries. In some developing countries, however, fertility initially increases as educational levels and socioeconomic status rise. With higher income, families are better able to afford the children they want. More money also means that women are healthier and therefore better able to conceive and carry a child to term. It may be a generation before this unmet desire for children abates.

In less-developed countries, where feeding and clothing children can be a minimal expense, adding one more child to a family usually doesn't cost much. By contrast, raising a child in a developed country can cost hundreds of thousands of dollars by the time the child finishes school and is independent. Under these circumstances, parents are more likely to choose to have one or two children on whom they can concentrate their time, energy, and financial resources.

Figure 4.11 shows U.S. birth rates between 1910 and 2000. As you can see, birth rates fell and rose in an irregular pattern. The period between 1910 and 1930 was a time of industrialization and urbanization. Women were getting more education than

ever before and entering the workforce in large numbers. The Great Depression in the 1930s made it economically difficult for families to have children, and birth rates were low. The birth rate increased at the beginning of World War II (as it often does in wartime). For reasons that are unclear, a higher percentage of boys are usually born during war years.

A "baby boom" followed World War II, as couples were reunited and new families started (fig. 4.12). During this time, the government encouraged women to leave their wartime jobs and stay home. A high birth rate persisted through the times of prosperity and optimism of the 1950s but began to fall in the 1960s. Part of this decline was caused by the small number of

FIGURE 4.12 A returning World War II vet and his family. The baby boom interrupted the demographic transition underway before the Great Depression. Over half a century later, the country has again resumed its trajectory toward a stable population.

babies born in the 1930s, which resulted in fewer young adults to give birth in the 1960s. Part was due to changed perceptions of the ideal family size. While in the 1950s women typically wanted four children or more, the norm dropped to one or two (or no) children in the 1970s. A small "echo boom" occurred in the 1980s, as baby boomers began to have babies, but changing economics and attitudes seem to have permanently altered our view of ideal family size in the United States.

A DEMOGRAPHIC TRANSITION CAN LEAD TO STABLE POPULATION SIZE

In 1945 demographer Frank Notestein pointed out that a typical pattern of falling death rates and birth rates due to improved living conditions usually accompanies economic development. He called this pattern the **demographic transition** from high birth and death rates to lower birth and death rates. Figure 4.13 shows an idealized model of a demographic transition. This model is often used to explain connections between population growth and economic development.

Economic and social conditions change mortality and births

The left side of figure 4.13 represents the conditions in a pre-modern society. Food shortages, malnutrition, lack of sanitation and medicine, accidents, and other hazards generally keep death rates in such a society around 30 per 1,000 people. Birth rates are correspondingly high to keep population densities relatively constant. As economic development brings better jobs, medical care, sanitation, and a generally improved standard of living, death rates often fall very rapidly. Birth rates may actually rise at first as more money and better nutrition allow

people to have the children they always wanted. In a couple of generations, however, birth rates fall as people see that all their children are more likely to survive and that the whole family benefits from concentrating more resources on fewer children. Note that populations grow rapidly during the time that death rates have already fallen but birth rates remain high. Depending on how long it takes to complete the transition, the population may go through one or more rounds of doubling before coming into balance again.

The right-hand side of each curve in figure 4.13 represents conditions in many developed countries, where the transition is complete and both birth rates and death rates are low, often a third or less than those in the predevelopment era. The population comes into a new equilibrium in this phase, but at a much larger size than before. Most of the countries of northern and western Europe went through a demographic transition in the nineteenth or early twentieth century similar to the curves shown in figure 4.13. In countries such as Italy, where fertility levels have fallen below replacement rates, there are now fewer births than deaths, and the total population curve has started to decline.

A huge challenge facing countries in the final stage of the demographic transition is the imbalance between people in their most productive years and people who are retired or in their declining years. The debate in the U.S. Congress about how to fund the social security system is due to the fact that, when this program was established, the United States was in the middle of the demographic transition, with a large number of young people relative to older people. In ten to fifteen years, that situation will change, with many more elderly people living longer, and fewer younger workers to support them.

Many of the most rapidly growing countries in the world, such as Kenya, Yemen, Libya, and Jordan, now are in the middle phase of this demographic transition. Their death rates have fallen close to the rates of the fully developed countries, but birth rates have not fallen correspondingly. In fact, both their birth rates and total population are higher than those in most European countries when industrialization began 300 years ago. The large disparity between birth and death rates means that many developing countries now are growing at 3 to 4 percent per year. Such high growth rates in developing countries could boost total world population to over 9 billion by the end of the twenty-first century. This raises what may be the two most important questions in this entire chapter: why are birth rates not yet falling in these countries, and what can be done about it?

Many countries are in a demographic transition

Some demographers claim that a demographic transition already is in progress in most developing nations. They believe that problems in taking censuses and a normal lag between falling death and birth rates may hide this for a time but that the world population should stabilize sometime in the twenty-first century. Some evidence supports this view. As mentioned earlier in this

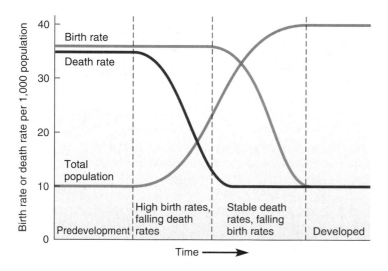

FIGURE 4.13 Theoretical birth, death, and population growth rates in a demographic transition accompanying economic and social development.

Principles of Environmental Science http://www.mhhe.com/cunningham4e

chapter, fertility rates have fallen dramatically nearly everywhere in the world over the past half century.

Some countries have had remarkable success in population control. In Thailand, Indonesia, and Colombia, for instance, total fertility dropped by more than half in 20 years. Morocco, the Dominican Republic, Jamaica, Peru, and Mexico all have seen fertility rates fall between 30 and 40 percent in a single generation. Surprisingly, one of the most successful family planning programs in recent years has been in Iran (see What Do You Think? p. 89).

The following factors help stabilize populations:

- Growing prosperity and social reforms that accompany development reduce the need and desire for large families in most countries.

- Technology is available to bring advances to the developing world much more rapidly than was the case a century ago, and the rate of technology exchange is much faster than it was when Europe and North America were developing.

- Less-developed countries have historic patterns to follow. They can benefit from the mistakes of more-developed countries and chart a course to stability relatively quickly.

- Modern communications (especially television) have caused a revolution of rising expectations that act as stimuli to spur change and development.

Two ways to complete the demographic transition

Kerala and Andra Pradesh exemplify two approaches to population growth. In Kerala, providing a fair share of social benefits to everyone was the key to completing the demographic transition. People adhering to this social justice view believe that the world has enough resources for everyone, but inequitable social and economic systems cause maldistributions of those resources. Hunger, poverty, violence, environmental degradation, and overpopulation are symptoms of a lack of justice, rather than a lack of resources. Although overpopulation exacerbates other problems, a focus on overpopulation alone encourages racism and hatred of the poor. Proponents of the social justice view argue that people in the richer nations should recognize their excess consumption levels and the impact that this consumption has on others. Figure 4.14 expresses the opinion of many people in less-developed countries about the relationship between resources and population.

The leaders of Andra Pradesh, on the other hand, adopted a view like that of economist Lester Brown of the Worldwatch Institute. Mr. Brown warns that many of the poorer countries of the world appear to be caught in a "demographic trap" that prevents them from escaping from the middle phase of the demographic transition. Their populations are growing so rapidly that

FIGURE 4.14 Do we reduce pressure on the environment by achieving population control in developing countries, or by limiting resource use in developed countries? It depends on whom you ask.

Used with permission of the Asian Cultural Forum on Development.

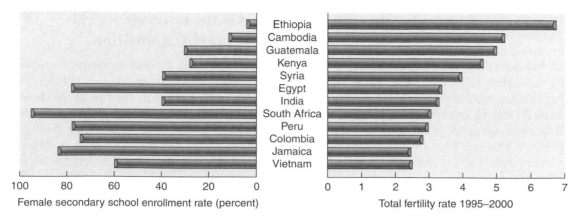

FIGURE 4.15 Total fertility declines as women's education increases. **Source:** Data from Worldwatch Institute, 2003.

human demands exceed the sustainable yield of local forests, grasslands, croplands, and water resources. The resulting resource shortages, environmental deterioration, economic decline, and political instability may prevent these countries from ever completing modernization. Their populations may continue to grow until catastrophe intervenes.

Many people believe that the only way to break out of the demographic trap is to reduce population growth immediately and drastically by whatever means are necessary, as occurred in Andra Pradesh and with China's one-child policy. They argue strongly for birth control education and bold national policies to encourage lower birth rates. Some agree with Malthus that helping the poor will simply increase their reproductive success and further threaten the resources on which we all depend. Author Garret Hardin described this view as lifeboat ethics:

> Each rich nation amounts to a lifeboat full of comparatively rich people. The poor of the world are in other much more crowded lifeboats. Continuously, so to speak, the poor fall out of their lifeboats and swim for a while, hoping to be admitted to a rich lifeboat, or in some other way to benefit from the goodies on board. . . . We cannot risk the safety of all the passengers by helping others in need. What happens if you share space in a lifeboat? The boat is swamped and everyone drowns. Complete justice, complete catastrophe.

Improving women's lives helps reduce birth rates

Kerala's leaders believed that women were key to achieving population stability. The 1994 International Conference on Population and Development in Cairo, Egypt, signaled this approach to population issues. A broad consensus reached by the 180 participating countries agreed that responsible economic development, education and empowerment of women, and high-quality health care (including family planning services) must be accessible to everyone if population growth is to be slowed. Child survival is one of the most critical factors in stabilizing population. When infant and child mortality rates are high, as they are in much of

the developing world, parents tend to have high numbers of children to ensure that some will survive to adulthood. There has never been a sustained drop in birth rates that was not first preceded by a sustained drop in infant and child mortality.

Increasing family income does not always translate into better welfare for children, since men in many cultures control most financial assets. As the UN Conference in Cairo noted, the best way to improve child survival often is to ensure the rights of mothers. Opportunities for women's education, for instance, as well as land reform, political rights, opportunities to earn an independent income, and improved health status of women often are better indicators of family welfare than is rising gross national product (fig. 4.15).

DECISIONS ABOUT FERTILITY START WITH FAMILIES

Family planning allows couples to determine the number and spacing of their children. It doesn't necessarily mean fewer children—people may use family planning to have the maximum number of children possible—but it does imply that the parents will control their reproductive lives and make rational, conscious decisions about how many children they will have and when those children will be born, rather than leaving it to chance (fig. 4.16).

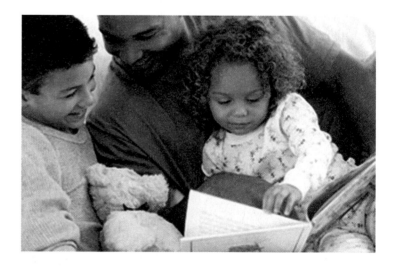

FIGURE 4.16 The number of children a family chooses to have is determined by many factors, including access to a variety of birth control methods.

Principles of Environmental Science

How Iran Shifted Its Perspective and Slowed Its Growth

After the Islamic Revolution in 1979, Iran had one of the world's highest population growth rates. In spite of civil war, large-scale emigration, and economic austerity, the country surged from 34 million to 63 million in just 20 years. A crude birth rate of 43.4 per 1,000 people and a total fertility rate of 5.1 per woman during this time resulted in an annual population growth of 3.9 percent and a doubling time of less than 18 years. Religious authorities exhorted couples to have as many children as Allah would give them. Any mention of birth control or family planning was forbidden, and the marriage age for girls was dropped to 9 years old. When a devastating war with Iraq in the 1980s killed at least 1 million young soldiers, producing more children to rebuild the army became a civic as well as religious duty.

In the late 1990s, however, the Iranian government became aware of the costs of such rapid population growth. With religious moderates gaining greater political power, public policy changed abruptly. Now the Iranian government is spending millions of dollars to lower birth rates. Couples must pass a national family planning course before they are allowed to marry. While it took a few years to convince people that this change will be long-lasting, most Iranian citizens are now eager for access to birth control information. Family planning classes are sought out both by engaged couples and those already married. A wide range of birth control methods are available. Implantable or injectable slow-release hormones, condoms, intrauterine devices (IUDs), pills, and male or female sterilization are free to all. Billboards, newspapers, television, and even water towers advertise this national program. Religious leaders have issued a *fatwah,* or command, that all faithful Muslims participate in family planning.

As a consequence, Iran has been remarkably successful in stemming its population growth. Between 1986 and 1996 the fertility rates for urban residents dropped by almost half, to less than three children per women, and the crude birth rate dropped from 43 to 18 per 1,000 people. By 2000 the average annual growth rate had fallen to 1.4 percent. While the population is still increasing, another decade of such progress would bring the country to a stable or even declining rate of growth.

Several societal changes have contributed to this rapid birth reduction. While the minimum marriage age has been returned to 15, couples are encouraged to wait until at least age 20 to begin their families. The educational benefits of concentrating the family resources on just one or two children are being promoted. Although women's roles are still highly restricted in the Islamic Republic, greater gender equity has given women more control over their reproductive lives. Access to modern, information-age jobs gives people an incentive to seek out education both for themselves and for their children. The demographic transition hasn't spread to all levels of Iranian society, however. Rural families, ethnic minorities, and some urban poor still tend to have many children.

Does it seem surprising that a religious, fundamentalist society would embrace modern birth control? Could the Iranian approach for reducing fertility rates work in other countries experiencing rapid growth, such as in Latin America or Africa?

As the desire for smaller families becomes more common, birth control often becomes an essential part of family planning. In this context, **birth control** usually means any method used to reduce births, including celibacy, delayed marriage, contraception, methods that prevent embryo implantation, and induced abortions.

Humans have always regulated their fertility

The high human birth rate of the last two centuries is not the norm compared to previous millennia of human existence. Evidence suggests that people in every culture and every historic period used a variety of techniques to control population size. Studies of hunting and gathering people, such as the !Kung, or San, of the Kalahari Desert in southwest Africa, indicate that our early ancestors had stable population densities, not because they killed each other or starved to death regularly but because they controlled fertility.

For instance, San women breast-feed children for three or four years. When calories are limited, lactation depletes body fat stores and suppresses ovulation. Coupled with taboos against intercourse while breast-feeding, this is an effective way of spacing children. (However, breast-feeding among well-nourished women in modern societies doesn't necessarily suppress ovulation or prevent conception.) Other ancient techniques to control population size include celibacy, folk medicines, abortion, and infanticide. We may find some or all of these techniques unpleasant or morally unacceptable, but we shouldn't assume that other people are too ignorant or too primitive to make decisions about fertility.

Today there are many options

Modern medicine gives us many more options for controlling fertility than were available to our ancestors. The major categories of birth control techniques include (1) avoidance of sex during fertile periods (for example, celibacy or the use of changes in body temperature or cervical mucus to judge when ovulation will occur), (2) mechanical barriers that prevent contact between sperm and egg (for example, condoms, spermicides, diaphragms, cervical caps, and vaginal sponges), (3) surgical methods that prevent release of sperm or egg (for example, tubal ligations in females and vasectomies in males), (4) hormone-like chemicals that prevent maturation or release of sperm or eggs or that prevent embryo implantation in the uterus (for example, estrogen plus progesterone, or progesterone alone, for females; gossypol for males), (5) physical barriers to implantation (for example, intrauterine devices), and (6) abortion.

Not surprisingly, the most effective birth control methods are also the ones most commonly used (table 4.4). In the United States, the majority of women younger than 30 who eventually want to become pregnant use the Pill. Most women over 35, with their child-bearing years behind them, choose sterilization. Male condom use is more effective than the remaining techniques in the table, and increases in effectiveness when used with a spermicide. Only 2 to 6 women in a hundred become pregnant in a year using this combination method. Condoms have the added advantage of protecting partners against sexually transmitted

TABLE 4.4

Some Birth Control Methods and Pregnancy Prevention Rates

Method	Number of Women in 100 Who Become Pregnant
Sterilization (male, female)	<1
IUD	<1
Oral contraceptive (the Pill)	1–2
Hormones (implant, patch, injection, etc.)	1–2
Male condom	11
Sponge and spermicide	14–28
Female condom (e.g., cervical cap)	15–23
Diaphragm together with spermicide	17
Abstinence during fertile periods	20
Morning-after-pill (e.g., Preven)	20
Spermicide alone	20–50
Actively seeking pregnancy	85

Source: U.S. Food and Drug Administration, *Birth Control Guide*, 2003 Revision.

diseases, including AIDS, if they are made of latex and used correctly. That may partly explain why their use in the United States went from 3.5 million users in 1980 to 8 million in 2000. Condoms are an ancient birth control method; the Egyptians used them some 3,000 years ago.

More than 100 new contraceptive methods are now being studied, and some appear to have great promise. Nearly all are biologically based (e.g., hormonal), rather than mechanical

(e.g., condom, IUD). Recently, the U.S. Food and Drug Administration approved five new birth control products. Four of these use various methods to administer female hormones that prevent pregnancy. Other methods are years away from use, but take a new direction entirely. Vaccines for women are being developed that will prepare the immune system to reject the hormone chorionic gonadotropin, which maintains the uterine lining and allows egg implant, or that will cause an immune reaction against sperm. Injections for men are focused on reducing sperm production, and have proven effective in mice. Without a doubt, the contemporary couple has access to many more birth control options than their grandparents had.

WHAT KIND OF FUTURE ARE WE CREATING NOW?

Because there is a lag between the time when a society reaches replacement birth rate and the end of population growth, we are deciding now what the world will look like in a hundred years. How many people will be in the world a century from now? Most demographers believe that world population will stabilize sometime during the twenty-first century. When we reach that equilibrium, the total number of humans is likely to be somewhere around 8 to 10 billion, depending on the success of family planning programs and the multitude of other factors affecting human populations. The United Nations Population Division projects three population scenarios (fig. 4.17). The optimistic (low) projection suggests that world population might stabilize by about 2030 and then drop back below current levels. This is not likely. The medium projection suggests that growth might continue to rise slowly until at least 2050 and then stabilize at about 9 billion shortly afterwards. The most pessimistic (high) projection indicates that population growth will climb at present rates for 150 years or more, quadrupling the world's population to 25 billion.

Which of these scenarios will we follow? As you have seen in this chapter, population growth is a complex subject. Stabilizing or reducing human populations will require substantial changes from business as usual. One impediment is that the United States refuses to release payments to the United Nations Family Planning Fund because some of the 135 countries that receive UN aid include abortion as part of population control programs. The UNFPA is the world's leading provider of funding for family planning and reproductive health programs. Since UNFPA began

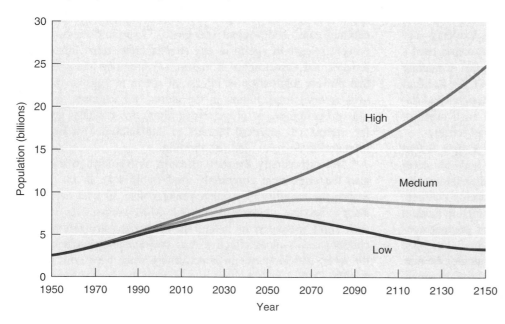

FIGURE 4.17 Population projections for different growth scenarios indicate that population growth is slowing. The medium projection is for a peak of 9.2 billion in 2075, compared to over 10 billion in previous estimates.
Source: Data from United Nations Population Division, 2004.

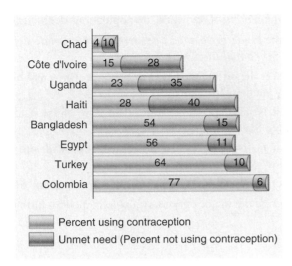

FIGURE 4.18 "Unmet need" is measured as percent of married women who want to control their fertility but do not use contraception. Globally, more than 100 million women in developing countries would prefer to avoid pregnancy but do not have access to family planning.
Source: Data from Population Reference Bureau, 2003.

FIGURE 4.19 Average total fertility rates for less-developed countries fell by more than half over the past 50 years, mostly due to China's one-child policy. By 2050 even the least-developed countries should approach the replacement rate of 2.1 children per reproductively fertile woman.
Source: Data from UN Population Division, *World Population Prospects, 1996,* and Population Reference Bureau, 2004.

operations in 1969, fertility rates in developing countries have fallen from six children per woman to three.

An encouraging sign is that worldwide contraceptive use has increased sharply in recent years. About half of the world's married couples used some family planning techniques in 2000, compared with only 10 percent 30 years earlier, but another 100 million couples say they want, but do not have access to, family planning. If given a choice, people prefer smaller families.

The unmet need for family planning among married women in some countries is high (fig. 4.18). When people in developing countries are asked what they want most, men say they want better jobs, but the first choice for a vast majority of women is family planning assistance. In general, a 15 percent increase in contraceptive use equates to about 1 fewer births per woman per lifetime. In Chad, for example, where only 4 percent of all women use contraceptives, the average fertility is 6.6 children per woman. In Colombia, by contrast, where 77 percent of the women who would prefer not to be pregnant use contraceptives, the average fertility is 2.6 children.

Successful family planning programs often require significant societal changes. Among the most important of these are (1) improved social, educational, and economic status for women (birth control and women's rights are often linked); (2) improved

status for children (fewer children are born if they are not needed as a cheap labor source); (3) acceptance of calculated choice as a valid element in life in general and in fertility in particular (the belief that we have no control over our lives discourages a sense of responsibility); (4) social security and political stability that give people the means and the confidence to plan for the future; and (5) the knowledge, availability, and use of effective and acceptable means of birth control. Concerted efforts to bring about these types of societal changes can be effective. Twenty years of economic development and work by voluntary family planning groups in Zimbabwe, for example, have lowered total fertility rates from 8.0 to 5.5 children per woman on average. Surveys show that desired family sizes have fallen nearly by half (9.0 to 4.6) and that nearly all women and 80 percent of men in Zimbabwe use contraceptives.

The current world average fertility rate of 2.7 births per woman is less than half what it was 50 years ago. If similar progress could be sustained for the next half century, fertility rates could fall to the replacement rate of 2.1 children per woman (fig. 4.19). Whether this scenario comes true or not depends on choices that all of us make. In their approach to population issues, the people of Kerala may have indicated for the rest of the world an effective yet supremely humane direction to take.

REVIEW QUESTIONS

1. About how many years of human existence passed before the world population reached its first billion? What factors restricted population before that time, and what factors contributed to growth after that point?

2. Describe the pattern of human population growth over the past 200 years. What is the shape of the growth curve (recall chapter 3)?

3. In what year might the human population of the planet stabilize? Is this level of population sustainable? What other factors besides population size determine the impact of a country on the environment?

4. Define *ecological footprint*. Why is it helpful, but why might it also be inaccurate?

5. How might growing populations lead to solutions to society's problems?

6. Why do some economists consider human resources more important than natural resources in determining a country's future?

7. In which regions of the world will most population growth occur during the twenty-first century? What conditions contribute to rapid population growth in these locations?

8. In which parts of the world are populations declining?

9. Define *crude birth rate, total fertility rate, crude death rate,* and *zero population growth.*

10. What is the difference between life expectancy and life span? Why are they different?

11. What is the dependency ratio, and how might it affect the United States in the future?

12. What factors increase or decrease people's desire to have babies?

13. Describe the conditions that lead to a demographic transition.

14. Describe the major options available in modern birth control.

CRITICAL THINKING AND DISCUSSION

Apply the principle you have learned in this chapter to discuss these questions with other students.

1. Suppose that you were head of a family planning agency in India. How would you design a scientific study to determine the effectiveness of different approaches to population stabilization? How would you account for factors such as culture, religion, education, and economics?

2. Why do you suppose that the United Nations gives high, medium, and low projections for future population growth? Why not give a single estimate? What factors would you consider in making these projections?

3. Some demographers claim that the total world population has already begun to slow, while others dispute this claim. How would you recognize a true demographic transition, as opposed to mere random fluctuations in birth and death rates?

4. Why do we usually express crude birth and death rates per thousand people? Why not give the numbers per person or for the entire population?

5. In northern Europe, the demographic transition began in the early 1800s, a century or more before the invention of modern antibiotics and other miracle drugs. What factors do you think contributed to this transition? How would you use historical records to test your hypothesis?

6. In chapter 3, we discussed carrying capacities. What do you think the maximum and optimum carrying capacities for humans are? Why is this a more complex question for humans than it might be for other species? Why is designing experiments in human demography difficult?

SUGGESTED READINGS

Ashford, Lori. 2003. *Unmet Need for Family Planning: Recent Trends and Their Implications for Programs.* Population Reference Bureau. Published online at http://www.prb.org.

Cohen, Joel E. 1995. *How Many People Can the Earth Support?* W.W. Norton.

Dasgupta, Partha S. 1995. Population, poverty and the local environment. *Scientific American,* February, 40–45.

Diamond, Jared. 2005. *Collapse: How Societies Choose to Fail or Succeed.* Penguin Books.

Ehrlich, Paul R. 2000. *Human Nature: Genes, Cultures, and the Human Prospect.* Island Press.

Hartmann, Betsy. 1999. Population, environment, and security: A new trinity. In *Dangerous Intersections: Feminist Perspectives on Population, Environment, and Development.* Jael Silliman and Ynestra King, eds. South End Press, pp. 1–23.

Lee, Ronald. 2003. The demographic transition: Three centuries of fundamental change. *Journal of Economic Perspectives* 17:167–290.

Pimentel, David, et al. 1999. Will limits of the earth's resources control human numbers? *Environment, Development and Sustainability* 1:19–39.

Sen, Amaryta. 1994. Population and reasoned agency: Food, fertility, and economic development. In *Population, Economic Development, and the Environment.* Kerstin Lindahl-Kiessling and Hans Landberg, eds. Oxford University Press, pp. 51–78.

Wackernagel, M., and W. E. Rees. 1996. *Our Ecological Footprint: Reducing Human Impact on the Earth.* New Society Publishers.

DATA analysis   Interpreting Graphs

An important part of any scientific investigation is presenting the data. Graphs and other pictorial representations of information are particularly good for helping others to understand what we have to say. It is one thing to read that "the human population is rising explosively," but it is considerably more compelling to see the J curve along with concrete numbers as evidence. The combination of picture and numbers often conveys powerful impressions beyond the numbers themselves. But it is important to recognize that graphs can be as truthful or misleading as the person who creates them intends to be.

Central questions addressed in this chapter are how serious is the worldwide population problem and does it demand action. The graphs in figures 1 and 2 present information relevant to human population studies, yet the impressions these figures convey vary considerably, depending on the type of data used and the design of the graph.

Reexamine figure 4.2, which presents a historical perspective on the growth in human numbers. What conclusions do you draw from it? Does it suggest that we are experiencing an explosive rise in human numbers? Does it indicate that the current increase is unprecedented in our species history? Does it also imply, perhaps more indirectly, that the earth's population problem is serious and urgent? In light of the concepts of carrying capacity, and overshoot and dieback, many people would conclude that the current population growth, highlighted by the graph, is unsustainable.

Now examine figure 1. It plots the same variables as figure 4.2 but suggests that population is rising at a modest rate. It does not produce the same sense of explosiveness and urgency as figure 4.2. How can the impressions be so different?

Notice that the graph in figure 1 covers a much shorter time period. This greatly changes the time interval lengths on the horizontal scale. In figure 4.2, 1 mm represents about 50 years, but in figure 1, it represents less than 1 year. This changes the line's slope from nearly vertical in figure 4.2 to a modest incline in figure 1. Slope impacts the visual impression created by a graph and, therefore, its interpretation. How could you change the time axis in figure 1 to flatten the slope even more?

Next examine figure 2. This graph plots the stabilization ratio, rather than population size over time. The graph reveals that the growth rate for most regions except Africa has been declining since about 1970. Do the graph's downward-trending lines suggest that there really isn't much of a population problem, or at least not much reason for concern?

All three graphs present information relevant to a discussion of worldwide population growth, yet each gives a different impression of the seriousness of the problem. So, how can we analyze graphs and their messages in a thoughtful way? Certainly, no single formula is applicable to all situations, but the following questions are worth keeping in mind:

1. Is the graph's time frame of reference appropriate, or is it too restricted to allow a valid, comprehensive assessment of the issue?

2. Are the unit intervals on the graph appropriately sized?

3. Is the impression the graph creates real or simply an artifact of the graph's format?

4. Do the data the graph presents provide only a partial, perhaps misleading, view of the whole?

Because graphs can create powerful impressions, we need to use and interpret them with care.

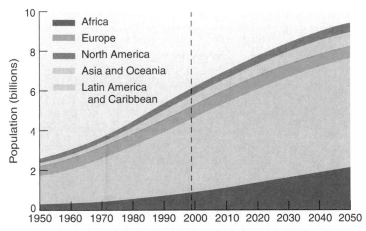

FIGURE 1 The UN Population Division projects continued population growth, although at a gradually slowing rate, over the next 50 years.
Source: World Resources Institute, World Resources 1998–99.

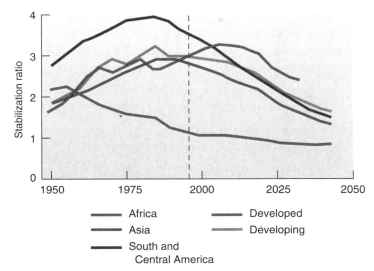

FIGURE 2 The stabilization ratio is measured by dividing crude birth rate by crude death rate. A ratio of 1 indicates zero population growth.
Source: United Nations (UN) Population Division, World Population Prospects, 1950–2050 (The 1996 Revision). The United Nations, New York, 1996.

For Additional Help in Studying This Chapter, please visit our website at www.mhhe.com/cunningham4e. You will find practice quizzes, a chapter summary, key terms, answers to review questions, additional case studies, regional examples, and an extensive reading list, all of which will help you understand the material in this chapter.

Wolves reintroduced to Yellowstone National Park
may be helping to restore biodiversity.

C H A P T E R

5

Biomes and Biodiversity

In the end, we conserve only what we love.
We will love only what we understand.
We will understand only what we are taught.

–Baba Dioum–

OBJECTIVES

Studying this chapter will help you answer the following questions:

- What are nine major terrestrial biomes, and what environmental conditions control their distribution?
- How does vertical stratification differentiate life zones in oceans?
- Why are coral reefs, mangroves, estuaries, and wetlands biologically important?

- What do we mean by *biodiversity?* List several regions of high biodiversity.
- What are four major benefits of biodiversity?
- What are the major human-caused threats to biodiversity?
- How can we reduce these threats to biodiversity?

Case Study Predators Help Restore Biodiversity in Yellowstone

In the past dozen years, Yellowstone National Park has been the site of a grand experiment in managing biodiversity. The park is renowned for its wildlife and wilderness. But between about 1930 an 1990, park managers and ecologists observed that growing elk populations were overbrowsing willow and aspen. Plant health and diversity were declining, and populations of smaller mammals, such as beaver, were gradually dwindling. Most ecologists blamed these changes on the eradication of one of the region's main predators, the gray wolf (*Canis lupus*). Now, a project to restore wolves has provided a rare opportunity to watch environmental change unfold.

Wolves were once abundant in the Greater Yellowstone ecosystem. In the 1890s, perhaps 100,000 wolves roamed the western United States. Farmers and ranchers, aided by government programs to eradicate predators, gradually poisoned, shot, and trapped nearly all the country's wolves. In the northern Rocky Mountains, elk and deer populations expanded rapidly without predators to restrain their numbers. Yellowstone's northern range elk herd grew to about 20,000 animals. Some ecologists warned that without a complete guild of predators, the elk, bison, and deer could damage one of our best-loved parks. After much controversy, 31 wolves were captured in the Canadian Rockies and relocated to Yellowstone National Park in 1995 and 1996.

Wolves expanded quickly in their new home. With abundant elk for prey, as well as occasional moose and bison, the wolves tripled their population in just three years. The population now exceeds 100 animals in over 20 packs, well beyond the recovery goal of 10 packs.

Following the wolf reintroductions, elk numbers have fallen, and regenerating willow and aspen show signs of increased stature and abundance. Elk carcasses left by wolves provide a feast for scavengers, and wolves appear to have driven down numbers of coyotes, which prey on small mammals. Many ecologists see evidence of more songbirds, hawks, foxes, voles, and ground squirrels, which they attribute to wolf reintroductions. Other ecologists caution that Yellowstone is a complex system, and wolves are one of many factors

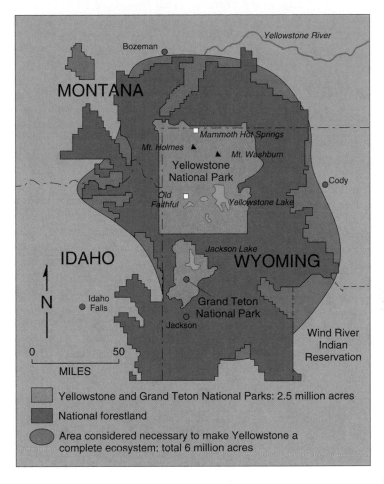

This map shows the Yellowstone ecosystem complex, or biogeographical region, which extends far beyond the park's boundaries. Park managers and ecologists believe that it is necessary to manage the entire region if the park itself is to remain biologically viable.

that may influence regeneration. For example, recent mild winters may have allowed elk to find alternate food sources, reducing browsing pressure on willows and aspen, and drought may have contributed to declining elk numbers. Debate on the exact impact of wolves underscores the difficulty of interpreting change in complex ecosystems. Whether or not wolves are driving ecological change, most ecologists agree that wolves contribute to a diverse and healthy ecosystem.

New Technology

How do ecologists study these changes? For years, ecologists have skiied and snowshoed out to remote observation posts to record wolf kills and observe wolf behavior. But the difficulty of this field work has limited the amount of data that could be collected. One of the exciting new alternatives involves a remotely controlled video camera that allows researchers to monitor wolf ecology and behavior without intruding on the wolves.

Ecologist Dan MacNulty, of the University of Minnesota, together with Glenn Plumb of the National Park Service, has installed a pair of computer-controlled video cameras in an open meadow near hot springs where bison gather to seek exposed grass in the winter. Wolves, hoping to prey on the bison, frequent the area as well. Instead of skiing long, cold miles to the site, MacNulty can now turn on a camera from his office desktop. The camera scans an area of 17 km^2, and it can

Dan MacNulty observes wolves from his office.

Case Study *continued*

zoom in to identify individual animals up to 9.5 km away. In the past, his team was lucky to get a few weeks of winter wolf observation in a year, but now MacNulty and his 12 undergraduate research assistants can watch for wolves year-round, any time it's light enough to see them. Working in shifts, the group records sightings and behavior of wolves, bears, elk, bison, coyotes, and fox. (For discussion of additional research methods, see Exploring Science, p. 113.)

Because they can observe the animals at length without disturbing them, MacNulty's group has made some intriguing observations. Wolves are masterful at detecting when a bison is vulnerable. They almost never attack a large bull, unless it has wandered from the group or gotten into deep snow, where bison quickly get bogged down. Bison

will also work together to fend off wolves—if they can avoid leaving the snow-free areas. Wolves can also be surprisingly persistent: in one case McNulty says wolves harassed a vulnerable bison for 36 hours before finally killing it. All these observations help ecologists understand how wolves hunt, select their prey, and avoid injury in the chase.

Yellowstone's wolf restoration is exciting because it is almost unique as a large, biodiversity restoration project. As a natural laboratory, the park has allowed researchers to explore many questions about ecosystem function, stability, and resilience. In this chapter, we'll examine the major living system that harbor the earth's biodiversity. We'll also consider the benefits of biodiversity, activities that threaten diversity, and opportunities to preserve it.

TERRESTRIAL BIOMES

Although all local environments are unique, it is helpful to understand them in terms of a few general groups with similar climate conditions, growth patterns, and vegetation types. We call these broad types of biological communities **biomes.** Understanding the global distribution of biomes, and knowing the differences in what grows where and why, is essential to the study of global environmental science. Biological productivity—and ecosystem resilience—varies greatly from one biome to another. Human use of biomes depends largely on those levels of productivity. Our ability to restore ecosystems, a topic of chapter 6, and nature's ability to restore itself, depend largely on biome productivity. Clear-cut forests regrow relatively quickly in New England but very slowly in Siberia, where current logging is expanding. Some grasslands rejuvenate quickly after grazing, and some are slower to recover. Why these differences? The sections that follow seek to answer this question.

Temperature and precipitation are among the most important determinants in biome distribution on land (fig. 5.1). If we know the general temperature range and precipitation level, we can predict what kind of biological community is likely to occur there, in the absence of human disturbance. Landforms, especially mountains, and prevailing winds also influence biological communities.

Many temperature-controlled biomes occur in latitudinal bands (fig. 5.2). For example, a band of boreal (northern) forests crosses Canada and Siberia, tropical forests occur near the equator, and expansive grasslands lie near—or just beyond—the

tropics. Many biomes are even named for their latitudes: tropical rainforests occur between the Tropic of Cancer (23° north) and the Tropic of Capricorn (23° south); arctic tundra lies near or above the Arctic Circle (66.6° north).

Temperature and precipitation change with elevation as well as with latitude. In mountainous regions, temperatures are cooler and precipitation is usually greater at high elevations. Mountains are cooler and often wetter, than low elevations. **Vertical zonation** is a term applied to vegetation zones defined by altitude. A 100 km transect from California's Central Valley up to Mount Whitney, for example, crosses as many vegetation zones as you would find on a journey from southern California to northern Canada (fig. 5.3).

FIGURE 5.1 Biomes most likely to occur in the absence of human disturbance or other disruptions, according to average annual temperature and precipitation. *Note:* This diagram does not consider soil type, topography, wind speed, or other important environmental factors. Still, it is a useful general guideline for biome location.

Source: Whitaker, Robert, C., *Communities & Ecosystems*, 2e. © 1975. Adapted by permission of Pearson Education, Inc., Upper Saddle River, NJ.

Principles of Environmental Science

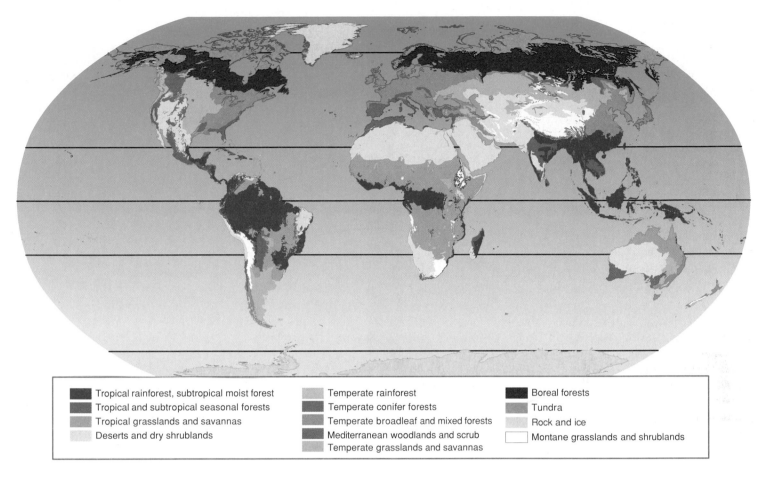

Tropical rainforest, subtropical moist forest
Tropical and subtropical seasonal forests
Tropical grasslands and savannas
Deserts and dry shrublands

Temperate rainforest
Temperate conifer forests
Temperate broadleaf and mixed forests
Mediterranean woodlands and scrub
Temperate grasslands and savannas

Boreal forests
Tundra
Rock and ice
Montane grasslands and shrublands

FIGURE 5.2 Major world biomes. Compare this map with figure 5.1 for generalized temperature and moisture conditions that control biome distribution. Also compare it with the satellite image of biological productivity (fig. 5.13).
Source: WWF Ecoregions.

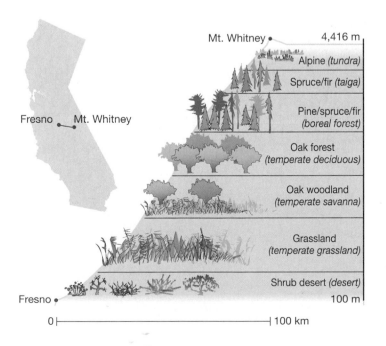

Mt. Whitney • 4,416 m

Alpine (tundra)
Spruce/fir (taiga)
Pine/spruce/fir (boreal forest)
Oak forest (temperate deciduous)
Oak woodland (temperate savanna)
Grassland (temperate grassland)
Shrub desert (desert)

Fresno • 100 m

0 ├────────────────┤ 100 km

As you consider the terrestrial biomes, compare the annual temperatures and precipitation levels that occur there. To begin, examine the three climate graphs in figure 5.4. These graphs show annual trends in temperature and precipitation (rainfall and snowfall). They also indicate the relationship between potential evaporation, which depends on temperature, and precipitation. When evaporation exceeds precipitation, dry conditions result (marked yellow). Moist climates may vary in precipitation rates, but evaporation rarely exceeds precipitation. Months above freezing temperature (marked brown) have most evaporation. Comparing these climate graphs helps us understand the different seasonal conditions that control plant and animal growth in the different biomes.

FIGURE 5.3 Vegetation changes with elevation because temperatures are lower and precipitation is greater high on a mountain side. A 100 km transect from Fresno, California, to Mount Whitney (California's highest point) crosses vegetation zones similar to about seven different biome types.

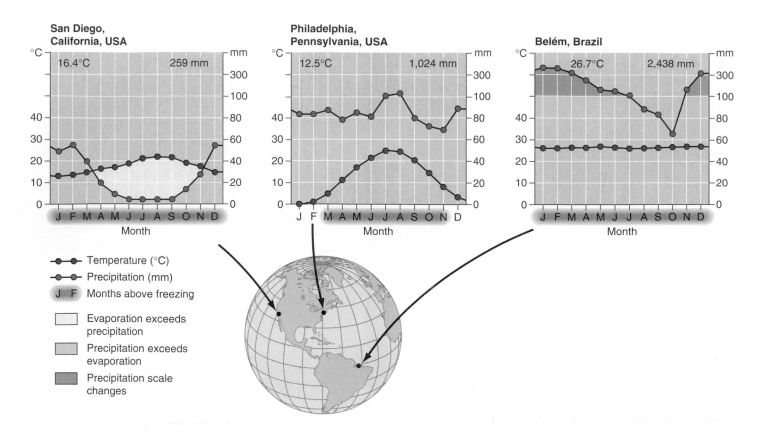

FIGURE 5.4 Moisture availability depends on temperature as well as precipitation. The horizontal axis on these climate diagrams shows months of the year; vertical axes show temperature (*left side*) and precipitation (*right*). The number of dry months (*shaded yellow*) and wetter months (*blue*) varies with geographic location. Mean annual temperature (°C) and precipitation (mm) are shown at the top of each graph.

Tropical moist forests are warm and wet year-round

The humid tropical regions support one of the most complex and biologically rich biome types in the world (fig. 5.5). Although there are several kinds of moist tropical forests, all have ample rainfall and uniform temperatures. Cool **cloud forests** are found high in the mountains where fog and mist keep vegetation wet all the time. **Tropical** rainforests occur where rainfall is abundant—more than 200 cm (80 in.) per year—and temperatures are warm to hot year-round.

The soil of both these tropical moist forest types tends to be thin, acidic, and nutrient-poor, yet the number of species present can be mind-boggling. For example, the number of insect species in the canopy of tropical rainforests has been estimated to be in the millions! It is estimated that one-half to two-thirds of all species of terrestrial plants and insects live in tropical forests.

The nutrient cycles of these forests also are distinctive. Almost all (90 percent) of the nutrients in the system are contained in the bodies of the living organisms. This is a striking contrast to temperate forests, where nutrients are held within the soil and made available for new plant growth. The luxuriant growth in tropical rainforests depends on rapid decomposition and recycling of dead organic material. Leaves and branches that

FIGURE 5.5 Tropical rainforests have luxuriant and diverse plant growth. Heavy rainfall in most months, shown in the climate graph, supports this growth.

Principles of Environmental Science

Comparing Biome Climates

Look back at the climate graphs for San Diego, California, an arid region, and Belém, Brazil, in the Amazon rainforest (fig. 5.4). How much colder is San Diego than Belém in January? In July? Which location has the greater range of temperature through the year? How much do the two locations differ in precipitation during their wettest months?

Compare the temperature and precipitation in these two places with those in the other biomes shown in the pages that follow. How wet are the wettest biomes? Which biomes have distinct dry seasons? How do rainfall and length of warm seasons explain vegetation conditions in these biomes?

Answers: San Diego is about 13°C colder in January, about 6°C colder in July; San Diego has the greater range of temperature; there is about 250 mm difference in precipitation in December–February.

Tropical savannas and grasslands are dry most of the year

Where there is too little rainfall to support forests, we find open **grasslands** or grasslands with sparse tree cover, which we call **savannas** (fig. 5.6). Like tropical seasonal forests, most tropical savannas and grasslands have a rainy season, but generally the rains are less abundant or less dependable than in a forest. During dry seasons, fires can sweep across a grassland, killing off young trees and keeping the landscape open. Savanna and grassland plants have many adaptations to survive drought, heat, and fires. Many have deep, long-lived roots that seek groundwater and that persist when leaves and stems above the ground die back. After a fire, or after a drought, fresh green shoots grow quickly from the roots. Migratory grazers, such as wildebeest, antelope, or bison thrive on this new growth. Grazing pressure from domestic livestock is an important threat to both the plants and animals of tropical grasslands and savannas.

Deserts are hot or cold, but always dry

You may think of deserts as barren and biologically impoverished. Their vegetation is sparse, but it can be surprisingly diverse, and most desert plants and animals are highly adapted to survive long droughts, extreme heat, and often extreme cold. **Deserts** occur where precipitation is uncommon and slight, usually with less than

fall to the forest floor decay and are incorporated almost immediately back into living biomass.

When the forest is removed for logging, agriculture, and mineral extraction, the thin soil cannot support continued cropping and cannot resist erosion from the abundant rains. And if the cleared area is too extensive, it cannot be repopulated by the rainforest community.

Tropical seasonal forests have annual dry seasons

Many tropical regions are characterized by distinct wet and dry seasons, although temperatures remain hot year-round. These areas support **tropical seasonal forests:** drought-tolerant forests that look brown and dormant in the dry season but burst into vivid green during rainy months. These forests are often called dry tropical forests because they are dry much of the year; however, there must be some periodic rain to support tree growth. Many of the trees and shrubs in a seasonal forest are drought-deciduous: they lose their leaves and cease growing when no water is available. Seasonal forests are often open woodlands that grade into savannas.

Tropical dry forests have typically been more attractive than wet forests for human habitation and have suffered greater degradation. Clearing a dry forest with fire is relatively easy during the dry season. Soils of dry forests often have higher nutrient levels and are more agriculturally productive than those of a rainforest. Finally, having fewer insects, parasites, and fungal diseases than a wet forest makes a dry or seasonal forest a healthier place for humans to live. Consequently, these forests are highly endangered in many places. Less than 1 percent of the dry tropical forests of the Pacific coast of Central America or the Atlantic coast of South America, for instance, remain in an undisturbed state.

FIGURE 5.6 Tropical savannas and grasslands experience annual drought and rainy seasons and year-round warm temperatures. Thorny acacias and abundant grazers thrive in this savanna. Yellow areas show moisture deficit.

30 cm of rain per year. Adaptations to these conditions include water-storing leaves and stems, thick epidermal layers to reduce water loss, and salt tolerance. As in other dry environments, many plants are drought-deciduous. Most desert plants also bloom and set seed quickly when a spring rain does fall.

Warm, dry, high-pressure climate conditions (chapter 9) create desert regions at about 30° north and south. Extensive deserts occur in continental interiors (far from oceans, which evaporate the moisture for most precipitation) of North America, Central Asia, Africa, and Australia (fig. 5.7). The rain shadow of the Andes produces the world's driest desert in coastal Chile. Deserts can also be cold. Antarctica is a desert; some inland valleys apparently get almost no precipitation at all.

Like plants, animals in deserts are specially adapted. Many are nocturnal, spending their days in burrows to avoid the sun's heat and desiccation. Pocket mice, kangaroo rats, and gerbils can get most of their moisture from seeds and plants. Desert rodents also have highly concentrated urine and nearly dry feces, which allow them to eliminate body waste without losing precious moisture.

Deserts are more vulnerable than you might imagine. Sparse, slow-growing vegetation is quickly damaged by off-road vehicles. Desert soils recover slowly. Tracks left by army tanks practicing in California deserts during World War II can still be seen today.

Deserts are also vulnerable to overgrazing. In Africa's vast Sahel (the southern edge of the Sahara Desert), livestock are destroying much of the plant cover. Bare, dry soil becomes drifting sand, and restabilization is extremely difficult. Without plant roots and organic matter, the soil loses its ability to retain what rain does fall, and the land becomes progressively drier and more bare. Similar depletion of dryland vegetation is happening in many desert areas, including Central Asia, India, and the American Southwest and Plains states.

Temperate grasslands have rich soils

As in tropical latitudes, temperate (midlatitude) grasslands occur where there is enough rain to support abundant grass but not enough for forests (fig. 5.8). Usually grasslands are a complex, diverse mix of grasses and flowering herbaceous plants, generally known as forbs. Myriad flowering forbs make a grassland colorful and lovely in summer. In dry grasslands, vegetation may be less than a meter tall. In more humid areas, grasses can exceed 2 m. Where scattered trees occur in a grassland, we call it a savanna.

Deep roots help plants in temperate grasslands and savannas survive drought, fire, and extreme heat and cold. These roots, together with an annual winter accumulation of dead leaves on the surface, produce thick, organic-rich soils in temperate grasslands. Because of this rich soil, many grasslands have been converted to farmland. The legendary tallgrass prairies of the central United States and Canada are almost completely replaced by corn, soybeans, wheat, and other crops. Most remaining grasslands in this region are too dry to support agriculture, and their greatest threat

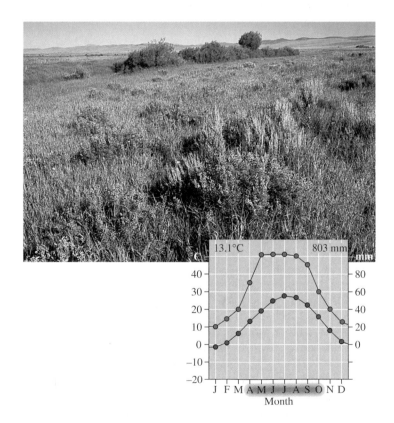

FIGURE 5.7 Deserts generally receive less than 300 mm (30 cm) of precipitation per year. Hot deserts, as in the American Southwest, endure year-round drought and extreme heat in summer.

FIGURE 5.8 Grasslands occur at midlatitudes on all continents. Kept open by extreme temperature, dry conditions, and periodic fires, grasslands can have surprisingly high plant and animal diversity.

Principles of Environmental Science

is overgrazing. Excessive grazing eventually kills even deep-rooted plants. As ground cover dies off, soil erosion results, and unpalatable weeds, such as cheatgrass or leafy spurge, spread.

Temperate shrublands have summer drought

Often, dry environments support drought-adapted shrubs and trees, as well as grass. These mixed environments can be highly variable. They can also be very rich biologically. Such conditions are often described as Mediterranean (where the hot season coincides with the dry season producing hot, dry summers and cool, moist winters). Evergreen shrubs with small, leathery, sclerophyllous (hard, waxy) leaves form dense thickets. Scrub oaks, drought-resistant pines, or other small trees often cluster in sheltered valleys. Periodic fires burn fiercely in this fuel-rich plant assemblage and are a major factor in plant succession. Annual spring flowers often bloom profusely, especially after fires. In California, this landscape is called **chaparral,** Spanish for thicket. Resident animals are drought tolerant species, such as jackrabbits, kangaroo rats, mule deer, chipmunks, lizards, and many bird species. Very similar landscapes are found along the Mediterranean coast as well as southwestern Australia, central Chile, and South Africa. Although this biome doesn't cover a very large total area, it contains a high number of unique species and is often considered a "hot-spot" for biodiversity. It also is highly desired for human habitation, often leading to conflicts with rare and endangered plant and animal species.

Temperate forests can be evergreen or deciduous

Temperate, or midlatitudes, forests occupy a wide range of precipitation conditions but occur mainly between about 30 and 55 degrees latitude (see fig. 5.2). In general we can group these forests by tree type, which can be broad-leaved **deciduous** (losing leaves seasonally) or evergreen **coniferous** (cone-bearing).

Deciduous forests

Broad-leaf forests occur throughout the world where rainfall is plentiful. In midlatitudes, these forests are deciduous and lose their leaves in winter. The loss of green chlorophyll pigments can produce brilliant colors in these forests in autumn (fig. 5.9). At lower latitudes, broad-leaf forests may be evergreen or drought-deciduous. Southern live oaks, for example, are broad-leaf evergreen trees.

Although these forests have a dense canopy in summer, they have a diverse understory that blooms in spring, before the trees leaf out. Spring ephemeral (short-lived) plants produce lovely flowers, and vernal (springtime) pools support amphibians and insects. These forests also shelter a great diversity of songbirds.

North American deciduous forests once covered most of what is now the eastern half of the United States and southern Canada. Most of western Europe was once deciduous forest but was cleared a thousand years ago. When European settlers first came to North America, they quickly settled and cut most of the eastern deciduous forests for firewood, lumber, and industrial uses, as well as to

FIGURE 5.9 Temperature deciduous forests have year-round precipitation and winters near or below freezing.

clear farmland. Many of those regions have now returned to deciduous forest, though the dominant species have changed.

Deciduous forests can regrow quickly because they occupy moist, moderate climates. But most of these forests have been occupied so long that human impacts are extensive, and most native species are at least somewhat threatened. The greatest threat to broad-leaf deciduous forests is in eastern Siberia, where deforestation is proceeding rapidly. Siberia may have the highest deforestation rate in the world. As forests disappear, so do Siberian tigers, bears, cranes, and a host of other endangered species.

Coniferous forests

Coniferous forests grow in a wide range of temperature and moisture conditions. Often they occur where moisture is limited. In cold climates, moisture is unavailable (frozen) in winter; hot climates may have seasonal drought; sandy soils hold little moisture, and they are often occupied by conifers. Thin, waxy leaves (needles) help these trees reduce moisture loss. Coniferous forests provide most wood products in North America. Dominant wood production regions include the southern Atlantic and Gulf coast states, the mountain West, and the Pacific Northwest (northern California to Alaska), but coniferous forests support forestry in many regions.

The coniferous forests of the Pacific coast grow in extremely wet conditions. The wettest coastal forests are known as **temperate rainforest,** a cool, rainy forest often enshrouded in fog

FIGURE 5.11 Boreal forests have moderate precipitation but are often moist because temperatures are cold most of the year. Cold-tolerant and drought-tolerant conifers dominate boreal forests and taiga, a the forest fringe.

FIGURE 5.10 Temperate rainforests have abundant but often seasonal precipitation that supports magnificent trees and luxuriant understory vegetation. Oten these forests experience dry summers.

(fig. 5.10). Condensation in the canopy (leaf drip) is a major form of precipitation in the understory. Mild year-round temperatures and abundant rainfall, up to 250 cm (100 in.) per year, result in luxuriant plant growth and giant trees such as the California redwoods, the largest trees in the world and the largest aboveground organism ever known to have existed. Redwoods once grew along the Pacific coast from California to Oregon, but logging has reduced them to a few small fragments.

Boreal forests lie north of the temperate zone

Because conifers can survive winter cold, they tend to dominate the **boreal forest,** or northern forests, that lie between about 50° and 60° north (fig. 5.11). Mountainous areas at lower latitudes may also have many characteristics and species of the boreal forest. Dominant trees are pines, hemlocks, spruce, cedar, and fir. Some deciduous trees are also present, such as maples, birch, aspen, and alder. These forests are slow-growing because of the cold temperatures and short frost-free growing season, but they are still an expansive resource. In Siberia, Canada, and the western United States, large regional economies depend on boreal forests.

The extreme, ragged edge of the boreal forest, where forest gradually gives way to open tundra, is known by its Russian name, **taiga.** Here extreme cold and short summer limits the growth rate of trees. A 10 cm diameter tree may be over 200 years old in the far north.

Tundra can freeze in any month

Where temperatures are below freezing most of the year, only small, hardy vegetation can survive. **Tundra,** a treeless landscape that occurs at high latitudes or on mountaintops, has a growing season of only two to three months, and it may have frost any month of the year. Some people consider tundra a variant of grasslands because it has no trees; others consider it a very cold desert because water is unavailable (frozen) most of the year.

Arctic tundra is an expansive biome that has low productivity because it has a short growing season. During midsummer, however, 24-hour sunshine supports a burst of plant growth and an explosion of insect life. Tens of millions of waterfowl, shorebirds, terns, and songbirds migrate to the arctic every year to feast on the

FIGURE 5.12 Tundra vegetation grows slowly in the short growing season that occurs at high latitudes and high altitudes (alpine tundra). Plant diversity is relatively low, and frost can occur even in summer.

abundant invertebrate and plant life and to raise their young on the brief bounty. These birds then migrate to wintering grounds, where they may be eaten by local predators—effectively they carry energy and protein from high latitudes to low latitudes. Arctic tundra is essential for global biodiversity, especially for birds.

Alpine tundra, occurring on or near mountaintops, has environmental conditions and vegetation similar to arctic tundra (fig. 5.12). These areas have a short, intense growing season. Often one sees a splendid profusion of flowers in alpine tundra; everything must flower at once in order to produce seeds in a few weeks before the snow comes again. Many alpine tundra plants also have deep pigmentation and leathery leaves to protect against the strong ultraviolet light in the thin mountain atmosphere.

Compared to other biomes, tundra has relatively low diversity. Dwarf shrubs, such as willows, sedges, grasses, mosses, and lichens tend to dominate the vegetation. Migratory musk-ox, caribou, or alpine mountain sheep and mountain goats can live on the vegetation because they move frequently to new pastures.

Because these environments are too cold for most human activities, they are not as badly threatened as other biomes. There are important problems, however. Global climate change may be altering the balance of some tundra ecosystems, and air pollution from distant cities tends to accumulate at high latitudes (chapter 9). In eastern Canada, coastal tundra is being badly depleted by overabundant populations of snow geese, whose numbers have exploded due to winter grazing on the rice fields of Arkansas and

Louisiana. Oil and gas drilling—and associated truck traffic—threatens tundra in Alaska and Siberia. Clearly, this remote biome is not independent of human activities at lower latitudes.

MARINE ECOSYSTEMS

The biological communities in oceans and seas are poorly understood, but they are probably as diverse and as complex as terrestrial biomes. In this section, we will explore a few facets of these fascinating environments. Oceans cover nearly three-fourths of the earth's surface, and they contribute in important, although often unrecognized, ways to terrestrial ecosystems. Like land-based systems, most marine communities depend on photosynthetic organisms. Often it is algae or tiny, free-floating photosynthetic plants (**phytoplankton**) that support a marine food web, rather than the trees and grasses we see on land. In oceans, photosynthetic activity tends to be greatest near coastlines, where nitrogen, phosphorus, and other nutrients wash offshore and fertilize primary producers. Ocean currents also contribute to the distribution of biological productivity, as they transport nutrients and phytoplankton far from shore (fig. 5.13).

As plankton, algae, fish, and other organisms die, they sink toward the ocean floor. Deep-ocean ecosystems, consisting of crabs, filter-feeding organisms, strange phosphorescent fish, and many other life-forms, often rely on this "marine snow" as a primary nutrient source. Surface communities also depend on this material. Upwelling currents circulate nutrients from the ocean floor back to the surface. Along the coasts of South America, Africa, and Europe, these currents support rich fisheries.

Vertical stratification is a key feature of aquatic ecosystems. Light decreases rapidly with depth, and communities below the photic zone (light zone, often reaching about 20 m deep) must rely on energy sources other than photosynthesis to persist. Temperature also decreases with depth. Deep-ocean species often grow slowly in part because metabolism is reduced in cold conditions. In contrast, warm, bright, near-surface communities, such as coral reefs and estuaries, are among the world's most biologically productive environments. Temperature also affects the amount of oxygen and other elements that can be absorbed in water. Cold water holds abundant oxygen, so productivity is often high in cold oceans, as in the North Atlantic, North Pacific, and Antarctic.

Open ocean communities vary from surface to hadal zone

Ocean systems can be described by depth and proximity to shore (fig. 5.14). In general, **benthic** communities occur on the bottom, and **pelagic** (from "sea" in Greek) zones are the water column. The epipelagic zone (*epi* = on top) has photosynthetic organisms. Below this are the mesopelagic (*meso* = medium) and bathypelagic (*bathos* = deep) zones. The deepest layers are the abyssal zone (to 4,000 m) and hadal zone (deeper than 6,000 m). Shorelines are known as littoral zones, and the area exposed by low tides is known as the intertidal zone. Often there is a broad,

FIGURE 5.13 Satellite measurements of chlorophyll levels in the oceans and on land. Dark green to blue land areas have high biological productivity. Dark blue oceans have little chlorophyll and are biologically impoverished. Light green to yellow ocean zones are biologically rich.

FIGURE 5.14 Light penetrates only the top 10–20 m of the ocean. Below this level, temperatures drop and pressure increases. Nearshore environments include the intertidal zone and estuaries.

relatively shallow region along a continent's coast, which may reach a few kilometers or hundreds of kilometers from shore. This undersea area is the continental shelf.

We know relatively little about marine ecosystems and habitats, and much of what we know we have learned only recently. The open ocean has long been known as a biological desert, because it has relatively low productivity, or biomass production. Fish and plankton abound in many areas, however. Sea mounts, or undersea mountain chains and islands, support many commercial fisheries and much newly discovered biodiversity. In the equatorial Pacific and Antarctic oceans, currents carry nutrients far from shore, supporting biological productivity (fig. 5.15). The Sargasso Sea, a large region of the Atlantic near Bermuda, is known for its free-floating mats of brown algae. These algae mats support a phenomenal diversity of animals, including sea turtles, fish, and other species. Eels that hatch amid the algae eventually migrate up rivers along the Atlantic coasts of North America and Europe.

Deep-sea thermal vent communities are another remarkable type of marine system that was completely unknown until 1977, when the deep-sea submarine *Alvin* descended to the deep-ocean floor. These communities are based on microbes that capture chemical energy, mainly from sulfur compounds released from thermal vents—jets of hot water and minerals on the ocean floor. Magma below the ocean crust heats these vents. Tube worms, mussels, and microbes on the vents are adapted to survive both extreme temperatures, often above 350°C (700°F), and intense water pressure at depths of 7,000 m (20,000 ft) or more. Oceanographers have discovered thousands of different types of

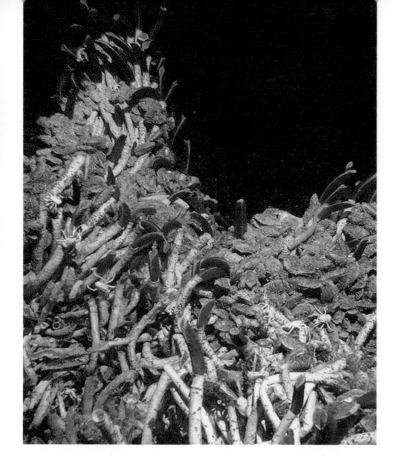

FIGURE 5.15 Deep-ocean thermal vent communities were discovered only recently. The have great diversity and are unusual because they rely on chemosynthesis, not photosynthesis, for energy.

organisms, most of them microscopic, in these communities. Some estimate that the total mass of microbes on the seafloor represents one-third of all biomass on the planet. For further discussion of these mysterious communities, see chapter 2.

Tidal shores support rich, diverse communities

As in the open ocean, shoreline communities vary with depth, light, and temperature. Some shoreline communities, such as estuaries, have high biological productivity and diversity because they are enriched by nutrients washing from the land. Others, such as coral reefs, occur where there is little runoff from shore, but where shallow, clear warm water supports photosynthesis.

Coral reefs are among the best-known marine systems, because of their extraordinary biological productivity and their diverse and beautiful organisms (fig. 5.16a). Reefs are colonies of minute, colonial animals ("coral polyps") that live symbiotically with photosynthetic algae. Calcium-rich coral skeletons shelter the algae, and algae nourish the coral animals. The complex structure of a reef also shelters countless species of fish, worms, crustaceans, and other life-forms. Reefs occur where the water is shallow and clear enough for sunlight to reach the photosynthetic algae. They cannot tolerate abundant nutrients in the water, as nutrients support tiny floating plants and animals called plankton, which block sunlight.

Reefs are among the most endangered biological communities. Sediment from coastal development, farming, sewage, or other pollution can reduce water clarity and smother coral. Destructive fishing practices, including dynamite and cyanide poison, have destroyed many Asian reefs. Reefs can also be damaged or killed by changes in temperature, by invasive fish, and by diseases. **Coral bleaching,** the whitening of reefs due to stress, often followed by coral death, is a growing and spreading problem that worries marine biologists (see chapter 1).

Sea-grass beds, or eel-grass beds, occupy shallow, warm, sandy coastlines. Like reefs, these support rich communities of grazers, from snails to turtles to Florida's manatees.

Mangroves are a diverse group of salt-tolerant trees that grow along warm, calm marine coasts around the world (fig. 5.16b). Growing in shallow, tidal mudflats, mangroves help stabilize shorelines, blunt the force of storms, and build land by trapping sediment and organic material. After the devastating Indonesian tsunami of 2004, studies showed that mangroves, where they still stood, helped reduce the speed, height, and turbulence of the tsunami waves. Detritus, including fallen leaves, collects below mangroves and provides nutrients for a diverse community of animals and plants. Both marine species (such as crabs and fish) and terrestrial species (such as birds and bats) rely on mangroves for shelter and food.

Like coral reefs and sea-grass beds, mangrove forests provide sheltered nurseries for juvenile fish, crabs, shrimp, and other marine species on which human economies depend. However, like reefs and sea-grass beds, mangroves have been devastated by human activities. More than half of the world's mangroves that stood a century ago, perhaps 22 million ha, have been destroyed or degraded. They are clear-cut for timber or cleared to make room for fish and shrimp ponds. They are also poisoned by sewage and industrial waste near cities. Some parts of Southeast Asia and South America have lost 90 percent of their mangrove forests. Most have been cleared for fish and shrimp farming.

Estuaries are bays where rivers empty into the sea, mixing fresh water with salt water. **Salt marshes,** shallow wetlands flooded regularly or occasionally with seawater, occur on shallow coastlines, including estuaries (fig. 5.16c). Usually calm, warm, and nutrient-rich, estuaries and salt marshes are biologically diverse and productive. Rivers provide nutrients and sediments, and a muddy bottom supports emergent plants (whose leaves emerge above the water surface), as well as the young forms of crustaceans, such as crabs and shrimp, and mollusks, such as clams and oysters. Nearly two-thirds of all marine fish and shellfish rely on estuaries and saline wetlands for spawning and juvenile development.

Estuaries near major American cities once supported an enormous wealth of seafood. Oyster beds and clam banks in the waters adjacent to New York, Boston, and Baltimore provided free and easy food to early residents. Sewage and other contaminants long

(a) Coral reefs

(b) Mangroves

(c) Estuary and salt marsh

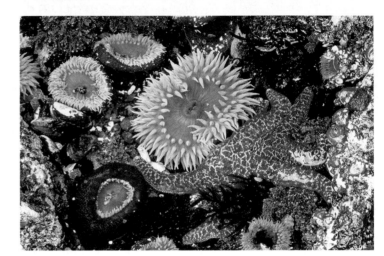

(d) Tide pool

FIGURE 5.16 Coastal environments support incredible diversity and help stabilize shorelines. Coral reefs (a), mangroves (b), and estuaries (c) also provide critical nurseries for marine ecosystems. Tide pools (d) also shelter highly specialized organisms.

ago eliminated most of these resources, however. Recently, major efforts have been made to revive Chesapeake Bay, America's largest and most productive estuary. These efforts have shown some success, but many challenges remain (see related story "Restoring the Chesapeake" at www.mhhe.com/cunningham4e).

In contrast to the shallow, calm conditions of estuaries, coral reefs, and mangroves, there are violent, wave-blasted shorelines that support fascinating life-forms in **tide pools.** Tide pools are depressions in a rocky shoreline that are flooded at high tide but retain some water at low tide. These areas remain rocky where wave action prevents most plant growth or sediment (mud) accumulation. Extreme conditions, with frigid flooding at high tide and hot, dessicating sunshine at low tide, make life impossible for most species. But the specialized animals and plants that do

occur in this rocky intertidal zone are astonishingly diverse and beautiful (fig. 5.16d).

FRESHWATER ECOSYSTEMS

Freshwater environments are far less extensive than marine environments, but they are centers of biodiversity. Most terrestrial communities rely, to some extent, on freshwater environments. In deserts, isolated pools, streams, and even underground water systems support astonishing biodiversity as well as provide water to land animals. In Arizona, for example, many birds are found in trees and bushes surrounding the few available rivers and streams.

Principles of Environmental Science http://www.mhhe.com/cunningham4e

Lakes have open water

Freshwater lakes, like marine environments, have distinct vertical zones (fig. 5.17). Near the surface a subcommunity of plankton, mainly microscopic plants, animals, and protists (single-celled organisms, such as amoebae), float freely in the water column. Insects such as water striders and mosquitoes also live at the air-water interface. Fish move through the water column, sometimes near the surface and sometimes at depth.

Finally, the bottom, or *benthos,* is occupied by a variety of snails, burrowing worms, fish, and other organisms. These make up the benthic community. Oxygen levels are lowest in the benthic environment, mainly because there is little mixing to introduce oxygen to this zone. Anaerobic (not using oxygen) bacteria may live in low-oxygen sediments. In the littoral zone, emergent plants, such as cattails and rushes, grow in the bottom sediment. These plants create important functional links between layers of an aquatic ecosystem, and they may provide the greatest primary productivity to the system.

Lakes, unless they are shallow, have a warmer upper layer that is mixed by wind and warmed by the sun. This layer is the *epilimnion*. Below the epilimnion is the hypolimnion (*hypo* = below), a colder, deeper layer that is not mixed. If you have gone swimming in a moderately deep lake, you may have discovered the sharp temperature boundary, known as the **thermocline,** between these layers. Below this boundary, the water is much colder. This boundary is also called the mesolimnion.

Local conditions that affect the characteristics of an aquatic community include (1) nutrient availability (or excess), such as nitrates and phosphates; (2) suspended matter, such as silt, that affects light penetration; (3) depth; (4) temperature; (5) currents; (6) bottom characteristics, such as muddy, sandy, or rocky floor; (7) internal currents; and (8) connections to, or isolation from, other aquatic and terrestrial systems.

FIGURE 5.17 The layers of a deep lake are determined mainly by gradients of light, oxygen, and temperature. The epilimnion is affected by surface mixing from wind and thermal convections, while mixing between the hypolimnion and epilimnion is inhibited by a sharp temperature and density difference at the thermocline.

Wetlands are shallow and productive

Wetlands are shallow ecosystems in which the land surface is saturated or submerged at least part of the year. Wetlands have vegetation that is adapted to grow under saturated conditions. These legal definitions are important because, although wetlands make up only a small part of most countries, they are disproportionately important in conservation debates and are the focus of continual legal disputes in North America and elsewhere around the world. Beyond these basic descriptions, defining wetlands is a matter of hot debate. How often must a wetland be saturated, and for how long? How large must it be to deserve legal protection? Answers can vary, depending on political, as well as ecological, concerns.

These relatively small systems support rich biodiversity, and they are essential for both breeding and migrating birds. Although wetlands occupy less than 5 percent of the land in the United States, the Fish and Wildlife Service estimates that one-third of all endangered species spend at least part of their lives in wetlands. Wetlands retain storm water and reduce flooding by slowing the rate at which rainfall reaches river systems. Floodwater storage is worth $3 billion to $4 billion per year in the United States. As water stands in wetlands, it also seeps into the ground, replenishing groundwater supplies. Wetlands filter, and even purify, urban and farm runoff, as bacteria and plants take up nutrients and contaminants in water. They are also in great demand for filling and development. They are often near cities or farms, where land is valuable, and, once drained, wetlands are easily converted to more lucrative uses. At least half of all the existing wetlands in the United States when Europeans first arrived have been drained, filled, or degraded. In some major farming states, losses have been even greater. Iowa, for example, has lost 99 percent of its original wetlands.

Wetlands are described by their vegetation. **Swamps,** also called forested wetlands, are wetlands with trees. **Marshes** are wetlands without trees (fig. 5.18). **Bogs** are areas of water-saturated ground, and usually the ground is composed of deep layers of accumulated, undecayed vegetation known as peat. **Fens** are similar to bogs except that they are mainly fed by groundwater, so that they have mineral-rich water and specially adapted plant species. Bogs are fed mainly by precipitation. Swamps and marshes have high biological productivity. Bogs and fens, which are often nutrient-poor, have low biological productivity. They may have unusual and interesting species, though, such as sundews and pitcher plants, which are adapted to capture nutrients from insects rather than from soil.

The water in marshes and swamps usually is shallow enough to allow full penetration of sunlight and seasonal warming. These mild conditions favor great photosynthetic activity, resulting in high productivity at all trophic levels. In short, life is abundant and varied. Wetlands are major breeding, nesting, and migration staging areas for waterfowl and shorebirds.

(a) Swamp, or wooded wetland

(b) Marsh

(c) Bog

FIGURE 5.18 Wetlands provide irreplaceable ecological services, including water filtration, water storage and flood reduction, and habitat. Forested wetlands (a) are often called swamps; marshes (b) have no trees; bogs (c) are acidic and accumulate peat.

Streams and rivers are open systems

Streams form wherever precipitation exceeds evaporation and surplus water drains from the land. Within small streams, ecologists distinguish areas of riffles, where water runs rapidly over a rocky substrate, and pools, which are deeper stretches of slowly moving current. Water tends to be well mixed and oxygenated in riffles; pools tend to collect silt and organic matter. If deep enough, pools can have vertical zones similar to those of lakes. As streams collect water and merge, they form rivers, although there isn't a universal definition of when one turns into the other. Ecologists consider a river system to be a continuum of constantly changing environmental conditions and community inhabitants from the headwaters to the mouth of a drainage or watershed. The biggest distinction between stream and lake ecosystems is that, in a stream, materials, including plants, animals, and water, are continually moved downstream by flowing currents. This downstream drift is offset by active movement of animals upstream, productivity in the stream itself, and input of materials from adjacent wetlands or uplands.

BIODIVERSITY

The biomes you've just learned about shelter an astounding variety of living organisms. From the driest desert to the dripping rainforests, from the highest mountain peaks to the deepest ocean trenches, life occurs in a marvelous spectrum of sizes, colors, shapes, life cycles, and interrelationships. The varieties of organisms and complex ecological relationships give the biosphere its unique, productive characteristics. **Biodiversity,** the variety of living things, also makes the world a more beautiful and exciting place to live. Three kinds of biodiversity are essential to preserve ecological systems and functions: (1) *genetic diversity* is a measure of the variety of versions of the same genes within individual species; (2) *species diversity* describes the number of different kinds of organisms within individual communities or ecosystems; and (3) *ecological diversity* means the richness and complexity of a biological community, including the number of niches, trophic levels, and ecological processes that capture energy, sustain food webs, and recycle materials within this system.

Increasingly, we identify species by genetic similarity

Species are a fundamental concept in understanding biodiversity, but what is a species? In general, species are distinct organisms that persist because they can produce fertile offspring. But many organisms reproduce asexually; others don't reproduce in nature just because they don't normally encounter one another. Because of such ambiguities, evolutionary biologists favor the **phylogenetic species concept,** which identifies genetic similarity. Alternatively, the **evolutionary species concept** defines species according to evolutionary history and common ancestors. Both of these approaches rely on DNA analysis to define similarity among organisms.

TABLE 5.1

Current Estimates of Known Living Species by Taxonomic Group

Mammals	5,416
Birds	9,917
Reptiles	8,163
Amphibians	5,743
Fishes	28,500
Insects	950,000
Mollusks	70,000
Crustaceans	40,000
Other invertebrates	130,200
Ferns, mosses, and lichens	38,025
Gymnosperms	980
Grasses	59,300
Flowering plants	199,350
Total	1,545,594

Source: IUCN, 2005.

How many organisms are there? Biologists have identified about 1.5 million species, but these probably represent only a small fraction of the actual number (table 5.1). Based on the rate of new discoveries by research expeditions—especially in the tropics—taxonomists estimate that somewhere between 3 million and

FIGURE 5.19 Insects and other invertebrates make up more than half of all known species. Many, like this blue morpho butterfly, are beautiful as well as ecologically important.

50 million different species may be alive today. About 70 percent of all known species are invertebrates (animals without backbones, such as insects, sponges, clams, and worms) (fig. 5.19). This group probably makes up the vast majority of organisms yet to be discovered and may constitute 90 percent of all species.

Biodiversity hot spots are rich and threatened

Most of the world's biodiversity concentrations are near the equator, especially tropical rainforests and coral reefs (fig. 5.20). Of all the world's species, only 10 to 15 percent live in North America

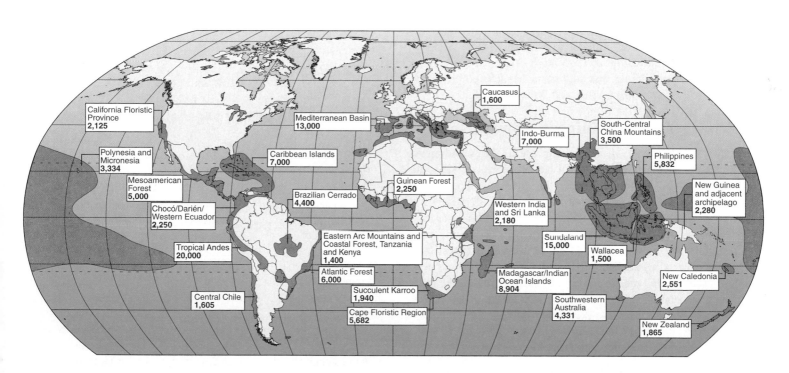

FIGURE 5.20 Biodiversity "hot spots" identified by Conservation International tend to be in tropical or Mediterranean climates and on islands, coastlines, or mountains where many habitats exist and physical barriers encourage speciation. Numbers represent estimated endemic (locally unique) species in each area.
Source: Data from Conservation International.

and Europe. Many of the organisms in megadiversity countries have never been studied by scientists. The Malaysian Peninsula, for instance, has at least 8,000 species of flowering plants, while Britain, with an area twice as large, has only 1,400 species. There may be more botanists in Britain than there are species of higher plants. South America, on the other hand, has fewer than 100 botanists to study perhaps 200,000 species of plants.

Areas isolated by water, deserts, or mountains can also have high concentrations of unique species and biodiversity. Madagascar, New Zealand, South Africa, and California are all midlatitude areas isolated by barriers that prevent mixing with biological communities from other regions and produce rich, unusual collections of species.

BENEFITS OF BIODIVERSITY

We benefit from other organisms in many ways, some of which we don't appreciate until a particular species or community disappears. Even seemingly obscure and insignificant organisms can play irreplaceable roles in ecological systems or be the source of genes or drugs that someday may be indispensable.

All of our food comes from other organisms

Many wild plant species could make important contributions to human food supplies either as they are or as a source of genetic material to improve domestic crops. Noted tropical ecologist Norman Myers estimates that as many as 80,000 edible wild plant species could be utilized by humans. Villagers in Indonesia, for instance, are thought to use some 4,000 native plant and animal species for food, medicine, and other valuable products. Few of these species have been explored for possible domestication or more widespread cultivation. A 1975 study by the U.S. National Academy of Science found that Indonesia has 250 edible fruits, only 43 of which have been cultivated widely (fig. 5.21).

Rare species provide important medicines

Living organisms provide us with many useful drugs and medicines (table 5.2). More than half of all prescriptions contain some natural products. The United Nations Development Programme estimates the value of pharmaceutical products derived from developing world plants, animals, and microbes to be more than $30 billion per year.

Consider the success story of vinblastine and vincristine. These anticancer alkaloids are derived from the Madagascar periwinkle (*Catharanthus roseus*). They inhibit the growth of cancer cells and are very effective in treating certain kinds of cancer. Twenty years ago, before these drugs were introduced, childhood leukemias were invariably fatal. Now the remission rate for some childhood leukemias is 99 percent. Hodgkin's disease was 98 percent fatal a few years ago but is now only 40 percent fatal, thanks to these compounds. The total value of the periwinkle crop is roughly $15 million per year, although Madagascar gets little of those profits.

FIGURE 5.21 Mangosteens from Indonesia have been called the world's best-tasting fruit, but they are practically unknown beyond the tropical countries where they grow naturally. There may be thousands of other traditional crops and world food resources that could be equally valuable but are threatened by extinction.

Biodiversity can support ecosystem stability

Human life is inextricably linked to ecological services provided by other organisms. Soil formation, waste disposal, air and water purification, nutrient cycling, solar energy absorption, and food production all depend on biodiversity (see chapters 2 and 7). Total value

TABLE 5.2

Some Natural Medicinal Products

Product	Source	Use
Penicillin	Fungus	Antibiotic
Bacitracin	Bacterium	Antibiotic
Tetracycline	Bacterium	Antibiotic
Erythromycin	Bacterium	Antibiotic
Digitalis	Foxglove	Heart stimulant
Quinine	Chincona bark	Malaria treatment
Diosgenin	Mexican yam	Birth control drug
Cortisone	Mexican yam	Anti-inflammation treatment
Cytarabine	Sponge	Leukemia cure
Vinblastine, vincristine	Periwinkle plant	Anticancer drugs
Reserpine	Rauwolfia	Hypertension drug
Bee venom	Bee	Arthritis relief
Allantoin	Blowfly larva	Wound healer
Morphine	Poppy	Analgesic

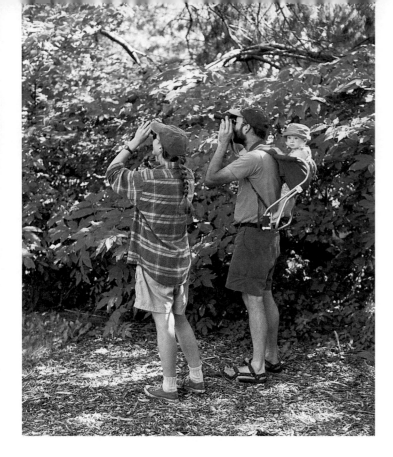

FIGURE 5.22 Birdwatching and other wildlife observation contribute more than $29 million each year to the U.S. economy.

of these ecological services is at least $33 trillion per year, or more than double total world GNP. In many environments, high diversity may help biological communities withstand environmental stress better and recover more quickly than those with fewer species.

Because we don't fully understand the complex interrelationships between organisms, we often are surprised and dismayed at the effects of removing seemingly insignificant members of biological communities. For instance, it is estimated that 95 percent of the potential pests and disease-carrying organisms in the world are controlled by natural predators and competitors. Maintaining biodiversity is essential to preserving these ecological services.

Biodiversity has aesthetic and cultural benefits

Millions of people enjoy hunting, fishing, camping, hiking, wildlife watching, and other nature-based activities. These activities provide invigorating physical exercise, and contact with nature can be psychologically and emotionally restorative. In many cultures, nature carries spiritual connotations, and a particular species or landscape may be inextricably linked to a sense of identity and meaning. Observing and protecting nature has religious or moral significance for many people. Some religious organizations call for the protection of nature simply because it is God's creation.

Nature appreciation is economically important. The U.S. Fish and Wildlife Service estimates that Americans spend $104 billion every year on wildlife-related recreation (fig. 5.22). This compares

to $81 billion spent each year on new automobiles. Forty percent of all adults enjoy wildlife, including 39 million who hunt or fish and 76 million who watch, feed, or photograph wildlife. Many communities are finding that local biodiversity can bring cash to remote areas through ecotourism (see chapter 6).

For many people, the value of wildlife goes beyond the opportunity to photograph or shoot a particular species. They argue that existence value, the value of simply knowing that a species exists, is reason enough to protect and preserve it. Even if we never see a whooping crane or a redwood tree, many of us are happy to know that these species have not been destroyed.

WHAT THREATENS BIODIVERSITY?

Extinction, the elimination of a species, is a normal process of the natural world. Species die out and are replaced by others, often their own descendants, as part of evolutionary change. In undisturbed ecosystems, the rate of extinction appears to be about one species lost every decade. Over the past century, however, human impacts on populations and ecosystems have accelerated that rate, possibly causing thousands of species, subspecies, and varieties to become extinct every year. Ecologist E. O. Wilson estimates that we are losing 10,000 species or subspecies a year—that makes more than 27 per day! If present trends continue, we may destroy millions of kinds of plants, animals, and microbes in the next few decades. In this section, we will look at some ways we threaten biodiversity.

In geologic history, extinctions are common. Studies of the fossil record suggest that more than 99 percent of all species that ever existed are now extinct. Most of those species were gone long before humans came on the scene. Periodically, mass extinctions have wiped out vast numbers of species and even whole families (table 5.3). The best studied of these events occurred at the end of the Cretaceous Period, when dinosaurs disappeared, along with at least 50 percent of existing species. An even greater disaster occurred at the end of the Permian Period, about 250 million years

TABLE 5.3

Mass Extinctions

Historic Period	Time (Before Present)	Percent of Species Extinct
Ordovician	444 million	85
Devonian	370 million	83
Permian	250 million	95
Triassic	210 million	80
Cretaceous	65 million	76
Quaternary	Present	33–66

Source: Data from W. W. Gibbs, 2001. "On the termination of species." *Scientific American* 285(5): 40–49.

ago, when 90 percent of species and half of all families died out over a period of about 10,000 years—a mere moment in geologic time. Current theories suggest that these catastrophes were caused by climate changes, perhaps triggered when large asteroids struck the earth. Many ecologists worry that global climate change caused by our release of greenhouse gases in the atmosphere could have similarly catastrophic effects (see chapter 9).

Human activities have sharply increased extinctions

The rate at which species are disappearing has increased dramatically over the past 150 years. Between A.D. 1600 and 1850, human activities appear to have eliminated two or three species per decade, about double the natural extinction rate. In the past 150 years, the extinction rate has increased to thousands per decade. If present trends continue, biologist Paul Ehrlich warns, somewhere between one-third to two-thirds of all current species could be extinct by the middle of the twenty-first century. Conservation biologists call this the sixth mass extinction but note that this time it's not asteroids or volcanoes but human impacts that are responsible. E. O. Wilson summarizes human threats to biodiversity with the acronym **HIPPO,** which stands for Habitat destruction, Invasive species, Pollution, Population (human), and Overharvesting. Let's look in more detail at each of these issues.

Habitat destruction is the main threat for many species

The most important extinction threat for most species—especially terrestrial ones—is habitat loss. Perhaps the most obvious example of habitat destruction is clear-cutting of forests and conversion of grasslands to crop fields (fig. 5.23). Over the past 10,000 years, humans have transformed billions of hectares of former forests and grasslands to croplands, cities, roads, and other uses. These human-dominated spaces aren't devoid of wild organisms, but they generally favor weedy species adapted to coexist with us.

Today, forests cover less than half the area they once did, and only around one-fifth of the original forest retains its old-growth characteristics. Species, such as the northern spotted owl (*Strix occidentalis caurina*), that depend on the varied structure and resources of old-growth forest vanish as their habitat disappears (chapter 6). Grasslands currently occupy about 4 billion ha (roughly equal to the area of closed-canopy forests). Much of the most highly productive and species-rich grasslands—for example, the tallgrass prairie that once covered the U.S. corn belt—has been converted to cropland. Much more may need to be used as farmland or pasture if human populations continue to expand.

Sometimes we destroy habitat as side effects of resource extraction, such as mining, dam-building, and indiscriminate fishing methods. Surface mining, for example, strips off the land covering along with everything growing on it. Waste from mining operations can bury valleys and poison streams with toxic material. Dam-building floods vital stream habitat under deep reservoirs and eliminates food sources and breeding habitat for some

FIGURE 5.23 Decrease in wooded area of Cadiz Township in southern Wisconsin during European settlement. Green areas represent the amount of land in forest each year.

aquatic species. Our current fishing methods are highly unsustainable. One of the most destructive fishing techniques is bottom trawling, in which heavy nets are dragged across the ocean floor, scooping up every living thing and crushing the bottom structure to lifeless rubble. Marine biologist Jan Lubechenco says that trawling is "like collecting forest mushrooms with a bulldozer."

Fragmentation reduces habitat to small, isolated patches

In addition to the total area of lost habitat, another serious problem is habitat **fragmentation**—the reduction of habitat into small, isolated patches. Breaking up habitat reduces biodiversity because many species, such as bears and large cats, require large territories to subsist. Other species, such as forest interior birds, reproduce successfully only in deep forest far from edges and human settlement. Predators and invasive species often spread quickly into new regions following fragment edges.

Fragmentation also divides populations into isolated groups, making them much more vulnerable to catastrophic events, such as storms or diseases. A very small population may not have enough breeding adults to be viable even under normal circumstances. An important question in conservation biology is what is the **minimum viable population** size for various species.

Understanding wolf recovery in Yellowstone (opening Case Study) requires careful monitoring of these animals. Day-to-day movements provide some of the most important clues to an animal's habitat needs, behavior, and biology: How much territory do they need? How fast do they move? What kinds of habitat do they prefer? How readily will they risk crossing roads or human settlements? But wolves are fast-moving, wide-ranging animals, and biologists only spot them occasionally—outside of the camera's range. Beyond special locations such as a camera-monitored meadow, how do they monitor such elusive animals?

Many biologists rely on **telemetry,** or locating animals from a distance, to track the locations of wolves, deer, birds, or other mobile species. The most common method is radio telemetry, which involves placing a radio transmitter on an animal, usually on a collar (fig. 1). The collar radio sends out pulses of radio waves, and the researcher detects the collar's location using a radio receiver. The radio receiver can detect direction, but not distance, to the animal. To pinpoint the animal's location, the researcher must locate the signal from two directions, then triangulate the actual position. By locating an animal day after day, or week after week, biologists can study its behavior, habitat requirements, winter survival, and many other questions.

Radio collars were first developed for large animals, such as bears, deer, and wolves. As radio technology has improved, transmitters have gotten small enough to attach to the backs of songbirds or small rodents. Tiny transmitters are even available that can be glued to the backs of beetles and cockroaches! Depending on the size of the animal, and the weight of batteries that it can easily carry, radios might transmit for a few weeks or more than a year. Depending on vegetation and terrain and the strength of the transmitter, signals may be detectable just within a few hundred meters or many kilometers away.

For animals that move too far or too fast to relocate with a radio receiver, such as migrating birds, polar bears, or sea turtles, biolo-gists are increasingly using satellite telemetry. With satellite telemetry, the collar sends a radio signal that is picked up by a group of satellites that orbit the earth about 850 km above the ground surface. When two or more satellites detect the radio signal, they can triangulate the collar's location, essentially as a biologist would do on the ground. The satellites send data to a receiving station on the ground, which distributes data to researchers for mapping and analysis (fig. 2).

This technology has allowed researchers to map hundreds of species, including detailed records of Yellowstone's wolf packs, and how their territories shift from year to year. Ornithologists have mapped migratory routes for cranes, mapped habitat use by songbirds, and located never-before-discovered arctic breeding grounds for waterfowl. Marine biologists have mapped the seasonal movements of sea turtles, sharks, and seals. The field of telemetry has made stunning advances in recent years and provided entirely new answers to old and difficult questions in wildlife studies.

FIGURE 1   An ecologist carries a radio receiver and antenna to pick up the signal from a radio collar.

FIGURE 2   With satellite telemetry, a satellite sends data to a ground receiver, which sends data to the researcher for mapping.

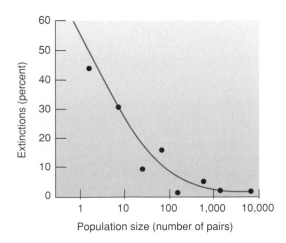

FIGURE 5.24 Extinction rates of bird species on the California Channel Islands as a function of population size over 80 years.
Source: Data from H. L. Jones and J. Diamond, "Short-term-base Studies of Turnover in Breeding Bird Populations on the California Coast Island," in *Condor*, vol. 78:526–549, 1976.

Much of our understanding of fragmentation was elegantly outlined in the theory of **island biogeography,** developed by R. H. MacArthur and E. O. Wilson in the 1960s. Noticing that small islands far from a mainland have fewer terrestrial species than larger, nearer islands, MacArthur and Wilson proposed that species diversity is a balance between colonization and extinction rates. An island far from a population source naturally has a lower rate of colonization than a nearer island because it is harder for terrestrial organisms to reach. At the same time, fewer resources on small islands means that the population of any single species is likely to be small and more vulnerable to extinction. By contrast, a large island can support more individuals of a given species and is, therefore, less vulnerable to natural disasters, genetic problems, or demographic uncertainty (the chance that all the members of a single generation will be of the same sex).

Island biogeographic effects have been observed in many places. Cuba, for instance is 100 times as large and has about 10 times as many amphibian species as its Caribbean neighbor, Montserrat. Similarly, in a study of bird species on the California Channel Islands, Jared Diamond observed that, on islands with fewer than 10 breeding pairs, 39 percent of the populations went extinct over an 80-year period, while only 10 percent of populations between 10 and 100 pairs went extinct in the same time (fig. 5.24). Only one species numbering between 100 and 1,000 pairs went extinct and no species with over 1,000 pairs disappeared over this time.

Many of the parks and wildlife refuges we establish are effectively islands of good habitat surrounded by oceans of inhospitable territory. Like small, remote oceanic islands, they are too isolated to be reached by new migrants, and they can't support large enough populations to survive catastrophic events or genetic problems. Large predators, such as tigers or wolves, need large expanses of contiguous range relatively free of human incursion

to survive. Glacier National Park in Montana, for example, is excellent habitat for grizzly bears. It can support only about 100 bears, however, and if there isn't migration at least occasionally from other areas, this probably isn't a large enough population to survive in the long run.

The challenge of monitoring wildlife in fragmented habitat has led to important new techniques in wildlife research (see Exploring Science, p. 113).

Invasive species are a growing threat

A major threat to native biodiversity in many places is from accidentally or deliberately introduced species. Called a variety of names—alien, exotic, non-native, nonindigenous, unwanted, disruptive, or pests—**invasive species** are organisms that thrive in new territory where they are free of predators, diseases, or resource limitations that may have controlled their population in their native habitat. Although humans have probably transported organisms into new habitats for thousands of years, the rate of movement has increased sharply in recent years with the huge increase in speed and volume of travel by air, water, and land. We move species around the world in a variety of ways. Some are deliberately released because people believe they will be aesthetically pleasing or economically beneficial. Others hitch a ride in ship ballast water, in the wood of packing crates, inside suitcases or shipping containers, in the soil of potted plants, even on people's shoes.

Over the past 300 years, approximately 50,000 non-native species have become established in the United States. Many of these introductions, such as corn, wheat, rice, soybeans, cattle, poultry, and honeybees, have proved to be both socially and economically beneficial. At least 4,500 of these species have established free-living populations, of which 15 percent cause environmental or economic damage (fig. 5.25). Invasive species are estimated to cost the United States $138 billion annually and are forever changing a variety of ecosystems (see What Can You Do? p. 116).

A few important examples of invasive species include the following:

- Eurasian milfoil (*Myriophyllum spicatum* L.) is an exotic aquatic plant native to Europe, Asia, and Africa. Scientists believe that milfoil arrived in North America during the late nineteenth century in shipping ballast. It grows rapidly and tends to form a dense canopy on the water surface, which displaces native vegetation, inhibits water flow, and obstructs boating, swimming, and fishing. Humans spread the plant between water body systems from boats and boat trailers carrying the plant fragments. Herbicides and mechanical harvesting are effective in milfoil control but can be expensive (up to $5,000 per hectare per year). There is also concern that the methods may harm nontarget organisms. A native milfoil weevil, *Euhrychiopsis lecontei,* is being studied as an agent for milfoil biocontrol.

- Water hyacinth (*Eichhornia crassipes*) is a free-floating aquatic plant that grows up to a meter in height. It has

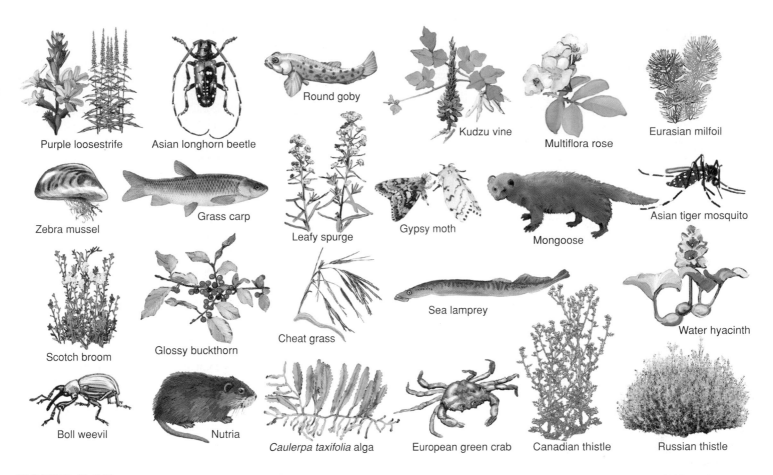

FIGURE 5.25 A few of the approximately 50,000 invasive species in North America. Do you recognize any that occur where you live? What others can you think of?

Image labels: Purple loosestrife, Asian longhorn beetle, Round goby, Kudzu vine, Multiflora rose, Eurasian milfoil, Zebra mussel, Grass carp, Leafy spurge, Gypsy moth, Mongoose, Asian tiger mosquito, Scotch broom, Glossy buckthorn, Cheat grass, Sea lamprey, Water hyacinth, Boll weevil, Nutria, *Caulerpa taxifolia* alga, European green crab, Canadian thistle, Russian thistle

thick, waxy, dark green leaves with bulbous, spongy stalks. It grows a tall spike of lovely flowers with blue, lavender, or pink petals, some of which have yellow splotches. This South American native was introduced into the United States in the 1880s. Its growth rate is among the highest of any plant known: hyacinth populations can double in as little as 12 days. Many lakes and ponds are covered from shore to shore with up to 500 tons of hyacinths per hectare. Besides blocking boat traffic and preventing swimming and fishing, water hyacinth infestations also prevent sunlight and oxygen from getting into the water. Thus, water hyacinth infestations reduce fisheries, shade-out submersed plants, crowd-out immersed plants, and diminish biological diversity. Water hyacinth is controlled by using herbicides, machines, and biocontrol insects.

- Kudzu vine (*Pueraria lobata*) has blanketed large areas of the southeastern United States. Long cultivated in Japan for edible roots, medicines, and fibrous leaves and stems used for paper production, kudzu was introduced by the U.S. Soil Conservation Service in the 1930s to control erosion. Unfortunately, it succeeded too well. In the ideal conditions of its new home, kudzu can grow 18 to 30 m in a single season. Smothering everything in its path, it kills trees, pulls down utility lines, and causes millions of dollars in damage every year.

- Asian tiger mosquitoes (*Aedes albopictus*) are unusually aggressive species that now infest many coastal states in the United States. These species have apparently arrived on container ships carrying used tires, a notorious breeding habitat for mosquitoes. Asian tiger mosquitoes spread West Nile virus (another species introduced with the mosquitoes), which is deadly to many wild birds and occasionally to people and livestock.

- Purple loosestrife (*Lythrum salicaria*) grows in wet soil. Originally cultivated by gardeners for its bright purple flower spikes, this tall wetland plant escaped into New England marshes about a century ago. Spreading rapidly across the Great Lakes, it now fills wetlands across much of the northern United States and southern Canada. Because it crowds out indigenous vegetation and has few native predators or symbionts, it tends to reduce biodiversity wherever it takes hold.

- Zebra mussels (*Dreissena polymorpha*) probably made their way from their home in the Caspian Sea to the Great

What Can You Do?

You Can Help Preserve Biodiversity

Our individual actions are some of the most important obstacles—and most important opportunities—in conserving biodiversity.

Pets and Plants

- Help control invasive species. Never release fish or vegetation from fish tanks into waterways or sewers. Pet birds, cats, dogs, snakes, lizards, and other animals, released by well-meaning owners, are widespread invasive predators.
- Keep your cat indoors. House cats are major predators of woodland birds and other animals.
- Plant native species in your garden. Exotic nursery plants often spread from gardens, compete with native species, and introduce parasites, insects, or diseases that threaten ecosystems. Local-origin species are an excellent, and educational, alternative.
- Don't buy exotic birds, fish, turtles, reptiles, or other pets. These animals are often captured, unsustainably, in the wild. The exotic pet trade harms ecosystems and animals.
- Don't buy rare or exotic house plants. Rare orchids, cacti, and other plants are often collected and sold illegally and unsustainably.

Food and Products

- When buying seafood, inquire about the source. Try to buy species from stable populations. Farm-raised catfish, tilapia, trout, Pacific pollack, Pacific salmon, mahimahi, squid, crabs, and crayfish are some of the stable or managed species that are good to buy. Avoid slow-growing top predators, such as swordfish, marlin, bluefin tuna, and albacore tuna.
- Buy shade-grown coffee and chocolate. These are also organic and often "fair trade" varieties that support workers' families as well as biodiversity in growing regions.
- Buy sustainably harvested wood products. Your local stores will start carrying sustainable wood products if you and your friends ask for them. Persistent consumers are amazingly effective forces of change!

Lakes in ballast water of transatlantic cargo ships, arriving sometime around 1985. Attaching themselves to any solid surface, zebra mussels reach enormous densities—up to 70,000 animals per square meter—covering fish-spawning beds, smothering native mollusks, and clogging utility intake pipes. Found in all the Great Lakes, zebra mussels have moved into the Mississippi River and its tributaries. Public and private costs for zebra mussel removal now amount to some $400 million per year. On the good side, mussels have improved water clarity in Lake Erie at least four-fold by filtering out algae and particulates.

Disease organisms, or pathogens, may also be considered predators. To be successful over the long term, a pathogen must establish a balance in which it is vigorous enough to reproduce, but not so lethal that it completely destroys its host. When a disease is introduced into a new environment, however, this balance may be lacking and an epidemic may sweep through the area.

The American chestnut was once the heart of many eastern hardwood forests. In the Appalachian Mountains, at least one of every four trees was a chestnut. Often over 45 m (150 ft) tall, 3 m (10 ft) in diameter, fast growing, and able to sprout quickly from a cut stump, it was a forester's dream. Its nutritious nuts were important for birds (such as the passenger pigeon), forest mammals, and humans. The wood was straight-grained, light, and rot-resistant, and it was used for everything from fence posts to fine furniture. In 1904 a shipment of nursery stock from China brought a fungal blight to the United States, and within 40 years, the American chestnut had all but disappeared from its native range. Efforts are now underway to transfer blight-resistant genes into the few remaining American chestnuts that weren't reached by the fungus or to find biological controls for the fungus that causes the disease.

The flow of organisms isn't just into the United States; Americans also send exotic species to other places. The Leidy's comb jelly, for example, which is native to the western Atlantic coast, has devastated the Black Sea, now making up more than 90 percent of all biomass at certain times of the year. Similarly, the bristle worm from North America has invaded the coast of Poland and now is almost the only thing living on the bottom of some bays and lagoons. A tropical seaweed named *Caulerpa taxifolia,* originally grown for the aquarium trade, has escaped into the northern Mediterranean, where it covers the shallow seafloor with a dense, meter-deep shag carpet from Spain to Croatia. Producing more than 5,000 leafy fronds per square meter, this aggressive weed crowds out everything in its path. Rarely growing in more than scattered clumps less than 25 cm (10 in.) high in its native habitat, this alga was transformed by aquarium breeding into a supercompetitor that grows over everything and can withstand a wide temperature range. Getting rid of these alien species once they dominate an ecosystem is difficult if not impossible.

Pollution poses many different types of risk

We have known for a long time that toxic pollutants can have disastrous effects on local populations of organisms. Pesticide-linked declines of fish-eating birds and falcons were well documented in the 1970s (fig. 5.26). Marine mammals, alligators, fish, and other declining populations suggest complex interrelations between pollution and health (chapter 8). Mysterious, widespread deaths of thousands of seals on both sides of the Atlantic in recent years are thought to be linked to an accumulation of persistent chlorinated hydrocarbons, such as DDT, PCBs, and dioxins, in fat, causing weakened immune systems that make animals vulnerable to infections. Similarly, mortality of Pacific sea lions, beluga whales in the St. Lawrence estuary, and striped dolphins in the Mediterranean is thought to be caused by accumulation of toxic pollutants.

FIGURE 5.26 Brown pelicans, and other bird species at the top of the food chain, were decimated by DDT in the 1960s. Pelicans and other species have largely recovered since DDT was banned in the United States.

Lead poisoning is another major cause of mortality for many species of wildlife. Bottom-feeding waterfowl, such as ducks, swans, and cranes, ingest spent shotgun pellets that fall into lakes and marshes. They store the pellets, instead of stones, in their gizzards and the lead slowly accumulates in their blood and other tissues. The U.S. Fish and Wildlife Service (USFWS) estimates that 3,000 metric tons of lead shot are deposited annually in wetlands and that between 2 and 3 million waterfowl die each year from lead poisoning.

Human population growth threatens biodiversity

If our consumption patterns remain constant, with more people we will need to harvest more timber, catch more fish, plow more land for agriculture, dig up more fossil fuels and minerals, build more houses, and use more water. All of these demands impact wild species. Unless we find ways to dramatically increase the crop yield per unit area, it will take much more land than is currently domesticated to feed everyone if our population grows to 8 to 10 billion, as current projections predict. This will be especially true if we abandon intensive (but highly productive) agriculture and introduce more sustainable practices. The human population growth curve is leveling off (chapter 4), but it remains unclear whether we can reduce global inequality and provide a tolerable life for all humans while also preserving healthy natural ecosystems and a high level of biodiversity.

Overharvesting has depleted or eliminated many species

Overharvesting involves taking more individuals than reproduction can replace. A classic example is the extermination of the American passenger pigeon (*Ectopistes migratorius*). Even though it inhabited only eastern North America, 200 years ago this was

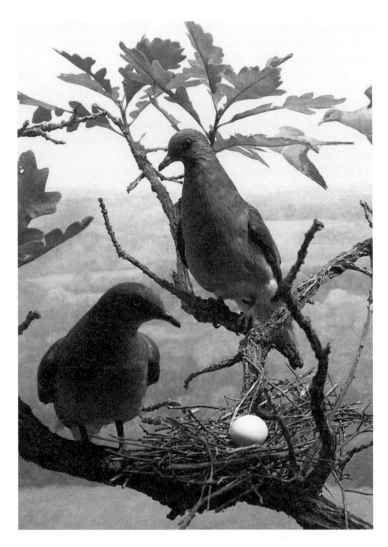

FIGURE 5.27 A pair of stuffed passenger pigeons (*Ectopistes migratorius*). The last member of this species died in the Cincinnati Zoo in 1914.
Courtesy of Bell Museum, University of Minnesota.

the world's most abundant bird, with a population of between 3 and 5 billion animals (fig. 5.27). It once accounted for about one-quarter of all birds in North America. In 1830 John James Audubon saw a single flock of birds estimated to be ten miles wide, hundreds of miles long, and thought to contain perhaps a billion birds. In spite of this vast abundance, market hunting and habitat destruction caused the entire population to crash in only about 20 years between 1870 and 1890. The last known wild bird was shot in 1900 and the last existing passenger pigeon, a female named Martha, died in 1914 in the Cincinnati Zoo.

At about the same time that passenger pigeons were being extirpated, the American bison, or buffalo (*Bison bison*), was being hunted to near extinction on the Great Plains. In 1850 some 60 million bison roamed the western plains. Many were killed only for their hides or tongues, leaving millions of carcasses to rot. Much of the bison's destruction was carried out by the U.S. Army to deprive native peoples who depended on bison for food, clothing, and shelter of these resources, thereby forcing them onto reservations. After 40 years, there were only about 150 wild bison left and another 250 in captivity.

Fish stocks have been seriously depleted by overharvesting in many parts of the world. A huge increase in fishing fleet size and efficiency in recent years has led to a crash of many oceanic populations. Worldwide, 13 of 17 principal fishing zones are now reported to be commercially exhausted or in steep decline. At least three-quarters of all commercial oceanic species are over-harvested. Canadian fisheries biologists estimate that only 10 percent of the top predators, such as swordfish, marlin, tuna, and shark, remain in the Atlantic Ocean. Groundfish, such as cod, flounder, halibut, and hake, also are severely depleted. You can avoid adding to this overharvest by eating only abundant, sustainably harvested varieties.

Perhaps the most destructive example of harvesting terrestrial wild animal species today is the African bushmeat trade. Wildlife biologists estimate that 1 million tons of bushmeat, including antelope, elephants, primates, and other animals, are sold in African markets every year. For many poor Africans, this is the only source of animal protein in their diet. If we hope to protect the animals targeted by bushmeat hunters, we will need to help them find alternative livelihoods and replacement sources of high-quality protein. The emergence of SARS in 2003 (chapter 8) resulted from the wild food trade in China and Southeast Asia, where millions of civets, monkeys, snakes, turtles, and other animals are consumed each year as luxury foods.

Commercial collection serves medicinal and pet trades

In addition to harvesting wild species for food, we also obtain a variety of valuable commercial products from nature. Much of this represents sustainable harvest, but some forms of commercial exploitation are highly destructive, however, and represent a serious threat to certain rare species (fig. 5.28). Despite international bans on trade in products from endangered species, smuggling of furs, hides, horns, live specimens, and folk medicines amounts to millions of dollars each year.

Developing countries in Asia, Africa, and Latin America with the richest biodiversity in the world are the main sources of wild animals and animal products, while Europe, North America, and some of the wealthy Asian countries are the principal importers. Japan, Taiwan, and Hong Kong buy three-quarters of all cat and snake skins, for instance, while European countries buy a similar percentage of live birds. The United States imports 99 percent of all live cacti and 75 percent of all orchids sold each year.

The profits to be made in wildlife smuggling are enormous. Tiger or leopard fur coats can bring $100,000 in Japan or Europe. The population of African black rhinos dropped from approximately 100,000 in the 1960s to about 3,000 in the 1980s because of a demand for their horns. In Asia, where it is prized for its supposed medicinal properties, powdered rhino horn fetches $28,000 per kg. In Yemen, a rhino horn dagger handle can sell for up to $1,000.

FIGURE 5.28 Parts from rare and endangered species for sale on the street in China. Use of animal products in traditional medicine and prestige diets is a major threat to many species.

Plants also are threatened by overharvesting. Wild ginseng has been nearly eliminated in many areas because of the Asian demand for the roots, which are used as an aphrodisiac and folk medicine. Cactus "rustlers" steal cacti by the ton from the American Southwest and Mexico. With prices as high as $1,000 for rare specimens, it's not surprising that many are now endangered.

The trade in wild species for pets is an enormous business. Worldwide, some 5 million live birds are sold each year for pets, mostly in Europe and North America. Currently, pet traders import (often illegally) into the United States some 2 million reptiles, 1 million amphibians and mammals, 500,000 birds, and 128 million tropical fish each year. About 75 percent of all saltwater tropical aquarium fish sold come from coral reefs of the Philippines and Indonesia.

Many of these fish are caught by divers using plastic squeeze bottles of cyanide to stun their prey (fig. 5.29). Far more fish die with this technique than are caught. Worst of all, it kills the coral animals that create the reef. A single diver can destroy all of the life on 200 m² of reef in a day. Altogether, thousands of divers currently destroy about 50 km² of reefs each year. Net fishing would prevent this destruction, and it could be enforced if pet owners would insist on net-caught fish. More than half the world's coral reefs are potentially threatened by human activities, with up to 80 percent at risk in the most populated areas.

Predator and pest control is expensive but widely practiced

Some animal populations have been greatly reduced, or even deliberately exterminated, because they are regarded as dangerous to humans or livestock or because they compete with our use of resources. Every year, U.S. government animal control agents trap, poison, or shoot thousands of coyotes, bobcats, prairie dogs, and other species considered threats to people, domestic livestock, or crops.

FIGURE 5.29 A diver uses cyanide to stun tropical fish being caught for the aquarium trade. Many fish are killed by the method itself, while others die later during shipment. Even worse is the fact that cyanide kills the coral reef itself.

This animal control effort costs about $20 million in federal and state funds each year and kills some 700,000 birds and mammals, about 100,000 of which are coyotes. Defenders of wildlife regard this program as cruel, callous, and mostly ineffective in reducing livestock losses. Protecting flocks and herds with guard dogs or herders or keeping livestock out of areas that are home range of wild species would be a better solution, they believe. Ranchers and trappers, on the other hand, argue that, without predator control, western livestock operations would be uneconomical.

ENDANGERED SPECIES MANAGEMENT AND BIODIVERSITY PROTECTION

Over the years, we have gradually become aware of the harm we have done—and continue to do—to wildlife and biological resources. Slowly, we are adopting national legislation and international treaties to protect these irreplaceable assets. Parks, wildlife refuges, nature preserves, zoos, and restoration programs have been established to protect nature and rebuild depleted populations. There has been encouraging progress in this area, but much remains to be done. While most people favor pollution control or protection of favored species, such as whales or gorillas, surveys show that few understand what biological diversity is or why it is important.

Hunting and fishing laws protect reproductive populations

In 1874 a bill was introduced in the United States Congress to protect the American bison, whose numbers were already falling to dangerous levels. This initiative failed, however, because

most legislators believed that all wildlife—and nature in general—was so abundant and prolific that it could never be depleted by human activity. As we discussed earlier in this chapter, however, by the end of the nineteenth century, bison had plunged from some 60 million to only a few hundred animals.

By the 1890s most states had enacted some hunting and fishing restrictions. The general idea behind these laws was to conserve the resource for future human use rather than to preserve wildlife for its own sake. The wildlife regulations and refuges established since that time have been remarkably successful for many species. A hundred years ago, there were an estimated half a million white-tailed deer in the United States; now there are some 14 million—more in some places than the environment can support. Wild turkeys and wood ducks were nearly gone 50 years ago. By restoring habitat, planting food crops, transplanting breeding stock, building shelters or houses, protecting these birds during breeding season, and using other conservation measures, we have restored populations of these beautiful and interesting birds to several million each. Snowy egrets, which were almost wiped out by plume hunters 80 years ago, are now common again.

The endangered species act protects habitat and species

Establishment of the U.S. Endangered Species Act (ESA) of 1973 and the Committee on the Status of Endangered Wildlife in Canada (COSEWIC) in 1976 represented powerful new approaches to wildlife protection. Where earlier regulations had been focused almost exclusively on "game" animals, these programs seek to identify all endangered species and populations and to save as much biodiversity as possible, regardless of its usefulness to humans. **Endangered species** are those considered in imminent danger of extinction, while **threatened species** are those that are likely to become endangered—at least locally—within the foreseeable future. Bald eagles, gray wolves, brown (or grizzly) bears, and sea otters, for instance, together with a number of native orchids and other rare plants, are considered to be locally threatened, even though they remain abundant in other parts of their former range. **Vulnerable species** are naturally rare or have been locally depleted by human activities to a level that puts them at risk. They often are candidates for future listing. For vertebrates, protected categories include species, subspecies, and local races or ecotypes.

The ESA regulates a wide range of activities involving endangered species, including "taking" (harassing, harming, pursuing, hunting, shooting, trapping, killing, capturing, or collecting) either accidentally or on purpose; importing into or exporting out of the United States; possessing, selling, transporting, or shipping; and selling or offering for sale any endangered species. Prohibitions apply to live organisms, body parts, and products made from endangered species. Violators of the ESA are subject to fines up to $100,000 and one-year imprisonment. Vehicles and equipment used in violations may be subject to forfeiture.

TABLE 5.4

Endangered and Threatened Species Worldwide

Mammals	1,101
Birds	1,213
Reptiles	304
Amphibians	1,770
Fish	800
Insects and other invertebrates	1,992
Total fauna	7,180
Plants	8,321

Source: Data from *IUCN Red List*, 2005.

In 1995 the Supreme Court ruled that critical habitat—habitat essential for a species' survival—must be protected, whether on public or private land.

Currently, the United States has 1,300 species on its endangered and threatened species lists and about 250 candidate species waiting to be considered. The number of listed species in different taxonomic groups reflects much more about the kinds of organisms that humans consider interesting and desirable than the actual number in each group. In the United States, invertebrates make up about three-quarters of all known species but only 9 percent of those deemed worthy of protection. Worldwide, the International Union for Conservation of Nature and Natural Resources (IUCN) lists a total of 16,496 endangered and threatened species, including nearly one-quarter of all known bird varieties (table 5.4).

Listing of new species in the United States has been very slow, generally taking several years from the first petition to final determination. Limited funding, political pressures, listing moratoria, and changing administrative policies have created long delays. In December 2000, flooded by lawsuits over species protection and badly underfunded, the USFWS stopped reviewing new cases altogether. Hundreds of species are classified as "warranted (deserving of protection) but precluded" for lack of funds or local support. At least 18 species have gone extinct since being nominated for protection. Advocates for listing insist that early action is far cheaper than last-ditch rescue efforts. Both environmentalists and property-rights advocates, however, tend to believe that the "single-species" approach is flawed and that efforts should be focused on multiple species and habitat protection.

When Congress passed the original ESA, it probably intended to protect only a few charismatic species, such as birds and big game animals. Sheltering obscure species, such as the Delhi Sands flower-loving fly, the Coachella Valley fringe-toed lizard, Mrs. Furbisher's lousewort, or the orange-footed pimple-back mussel, most likely never occurred to those who voted for the bill. This raises some interesting ethical questions about the rights and values of seemingly minor species. Although uncelebrated, these species may be indicators of environmental health. Protecting them usually preserves habitat and a host of unlisted species.

Recovery plans

Once a species is officially listed as endangered, the Fish and Wildlife Service is required to prepare a recovery plan detailing how populations will be rebuilt to sustainable levels. It usually takes years to reach agreement on specific recovery plans. Among the difficulties are costs, politics, interference on local economic interests, and the fact that, once a species is endangered, much of its habitat and ability to survive is likely compromised. The total cost of recovery plans for all currently listed species is estimated to be nearly $5 billion.

The United States currently spends about $150 million per year on endangered species protection and recovery. About half that amount is spent on a dozen charismatic species, such as the California condor and the Florida panther and grizzly bear, which receive around $13 million per year. By contrast, the 137 endangered invertebrates and 532 endangered plants get less than $5 million per year altogether. Our funding priorities often are based more on emotion and politics than biology. A variety of terms are used for rare or endangered species thought to merit special attention:

- *Keystone species* are those with major effects on ecological functions and whose elimination would affect many other members of the biological community; examples are prairie dogs (*Cynomys ludovicianus*) and bison (*Bison bison*).

- *Indicator species* are those tied to specific biotic communities or successional stages or environmental conditions. They can be reliably found under certain conditions but not others; an example is brook trout (*Salvelinus fontinalis*).

- *Umbrella species* require large blocks of relatively undisturbed habitat to maintain viable populations. Saving this habitat also benefits other species. Examples of umbrella species are the northern spotted owl (*Strix occidentalis caurina*) and tiger (*Panthera tigris*) (fig. 5.30).

- *Flagship species* are especially interesting or attractive organisms to which people react emotionally. These species can motivate the public to preserve biodiversity and contribute to conservation; an example is the giant panda (*Ailuropoda melanoleuca*).

Some recovery plans have been gratifyingly successful. The American alligator was listed as endangered in 1967 because hunting (for meat, skins, and sport) and habitat destruction had reduced populations to precarious levels. Protection has been so effective that the species is now plentiful throughout its entire southern range. Florida alone estimates that it has at least 1 million alligators.

Twenty years ago, due mainly to DDT poisoning, only 800 bald eagles (*Haliaeetus leucocephalis*) remained in the contiguous United States. By 1994 the population had rebounded to more than 8,000 birds and the eagle's status was reduced from

FIGURE 5.30 Umbrella or flagship species generally are highly charismatic and can mobilize support to protect large blocks of habitat that benefit other creatures as well.

"DAMN SPOTTED OWL!"

FIGURE 5.31 Endangered species often serve as a barometer for the health of an entire ecosystem and as surrogate protector for a myriad of less well-known creatures.
Copyright 1990 by Herblock in the Washington Post.

endangered to threatened. Bald eagles have been proposed for complete delisting, but disagreement over how they will be managed has delayed this step. Similarly, peregrine falcons, which had been down to only 39 breeding pairs in the 1970s, had rebounded to 1,650 pairs by 1999 and were taken off the endangered species list. Declaring the ESA a success, former Interior Secretary Bruce Babbit announced that 29 species, including mammals, fish, reptiles, birds, plants, and even one insect (the Tinian monarch) in addition to eagles and falcons have been removed or downgraded from the endangered species list.

Opponents of the ESA have repeatedly tried to require that economic costs and benefits be incorporated into endangered species planning. An important test of the ESA occurred in 1978 in Tennessee, where construction of the Tellico Dam threatened a tiny fish called the snail darter. As a result of this case, a federal committee (the so-called God Squad) was given power to override the ESA for economic reasons. Another important example is the case of the northern spotted owl (chapter 6), whose protection depends on preserving old-growth forest in the Pacific Northwest. Timber-industry economists estimate that saving a population of 1,600 to 2,400 owls would cost $33 billion, with most of the losses borne by local companies and residents of Washington and Oregon. Conservationists dispute these numbers and claim the owl is an umbrella species whose protection would aid many other organisms and resources (fig. 5.31).

An even more costly recovery program may be required for Columbia River salmon and steelhead endangered by hydropower dams and water storage reservoirs that block their migration to the sea. Opening floodgates to allow young fish to run downriver and adults to return to spawning grounds would have high economic costs to barge traffic, farmers, and electric rate payers who have come to depend on abundant water and cheap electricity. On the other hand, commercial and sport fishing for salmon is worth $1 billion per year and employs about 60,000 people directly or indirectly.

Reauthorizing the ESA has been contentious

The ESA officially expired in 1992. Since then, Congress has debated many alternative proposals ranging from outright elimination to substantial strengthening of the act. Perhaps no other environmental issue divides Americans more strongly than the ESA. In the western United States, where traditions of individual liberty and freedom are strong and the federal government is viewed with considerable suspicion and hostility, the ESA seems to many to be a diabolical plot to take away private property and trample on individual rights. Many people believe that the law puts the welfare of plants and animals above that of humans. Farmers, loggers, miners, ranchers, developers, and other ESA opponents repeatedly have tried to scuttle the law or greatly reduce its power. Environmentalists, on the other hand, see the ESA as essential to protecting nature and maintaining the viability of the planet. They regard it as the single most effective law in their arsenal and want it enhanced and improved.

Proposals for a new ESA generally fall into one of two general categories. Environmentalists have repeatedly introduced versions that encourage an ecosystem and habitat protection approach rather than focusing on individual species. Reinforcing critical habitat protection and placing deadlines for listing and recovery plan completion, these bills would require that proposed recovery plans be independently peer-reviewed by qualified scientists. They also would expand citizen ability to enforce the law and increase civil and criminal penalties.

On the other side, ESA opponents want to protect private property rights and economic interests from what they see as unwarranted government actions. All recovery plan teams in their versions of the ESA would be required to include individuals with vested economic interests. Under these proposals, only the least costly, most cost-effective, or least burdensome measures would be taken to protect endangered organisms. Federal agencies would be allowed to do "self-consultation"—that is, to avoid consulting wildlife specialists from the Fish and Wildlife Service before undertaking projects. ESA opponents also would require notification and hearings in all states with resident candidate species and would give small-parcel landowners special exemptions from ESA regulations.

Habitat protection is a source of hope and controversy

Over the past decade, growing numbers of scientists, land managers, policymakers, and developers have been making the case that it is time to focus on a rational, continentwide preservation of ecosystems that supports maximum biological diversity rather than a species-by-species battle for the rarest or most popular organisms. By focusing on populations already reduced to only a few individuals, we spend most of our conservation funds on species that may be genetically doomed no matter what we do. Furthermore, by concentrating on individual species, we spend millions of dollars to breed plants or animals in captivity that have no natural habitat where they can be released. While flagship species, such as mountain gorillas and Indian tigers, are reproducing well in zoos and wild animal parks, the ecosystems that they formerly inhabited have largely disappeared.

A leader of this new form of conservation is J. Michael Scott, who was project leader of the California condor recovery program in the mid-1980s and had previously spent ten years working on endangered species in Hawaii. In making maps of endangered species, Scott discovered that even Hawaii, where more than 50 percent of the land is federally owned, has many vegetation types completely outside of natural preserves (fig. 5.32). The gaps between protected areas may contain more endangered species than are preserved within them.

This observation has led to an approach called **gap analysis,** in which conservationists and wildlife managers look for unprotected landscapes, or gaps in the network of protected lands, that are rich in species. Gap analysis uses GIS to overlay protected conservation areas with high-biodiversity areas. This overlay makes it easy to identify priority spots for conservation efforts.

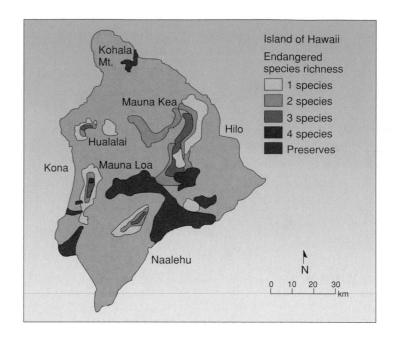

FIGURE 5.32 Protected lands (*green*) are often different from biologically diverse areas (*red shades*), as shown here on the island of Hawaii.

Maps also help biologists and land-use planners communicate about threats to biodiversity. This broad-scale, holistic approach seems likely to save more species than a piecemeal approach.

Conservation biologist R. E. Grumbine suggests four remanagement principles for protecting biodiversity in a large-scale, long-range approach:

1. Protect enough habitat for viable populations of all native species in a given region.

2. Manage at regional scales large enough to accommodate natural disturbances (fire, wind, climate change, etc.).

3. Plan over a period of centuries, so that species and ecosystems can continue to evolve.

4. Allow for human use and occupancy at levels that do not result in significant ecological degradation.

International wildlife treaties protect global biodiversity

The 1975 Convention on International Trade in Endangered Species (CITES) was a significant step toward worldwide protection of endangered flora and fauna. It regulated trade in living specimens and products derived from listed species, but it has not been foolproof. Species are smuggled out of countries where they are threatened or endangered, and documents are falsified to make it appear they have come from areas where the species are still common. Investigations and enforcement are especially difficult in developing countries where wildlife is disappearing most rapidly. Still, eliminating markets for endangered wildlife is an effective way of stopping poaching. Appendix I of CITES lists 700 species threatened with extinction by international trade.

Principles of Environmental Science

REVIEW QUESTIONS

1. Why did ecologists want to reintroduce wolves to Yellowstone Park? What goals did they have, and have their goals been achieved?

2. Describe nine major types of terrestrial biomes.

3. Explain how climate graphs (as in fig. 5.4) should be read.

4. Describe conditions under which coral reefs, mangroves, estuaries, and tide pools occur.

5. Throughout the central portion of North America is a large biome once dominated by grasses. Describe how physical conditions and other factors control this biome.

6. Explain the difference between swamps, marshes, and bogs.

7. How do elevation (on mountains) and depth (in water) affect environmental conditions and life-forms?

8. Figure 5.13 shows chlorophyll (plant growth) in oceans and on land. Explain why green, photosynthesizing organisms occur in long bands at the equator and along the edges of continents. Explain the very dark green areas and yellow/orange areas on the continents.

9. Define *biodiversity* and give three types of biodiversity essential in preserving ecological systems and functions.

10. What is a biodiversity "hot spot"? List several of them (see fig. 5.20).

11. How do humans benefit from biodiversity?

12. What are some of the main threats to biodiversity?

13. Have extinctions occurred in the past? Is there anything unusual about current extinctions?

14. Why are exotic or invasive species a threat to biodiversity? Give several examples of exotic invasive species (see fig. 5.25).

15. What is the Endangered Species Act? Describe some of the main arguments of its proponents and opponents.

16. What is a flagship or umbrella species? Why are they often important, even though they are costly to maintain?

CRITICAL THINKING AND DISCUSSION

Apply the principles you have learned in this chapter to discuss these questions with other students.

1. Many poor tropical countries point out that a hectare of shrimp ponds can provide 1,000 times as much annual income as the same area in an intact mangrove forest. Debate this point with a friend or classmate. What are the arguments for and against saving mangroves?

2. Genetic diversity, or diversity of genetic types, is believed to enhance stability in a population. Most agricultural crops are genetically very uniform. Why might the usual importance of genetic diversity *not* apply to food crops? Why *might* it apply?

3. Scientists need to be cautious about their theories and assumptions. What arguments could you make for *and* against the statement that humans are causing extinctions unlike any in the history of the earth?

4. A conservation organization has hired you to lead efforts to reduce the loss of biodiversity in a tropical country. Which of the following problems would you focus on first and why: habitat destruction and fragmentation, hunting and fishing activity, harvesting of wild species for commercial sale, or introduction of exotic organisms?

5. Many ecologists and resource scientists work for government agencies to study resources and resource management. Do these scientists serve the public best if they try to do pure science, or if they try to support the political positions of democratically elected representatives, who, after all, represent the positions of their constituents?

6. You are a forest ecologist living and working in a logging community. An endangered salamander has recently been discovered in your area. What arguments would you make for and against adding the salamander to the official endangered species list?

7. A variety of environments are adapted to disturbance, including fire-adapted grasslands, fire-dependent forest types, and floodplains. How do you explain the difference between disturbance-adapted ecosystems and ecosystems that simply never mature and never reach climax conditions?

SUGGESTED READINGS

du Toit, Johan T., et al. 2003. *The Kruger Experience: Ecology and Management of Savanna Heterogeneity.* Island Press.

Field, John G., et al., eds. 2002. *Oceans 2020: Science, Trends, and the Challenge of Sustainability.* Island Press.

Galindo-Leal, Carlos, and Ibsen de Gusmao Camara, eds. 2003. *The Atlantic Forest of South America: Biodiversity Status, Threats, and Outlook.* Island Press.

Jones, A. G., et al. 2003. A review of conservation threats on Gough island: A case study for terrestrial conservation in the Southern Oceans. *Biological Conservation* 113(1):75–87.

Kalluri, S., et al. 2003. The potential of remote sensing data for decision makers at the state, local and tribal level: Experiences from NASA's Synergy program. *Environmental Science and Policy* 6(6):487–500.

Quammen, David. 1997. *The Song of the Dodo: Island Biogeography in an Age of Extinctions.* Touchstone Books.

Tidwell, Mike. 2003. *Bayou Farewell: The Rich Life and Tragic Death of Louisiana's Cajun Coast.* Pantheon Books.

Woodward, Colin. 2003. Saving Maine. *OnEarth* 25(2):14–22.

DATA analysis   Confidence Limits in the Breeding Bird Survey

A central principle of science is the recognition that all knowledge involves uncertainty. No study can observe every possible event in the universe, so there is always missing information. Scientists try to define the limits of their uncertainty, in order to allow a realistic assessment of their results. A corollary of this principle is that the more data we have, the less uncertainty we have. More data increase our confidence that our observations represent the range of possible observations.

One of the most detailed records of wildlife population trends in North America is the Breeding Bird Survey (BBS). Every June, more than a thousand volunteers drive established routes and count every bird they see or hear. The accumulated data from thousands of routes, over many years, indicates population *trends,* telling which populations are increasing, decreasing, or expanding into new territory.

Because many scientists use BBS data, it is essential to communicate how much confidence there is in the data. The online BBS database reports measures of data quality, including:

- **N:** the number of survey routes from which population trends are calculated.

- **Confidence limits:** because the reported trend is an average of a small *sample* of year-to-year changes on routes, confidence limits tell us how close the sample's average probably is to the average for the *entire population* of that species. Statistically, 95 percent of all samples should fall in-between the confidence limits. In effect, we can be 95 percent sure that the entire population's actual trend falls between the upper and lower confidence limits.

1. Examine the table at right, which shows 10 species taken from the online BBS database. How many species have a positive population trend (>0)? If a species had a trend of 0, how much would it change from year to year?

2. Which species has the greatest decline per year? For every 100 birds this year, how many fewer will there be next year?

3. If the distance between upper and lower confidence limits (the *confidence interval*) is narrow, then we can be reasonably sure the trend in our *sample* is close to the trend for the *total population* of that species. What is the reported trend for ring-necked pheasant? What is the number of routes (N) on which this trend is based? What is the range in which the pheasant's true population trend probably

falls? Is there a reasonable chance that the pheasant population's average annual change is 0? That the population trend is actually −5?

Now look at the ruffed grouse, on either the table or the graph. Does the trend show that the population is increasing

Species	Trend 1966-2004	N	Lower limit	Upper limit
Ring-necked pheasant	−0.5	397	−1.2	0.2
Ruffed grouse	0.7	26	−19.7	21.1
Sage grouse	−2.4	23	−5.6	0.9
Sharp-tailed grouse	0	104	−3.4	3.5
Greater prairie-chicken	−6.8	41	−12.8	−0.8
Wild turkey	8.2	308	5.1	11.2
Northern bobwhite	−2.2	541	−2.7	−1.8
Clapper rail	−0.7	8	−7.3	5.8
King rail	−6.8	22	−11.7	−1.8
Virginia rail	4.7	21	1.4	7.9

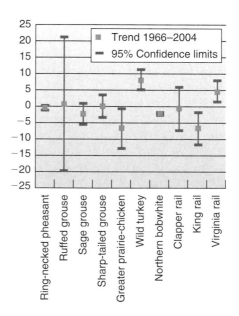

or decreasing? Can you be certain that the actual trend is not 7, or 17? On how many routes is this trend based?

4. In general, confidence limits depend on the number of observations (N), and how much all the observed values (trends on routes, in this case) differ from the average value. If the values vary greatly, the confidence interval will be wide. Examine the table and graph. Does a *large* N tend to widen or narrow the confidence interval?

5. A trend of 0 would mean no change at all. When 0 falls within the confidence interval, we have little certainty that the trend is *not* 0. In this case, we say that the trend is not significant. How many species have trends that are not significant (at 95 percent certainty)?

   Can we be certain that the sharp-tailed grouse and greater prairie-chicken are changing at different rates? How about the sharp-tailed grouse and wild turkey?

6. Does uncertainty in the data mean results are useless? Does reporting of confidence limits increase or decrease your confidence in the results?

For further information on the Breeding Bird Survey, see http://www.mbr-pwrc.usgs.gov/bbs/.

Elephants gather in a clearing in Nouabalé-Ndoki
National Park, Congo.

Environmental Conservation

6

Forests, Grasslands, Parks, and Nature Preserves

*What a country chooses to save is what a country
chooses to say about itself.*

–Mollie Beatty, former director, U.S. Fish and Wildlife Service–

OBJECTIVES

Studying this chapter will help you answer the following questions:

- What portion of the world's original forests remain?
- What activities threaten global forests? What steps can be taken to preserve them?
- Why is road construction a challenge to forest conservation?
- Where are the world's most extensive grasslands?
- How are the world's grasslands distributed, and what activities degrade grasslands?

- What are the original purposes of parks and nature preserves in North America?
- What is a *wilderness*? Why are wilderness areas both important and controversial?
- What are some steps to help restore natural areas?

Case Study   Saving an African Eden

When conservation biologist Paul Elkan first entered the Goualougo tract in remote northern Republic of Congo on a wildlife reconnaissance expedition in 1997, he was astonished at how little fear the animals showed. Generally, when wild monkeys or chimpanzees spot humans, they shriek and run away. Conditioned by years of hunting, wild animals fear humans. In the pristine forests of Goualougo, however, Elkan found that the animals followed him for hours, seemingly curious about this strange new species in their forest home. Bordered by two untamed rivers and many kilometers of dense, flooded forests, the Goualougo shows no signs of human intrusion. Even the local Bambendjelle people (called forest Pygmies by Europeans) had no memories of ever hunting there. "This is one of the last great wild places in the world," Elkan said.

The Congo basin holds about one-fourth of the world's tropical forests and is the largest stretch of lowland rainforest in the world, aside from the Amazon. Those forests are home to some of the most important wildlife populations remaining in Central Africa, including forest elephants (*Loxodonta africana cyclotis*), lowland gorillas (*Gorilla gorilla*), chimpanzees (*Pan troglodytes troglodytes*), bongo (*Tragelaphus eurycerus*), buffalo (*Syncerus caffer nanus*), leopard (*Panthera pardus*), six species of small antelope, and eight species of monkeys. Logging and hunting are a growing threat to the forests and the wildlife they shelter. It has long been common practice for loggers to subsist on the local wildlife, or bushmeat. And logging roads provide access deep into the forest for hunters and settlers. Up to 1 million tons of bushmeat may be consumed in Central Africa each year.

In 1982, all the lands in the northern Republic of Congo were divided into large logging concessions. In the late 1980s, wildlife explorations revealed that the Nouabalé-Ndoki region held one of the densest assemblages of wildlife and the richest collection of primates on earth. Recognizing its ecological importance, the government of Congo upgraded the Noubalé-Ndoki forest from a forest management unit to national park in 1993. Adjacent parks in the Central African Republic and Cameroon create a contiguous area of irreplaceable lowland tropical rainforest.

Unfortunately, when the parks were being established, the most pristine area was cut in two, and half of it, Goualougo triangle, was slated for logging. Logging rights were sold to a German-owned company, Congolaise Industrielle des Bois (C.I.B.). Biologist/explorer Mike Fay, who publicized the region's plight with his "megatransect" across the Congo basin, calls the Gualougo triangle the Inner Sanctum

of this great forest. Elkan and others have confirmed the richness of this African Eden.

In July 2002, C.I.B., the Congolese government, and the Wildlife Conservation Society (WCS) announced that a 26,000 ha (100 mi^2) tract of the Goualougo triangle would be added to the Nouabalé-Ndoki National Park. This was the first time any logging company in the entire Congo Basin had voluntarily given up rights to valuable timberland. The C.I.B. estimated that it would forgo logging timber worth an estimated $40 million (U.S.). In addition, C.I.B. promised to practice sustainable forest management on its remaining timber concessions and to participate in a wildlife protection program run by the WCS.

C.I.B. promises to prohibit hunting and transporting of bushmeat by company employees and promises to develop alternative protein sources (imported beef, fish, and poultry farms). The company will provide garden supplies and training to local farmers and a conservation awareness program for villagers and company employees. The forest outside the park will be zoned with community hunting areas near villages and no-hunting zones immediately surrounding the park. Protected areas will be patrolled by six government officers and 40 local ecoguards funded and managed by the WCS.

Why did the C.I.B. give up such lucrative logging concessions? Biologists played a vital role by revealing the incredible richness of the wildlife and the absence of human encroachment. Consumers who voiced concern about tropical forests were also key. More and more, consumers of paper products, furniture, and lumber are recognizing their links to tropical forests, and public pressure often leads to better policies. European environmental groups provided leadership and helped educate the public. C.I.B. gave up the Goualougo triangle as part of an effort to get its products certified as sustainably harvested, so that it could retain its European market.

Land-use policies are central to environmental quality, and they are influenced by decisions of national governments, corporations, citizen organizations, and individuals. In this chapter, we'll look at land uses, especially on forests and grasslands, and we'll consider how each of us plays a role, as individuals or as members of a community, in land-use decisions affecting our common environment.

Adjacent parks in Cameroon, Central African Republic, and Congo protect an irreplaceable expanse of tropical rainforest.

World Forests and Grasslands

Forests and grasslands together occupy almost 60 percent of global land cover (fig. 6.1). These ecosystems provide many of our essential resources, such as lumber, paper pulp, and grazing lands for livestock. They also provide essential ecological services, including regulating climate, controlling water runoff, providing wildlife habitat, purifying air and water, and supporting rainfall. Forests and grasslands also have scenic, cultural, and historic values that deserve protection. Forests and grasslands are also among the most heavily disturbed ecosystems (chapter 5).

In many cases, these competing land uses and needs are incompatible. Most political debates over conservation have concerned protection or use of forests, prairies, and rangelands. This chapter examines the ways we use and abuse these biological communities, as well as some of the ways we can protect them and conserve their resources. We discuss forests first, followed by grasslands and then strategies for conservation, restoration, and preservation.

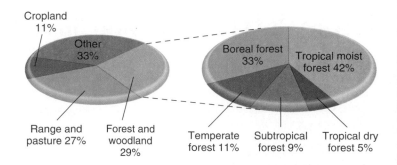

FIGURE 6.1 World land use and forest types. The "other" category includes tundra, desert, wetlands, and urban areas.
Source: UN Food and Agricultural Organization (FAO).

Boreal and tropical forests are most abundant

Forests are widely distributed, but most remaining forests are in the humid tropics and the cold boreal ("northern") or taiga regions (fig. 6.2). Assessing forest distribution is tricky, because

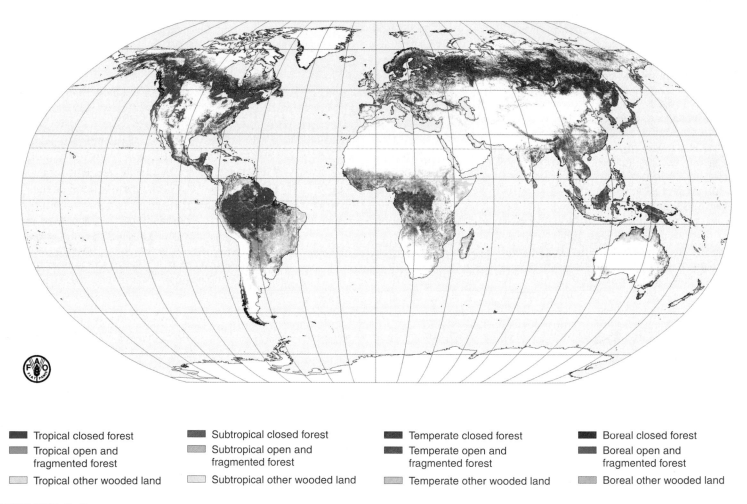

Tropical closed forest
Tropical open and fragmented forest
Tropical other wooded land

Subtropical closed forest
Subtropical open and fragmented forest
Subtropical other wooded land

Temperate closed forest
Temperate open and fragmented forest
Temperate other wooded land

Boreal closed forest
Boreal open and fragmented forest
Boreal other wooded land

FIGURE 6.2 Major forest types. Note that some of these forests are dense; others may have only 10 to 20 percent actual tree cover.
Source: Data from United Nations Food and Agriculture Organization, 2002.

forests vary in density and height, and many are inaccessible. The UN Food and Agriculture Organization (FAO) defines "forest" as any area where trees cover more than 10 percent of the land. This definition includes woodlands ranging from open **savannas,** whose trees cover less than 20 percent of the ground, to **closed-canopy forests,** in which tree crowns cover most of the ground. The largest tropical forest is in the Amazon River basin. The highest rates of forest loss are in Africa, especially tropical West Africa, the region discussed in the opening Case Study, and in South America (fig. 6.3). Some of the world's most biologically diverse regions are undergoing rapid deforestation, including Southeast Asia and Central America (see related story "Protecting Forests to Preserve Rain" at www.mhhe.com/cunningham4e).

Forests are a huge carbon sink, storing some 422 billion metric tons of carbon in standing biomass. Clearing and burning of forests releases much of this carbon into the atmosphere (chapter 9) and may contribute substantially to global climate change. Moisture released from forests also contributes to rainfall. For example, recent climate studies suggest that deforestation of the Amazon could reduce precipitation in the American Midwest.

Among the forests of greatest ecological importance are the remnants of primeval forests that are home to much of the world's biodiversity, endangered species, and indigenous human cultures. Sometimes called frontier forests, **old-growth forests** are those that cover a large enough area and have been undisturbed by human activities long enough that trees can live out a natural life cycle and ecological processes can occur in relatively normal fashion. That doesn't mean that all trees need be enormous or thousands of years old. In some old-growth forests, most trees live less than a century before being killed by disease or some natural disturbance, such as a fire. Nor does it mean that humans have never been present. Where human occupation entails relatively little impact, an old-growth forest may have been inhabited by people for a very long time. Even forests that have been logged or converted to cropland often can revert to old-growth characteristics if left alone long enough.

While forests still cover about half the area they once did worldwide, only one-quarter of those forests retain old-growth features. The largest remaining areas of old-growth forest are in Russia, Canada, Brazil, Indonesia, and Papua New Guinea. Together, these five countries account for more than three-quarters of all relatively undisturbed forests in the world. In general, remoteness rather than laws protect those forests. Although official data describe only about one-fifth of Russian old-growth forest as threatened, rapid deforestation—both legal and illegal—especially in the Russian Far East, probably put a much greater area at risk.

Forest products include fuel, paper, and lumber

Wood plays a part in more activities of the modern economy than does any other commodity. There is hardly any industry that does not use wood or wood products somewhere in its manufacturing and marketing processes. Think about the amount of junk mail, newspapers, photocopies, and other paper products that each of us in developed countries handles, stores, and disposes of in a single day. Total annual world wood consumption is about 4 billion m^3. This is more than steel and plastic consumption combined. International trade in wood and wood products amounts to more than $100 billion each year. Developed countries produce less than half of all industrial wood but account for about 80 percent of its consumption. Less-developed countries, mainly in the tropics, produce more than half of all industrial wood but use only 20 percent.

Paper pulp, the fastest growing type of forest product, accounts for nearly a fifth of all wood consumption. Most of the world's paper is used in the wealthier countries of North America, Europe, and Asia. Global demand for paper is increasing rapidly, however, as other countries develop. The United States, Russia, and Canada are the largest producers of both paper pulp and industrial wood (lumber and panels). Much industrial logging in Europe and North America occurs on managed plantations, rather than in untouched old-growth forest. However, paper production is increasingly blamed for deforestation in Southeast Asia, West Africa, and other regions.

Fuelwood accounts for nearly half of global wood use. More than half of the people in the world depend on firewood or charcoal as their principal source of heating and cooking fuel

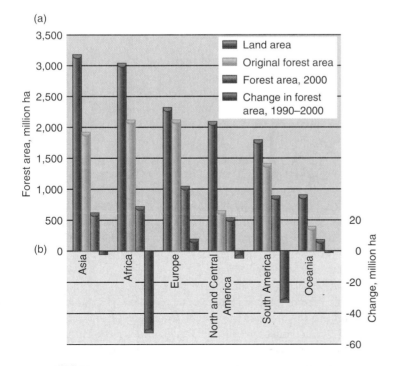

FIGURE 6.3 Regional distribution of original and current forests (a) and regional forest losses, 1990 to 2000 (b). Europe includes Russia's Siberian forests.
Source: Data from United Nations Food and Agriculture Organization, 2002.

Calculate Forest Area

Examine figures 6.2 and 6.3 to evaluate changes in global forest cover: Which region(s) originally had the greatest area of forest according to the FAO (fig. 6.3). What kinds of forest occupied those regions? Looking at figure 6.3, which region would you say had the greatest *percentage* forest cover? The lowest percent coverage?

Figures 6.2 and 6.3 show the extent of global forests in *map* form and *graph* form. You can also compare forest distribution *numerically.* You have just made a visual estimate of the percentage cover. Now try calculating percentage cover from the graph in figure 6.3. (Although we'll ask you to do some approximation of the data, the numbers you're comparing are so enormous that some generalization is acceptable for now.)

Starting with Asia, read the approximate size of the land area on the vertical axis (about 3,200 million ha). Then find the extent of original forest (about 1,900 million ha).

Now calculate the percentage of original forest, as follows:

$$\text{percentage} = \frac{\text{forest area}}{\text{land area}} = \frac{1,900 \text{ million ha}}{3,200 \text{ million ha}}$$

cancel out common terms and zeros:

$$\frac{1,900 \text{ million ha}}{3,200 \text{ million ha}} = \frac{19}{32}$$

Now divide the numbers by hand (notice that this is pretty close to 20/30, so you should expect a result not far from 2/3):

$$32\overline{)19} \rightarrow 32\overline{)19.0} \rightarrow \frac{\begin{array}{r} 0.5 \\ \hline 16.0 \end{array}}{} \rightarrow \frac{\begin{array}{r} 0.59 \\ 32\overline{)19.00} \\ \underline{160} \\ 300 \\ \underline{288} \\ 12 \end{array}}{} \rightarrow 0.59$$

To look at this percentage as a whole number, multiply your result by 100:

$$0.59 \times 100 = 59\%$$

Repeat this process for the other regions in figure 6.3, and fill in this table:

Original percentage forest cover

Asia	Africa	Europe	N., C. America	S. America	Oceania
59%					

Working with round numbers makes comparison easy. How important are the details lost when you read approximate numbers from the graph? What kinds of generalization might have gone into producing the FAO's original data? What kinds of generalization might have been unavoidable?

Answers: Approximate percentages should be: Africa 70%; N., C. America 29%; S. America 86%; Oceania 44%. Usually approximate numbers provide a quick, useful comparison, and further detail isn't needed until further analysis is done. Defining forests and forest types, or determining extent and "original" extent all involve considerable, usually unavoidable, generalization.

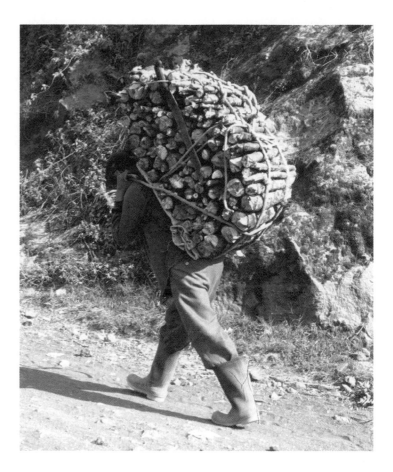

FIGURE 6.4 Firewood accounts for almost half of all wood harvested worldwide and is the main energy source for nearly half of all humans.

(fig. 6.4). The average amount of fuelwood used in less-developed countries is about 1 m³ per person per year, roughly equal to the amount that each American consumes each year as paper products alone. Demand for fuelwood, which is increasing at slightly less than the global population growth rate, is causing severe fuelwood shortages and depleting forests in some developing areas, especially around growing cities. About 1.5 billion people have less fuelwood than they need, and many experts expect shortages to worsen as poor urban areas grow. Because fuelwood is rarely taken from closed-canopy forest, however, it does not appear to be a major cause of deforestation. Some analysts argue that fuelwood could be produced sustainably in most developing countries, with careful management.

Approximately one-quarter of the world's forests are managed for wood production. Ideally, forest management involves scientific planning for sustainable harvests, with particular attention paid to forest regeneration. In temperate regions, according to the UN Food and Agriculture Organization, more land is being replanted or allowed to regenerate naturally than is being permanently deforested. Much of this reforestation, however, is in large plantations of single-species, single-use, intensive cropping called **monoculture forestry.** Although this produces rapid growth and easier harvesting than a more diverse forest, a dense, single-species stand often supports little biodiversity and does poorly in providing the ecological services, such as soil erosion control and clean water production, that may be the greatest value of native forests.

Some of the countries with the most successful reforestation programs are in Asia. China, for instance, cut down most of its forests 1,000 years ago and has suffered centuries of erosion and terrible floods as a consequence. Recently, however, timber cutting in the headwaters of major rivers has been outlawed, and a massive reforestation project has begun. In the 1990s, China planted 42 billion trees, mainly in Xinjiang Provence, to stop the spread of deserts. Korea and Japan also have had very successful forest restoration programs. After being almost totally denuded during World War II, both countries are now about 70 percent forested.

Tropical forests are being cleared for lumber and farming

Tropical forests, like the Ndoki forest in the opening Case Study, are among the richest and most diverse terrestrial systems. Although they now occupy less than 10 percent of the earth's land surface, these forests are thought to contain more than two-thirds of all higher plant biomass and at least half of all the plant, animal, and microbial species in the world.

A century ago, an estimated 12.5 million km² of tropical lands were covered with closed-canopy forest. This was an area larger than the entire United States. The FAO estimates that about 9.2 million ha, or about 0.6 percent, of the remaining tropical forest is cleared each year (fig. 6.6).

There is considerable debate about current rates of deforestation in the tropics. In 2003, satellite data showed more than 30,000 fires in a single month in Brazil. Remote sensing experts calculate that 3 million ha per year are now being cut and burned in the Amazon basin alone. However, there are different definitions of **deforestation.** Some scientists insist that it means a complete change from forest to agriculture, urban areas, or desert. Others include any area that has been logged, even if the cut was selective and regrowth will be rapid. Furthermore, savannas, open woodlands, and succession following natural disturbance are hard to distinguish from logged areas. Consequently, estimates for total tropical forest losses range from about 5 million to more than 20 million ha per year. The FAO estimates of 12.3 million ha deforested per year are generally the most widely accepted. To put that figure in perspective, it means that about 1 acre—or the area of a football field—is cleared every second, on average, around the clock.

By most accounts, Brazil has the highest total deforestation rate in the world, but it also has by far the largest tropical forests. Indonesia and Malaysia together may be losing as much primary forest each year as Brazil, even though their original amount was far lower. In 1997 forest fires on Borneo and Sumatra, exacerbated by severe drought, burned 20,000 km² (8,000 mi²). The fires reportedly were set both to clear land for agriculture and to hide illegal logging.

In Africa, the coastal forests of Senegal, Sierra Leone, Ghana, Madagascar, Cameroon, and Liberia already have been mostly destroyed. Haiti was once 80 percent forested; today, essentially all that forest has been destroyed, and the land lies barren and eroded. India, Burma, Kampuchea (Cambodia), Thailand, and Vietnam all have little old-growth lowland forest left. In Central America, nearly two-thirds of the original moist tropical forest has been destroyed, mostly within the past 30 years and primarily due to conversion of forest to cattle range. (See related story on "Disappearing Butterfly Forests" at www.mhhe.com/cunningham4e.)

Causes of deforestation

A variety of factors contribute to deforestation, and, as figure 6.5 shows, different forces predominate in various parts of the world. Logging for valuable tropical hardwoods, such as teak and mahogany, is generally the first step. Although loggers may take only one or two of the largest trees per hectare, the canopy of tropical forests is usually so strongly linked by vines and interlocking branches that felling one tree can bring down a dozen others. Building roads to remove logs kills more trees, but even more important, it allows entry to the forest by farmers, miners, hunters, and others who cause further damage (fig. 6.6).

In Africa, conversion of forest into small-scale agriculture accounts for nearly two-thirds of all tropical forest destruction. In Latin America, poor, landless farmers often start the deforestation but are bought out—or driven out—after a few years by large-scale farmers or ranchers (fig. 6.7). The thin, nutrient-poor tropical forest soils frequently are worn out after a few years of cropping, and much of this land becomes unproductive pasture or impoverished scrubland. Shifting cultivation (sometimes called

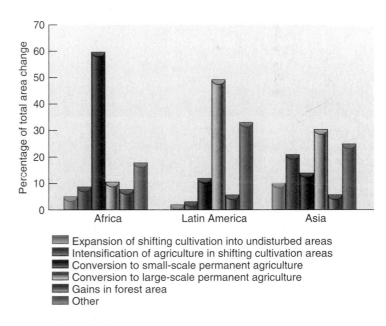

FIGURE 6.5 Tropical forest losses by different processes.
Source: Data from Food and Agriculture Organization of the United Nations, 2002.

1975   1989   2001

FIGURE 6.6 Forest destruction in Rondonia, Brazil, between 1975 and 2001. Construction of logging roads creates a feather-like pattern that opens forests to settlement by farmers.

FIGURE 6.7 Cattle graze on recently cleared tropical rainforest land. Millions of hectares have been cleared for pasture and cropland in recent decades. Unfortunately, the soil is poorly suited to grazing or farming, and new land must be cleared every few years.

"slash and burn" or milpa farming) is often blamed for forest destruction. This practice can be sustainable where population densities are low and individual plots are allowed to regenerate for a decade or two between cultivation periods. In some countries, however, growing populations and shrinking forests lead to short rotation cycles, and cropping intensification can lead to permanent deforestation.

As forests are cleared, rainfall patterns can change. A computer model created by Pennsylvania State University scientists suggests this phenomenon might create a kind of chain reaction. As forests are cut down, plant transpiration and rainfall decrease. Drought kills more vegetation, and fires become more numerous and extensive. In a worst-case scenario, an area as large as the entire Amazonian forest might be permanently damaged in just a few decades.

Forest protection

What can be done to stop this destruction and encourage tropical forest protection? While much of the news is discouraging, there are some hopeful signs for forest conservation in the tropics. Many countries now recognize that forests are valuable resources. Intensive scientific investigations are underway to identify the best remaining natural areas (see Exploring Science, p. 133).

About 12 percent of all world forests are in some form of protected status, but the effectiveness of that protection varies greatly. Costa Rica has one of the best plans for forest guardianship in the world. Attempts are being made there not only to rehabilitate the land (make an area useful to humans) but also to restore the ecosystems to naturally occurring associations. One of the best-known of these projects is Dan Janzen's work in Guanacaste National Park. Like many dry, tropical forests, the northwestern part of Costa Rica had been almost completely converted to ranchland. By controlling fires, however, Janzen and his coworkers are bringing back the forest. One of the keys to this success is involving local people in the project. Janzen also permits grazing in the park. The original forest evolved, he reasons, together with ancient grazing animals that are now extinct. Horses and cows can play a valuable role as seed dispersers.

Protecting areas such as the Nouabelé-Ndoki forest (opening Case Study) is hard because information is so elusive. Deep, remote, swampy, tropical jungles are difficult to enter, map, and assess for their ecological value. Yet without information about their ecological importance, most people have little reason to care about these remote, trackless forests. How can you conserve ecosystems if you don't know what's there?

For most of history, understanding the extent and conditions of a remote area required an arduous trek to see the place in person. Even on publicly owned lands, only those who could afford the time, or who could afford to pay surveyors, might understand the resources. Over time, maps improved, but maps usually show only a few features, such as roads, rivers, and some boundaries.

In recent years, details about public lands and resources have suddenly burst into public view through the use of geographic information systems (**GIS**). A GIS consists of spatial data, such as boundaries or road networks, and software to display and analyze the data. Spatial data can include variables that are hard to see on the ground—watershed boundaries, annual rainfall, land ownership, or historical land use. Data can also represent phenomena much larger than we can readily see—land surface slopes and elevation, forested regions, river networks, and so on. By overlaying these layers, GIS analysts can investigate completely new questions about conservation, planning, and restoration.

You have probably used a GIS. Online mapping programs such as MapQuest or Google Earth organize and display spatial data. They let you turn layers on and off, or zoom in and out to display different scales. You can also use an online mapping program to calculate distances and driving directions between places. An ecologist, meanwhile, might use a GIS to calculate the extent of habitat areas, to monitor changes in area, to calculate the size of habitat fragments, or to calculate the length of waterways in a watershed.

Identifying Priority Areas

Recently, a joint effort of several conservation organizations has used GIS and spatial data to identify priority areas for conservation in Central Africa, in the region surrounding the Nouabelé-Ndoki Park. The project was initiated because new data, including emerging GIS data, were showing dramatic increases in planned logging, in a region that contains the world's second greatest extent of tropical forest (fig. 1).

Researchers from the Wildlife Conservation Society, Worldwide Fund for Nature, World Resources Institute, USGS, and other agencies and groups, began collecting GIS data on a variety of variables. They identified the range of great apes and other rare or threatened species. They identified areas of

FIGURE 1 Gabon, Central Africa, has seen a steady increase in logging concessions.
Source: Wildlife Conservation Society.

Logging concessions
Remaining forest

Isolated forest
Forest within 30 km from a road

Existing conservation areas
Proposed

FIGURE 2 A few GIS layers used to identify priority conservation areas.
Source: Wildlife Conservation Society.

extreme plant diversity. They calculated the sizes of forest fragments to identify concentrations of intact, ancient forests. Using maps of logging roads, they calculated the area within a 30 km "buffer" around roads, since loggers, settlers, and hunters usually threaten biodiversity near roads. They also mapped existing and planned conservation areas (fig. 2).

By overlaying these and other layers, analysts identified priority conservation areas of extensive original forest, which have high biodiversity and rare species. Overlaying these priority areas with a map of protected lands and a map of timber concessions, they identified *threatened* priority areas (fig. 3).

Most of the unprotected priority areas may never be protected, but having this map provides two important guides for future conservation. First, it assesses the state of the problem. With this map, we know that most of the forest is unprotected but also that the region's primary forest is extensive. Second, this map provides priorities for conservation planning. In addition, maps are very effective tools for publicizing an issue. When a map like this is published, more

people become enthusiastic about joining the conservation effort.

GIS has become an essential tool for conserving forests, grasslands, ecosystems, and nature preserves. GIS has revolutionized the science of planning and conservation—examining problems using quantitative data—just as it may have revolutionized the way you plan a driving trip.

National parks
Original forest
Priority areas

FIGURE 3 Priority areas outside of national parks.
Source: Wildlife Conservation Society.

People also are working on the grassroots level to protect and restore forests in other countries. India, for instance, has a long history of nonviolent, passive resistance movements—called *satyagrahas*—to protest unfair government policies. These protests go back to the beginning of Indian culture and often have been associated with forest preservation. Gandhi drew on this tradition in his protests of British colonial rule in the 1930s and 1940s. During the 1970s, commercial loggers began large-scale tree felling in the Garhwal region in the state of Uttar Pradesh in northern India. Landslides and floods resulted from stripping the forest cover from the hills. The firewood on which local people depended was destroyed, and the way of life of the traditional forest culture was threatened. In a remarkable display of courage and determination, the village women wrapped their arms around the trees to protect them, sparking the Chipko Andolan movement (literally, movement to hug trees). They prevented logging on 12,000 km^2 of sensitive watersheds in the Alakanada basin. Today the Chipko Andolan movement has grown to more than 4,000 groups working to save India's forests.

Debt-for-nature swaps

Those of us in developed countries also can contribute toward saving tropical forests. Financing nature protection is often a problem in developing countries, where the need is greatest. One promising approach is called **debt-for-nature swaps.** Banks, governments, and lending institutions now hold nearly $1 trillion in loans to developing countries. There is little prospect of ever collecting much of this debt, and banks are often willing to sell bonds at a steep discount—perhaps as little as 10 cents on the dollar. Conservation organizations buy debt obligations on the secondary market at a discount and then offer to cancel the debt if the debtor country agrees to protect or restore an area of biological importance.

There have been many such swaps. Conservation International, for instance, bought $650,000 of Bolivia's debt for $100,000—an 85 percent discount. In exchange for canceling this debt, Bolivia agreed to protect nearly 1 million ha (2.47 million acres) around the Beni Biosphere Reserve in the Andean foothills. Ecuador and Costa Rica have had a different kind of debt-for-nature swap. They have exchanged debt for local currency bonds that fund activities of local private conservation organizations in the country. This has the dual advantage of building and supporting indigenous environmental groups while protecting the land. Critics, however, charge that these swaps compromise national sovereignty and do little to reduce the developing world's debt or to change the situations that led to environmental destruction in the first place.

Temperate forests have competing uses

Tropical countries aren't unique in harvesting forests at an unsustainable rate. Northern countries, such as the United States and Canada, also have allowed controversial forest management practices in many areas. For many years, the official policy of the U.S. Forest Service was "multiple use," which implied that the forests could be used for everything that we might want to do

FIGURE 6.8 The temperate rainforests of the Pacific Northwest have the highest biomass per hectare of any landscape in the world.

there simultaneously. Some uses are incompatible, however. Birdwatching, for example, isn't very enjoyable in a dirt bike racing course. And protecting species that need unbroken old-growth forest isn't feasible when you cut down the forest.

Old-growth forests

The most contentious forestry issues in the United States and Canada in recent years have centered on logging in old-growth forests in the Pacific Northwest. These forests have incredibly high levels of biodiversity, and they accumulate more total biomass in standing vegetation per unit area than any other ecosystem on earth (fig. 6.8). Many endemic species, such as the northern spotted owl (fig. 6.9), Vaux's swift, and the marbled murrelet, are so highly adapted to the unique conditions of these ancient forests that they live nowhere else.

Less than 10 percent of old-growth forest in the United States remains intact, and 80 percent of what remains is scheduled to be cut in the near future. Just a century ago, most of the coastal ranges of Washington, Oregon, northern California, British Columbia, and southeastern Alaska were clothed in a lush forest of huge trees, many a thousand years old or more. British Columbia has felled at least 60 percent of its richest and most productive ancient forests and is now cutting some 240 million ha (600 million acres) annually, about ten times the rate of old-growth harvest in the United States. At these rates, the only remaining ancient forests in North America in 50 years will be a fringe around the base of the mountains in a few national parks.

In 1989 wildlife conservationists sued the U.S. Forest Service over logging rates in Washington and Oregon, arguing that northern spotted owls are endangered and must be protected under the Endangered Species Act. Federal courts agreed and ordered some 1 million ha (2.5 million acres) of ancient forest set aside to preserve the last 2,000 pairs of owls. This is about half the remaining virgin forest in Washington and Oregon. The timber industry claimed that 40,000 jobs would be lost, although economic

FIGURE 6.9 Only about 2,000 pairs of northern spotted owls remain in the old-growth forests of the Pacific Northwest. Cutting old-growth forests threatens endangered species, but reduced logging threatens the jobs of many timber workers.

FIGURE 6.10 This huge clear-cut in Washington's Gifford Pinchot National Forest threatens species dependent on old-growth forest and exposes steep slopes to soil erosion. Restoring something like the original forest will take hundreds of years.

studies show that most logging jobs were lost as a result of log exports and mechanization. Still, outrage in the logging communities was loud and clear. Convoys of logging trucks converged on protest sites, while angry crowds burned conservationists in effigy. Bumper stickers urged, "Save a logger; eat an owl."

A compromise forest management plan was finally agreed upon that allows some continued logging but protects the most valuable forests, including selected old-growth preserves, riparian (streamside) buffer strips, and prime wildlife habitat. This plan may not be enough protection, however, to ensure the survival of endangered salmon and trout populations in northwestern rivers. An even bigger battle may be shaping up as specific fish populations are considered for listing as endangered species. Several salmon and steelhead trout populations were listed as endangered under the Clinton administration. These listings have been challenged by the timber industry, but the $1 billion-per-year fishing industry in the Pacific Northwest may be at least as powerful a constituency as were those concerned about spotted owls.

Harvest methods

Most lumber and pulpwood in the United States and Canada currently are harvested by **clear-cutting,** in which every tree in a given area is cut, regardless of size (fig. 6.10). This method is effective for producing even-age stands of sun-loving species, such as aspen or pines, but often increases soil erosion and eliminates habitat for many forest species when carried out on large blocks. It was once thought that good forest management required immediate removal of all dead trees and logging residue. Research has shown, however, that standing snags and coarse woody debris play important ecological roles, including soil protection, habitat for a variety of organisms, and nutrient recycling.

Some alternatives to clear-cutting include **shelterwood harvesting,** in which mature trees are removed in a series of two or more cuts, and **strip-cutting,** in which all the trees in a narrow corridor are harvested. For many forest types, the least disruptive harvest method is **selective cutting**, in which only a small percentage of the mature trees are taken in each 10- or 20-year rotation. Ponderosa pine, for example, are usually selectively cut to thin stands and improve growth of the remaining trees. A forest managed by selective cutting can retain many of the characteristics of age distribution and groundcover of a mature old-growth forest. (See related story "Forestry for the Seventh Generation" at www.mhhe.com/cunningham4e.)

How much should we subsidize logging on public lands?

An increasing number of people in the United States are calling for an end to all logging on federal lands. Many argue that ecological services, from maintaining river levels for fish and irrigation to recreation, generate more revenue at lower costs. Many remote communities depend on logging jobs, but these jobs depend on subsidies. The federal government builds roads, manages forests, fights fires, and sells timber for less than the administrative costs of the sales. What is the best policy?

Some argue that logging should be restricted to privately owned lands. Just 4 percent of the nation's timber comes from national forests, and this harvest adds about $4 billion to the American economy per year. In contrast, recreation, fish and wildlife, clean water, and other ecological services provided by the forest, by their calculations, are worth at least $224 billion each year. Timber industry officials, on the other hand, dispute these claims, arguing that logging not only provides jobs and supports rural communities but also keeps forests healthy. What do you think? Could we make up for decreased timber production

Lowering Your Forest Impacts

For most urban residents, forests—especially tropical forests—seem far away and disconnected from everyday life. There are things that each of us can do, however, to protect forests.

- Reuse and recycle paper. Make double-sided copies. Save office paper, and use the back for scratch paper.

- Use email. Store information in digital form, rather than making hard copies of everything.

- If you build, conserve wood. Use wafer board, particle board, laminated beams, or other composites, rather than plywood and timbers made from old-growth trees.

- Buy products made from "good wood" or other certified sustainably harvested wood.

- Don't buy products made from tropical hardwoods, such as ebony, mahogany, rosewood, or teak, unless the manufacturer can guarantee that the hardwoods were harvested from agroforestry plantations or sustainable-harvest programs.

- Don't patronize fast-food restaurants that purchase beef from cattle grazing on deforested rainforest land. Don't buy coffee, bananas, pineapples, or other cash crops if their production contributes to forest destruction.

- Do buy Brazil nuts, cashews, mushrooms, rattan furniture, and other nontimber forest products harvested sustainably by local people from intact forests. Remember that tropical rainforest is not the only biome under attack. Contact the Taiga Rescue Network (www.taigarescue.org) for information about boreal forests.

- If you hike or camp in forested areas, practice minimum-impact camping. Stay on existing trails, and don't build more or bigger fires than you absolutely need. Use only downed wood for fires. Don't carve on trees or drive nails into them.

- Write to your congressional representatives, and ask them to support forest protection and environmentally responsible government policies. Contact the U.S. Forest Service, and voice your support for recreation and nontimber forest values.

from public lands by more intensive management of private holdings and by substitution or recycling of wood products? Are there alternative ways you could suggest to support communities now dependent on timber harvesting?

Roads on public lands are another controversy. Over the past 40 years, the Forest Service has expanded its system of logging roads more than ten-fold, to a current total of nearly 550,000 km (343,000 mi), or more than ten times the length of the interstate highway system. Government economists regard road building as a benefit because it opens up the country to motorized recreation and industrial uses. Wilderness enthusiasts and wildlife supporters, however, see this as an expensive and disruptive program. In 2001 the Clinton administration announced a plan to protect 23.7 million ha (58.5 million acres) of de facto wilderness from

roads. Timber, mining, and oil companies protested this rule. In 2002 a federal judge in Idaho ruled that blocking road construction would do "irreparable harm" to the forest. He also ruled that the 600 public meetings held over a 12-month period to discuss this plan, and the record 1.6 million written comments gathered (90 percent of which supported increased protection), did not represent adequate public input. The Bush administration overturned the "roadless rule" and ordered resource managers to expedite logging, mining, and motorized recreation. What do you think? How much of the remaining old growth should be protected as ecological reserves?

Fire management

Following a series of disastrous fire years in the 1930s, in which hundreds of millions of hectares of forest were destroyed, whole towns burned to the ground, and several hundred people died, the U.S. Forest Service adopted a policy of aggressive fire control in which every blaze on public land was to be out before 10 A.M. Smokey Bear was adopted as the forest mascot and warned us that "only you can prevent forest fires." Recent studies, however, of fire's ecological role suggest that our attempts to suppress all fires may have been misguided. Many biological communities are fire-adapted and require periodic burning for regeneration. Furthermore, eliminating fire from these forests has allowed woody debris to accumulate, greatly increasing the chances of a very big fire (fig. 6.11).

Forests that once were characterized by 50 to 100 mature, fire-resistant trees per hectare and an open understory now have a thick tangle of up to 2,000 small, spindly, mostly dead saplings in the same area. The U.S. Forest Service estimates that 33 million ha (73 million acres), or about 40 percent of all federal forestlands, are at risk of severe fires. To make matters worse, Americans increasingly

FIGURE 6.11 By suppressing fires and allowing fuel to accumulate, we make major fires such as this more likely. The safest and most ecologically sound management policy for some forests may be to allow natural or prescribed fires, which don't threaten property or human life, to burn periodically.

What Do You Think?

Forest Thinning and Salvage Logging

For more than 70 years, firefighting has been a high priority for forest managers. Unfortunately, our efforts have been so successful that dead wood and brush have now built up to dangerous levels in many forests. Lands that would once have been cleaned out by frequent low-temperature ground fires now have so much accumulated fuel that a catastrophic wildfire is all but inevitable. To make matters worse, increasing numbers of people are building cabins and homes in remote, fire-prone areas, where they expect to be protected from unavoidable risks.

People living in or near national forests demand protection from wildfires. In response, federal agencies are starting thinning programs to remove excess fuel. To make it profitable for loggers to remove fire-prone dead wood, small trees, and brush, the government is allowing them to harvest large, valuable, and fire-resistant trees located in the backcountry, often miles away from the nearest communities. And to avoid what many loggers claim is "red tape and litigation" that have tied up forest managers in "analysis paralysis," most thinning projects are exempt from public comment and administrative appeals, as well as from environmental reviews under the National Environmental Policy Act (NEPA).

Environmental groups denounce these projects as merely logging without laws. Thinning, they argue, looks very much like clearcutting. Forest ecologists argue that thinning, unless it is repeated every few years, actually makes the forests more, rather than less, fireprone. Removing big, old trees opens up the forest canopy and encourages growth of brush and new tree seedlings. Furthermore, logging

Demonstration forest thinning (or fuel reduction) project in the Coconino National Forest, Arizona. "Exactly this kind of treatment has to happen across the western U.S.," said Forest Service chief Dale Bosworth. "We have only 20 years to treat 30 million acres."

compacts soil and introduces invasive species. Many forest ecologists maintain that small, prescribed burns can reduce fuel and produce healthier forests than commercial logging.

A 2005 study of an Oregon fire found that salvage logging actually increased fire susceptibility. Dry branches and debris left on the ground provided much better fuel than did standing dead trees. In addition, seedling growth among standing dead trees was robust, with three times the density of seedlings in salvage areas.

Other research calls into question a main justification for thinning: protecting housing that has proliferated around the edges of federal forests. The only thinning needed to protect houses, according to fire experts, is within about 60 m around the building. Regardless of how intense the fire is, building damage can be minimized by installing a metal roof, clearing pine needles and brush around the house, and removing trees immediately around the building.

After a fire has burned a forest, local residents often call for emergency salvage logging to utilize dead trees before they fall and become fuel for future fires. Foes say that salvage sales allow timber companies to cut many trees that survive the flames and, like thinning, leave roads and scars from heavy equipment that last far longer than any effects of the fires themselves. And, critics contend, the salvage contracts rarely require timber companies to remove the very fuel that stokes wildfires—the underbrush and downed timber for which there is little or no market. The result, some conservationists say, is the ecological equivalent of mugging a fire victim.

What do you think? How can we get out of the crisis we've created by preventing fires and letting dead wood accumulate? If you were a forest manager, would you authorize thinning and salvage logging, or can you think of other ways to remove forest fuels in an ecologically and economically sustainable manner? What research projects would you direct your staff to undertake in order to inform your decision in this dilemma?

---

live in remote areas where wildfires are highly likely. Because there haven't been fires in many of these places in living memory, many people assume there is no danger, but by some estimates, 40 million U.S. residents now live in areas with high wildfire risk.

A recent prolonged drought in the western United States has heightened fire danger. In 2002 more than 88,000 wildfires burned 2.8 million ha (6.9 million acres) of forests and grasslands in the United States. Federal agencies spent about $1.6 billion to fight these fires, nearly four times the previous ten-year average.

The dilemma is how to undo years of fire suppression and fuel buildup. Fire ecologists favor small, prescribed burns to clean out debris. Loggers decry this approach as a waste of valuable timber, and local residents of fire-prone areas fear that prescribed fires will escape and threaten them. Recently the Forest Service proposed a massive new program of forest

thinning and emergency salvage operations (removing trees and flammable material from mature or recently burned forests) on 16 million ha (40 million acres) of national forest. Carried out over a 20-year period, this program could cost as much as $12 billion and would open up much roadless, de facto wilderness to large-scale logging. Critics complain that this program is ill advised and environmentally destructive. Proponents argue that the only way to save the forest is to log it (see What Do You Think, p. 137).

Ecosystem management

In the 1990s the U.S. Forest Service began to shift its policies from a timber production focus to **ecosystem management,** which attempts to integrate sustainable ecological, economic, and

TABLE 6.1

Draft Criteria for Sustainable Forestry

1. Conservation of biological diversity
2. Maintenance of productive capacity of forest ecosystems
3. Maintenance of forest ecosystem health and vitality
4. Maintenance of soil and water resources
5. Maintenance of forest contribution to global carbon cycles
6. Maintenance and enhancement of long-term socioeconomic benefits to meet the needs of legal, institutional, and economic framework for forest conservation and sustainable management

Source: Data from USFS, 2002.

social goals in a unified, systems approach. Some of the principles of this new philosophy include

- Managing across whole landscapes, watersheds, or regions over ecological time scales
- Considering human needs and promoting sustainable economic development and communities
- Maintaining biological diversity and essential ecosystem processes
- Utilizing cooperative institutional arrangements
- Generating meaningful stakeholder and public involvement and facilitating collective decision making
- Adapting management over time, based on conscious experimentation and routine monitoring

Some critics argue that we don't understand ecosystems well enough to make practical decisions in forest management on this basis. They argue we should simply set aside large blocks of untrammeled nature to allow for chaotic, catastrophic, and unpredictable events. Others see this new approach as a threat to industry and customary ways of doing things. Still, elements of ecosystem management appear in the *National Report on Sustainable Forests* prepared by the U.S. Forest Service. Based on the Montreal Working Group criteria and indicators for forest health, this report suggests goals for sustainable forest management (table 6.1).

GRASSLANDS

After forests, grasslands are among the biomes most heavily used by humans. Prairies, savannas, steppes, open woodlands, and other grasslands occupy about one-quarter of the world's land surface. Much of the U.S. Great Plains and the Prairie Provinces of Canada fall in this category (fig. 6.12). The 3.8 billion ha (12 million mi^2) of pastures and grazing lands in this biome make up about twice the area of all agricultural crops. When you add to this about 4 billion ha of other lands (forest, desert, tundra, marsh, and thorn scrub) used for raising livestock, more than half of all land is used at least occasionally for grazing. More than 3 billion cattle, sheep, goats, camels, buffalo, and other domestic animals on these lands make a valuable contribution to human nutrition. Sustainable pastoralism can increase productivity while maintaining biodiversity in a grassland ecosystem.

Because grasslands, chaparral, and open woodlands are attractive for human occupation, they frequently are converted to cropland, urban areas, or other human-dominated landscapes. Worldwide the rate of grassland disturbance each year is three times that of tropical forest. Although they may appear to be uniform and monotonous to the untrained eye, native prairies can be highly productive and species-rich. According to the U.S. Department of Agriculture, more threatened plant species occur in rangelands than in any other major American biome.

Grazing can be sustainable or damaging

By carefully monitoring the numbers of animals and the condition of the range, ranchers and **pastoralists** (people who live by herding animals) can adjust to variations in rainfall, seasonal plant conditions, and the nutritional quality of forage to keep livestock healthy and avoid overusing any particular area. Conscientious management can actually improve the quality of the range.

When grazing lands are abused by overgrazing—especially in arid areas—rain runs off quickly before it can soak into the soil to nourish plants or replenish groundwater. Springs and wells dry up. Seeds can't germinate in the dry, overheated soil. The barren ground reflects more of the sun's heat, changing wind patterns, driving away moisture-laden clouds, and leading to further desiccation. This process of conversion of once fertile land to desert is called **desertification.**

This process is ancient, but in recent years it has been accelerated by expanding populations and the political conditions that force people to overuse fragile lands. According to the International Soil Reference and Information Centre in the Netherlands,

FIGURE 6.12 This short-grass prairie in northern Montana is too dry for trees but, nevertheless, supports a diverse biological community.

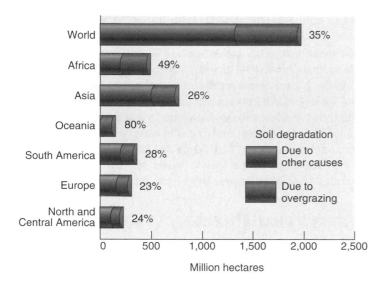

FIGURE 6.13 Rangeland soil degradation due to overgrazing and other causes. Notice that, in Europe, Asia, and the Americas, farming, logging, mining, urbanization, and so on are responsible for about three-quarters of all soil degradation. In Africa and Oceania, where more grazing occurs and desert or semiarid scrub make up much of the range, grazing damage is higher. **Source:** Data from World Resource Institute.

FIGURE 6.14 More than half of all publicity owned grazing land in the United States is in poor or very poor condition. Overgrazing and invasive weeds are the biggest problems.

nearly three-quarters of all rangelands in the world show signs of either degraded vegetation or soil erosion. Overgrazing is responsible for about one-third of that degradation (fig. 6.13). The highest percentage of moderate, severe, and extreme land degradation is in Mexico and Central America, while the largest total area is in Asia, where the world's most extensive grasslands occur. Can we reverse this process? In some places, people are reclaiming deserts and repairing the effects of neglect and misuse.

Overgrazing threatens many U.S. rangelands

The health of most public grazing lands in the United States is not good. Political and economic pressures encourage managers to increase grazing allotments beyond the carrying capacity of the range. Lack of enforcement of existing regulations and limited funds for range improvement have resulted in **overgrazing,** damage to vegetation and soil including loss of native forage species and erosion. The Natural Resources Defense Council claims that only 30 percent of public range-lands are in fair condition, and 55 percent are poor or very poor (fig. 6.14).

Overgrazing has allowed populations of unpalatable or inedible species, such as sage, mesquite, cheatgrass, and cactus, to build up on both public and private rangelands. Wildlife conservation groups regard cattle grazing as the most ubiquitous form of ecosystem degradation and the greatest threat to endangered species in the southwestern United States. They call for a ban on cattle and sheep grazing on all public lands, noting that it provides only 2 percent of the total forage consumed by beef cattle and supports only 2 percent of all livestock producers.

Like federal timber management policy, grazing fees charged for use of public lands often are far below market value and represent an enormous hidden subsidy to western ranchers. Holders of grazing permits generally pay the government less than 25 percent the amount of leasing comparable private land. The 31,000 permits on federal range bring in only $11 million in grazing fees but cost $47 million per year for administration and maintenance. The $36 million difference amounts to a massive "cow welfare" system of which few people are aware.

On the other hand, ranchers defend their way of life as an important part of western culture and history. Although few cattle go directly to market from their ranches, they produce almost all the beef calves subsequently shipped to feedlots. And without a viable ranch economy, they claim, even more of the western landscape would be subdivided into small ranchettes to the detriment of both wildlife and environmental quality. What do you think? How much should we subsidize extractive industries to preserve rural communities and traditional occupations?

Ranchers are experimenting with new methods

Where a small number of livestock are free to roam a large area, they generally eat the tender, best-tasting grasses and forbs first, leaving the tough, unpalatable species to flourish and gradually dominate the vegetation. In some places, farmers and ranchers find that short-term, intensive grazing helps maintain forage quality. As South African range specialist Allan Savory observed, wild ungulates (hoofed animals), such as gnus or zebras in Africa or bison (buffalo) in America, often tend to form dense herds that graze briefly but intensively in a particular location before moving on to the next area. Rest alone doesn't necessarily improve pastures and rangelands. Short-duration, **rotational grazing**—confining animals to a small area for a short time (often only a day or two) before shifting them to a new location—simulates the effects of wild herds (fig. 6.15). Forcing livestock to eat everything equally, to trample the ground thoroughly, and to fertilize heavily with manure before

FIGURE 6.15 Intensive, rotational grazing encloses livestock in a small area for a short time (often only one day) within a movable electric fence to force them to eat vegetation evenly and fertilize the area heavily.

FIGURE 6.16 Red deer (*Cervus elaphus*) are raised in New Zealand for antlers and venison.

moving on helps keep weeds in check and encourages the growth of more desirable forage species. This approach doesn't work everywhere, however. Many plant communities in the U.S. desert Southwest, for example, apparently evolved in the absence of large, hoofed animals and can't withstand intensive grazing.

Another approach to ranching in some areas is to raise wild species, such as red deer, impala, wildebeest, or oryx (fig. 6.16). These animals forage more efficiently, resist harsh climates, often are more pest- and disease-resistant, and fend off predators better than usual domestic livestock. Native species also may have different feeding preferences and needs for water and shelter than cows, goats, or sheep. The African Sahel, for instance, can

provide only enough grass to raise about 20 to 30 kg (44 to 66 lbs) of beef per hectare. Ranchers can produce three times as much meat with wild native species in the same area because these animals browse on a wider variety of plant materials.

In the United States, ranchers find that elk, American bison, and a variety of African species take less care and supplemental feeding than cattle or sheep and result in a better financial return because their lean meat can bring a better market price than beef or mutton. Media mogul Ted Turner has become both the biggest private landholder in the United States and the owner of more American bison than anyone other than the government.

ECOSYSTEM PRESERVATION

While most forests and grasslands serve useful, or utilitarian, purposes most societies also set aside some natural areas for aesthetic or recreational purposes. Natural preserves have existed for thousands of years. Ancient Greeks protected sacred groves for religious purposes. Royal hunting grounds have preserved forests in Europe for centuries. Although these areas were usually reserved for elite classes in society, they have maintained biodiversity and natural landscapes in regions where most lands are heavily used.

The first public parks open to ordinary citizens may have been the tree-sheltered agoras in planned Greek cities. But the idea of providing natural space for recreation, and to preserve natural environments, has really developed in the past century or so. While the first parks were intended mainly for the recreation of growing urban populations, parks have taken on many additional purposes. Today we see our national parks as playgrounds for rest and recreation, as havens for wildlife, as places to experiment with ecological management, and as opportunities to restore ecosystems.

National parks serve many purposes

Yellowstone National Park was the first national park in the world, established in 1872 by President Ulysses S. Grant. At the time, the United States was emerging from the Civil War, and railroads were changing the landscape. Public lands were rapidly disappearing, as they were sold or given to settlers, railroads, and timber companies. Establishing a large, remote, wilderness park just to preserve natural scenery was a revolutionary but sensible idea. Yellowstone encompassed about 800,000 ha (2 million acres) in the Wyoming, Montana, and Idaho territories. Although its purpose was to protect the natural curiosities of the geysers and hot springs (fig. 6.17), the park was large enough to encompass and preserve extensive wilderness. As the North American frontier closed, and as wild nature and scenic beauty continued to disappear, interest in national parks grew. Canada established Banff National Park in 1887. Yoho, Waterton Lakes, and St. Lawrence Islands soon followed. In just 15 years, between 1900 and 1915, the United States added the Grand Canyon, Glacier, Mesa Verde, and Rocky Mountain National Park to the federal park system.

The United States' national park system has grown to more than 376 parks, monuments, historic sites, and recreation areas, which comprise over 280,000 km² (108,000 mi²). About 300 million

FIGURE 6.17 Yellowstone National Park, established in 1872, is regarded as the first national park in the world. Although initial interest focused on the park's natural curiosities and wonders, Yellowstone has come to be appreciated for its beauty and wilderness values.

FIGURE 6.18 Golden Gate National Recreation Area, the largest urban park in the world, gets more than 13 million recreational visits per year.

people visit the parks each year. The most heavily visited units are the urban recreation areas, parkways, and historic sites (fig. 6.18). What most people imagine when they think of a national park, however, are the great wilderness parks of the West. While these parks preserve areas of great natural beauty, they often have far more rocks, ice, and snow than biodiversity or ecological complexity. When the U.S. public domain was being carved up over a century ago, the most productive land generally ended up in private hands.

Most of our national parks are intended to preserve an area as we believe it looked before any human habitation. Often this is an illusion because people have inhabited and affected the land for thousands of years. Furthermore, the idea of maintaining a static equilibrium in nature is based on ideas of climax communities discarded by most ecologists today. What would it mean for park management if we conceived of biological communities

FIGURE 6.19 Off-road vehicles cause severe, long-lasting environmental damage when driven through wetlands.

as constantly changing mosaics of chance associations of plants and animals (including humans)?

Users place conflicting demands on our parks

Originally the great wilderness parks of Canada and the United States were fortresses protected from development or exploitation by legal boundaries and diligent park rangers. Most were buffered from human impacts by their remote location and the wild lands surrounding them. Today the situation has changed. Many parks have become islands of nature surrounded and threatened by destructive land uses and burgeoning human populations that crowd park boundaries. Forests are clear-cut right up to the edges of some parks, while mine drainage contaminates streams and groundwater. Garish tourist traps clustered at park entrances detract from the beauty and serenity that most visitors seek.

Some popular parks, such as Yosemite and Grand Canyon, have become so crowded that managers have had to restrict automobile access and institute mass-transit systems to reduce traffic jams and pollution. Sometimes trails are so packed with tourists that they seem more like city streets than wilderness refuges. Visitors demand services, such as golf courses, laundries, video arcades, bars, grocery stores, and other facilities, that intrude on the solitude and natural beauty that parks were established to protect. Park rangers often spend more time on crime prevention and crowd control than natural history.

In many parks, dune buggies, dirt bikes, and off-road vehicles (ORV) run over fragile landscapes, disturbing vegetation and wildlife and destroying the aesthetic experience of those who come to enjoy nature (fig. 6.19). The U.S. Park Service has proposed plans to exclude snowmobiles and personal watercraft (jet skis and waverunners) from many parks, but these plans have been challenged by manufacturers and owners' groups. Yellowstone National Park is a case in point. On a typical winter weekend in 2000, more than 1,000 snowmobiles entered the park. Air pollution from their engines was so bad that rangers at the west Yellowstone entry point were forced to wear respirators. Although the snowmobile phase-out plan was based on 22 public hearings and 65,000 written comments (a majority of which favored a ban), Secretary of the Interior Gale Norton ordered the Park Service to continue to allow snowmobile use. A federal court judge then ruled, however, the secretary's ruling lacked merit and ordered the park to ban snowmobiles. At the time of this writing, the issue was still being contested. Ask your instructor what has happened recently.

Air pollution has become a serious threat to other parks as well. Sulfate concentrations in Shenandoah and Great Smoky Mountains National Parks are five times human health standards, and ozone levels in Acadia National Park in Maine exceed primary air quality standards by as much as 50 percent on some summer days. Visitors to the Grand Canyon once could see mountains 160 km (100 mi) away; now the air is so smoggy you can't see from one rim to the other during one-third of the year because of soot and sulfate aerosols from nearby power plants in Utah and Arizona.

Mining and oil interests continue to push for permission to dig and drill in the parks, especially on the 3 million acres of private **inholdings** (private lands) in the parks. Oil and gas wells recently have been drilled in Padre Island National Seashore, Big Cypress National Preserve, and the Upper Missouri River Breaks National Monument.

We expect parks to support wildlife

Wildlife is at the center of many arguments regarding whether the purpose of the parks is to preserve nature or to provide entertainment for visitors (fig. 6.20). In the early days of the parks, "bad" animals (such as wolves and mountain lions) were killed, so that populations of "good" animals (such as deer and elk) would be high. Rangers cut trees to improve views, put out salt blocks to lure animals to good viewing points close to roads, and otherwise manipulated nature to provide a more enjoyable experience for the guests.

Critics of this policy claim that favoring some species over others has unbalanced ecosystems and created a misleading illusion of a natural system. They claim that excessively large deer and elk populations in Yellowstone and Grand Teton National Parks, for instance, have degraded the range so badly that massive die-offs have occurred in severe winters. Park rangers tried hiring professional hunters to reduce the herds, but a storm of protest was raised. Sportsmen want to be able to hunt the animals themselves, animal lovers don't want the animals to be killed at all, and wilderness advocates don't like the precedent of hunting in national parks. Proposals to reduce elk and bison populations by reintroducing predators, such as wolves and mountain lions, have been highly controversial as discussed in the opening case study.

One of the biggest problems with managing parks and nature preserves is that boundaries usually are based on political rather than ecological considerations. Airsheds, watersheds, and animal territories or migration routes often extend far beyond official boundaries and yet profoundly affect communities that we are attempting to preserve. Yellowstone and Grand Teton National Parks in northwestern Wyoming are examples of this dilemma.

FIGURE 6.20 Wild animals have always been one of the main attractions in national parks. Many people lose all common sense when interacting with big, dangerous animals. This is not a petting zoo.

Although about 1 million ha (2.47 million acres) in total size, these parks probably cannot preserve viable populations of large predators, such as grizzly bears and mountain lions, completely within their borders. Management policies in the surrounding national forests and private lands seriously affect conditions in the park (see opening Case Study, chapter 5). The natural **biogeographical area** (an entire ecosystem and its associated land, water, air, and wildlife resources) must be managed as a unit if we are to preserve all its values.

The International Union for the Conservation of Nature and Natural Resources (IUCN) (also known as the World Conservation Union) divides protected areas into five categories, with increasing levels of protection and decreasing human impacts (table 6.2). Many of our parks have tried to meet all these goals simultaneously but may have to select those of highest value. Most parks limit the number of overnight visitors. The time may come when park permits will need to be reserved years in advance, and visits to certain parks will be limited to once in a lifetime! How would you feel about such a policy? Would you rather visit a pristine, uncrowded park only once or a less perfect place whenever you wanted?

One solution to congestion and overuse of our parks is to create new ones to distribute the load. This would also allow us to protect and enjoy other magnificent areas that deserve preservation in their own right. At the end of his term in office, President Bill Clinton created ten new parks and national monuments,

TABLE 6.2

IUCN Categories of Protected Areas

Category	Allowed Human Impact or Intervention
1. Ecological reserves and wilderness areas	Little or none
2. National parks	Low
3. Natural monuments and archaeological sites	Low to medium
4. Habitat and wildlife management areas	Medium
5. Cultural or scenic landscapes, recreation areas	Medium to high

Source: Data from World Conservation Union, 1990.

FIGURE 6.21 Corridors serve as routes of migration, linking isolated populations of plants and animals in scattered nature preserves. Although individual preserves may be too small to sustain viable populations, connecting them through river valleys and coastal corridors can facilitate interbreeding and provide an escape route if local conditions become unfavorable.

protecting nearly 2 million ha (about 5 million acres) in western states. Recreation and conservation groups celebrated protection of these areas; local residents, however, generally opposed what they feared would be increased restrictions on their use of the land and its resources. The Bush administration promised to give local residents and officials unprecedented input in devising management plans for these new parklands and monuments.

Species survival can depend on preserve size and shape

Many natural parks and preserves are increasingly isolated, remnant fragments of ecosystems that once extended over large areas. As park ecosystems are shrinking, however, they are also becoming more and more important for maintaining biological diversity. Principles of landscape design and landscape structure become important in managing and restoring these shrinking islands of habitat.

For years, conservation biologists have disputed whether it is better to have a *single large* or *several small* reserves (the SLOSS debate). Ideally, a reserve should be large enough to support viable populations of endangered species, keep ecosystems intact, and isolate critical core areas from damaging external forces. For some species with small territories, several small, isolated refuges can support viables populations, and having several small reserves provides insurance against a disease, habitat destruction, or other calamities that might wipe out a single population. But small preserves can't support species such as elephants or tigers, which need large amounts of space. Given human needs and pressures, however, big preserves aren't always possible. One proposed solution has been to create **corridors** of natural habitat that can connect to smaller habitat areas (fig. 6.21). Corridors could effectively create a large preserve from several small ones. Corridors could also allow populations to maintain genetic diversity or expand into

new breeding territory. The effectiveness of corridors probably depends on how long and wide they are, and on how readily a species will use them.

One of the reasons large preserves are considered better than small preserves is that they have more **core habitat,** areas deep in the interior of a habitat area, and that core habitat has better conditions for specialized species than do edges. **Edge effects** is a term generally used to describe habitat edges: for example, a forest edge is usually more open, bright, and windy than a forest interior, and temperatures and humidity are more varied. For a grassland, on the other hand, edges may be wooded, with more shade, and perhaps more predators, than in the core of the grassland area. As human disturbance fragments an ecosystem, habitat is broken into increasingly isolated islands, with less core and more edge. Small, isolated fragments of habitat often support fewer species, especially fewer rare species, than do extensive, uninterrupted ecosystems. The size and isolation of a wildlife preserve, then, may be critical to the survival of rare species.

Landscape ecology is a science that examines the relationship between these spatial patterns and ecological processes, such as species movement or survival. Landscape ecologists measure variables like habitat size, shape, and the relative amount of core habitat and edge (fig. 6.22). Landscape ecologists also examine the kind of land cover that surrounds habitat areas—is it a city, farm fields, or clear-cut forest? And they include utilitarian landscapes, such as farm fields, in their analysis, as well as pristine wilderness. By quantifying factors such as habitat shape and landscape complexity, and by monitoring the number of species or the size of popuations, landscape ecologists try to guide more effective design of nature preserves and parks.

(a) Patch size and shape: these influence core/edge ratio

(b) Number of habitat patches, and patch proximity and connectivity

(c) Contrast between habitat patches and surrounding landscape

(d) Complexity of the landscape mosaic

FIGURE 6.22 Some spatial variables examined in landscape ecology. Size, arrangement, context, and other factors often vary simultaneously. Together, they can strongly influence the ability of a wildlife preserve to support rare species.

FIGURE 6.23 How small can a nature preserve be? In an ambitious research project, scientists in the Brazilian rainforest are carefully tracking wildlife in plots of various sizes, either connected to existing forests or surrounded by clear-cuts. As you might expect, the largest and most highly specialized species are the first to disappear.

A dramatic experiment in reserve size, shape, and isolation is being carried out in the Brazilian rainforest. In a project funded by the World Wildlife Fund and the Smithsonian Institution, loggers left 23 test sites when they clear-cut a forest. Test sites range from 1 ha (2.47 acres) to 10,000 ha. Clear-cuts surround some, and newly-created pasture surrounds others (fig. 6.23); others remain connected to the surrounding forest. Selected species are regularly inventoried to monitor their survival after disturbance. As expected some species disappear very quickly, especially from small areas. Sun-loving species flourish in the newly-created forest edges, but deep-forest, shade-loving species disappear, particularly when the size or shape of a reserve reduces availability of core habitat. This experiment demonstrates the importance of maintaining core habitat in preserves.

WORLD PARKS AND PRESERVES

Global parks and preserves face innumerable challenges, but the statistics have been improving in recent years. The idea of setting aside nature preserves has spread rapidly over the past 50 years, as people around the world have become aware of the growing scarcity of wildlife and wild places. There are several reasons for a global increase in protected areas. International nongovernment organizations have developed a range of tactics, including debt-for-nature swaps to establish preserves. Consumer pressure has forced logging and mining companies to collaborate in preserving forests. And governments are increasingly interested in slowing resource depletion and in gaining status by protecting lands. Currently some 1.3 billion ha (about 10 percent of the earth's land area) is in some form of protected status.

Within the past several years, Brazil, for example, has pledged to protect 12 percent of its Amazon rainforest, some 500,000 km^2, and both Brazil and Bolivia are working to protect the Pantanal region, the world's largest freshwater wetland complex. Mexico has increased protection for its biosphere reserve at Laguna San Ignacio, the breeding ground for gray whales. Mozambique, despite extreme poverty, has created two new national parks to protect dugongs, sea turtles, elephants, and other wildlife.

Regions with the most dramatic increases in protected areas have been Asia, North America, and Latin America (fig. 6.24). Among the individual countries with the most admirable plans to protect natural resources are Costa Rica, Tanzania, Rwanda, Botswana, Benin, Senegal, Central African Republic, Zimbabwe, Bhutan, and Switzerland, each of which has designated 10 percent or more of its land as ecological protectorates.

So far, however, many of these areas are **"paper parks,"** or parks in name only. Lacking guards, visitor centers, administrative personnel, or even boundary fences, many are vulnerable to poaching by hunters and loggers, as well as to encroachment by settlers. In many countries it is easier to declare a new park than to protect and manage it. Often parks are created where people

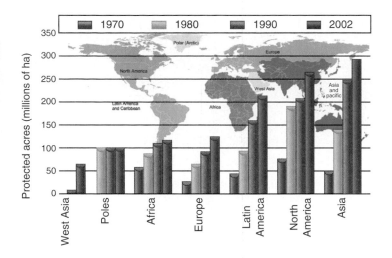

FIGURE 6.24 Change in amount of protected areas by region.
Source: Data from UN Environmental Program, 2003.

already live, farm, and hunt. The needs of these people must be accounted for if a protected area is to work on the ground, rather than just on paper.

Marine preserves aim to protect ocean species

As ocean fish stocks become increasingly depleted globally, biologists are calling for protected areas where marine organisms are sheltered from destructive harvest methods. Research has shown that "no-take" refuges not only protect the species living within them but also serve as nurseries for nearby areas. In a study of 100 of these refuges around the world, researchers found that, on average, the number of organisms inside refuges was twice as high as surrounding areas where harvest was allowed. In addition, the biomass of organisms was three times as great and individual animals were, on average, 30 percent larger inside the refuge, compared with outside. The size necessary for a safe haven to protect flora and fauna and replenish the area around it depends on the species involved, but some marine biologists have called on nations to protect at least 20 percent of their nearshore territory as marine refuges.

Coral reefs are among the most threatened marine ecosystems in the world (chapter 1). Recent remote sensing surveys show that living coral covers only about 285,000 km² (110,000 mi²), or an area about the size of Nevada. This is less than half of previous estimates, and 90 percent of all reefs face threats from sea temperature change, destructive fishing methods, coral mining, sediment runoff, and other human disturbance. In many ways, coral reefs are the old-growth rainforests of the ocean. Biologically rich, these sensitive habitats can take a century or more to recover from damage. If current trends continue, some researchers predict that in 50 years there will be no viable coral reefs anywhere in the world.

What can be done to reverse this trend? Among the national monuments created by President Bill Clinton are three that preserve coral reefs: the U.S. Virgin Islands Coral Reef National Monument, The Pacific Ocean Preserve northwest of Hawaii, and the Tortugas Refuge within the Florida Keys National Marine Sanctuary. Together, these refuges protect 90 percent of all coral in U.S. territory. All are closed to commercial fishing, and the Tortugas reefs are completely protected as a "no-take" zone. Unfortunately, most reefs lack such protection. A survey of marine biological resources identified the ten richest and most threatened hot spots, which are the Philippines, the Gulf of Guinea and Cape Verde Islands (off the west coast of Africa), Indonesia's Sunda Islands, the Mascarene Islands in the Indian Ocean, South Africa's coast, southern Japan and China, the western Caribbean, and the Red Sea and Gulf of Aden. We urgently need a system of underwater national parks and preserves like those established to protect terrestrial biomes.

Threats persist even in preserves

Even parks and preserves with protective systems in place are not always safe from exploitation or changes in political priorities. Many problems threaten natural resources and environmental quality in these areas. In Greece, the Pindus National Park is threatened by plans to build a hydroelectric dam in the center of the park. Furthermore, excessive stock grazing and forestry exploitation in the peripheral zone are causing erosion and loss of wildlife habitat. In Colombia, the Paramillo National Park also is threatened by dam building. Oil exploration along the border of the Yasuni National Park in Ecuador pollutes water supplies, while miners and loggers in Peru have invaded portions of Huascarán National Park. In Palau, coral reefs identified as a potential biosphere reserve are damaged by dynamiting, while on some beaches in Indonesia, every egg laid by endangered sea turtles is taken by egg hunters. These are just a few of the many problems around the world. Often countries with the most important biomes lack funds, trained personnel, and experience to manage some of the areas under their control.

The IUCN has developed a **world conservation strategy** for natural resources that includes the following three objectives: (1) to maintain essential ecological processes and life-support systems (such as soil regeneration and protection, nutrient recycling, and water purification) on which human survival and development depend; (2) to preserve genetic diversity essential for breeding programs to improve cultivated plants and domesticated animals; and (3) to ensure that any utilization of wild species and ecosystems is sustainable. These goals are further elaborated in the ecological plan of action adopted by the IUCN and shown in table 6.3.

Conservation and economic development can work together

Many of the most seriously threatened species and ecosystems of the world are in the developing countries, especially in the tropics. This situation concerns us all because these countries are the guardians of biological resources that may be vital to all of us. Unfortunately, where political and economic systems fail to

TABLE 6.3

IUCN Ecological Plan of Action

1. Launch a consciousness-raising exercise to bring the issue of biological resources to the attention of policymakers and the public at large.
2. Design national conservation strategies that take explicit account of the values at stake.
3. Expand our network of parks and preserves to establish a comprehensive system of protected areas.
4. Undertake a program of training in the fields relevant to biological diversity to improve the scientific skills and technological grasp of those charged with its management.
5. Work through conventions and treaties to express the interest of the community of nations in the collective heritage of biological diversity.
6. Establish a set of economic incentives to make species conservation a competitive form of land use.

Source: Data from International Union for Conservation and National Resources.

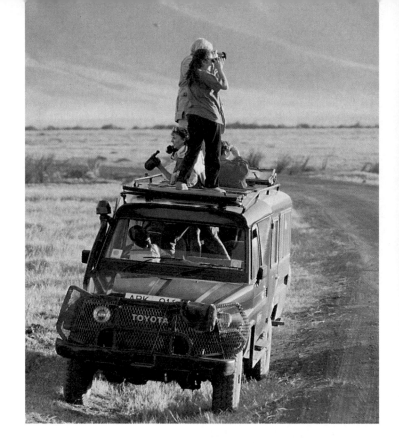

FIGURE 6.25 Ecotourism can be a sustainable resource use. If local communities share in the revenue, ecotourism gives them an incentive to value and protect biodiversity and natural beauty.

provide people with land, jobs, and food, disenfranchised citizens turn to legally protected lands, plants, and animals for their needs. Immediate human survival always takes precedence over long-term environmental goals. Clearly, the struggle to save species and unique ecosystems cannot be divorced from the broader struggle to achieve a new world order in which the basic needs of all are met.

People in some developing countries are beginning to realize that the biological richness of their environment may be their most valuable resource and that its preservation is vital for sustainable development. **Ecotourism** (tourism that is ecologically and socially sustainable) can be more beneficial to many of these countries over the long term than extractive industries, such as logging and mining (fig. 6.25). The What Can You Do? box (p. 147) suggests some ways to ensure that your vacations are responsible ecotourism.

At the 1982 World Congress on National Parks in Bali, 500 scientists, managers, and politicians discussed the design and location of biological reserves and the ecological, economic, and social factors that impinge on wildlife preservation. They concluded that conservation and rural development are not necessarily incompatible. In many cases, sustainable production of food, fiber, medicines, and water in rural areas depends on ecosystem services derived from adjacent conservation reserves. Tourism associated with wildlife watching and outdoor recreation can be a welcome source of income for underdeveloped countries. If local people share in the benefits of saving wildlife, they probably will cooperate, and the programs will be successful. To reformulate Thoreau's famous dictum, "In broadly shared economic progress is preservation of the wild."

Biosphere reserves can aid indigenous communities

Areas chosen for nature preservation are often traditional lands of indigenous people who cannot simply be ordered out. Finding ways to integrate human needs with those of wildlife is essential for local acceptance of conservation goals in many countries. In 1986 UNESCO (United Nations Educational, Scientific, and Cultural Organization) initiated its **Man and Biosphere (MAB) program,** which encourages the designation of **biosphere reserves,** protected areas divided into zones with different purposes. Critical ecosystem functions and endangered wildlife are protected in a central core region, where limited scientific study is the only human access allowed. Ecotourism and research facilities are located in a relatively pristine buffer zone around the core, while sustainable resource harvesting and permanent habitation are allowed in multiple-use peripheral regions (fig. 6.26).

Mexico's 545,000-ha (2,100-mi^2) Sian Ka'an Reserve on the Caribbean coast is a good example of a biosphere reserve. The core area includes 528,000 ha (1.3 million acres) of coral reef and adjacent bays, marshes, and lowland tropical forest. More than 335 bird species have been observed within the reserve, along with endangered manatees, five types of jungle cats, spider and howler monkeys, and four species of increasingly rare sea turtles. Approximately 25,000 people live in communities in peripheral regions around the reserve, and the resort developments of Cancún

What Can You Do?

Being a Responsible Ecotourist

1. *Pretrip preparation.* Learn about the history, geography, ecology, and culture of the area you will visit. Understand the do's and don'ts that will keep you from violating local customs and sensibilities.

2. *Environmental impact.* Stay on designated trails and camp in established sites, if available. Take only photographs and memories and leave only goodwill wherever you go.

3. *Resource impact.* Minimize your use of scarce fuels, food, and water resources. Do you know where your wastes and garbage go?

4. *Cultural impact.* Respect the privacy and dignity of those you meet and try to understand how you would feel in their place. Don't take photos without asking first. Be considerate of religious and cultural sites and practices. Be as aware of cultural pollution as you are of environmental pollution.

5. *Wildlife impact.* Don't harass wildlife or disturb plant life. Modern cameras make it possible to get good photos from a respectful, safe distance. Don't buy ivory, tortoise shell, animal skins, feathers, or other products taken from endangered species.

6. *Environmental benefit.* Is your trip strictly for pleasure, or will it contribute to protecting the local environment? Can you combine ecotourism with work on cleanup campaigns or delivery of educational materials or equipment to local schools or nature clubs?

7. *Advocacy and education.* Get involved in letter writing, lobbying, or educational campaigns to help protect the lands and cultures you have visited. Give talks at schools or to local clubs after you get home to inform your friends and neighbors about what you have learned.

FIGURE 6.26 A model biosphere reserve. Traditional parks and wildlife refuges have well-defined boundaries to keep wildlife in and people out. Biosphere reserves, by contrast, recognize the need for people to have access to resources. Critical ecosystem is preserved in the core. Research and tourism are allowed in the buffer zone, while sustainable resource harvesting and permanent habitations are situated in the multiple-use area around the perimeter.

are located just to the north. In addition to tourism, the economic base of the area includes lobster fishing, small-scale farming, and coconut cultivation.

The Amigos de Sian Ka'an, a local community organization, played a central role in establishing the reserve and is working to protect the resource base while it improves living standards for local people. New intensive farming techniques and sustainable harvesting of forest products enable people to make a living without destroying the resource base. Better lobster harvesting techniques developed at the reserve have improved the catch without depleting native stocks. Local people now see the preserve as a benefit rather than an imposition from outside. Unfortunately, the government has very limited funds to develop or patrol the reserve. (See related story "Ecotourism on the Roof of the World" at www.mhhe.com/cunningham4e.)

WILDERNESS AREAS

Although indigenous people had lived in the Americas for thousands of years before the first Europeans arrived, introduced diseases killed up to 90 percent of the existing population, so that the continent appeared a vast, empty wilderness to early explorers. As historian Frederick Jackson Turner pointed out in a series of articles and speeches at the beginning of the twentieth century, a belief that wilderness is not only a source of wealth but also the origin of strength, self-reliance, wisdom, and character is deeply embedded in our culture. The frontier was seen as a place for the continuous generation of democracy, social progress, economic growth, and national energy. A number of authors, including Henry Thoreau, Aldo Leopold, Sigurd Olson, Edward Abbey, and Wallace Stegner, have written about the physical, mental, and social benefits for modern people of rediscovering solitude and challenge in wilderness (fig. 6.27).

Beginning with designated primitive and roadless areas in a few national forests in the 1920s, the United States has established a system of 264 wilderness areas, encompassing nearly 40 million ha (100 million acres). These wilderness areas are found in national parks, national forests, wildlife refuges, and lands administered by the Bureau of Land Management (BLM).

The Wilderness Act legalized the concept of wilderness

The 1964 Wilderness Act, which is the basis for much of this system, defined **wilderness** as "an area of undeveloped land affected primarily by the forces of nature, where man is a visitor who does

FIGURE 6.27 Wilderness provides opportunities for solitude, self-propelled recreation, and undisturbed nature.

not remain; it contains ecological, geological, or other features of scientific or historic value; it possesses outstanding opportunities for solitude or a primitive and unconfined type of recreation; and it is an area large enough so that continued use will not change its unspoiled, natural conditions." Most of the areas meeting these standards are in the western states and Alaska.

Additional wilderness areas continue to be evaluated for protected status. Conservation groups argue that another 80 million ha (200 million acres), half of it in Alaska, should be added to the wilderness system. A prolonged battle has been waged over these de facto wilderness areas, pitting environmental groups who want more wilderness against loggers, miners, ranchers, and others who want less wilderness. The arguments for saving wilderness are that it provides (1) a refuge for endangered wildlife, (2) an opportunity for solitude and primitive recreation, (3) a baseline for ecological research, and (4) an area where we have chosen simply to leave things in their natural state.

Among the most common arguments against more wilderness are that timber, energy resources, and critical minerals contained on these lands are essential for economic development in local communities, while people want access for motorized recreation, such as off-road vehicles, snowmobiles, and motorboats. For many people living in remote areas, jobs, personal freedom, and local control of resources seem more important than abstract values of wilderness. They often see themselves as an embattled minority trying to protect an endangered, traditional way of life against a wealthy elite who want to lock up huge areas for recreation or aesthetic purposes. Wilderness proponents point out that 96 percent of the country already is open for resource exploitation; the remaining 4 percent is mostly land that developers didn't want, anyway.

For many people, especially those in developing countries, the idea of pristine wilderness untouched by humans is neither very important nor very interesting. In most places, all land is occupied fully—if sparsely—by indigenous people. To them, the area is home, no matter how empty it may look to outsiders. From this perspective, preserving biological diversity, scenic beauty, and other natural resources may be a good idea, but excluding humans and human features from the land does not necessarily make it more valuable. In fact, saving cultural heritage, working landscapes, and historical evidence of early human occupation can often be among the most important reasons to protect an area.

WILDLIFE REFUGES

In 1901 President Teddy Roosevelt established 51 national **wildlife refuges,** the first in an important but troubled system for wildlife preservation in the United States. There are now 511 refuges in this system, encompassing nearly 40 million ha of land and water and representing every major biome in North America. They range in size from less than 1 ha (2.47 acres) for the tiny Mille Lacs Refuge in Minnesota to 7.3 million ha (18 million acres) in the Arctic National Wildlife Refuge of Alaska. Altogether, about 1 percent of U.S. surface area is designated as wildlife refuge. Every state has at least one refuge. Ostensibly, wildlife refuges are our most direct efforts to protect wildlife. Many of these sanctuaries are managed, however, primarily to produce game species for hunters.

Over the years, a number of problematic uses have become accepted in wildlife refuges, including oil drilling, cattle grazing, snowmobiling, motorboating, off-road vehicle use, timber harvesting, hay cutting, trapping, and military exercises. A U.S. General Accounting Office report found that 60 percent of all refuges allow activities that are harmful to wildlife. The biggest current battle over U.S. wildlife refuges concerns proposals for oil and gas drilling in the Arctic National Wildlife Refuge on the north slope of Alaska's Brooks Range.

Refuges also face threats from external activities. More than three-quarters of all U.S. refuges have water pollution problems, two-thirds of which are serious enough to affect wildlife. A notorious example is the former Kesterson Wildlife Refuge in California, where selenium-contaminated irrigation water drained from farm fields turned the marsh into a death trap for wildlife, rather than a sanctuary. Eventually, the marsh had to be drained and capped with clean soil to protect wildlife. Subsequent research has shown that at least 20 wildlife refuges in western states have toxic metal pollution caused by agricultural and industrial activities outside the refuge.

Refuges are key in developing countries

As we saw earlier in this chapter, most developing countries rarely have separate systems of parks and wildlife refuges. Many nature preserves are set up primarily to protect wildlife, however. An outstanding example of both the promise and the problems in managing parks in the less-developed countries is seen in the Serengeti ecosystem in Kenya and Tanzania. This area of savanna, thorn woodland, and volcanic highland lying between Lake Victoria and the Great Rift Valley in East Africa is home to the highest density of ungulates (hoofed grazing animals) in the world. Over 1.5 million wildebeests (or gnus) graze on the savanna in the wet season, when grass is available, and then migrate through the woodlands into the northern highlands during the dry season. The ecosystem also supports hundreds of thousands of zebras, gazelles, impalas, giraffes, and other beautiful and intriguing animals. The herbivores, in turn, support lions and a variety of predators and scavengers, such as leopards, hyenas, cheetahs, wild dogs, and vultures. This astounding diversity and abundance is surely one of the greatest wonders of the world.

Tanzania's Serengeti National Park was established in 1940 to protect 15,000 km^2 (5,700 mi^2), an area about the size of Connecticut, or twice as big as Yellowstone National Park. It is bordered on the east by the much smaller Ngorongoro Conservation Area and Lake Manyara National Park. Kenya's Masai Mara National Reserve borders the Serengeti on the north. Rapidly growing human populations push against the boundaries of the park on all sides. Herds of domestic cattle compete with wild animals for grass and water. Agriculturalists clamor for farmland, especially in the temperate highlands along the Kenya-Tanzania border. So many tourists flock to these parks that the vegetation is ground to dust by hundreds of sight-seeing vans, and wildlife find it impossible to carry out normal lives.

Perhaps the worst problem in Africa is **poaching** by illegal hunters who massacre wildlife for valuable meat, horns, and tusks. Where there once were about 1 million rhinos in Africa, the population had dropped by the mid-1980s to fewer than 10,000 animals. Antipoaching efforts have allowed the population to recover to about 15,000 currently.

Elephants are under a similar assault. Thirty years ago, there were no elephants in the Serengeti, but perhaps 3 million in all of Africa (fig. 6.28). Since then, about 80 percent of the African elephants have been killed—mainly for their ivory—at a rate of 100,000 each year. The 2,000 elephants now in Serengeti National Park were driven there by hunting pressures elsewhere. Fortunately, the elephants find refuge in the park and add to the pleasure of tourists who go to see the wildlife; however, they are changing the ecosystem. Crowded into inadequate space, the ele-

FIGURE 6.28 Elephant populations have been decimated throughout the world. The biggest threat is from poachers, who kill for the elephants' valuable ivory.

phants smash down the acacia trees, turning the woodland and mixed savanna into continuous grassland. This is beneficial for some animals, but not for others.

Poachers continue to pursue the elephants and rhinos, even in the park. Armed with high-powered rifles and even machine guns and bazookas from the many African wars in the past decade, the poachers take a terrible toll on the wildlife. Park rangers try to stop the carnage, but they often are outgunned by the poachers. The parks are beginning to resemble war zones, with fierce, lethal firefights rather than peace and tranquility.

Still, as ecotourism grows in importance in national budgets, both local residents and governments are coming to recognize that protecting wildlife and natural resources makes economic sense. Although wildlife has been decimated in many areas, vast numbers remain an inspiring spectacle elsewhere. With careful stewardship, we may be able to protect much of what's left of nature's diversity and abundance.

REVIEW QUESTIONS

1. What portion of original forests still remain? What portion of remaining forests have old growth characteristics? Which forest types are the most abundant on earth (fig. 6.2)? Which continent has lost the most forest area recently (fig. 6.3)?

2. What do we mean by *closed-canopy* forest and *old-growth* forest?

3. Which commodity is used most heavily in industrial economies: steel, plastic, or wood? What portion of the world's population depends on wood or charcoal as the main energy supply?

4. What land use is responsible for most forest losses in Africa? In Latin America? In Asia? (fig. 6.5).

5. What is a *debt-for-nature swap*?

6. Why is fire suppression a controversial strategy? Why are forest thinning and salvage logging controversial?

7. Are pastures and rangelands always damaged by grazing animals? What are some results of overgrazing?

8. What portion of the United States' public rangelands are in poor or very poor condition due to overgrazing? Why do some groups say grazing fees amount to a "hidden subsidy"?

9. What is *rotational grazing,* and how does it mimic natural processes?

10. What was the first national park in the world, and when was it established? How have the purposes of this park and others changed?

11. How do the size and design of nature preserves influence their effectiveness? What do landscape ecologists mean by *interior habitat* and *edge effects*?

12. What percentage of the earth's land area has some sort of protected status? How has the amount of protected areas changed globally (fig. 6.24)?

13. What is a *paper park*?

14. What is *ecotourism*, and why is it important?

15. What is a *biosphere reserve*, and how does it differ from a wilderness area or wildlife preserve?

CRITICAL THINKING AND DISCUSSION

Apply the principles you have learned in this chapter to discuss these questions with other students.

1. Paper and pulp are the fastest growing sector of the wood products market, as emerging economies of China and India catch up with the growing consumption rates of North America, Europe, and Japan. What should be done to reduce paper use?

2. Conservationists argue that watershed protection and other ecological functions of forests are more economically valuable than timber. Timber companies argue that continued production supports stable jobs and local economies. If you were a judge attempting to decide which group was right, what evidence would you need on both sides? How would you gather this evidence?

3. Divide your class into a ranching group, a conservation group, and a suburban home-builders group, and debate the merits of subsidized grazing in the American West. What is the best use of the land? What landscapes are most desirable? Why? How do you propose to maintain these landscapes?

4. Calculating forest area and forest losses is complicated by the difficulty of defining exactly what constitutes a forest. Outline a definition for what counts as forest in your area, in terms of size, density, height, or other characteristics. Compare your definition to those of your colleagues. Is it easy to agree? Would your definition change if you lived in a different region?

5. There is considerable uncertainty about the extent of degradation on grazing lands. Suppose you were a range management scientist, and it was your job to evaluate degradation for the state of Montana. What data would you need? With an infinite budget, how would you gather the data you need? How would you proceed if you had a very small budget?

6. Why do you suppose dry topical forest and tundra are well represented in protected areas, while grasslands and wetlands are protected relatively rarely? Consider social, cultural, geographic, and economic reasons in your answer.

7. Oil and gas companies want to drill in several parks, monuments, and wildlife refuges. Do you think this should be allowed? Why or why not? Under what conditions would drilling be allowable?

SUGGESTED READINGS

Arno, Stephen F., and Steven Allison-Bunnell. 2002. *Flames in Our Forest: Disaster or Renewal?* Island Press.

Dombeck, Michael P., Christopher A. Wood, and Jack E. Williams. 2003. *From Conquest to Conservation: Our Public Lands Legacy.* Island Press.

Fischman, Robert L. 2004. *The National Wildlife Refuges: Coordinating a Conservation System Through Law.* Island Press.

Havlick, David G. 2002. *No Place Distant: Roads and Motorized Recreation on America's Public Lands.* Island Press.

Honey, Martha. 2002. *Ecotourism and Certification: Setting Standards in Practice.* Island Press.

Knight, Richard L., Wendell C. Gilgert, and Ed Marston. 2002. *Ranching West of the 100th Meridian: Culture Ecology and Economics.* Island Press.

Lawton, R. O., et al. 2001. Climate impact of tropical lowland deforestation on nearby montane cloud forests. *Science* 294:584–87.

Lindenmayer, David B., and Jerry F. Franklin, eds. 2003. *Towards Forest Sustainability: Regional, National, and Global Perspectives.* Island Press.

Mittermeir, R. A., et al. 2003. Wilderness and biodiversity conservation. *Proceedings of the National Academy of Sciences of the United States of America* 100(18):10309–13.

Terborgh, J. 2000. The fate of tropical forests: A matter of stewardship. *Conservation Biology* 14(5):1358–61.

Edge effects are a fundamental consideration in nature preserves. We usually expect to find dramatic edge effects in pristine habitat with many specialized species. But you may be able to find interior-edge differences on your own college campus, or in a park or other unbuilt area near you. Here are three testable questions you can examine using your own local patch of habitat: (1) Can an edge effect be detected or not? (2) Which species will indicate the difference between edge and interior conditions? (3) At what distance can you detect a difference between edge and interior conditions? To answer these questions, you can form a hypothesis and test it as follows:

1. Choose a study area. Find a distinct patch of habitat, such as woods, unmowed grass, or marshy but walkable wetland, about 50 m wide or larger. With other students, list some local, familiar plant species that you would expect to find in your study area. If possible, make this list on a visit to your site.

2. Form a hypothesis. Examine your list, and predict which species will occur most on edges, and (if you can) which you think will occur more in the interior. Form a hypothesis, or a testable statement, based on one of the three questions above. For example, "I will be able to detect an edge effect in my patch," or "I think an edge effect will be indicated by these species: _____," or "I think changes in species abundance

will indicate an edge-interior change at _____ m from the edge of the patch."

3. Gather data. Get a meter tape and lay it along the ground from the edge of your habitat patch toward the interior. (You can also use a string and pace distances: treat one pace as a meter.) This line is your *transect*. At the edge end of the tape (or string), count the number of different species you can see within 1 m^2 on either side of your line. Repeat this count at each 5 m interval, up to 25 m. Thus you will create a list of species at 0, 5, 10, 15, 20, and 25 m in from the edge.

4. Examine your lists, and determine whether your hypothesis was correct. Can you see a change in species presence/absence from 0 to 25 m? Were you correct in your prediction of which species disappeared with distance from the edge? If you can identify an edge effect, at what distance did it occur?

5. Consider ways that your test could be improved. Should you take more frequent samples? Larger samples? Should you compare abundance rather than presence/absence? Might you have gotten different results if you had chosen a different study site? Or if your class had examined many sites and averaged the results? How else might you modify your test to improve the quality of your results?

For Additional Help in Studying This Chapter, please visit our website at www.mhhe.com/cunningham4e. You will find practice quizzes, a chapter summary, key terms, answers to review questions, additional case studies, regional examples, and an extensive reading list, all of which will help you understand the material in this chapter.

Enormous farms have been carved out of Brazil's Cerrado (savanna), which once was the most biodiverse grassland and open tropical forest complex in the world.

Food and Agriculture

We can't solve problems by using the same kind of thinking we used when we created them.

–Albert Einstein–

OBJECTIVES

Studying this chapter will help you answer the following questions:

- Will there be enough food for everyone in the world?
- How many people are now chronically hungry, and why, in a world of surpluses, does hunger persist?
- What vitamins and nutrients are essential for normal growth and development? How might we provide these vital supplements?
- What is soil and why is it a living resource?

- How do we degrade soil?
- Which agricultural inputs are most likely to be in short supply in the future?
- How could agriculture contribute to sustainable energy?
- Are GMOs safe? Are they useful?
- What is low-input, sustainable, regenerative agriculture?

Case Study Farming the Cerrado

A soybean boom is sweeping across South America. Inexpensive land, availability of new crop varieties, and government policies that favor agricultural expansion have made South America the fastest growing agricultural area in the world. The center of this rapid expansion is the Cerrado, a huge area of grassland and tropical forest stretching from Bolivia and Paraguay across the center of Brazil almost to the Atlantic Ocean. Biologically, this rolling expanse of grasslands and tropical woodland is the richest savanna in the world, with at least 130,000 different plant and animal species, many of which are threatened by agricultural expansion.

Until recently, the Cerrado, which is roughly equal in size to the American Midwest, was thought to be unsuitable for cultivation. Its red iron-rich soils are highly acidic and poor in essential plant nutrients. Furthermore, the warm, humid climate harbors many destructive pests and pathogens. For hundreds of years, the Cerrado was primarily cattle country with many poor-quality pastures producing low livestock yields.

In the past few decades, however, Brazilian farmers have learned that modest applications of lime and phosphorus can quadruple yields of soybeans, maize, cotton, and other valuable crops. Researchers have developed more than 40 varieties of soybeans—mostly through conventional breeding, but some created with molecular techniques—specially adapted for the soils and climate of the Cerrado. Until about 30 years ago, soybeans were a relatively minor crop in Brazil. Since 1975, however, the area planted with soy has doubled about every four years, reaching more than 25 million ha (60 million acres) in 2005. Although that's a large area, it represents only one-eighth of the Cerrado, more than half of which is still occupied by pasture.

Brazil is now the world's top soy exporter, shipping some 27 million metric tons per year, or about 10 percent more than the United States. With two crops per year, cheap land, low labor costs, favorable tax rates, and yields per hectare equal to those in the American Midwest, Brazilian farmers can produce soybeans for less than half the cost in America. Agricultural economists predict that, by 2020, the global soy crop will double from the current 160 million metric tons per year, and that South America could be responsible for most of that growth. In addition to soy, Brazil now leads the world in beef, corn (maize), oranges, and coffee exports. This dramatic increase in

Brazil's Cerrado, 2 million ha of savanna (grassland) and open woodland is the site of the world's fastest growing soybean production. Cattle ranchers and agricultural workers, displaced by mechanized crop production, are moving northward into the "arc of destruction" at the edge of the Amazon rainforest, where the continent's highest rate of forest clearing is occurring.

South American agriculture helps answer the question of how the world may feed a growing human population.

A major factor in Brazil's current soy expansion is rising income in China. With more money to spend, the Chinese are consuming more soy, both directly as tofu and other soy products, and indirectly as animal feed. Between 2002 and 2004, China's soy imports doubled to more than 21 million metric tons, or about one-third of total global soy shipments. The outbreak of mad cow disease (or BSE, see chapter 8) in Europe, Canada, and Japan also fueled increased worldwide demand for soybeans. Rather than feed livestock meat-processing wastes that may convey mad cow disease, producers are turning to protein- and lipid-rich soy meal. With 175 million free-range, grass-fed (and presumably BSE-free) cattle, Brazil has become the world's largest beef exporter.

Increasing demand for both soybeans and beef create land conflicts in Brazil. The pressure for more cropland and pasture is a leading cause of deforestation and habitat loss, most of which is occurring in the "arc of destruction" between the Cerrado and the Amazon. Small family farms are being gobbled up, and farm workers, displaced by mechanization, often migrate either to the big cities or to frontier forest areas. Increasing conflicts between poor farmers and big landowners have led to violent confrontations. The Landless Workers Movement claims that 1,237 rural workers died in Brazil between 1985 and 2000 as a result of assassinations and clashes over land rights. In 2005, a 74-year-old Catholic nun, Sister Dorothy Stang, was shot by gunmen hired by ranchers who resented her advocacy for native people, workers, and environmental protection. Over the past 20 years, Brazil claims to have resettled 600,000 families from the Cerrado. Still, tens of thousands of landless farm workers and displaced families live in unauthorized squatter camps and shantytowns across the country awaiting relocation.

As you can see, rapid growth of beef and soy production in Brazil have both positive and negative aspects. On one hand, more high-quality food is now available to feed the world. The 2 million km^2 of the Cerrado represents one of the world's last opportunities to

Case Study *continued*

open a large area of new, highly productive cropland. On the other hand, the rapid expansion and mechanization of agriculture in Brazil is destroying biodiversity and creating social conflicts as people move into formerly pristine lands. The issues raised in this case study illustrate many of the major themes in this chapter. Will there be enough food for everyone in the world? What will be the environmental and social consequences of producing the nutrition we need? In this chapter, we'll look at world food supplies, agricultural inputs, and sustainable approaches that can help solve some of the difficult questions we face.

FOOD AND NUTRITION

Despite dire predictions that runaway population growth would soon lead to terrible famines (chapter 4), world food supplies have more than kept up with increasing human numbers over the past two centuries. The past 40 years have seen especially encouraging strides in reducing world hunger. While population growth averaged 1.7 percent per year during that time, world food production increased an average of 2.2 percent. New cropland, or more efficient use of existing cropland, as is now occurring in Brazil, coupled with increased use of irrigation, improved crop varieties, more readily available fertilizers, and distribution systems to transport food from regions with surpluses to those in need have brought improved nutrition to billions of people. In this section, we'll look at the causes and effects of remaining chronic and acute food shortages, as well as recommendations for a balanced, healthful diet.

Millions of people don't have enough to eat

In 1960 nearly 60 percent of the residents of developing countries were considered **chronically undernourished,** meaning their diet didn't provide the 2,200 kcal per day, on average, considered necessary for a healthy, productive life. Today, despite the fact that their population has doubled over the past 40 years, the proportion in these countries suffering from chronic caloric deficiency has fallen to less than 15 percent.

The UN Food and Agriculture Organization (FAO) expects agricultural production to continue to grow over the next few decades. Where the current world food supply would be sufficient, if equitably distributed, to provide an average of 2,800 kcal per person per day, the FAO predicts that, by 2030, there will be enough food available to supply 3,050 kcal per day to everyone, or about 30 percent more than most of us need. In countries such as the United States, the problem already has become what to do with surplus food. Farmers in these countries are paid billions of dollars per year not to grow crops.

Still, in a world of surplus food, about 850 million people don't have enough to eat. Figure 7.1 shows countries with the highest risk of food shortages. As you can see, most of sub-Saharan Africa, South and Southeast Asia, and parts of Latin America fall in this category. Ninety-five percent of the chronically undernourished are in developing countries, but an increasing number

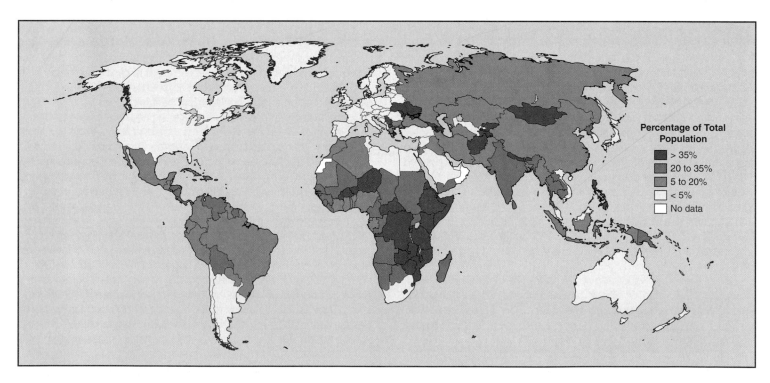

FIGURE 7.1 Hunger around the world. In 2005, the United Nations reported that 852 million people—815 million of them in developing countries—suffered from chronic hunger and malnutrition. Africa has the largest number of countries with food shortages.
Source: Hunger map from Food and Agriculture Organization of the United Nations website. Used by permission.

Principles of Environmental Science http://www.mhhe.com/cunningham4e

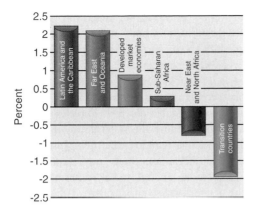

FIGURE 7.2 Changes in agricultural production in 2000. Transition countries are states from the former Soviet Union undergoing a change from socialism to capitalism.
Source: Data from Food and Agriculture Organization (FAO), 2002.

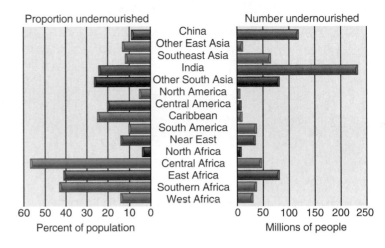

FIGURE 7.3 Number and percent of population chronically undernourished by region in the developing world.
Source: Data from Food and Agriculture Organization (FAO), 2002.

are in transition countries (primarily states of the former Soviet Union and its allies undergoing a change from socialism to market economies) where bad weather, poor management, and social crises have resulted in sharply falling agricultural production (fig. 7.2). Even in the richest countries, where excess calories are the greatest problem for the majority, some 11 million people don't have enough to eat.

Poverty is the greatest threat to **food security,** or the ability to obtain sufficient food on a day-to-day basis. The 1.5 billion people in the world who live on less than $1 per day all too often can't buy the food they need and don't have access to resources to grow it for themselves. Food security occurs at multiple scales. In the poorest countries, hunger may affect nearly everyone. In other countries, although the average food availability may be satisfactory, some individual communities or families may not have enough to eat. And within families, males often get both the largest share and the most nutritious food, while women and children—who need food most—all too often get the poorest diet. At least 6 million children under 5 years old die every year of diseases exacerbated by hunger and malnutrition. Providing a healthy diet might eliminate as much as 60 percent of all premature deaths worldwide.

Hungry people can't work their way out of poverty, the Nobel Prize–winning economist Robert Fogel points out. He estimates that in 1790 about 20 percent of the population of England and France was effectively excluded from the labor force because they were too weak and hungry to work. Improved nutrition, he calculates, accounted for about half of all European economic growth during the nineteenth century. Since many developing countries are as poor now (in relative terms) as Britain and France were in 1790, his analysis suggests that reducing hunger could yield more than $120 billion (U.S.) in economic growth produced by longer, healthier, more productive lives for several hundred million people.

The 2003 UN World Food Summit reaffirmed the goal set by previous conventions of reducing the number of chronically undernourished people to 400 million by 2015. We aren't on track to meet that goal, but some countries have made impressive

progress. China, alone, has reduced its number of undernourished people by 74 million over the past decade. Indonesia, Vietnam, Thailand, Nigeria, Ghana, and Peru each reduced chronic hunger by about 3 million people. In 47 other countries, however, the numbers of chronically underfed people have increased. India now has by far the largest number of persistently hungry people in the world, while central Africa has the highest percentage (fig. 7.3).

Recognizing the role of women in food production is an important step toward food security for all. Throughout the developing world, women do 50 to 70 percent of all farm work but control only a tiny fraction of the land and rarely have access to capital or developmental aid. In Nigeria, for example, home gardens occupy only 2 percent of all cropland but provide half the food families eat. Making land, credit, education, and access to markets available to women could contribute greatly to family nutrition.

Famines are acute food shortages

Famines are characterized by large-scale food shortages, massive starvation, social disruption, and economic chaos. Starving people eat their seed grain and slaughter their breeding stock in a desperate attempt to keep themselves and their families alive. Even if better conditions return, they have often sacrificed their productive capacity and will take a long time to recover. Famines often trigger mass migrations, as starving people travel to refugee camps in search of food and medical care (fig. 7.4). Many die on the way or fall prey to robbers.

In 2004 the United Nations reported that approximately 60 million people in 30 countries needed emergency food aid. What causes these emergencies? Environmental conditions are usually the immediate trigger, but politics and economics are often equally important in preventing people from getting the food they need. Drought, floods, insect outbreaks, and other natural disasters cause crop failures and create food shortages. But the Nobel Prize–winning work of Harvard economist Amartya K. Sen shows that these factors have often been around for a long time,

FIGURE 7.4 Children wait for their daily ration of porridge at a feeding station in Somalia. When people are driven from their homes by hunger or war, social systems collapse, diseases spread rapidly, and the situation quickly becomes desperate.

and local people usually have ways to get through hard times if they aren't thwarted by inept or corrupt governments and greedy elites. National politics, commodity hoarding, price gouging, poverty, wars, landlessness, and other social factors often conspire, so that poor people can neither grow their own food nor find jobs to earn money to buy the food they need. Professor Sen points out that armed conflict and political oppression almost always are at the root of famine. No democratic country with a relatively free press, he says, has ever had a major famine.

The aid policies of rich countries often serve more to get rid of surplus commodities and make us feel good about our generosity than to eliminate the root causes of starvation. Crowding people into feeding camps generally is the worst thing to do for them. The stress of getting there kills many of them, and the density and lack of sanitation in the camps expose them to epidemic diseases. There are no jobs in refugee camps, so people can't support themselves if they try. Social chaos and family breakdown expose the weakest to robbery and violence. Having left their land and tools behind, people can't replant crops when the weather returns to normal.

We need the right kinds of food

In addition to energy (calories), we also need specific nutrients in our diet, such as proteins, vitamins, and certain trace minerals. You might have more than enough calories and still suffer from **malnourishment,** a nutritional imbalance caused by a lack of specific dietary components or an inability to absorb or utilize essential nutrients. Those of us in richer countries often eat too much meat, salt, and fat and too little fiber, vitamins, trace minerals, and other components lost from highly processed foods. On average, we consume about one-third more calories than we need and get too little exercise. Food provides solace in stressful lives, and we're constantly bombarded with advertising tempting us to consume more. According to the U.S. surgeon general, 64 percent of all adult Americans are overweight, up from 40 percent only a decade

ago, and about one-third are seriously overweight or **obese**— generally considered to be a body mass greater than 30 kg/m^2, or roughly 30 pounds above normal for an average person.

Being overweight substantially raises your risk for hypertension, diabetes, heart attacks, stroke, gallbladder disease, osteoarthritis, respiratory problems, and certain cancers. Every year, about 400,000 in the United States die from illnesses related to obesity. This number is approaching the 435,000 annual deaths from diseases related to smoking. Paradoxically, food insecurity and poverty can contribute to obesity. In one study, more than half of the women who reported not having enough to eat were overweight, compared with one-third of the food-secure women. Lack of time for cooking and access to healthy food choices, along with ready availability of fast-food snacks and calorie-laden drinks, lead to dangerous dietary imbalances for many people.

This trend isn't limited to richer countries. Obesity is spreading around the world (fig. 7.5). For the first time in history, there are probably more overweight people (more than 1 billion) than underweight. As chapter 8 points out, Western diets and lifestyles are being adopted by many in the developing world, and diseases such as heart attack, stroke, diabetes, and depression, which once were thought to afflict only wealthy nations, are now becoming the most prevalent causes of death and disability everywhere.

Many poor people can't afford meat, fruits, and vegetables that would provide a balanced diet. The FAO estimates that nearly 3 billion people (nearly half the world population) suffer from vitamin, mineral, or protein deficiencies. This results in devastating illnesses and deaths, as well as reduced mental capacity, developmental abnormalities, and stunted growth. Altogether, these problems bring an incalculable loss of human potential and social capital.

Anemia (low hemoglobin levels in the blood, usually caused by dietary iron deficiency) is the most common nutritional problem in the world. According to the FAO, more than 2 billion people (52 percent are pregnant women and 39 percent are children under 5) suffer from iron deficiencies. The problem is most

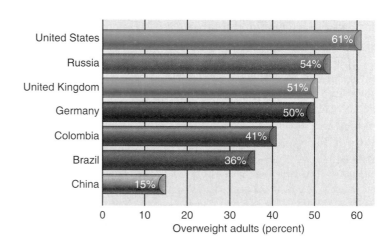

FIGURE 7.5 While nearly a billion people are chronically undernourished, people in wealthier countries are at risk from eating too much. **Source:** Data from Worldwatch Institute, 2001.

FIGURE 7.6 Goiter, a swelling of the thyroid gland at the base of the neck, is often caused by an iodine deficiency. It is a common problem in many parts of the world, particularly where soil iodine is low and seafood is unavailable.

severe in India, where more than 80 percent of all pregnant women are anemic. Anemia increases the risk of maternal deaths from hemorrhage in childbirth and affects childhood development. Red meat, eggs, legumes, and green vegetables all are good sources of dietary iron.

Iodine is essential for synthesis of thyroxin, an endocrine hormone that regulates metabolism and brain development, among other things. Chronic iodine deficiency causes goiter (a swollen thyroid gland, fig. 7.6), stunted growth, and reduced mental ability. The FAO estimates that 740 million people—mainly in South and Southeast Asia—suffer from iodine deficiency and that 177 million children have stunted growth and development. Adding a few pennies worth of iodine to our salt has largely eliminated this problem in developed countries.

Starchy foods, such as maize (corn), polished rice, and manioc (tapioca), which form the bulk of the diet for many poor people, tend to be low in several essential vitamins as well as minerals. Folic acid, for example, (found in dark green, leafy vegetables) is essential for early fetal development. Folic acid deficiencies have been linked to neurological problems in babies, including microencephaly (an abnormally small head) or even anencephaly (lacking a brain). According to the FAO, vitamin A deficiencies affect between 100 and 140 million children at any given time. More than a million of those children die and another 350,000 go blind every year from the effects of this vitamin shortage. There is good news in treating vitamin deficiencies. Dr. Alfred Sommer, an ophthalmologist from Johns Hopkins University, has shown that giving children just two cents' worth of vitamin A twice a year could prevent almost all the cases of childhood blindness and premature death associated with vitamin A deficiencies. Furthermore, vitamin A supplements reduced maternal mortality by nearly 40 percent in a study in Nepal. One promising way to deliver this crucial nutrient is through genetic engineering of common food species (see the related story, "Golden Rice" at www.mhhe.com/cunningham4e).

Protein also is essential for normal growth and development. The two most widespread human protein deficiency diseases are kwashiorkor and marasmus. **Kwashiorkor** is a West African word meaning "displaced child." (A young child is displaced—and deprived of nutritious breast milk—when a new baby is born.) This condition occurs mainly in young children who eat mostly cheap, starchy foods and don't get enough good-quality protein. Children with kwashiorkor often have reddish-orange hair; puffy, discolored skin; and a bloated belly. **Marasmus** (from the Greek "to waste away") is caused by a diet low in both calories and protein. A child suffering from severe marasmus is generally thin and shriveled, like a tiny, very old, starving person (fig. 7.7). Children with both these deficiencies have low resistance to infections and are likely to suffer from stunted growth, mental retardation, and other developmental problems. Altogether, the FAO estimates, the annual losses due to deaths and diseases caused by calorie and nutrient deficiencies are equivalent to 46 million years of productive life (see more in chapter 8 on disability-adjusted life years).

FIGURE 7.7 Marasmus is caused by combined energy (calorie) and protein deficiencies. Children with marasmus have the wizened look and dry, flaky skin of an old person.

Eating a balanced diet is essential for good health

What's the best way to be sure you're getting a healthy diet? Generally, it isn't necessary to take synthetic dietary supplements. Eating a good variety of foods should give you all the nutrients you need. For years, Americans were advised to eat daily servings of four major food groups: meat, dairy products, grains, and fruits and vegetables. These recommendations were revised in 1992 to emphasize only sparing servings of meat, dairy, fats, and sweets.

Some nutritionists believe that the 1992 recommendations for 6 to 11 servings a day of bread, cereal, rice, and pasta still provide too many simple sugars (or starches that are quickly converted to sugar). Based on observations of the health effects of Mediterranean diets, as well as a long-term study of 140,000 U.S. health professionals, Drs. Walter Willett and Meir Stampfer of Harvard University developed suggestions for a healthy diet (fig. 7.8a). They recommend that both red meat and starchy food, such as white rice, white bread, potatoes, and pasta, should be eaten sparingly. Nuts, legumes (beans, peas, and lentils), fruit, vegetables, and whole-grain foods form the basis of this diet. Unsaturated plant oils should make up 30 to 40 percent of dietary calories, according to this view. Trans fat (the kind found in hydrogenated margarine), on the other hand, is not recommended

(a) Harvard food pyramid

Colors represent food groups and oils

Vegetables (green)   Oils (yellow)   Meat & beans (purple)

Grains (orange)   Fruits (red)   Milk (blue)

(b) USDA recommendations

FIGURE 7.8 The Harvard food pyramid (a) emphasizes fruits, vegetables, and whole grains as the basis of a healthy diet. Red meat, white rice, pasta, and potatoes should be used sparingly. The new USDA food pyramid (b) is confusing to many people and seems to encourage more consumption of meat and milk.
Source: (a) Data from Willett and Stampfer, 2002. (b) USDA, 2005.

countries. These two grass species supply around 60 percent of the calories that humans consume directly.

Potatoes, barley, oats, and rye are staples in mountainous regions and high latitudes (northern Europe, North Asia) because they grow well in cool, moist climates. Cassava, sweet potatoes, and other roots and tubers grow well in warm, wet areas and are staples in Amazonia, Africa, Melanesia, and the South Pacific. Sorghum and millet are drought-resistant and are staples in the dry regions of Africa.

Fruits and vegetables—including vegetable oils—make a surprisingly large contribution to human diets. They are especially welcome because they typically contain high levels of vitamins, minerals, dietary fiber, and complex carbohydrates.

at all. Combined with regular, moderate exercise, this food selection should provide most people with all the nutrition they need.

In 2005, the U.S. Department of Agriculture unveiled an alternative food pyramid (fig. 7.8b). Rather than give specific recommendations, the USDA pyramid groups foods together and shows some consumption of each food group at every level. Many people find this graphic confusing. Critics charge that it was developed by consultants who simultaneously represented the food industry, and that it overemphasizes meat and dairy while downplaying whole grains, vegetables, fruits, and unsaturated oils.

KEY FOOD SOURCES

Of the thousands of edible plants and animals in the world, only about a dozen types of seeds and grains, three root crops, twenty or so common fruits and vegetables, six mammals, two domestic fowl, and a few fish and other forms of marine life make up almost all of the food humans eat. Table 7.1 shows annual production of some important foods in human diets. In this section, we will highlight sources and characteristics of those foods.

We depend on only a few crops for most of our food

The three crops on which humanity depends for the majority of its nutrients and calories are wheat, rice, and maize. Together, about 1,900 million metric tons of these three grains are grown each year. Wheat and rice are especially important, since they are the staple foods for most of the 5 billion people in developing

Meat and dairy are important protein sources

Protein-rich foods are prized by people nearly everywhere. In the past, the rich countries consumed a vast majority of the meat and dairy traded internationally. The four-fifths of the world's people

TABLE 7.1

Some Important Food Sources

Crop	2004 Yield (million metric tons)
Wheat	620
Rice (paddy)	610
Maize (corn)	643
Potatoes	310
Coarse grains*	1,013
Soybeans	221
Cassava and sweet potatoes	449
Sugar (cane and beet)	144
Pulses (beans, peas)	55
Oil seeds	375
Vegetables and fruits	1,206
Meat and milk	870
Fish and seafood	140

*Barley, oats, sorghum, rye, millet.
Source: Data from Food and Agriculture Organization (FAO), 2005.

FIGURE 7.9 Meat and dairy consumption has quadrupled in the past 40 years, and China represents about 40 percent of that increased demand.

in less-developed countries generally raised 60 percent of the 3 billion domestic ruminants and 6 billion poultry in the world but consumed only one-fifth of all commercial animal products. The FAO reports, however, that, as incomes rise in developing countries, food choices throughout the world are shifting toward higher-quality and more expensive foods. Meat consumption in developing countries, for example, has risen from only 10 kg per person annually in the 1960s to 26 kg currently (fig. 7.9). By 2030, average annual meat consumption in those countries is expected to be nearly 40 kg per person.

Much of that increase in meat production could be provided by improved productivity in tropical areas. As the opening Case Study for this chapter shows, there has been a huge increase in beef and soy production in South America. Modern breeding techniques that combine the best qualities of Indian zebu (humped) cattle and beef-producing varieties from Europe and North America have provided animals that resist heat and tropical diseases but have meat characteristics preferred by consumers in developed countries (fig. 7.10a). Over the past 40 years, the area devoted to soybeans in Brazil has increased more than 20-fold (fig. 7.10b), while the yield per hectare has doubled (fig. 7.10c). Carving farmland and pasture out of tropical forest and savanna is helping to feed the world, but it is also reducing biological diversity and causing social conflict.

Most of the livestock grown in North America is confined in concentrated animal feeding operations for part or all of the animals' lifetime. A diet rich in grain, oil, and protein fattens animals quickly and produces meat preferred by many consumers. Globally, some 660 million metric tons of cereals are used as livestock feed each year, representing just over a third of the world total cereal use. It's often pointed out that we could feed about ten times as many people if we ate grain directly rather than feeding it to livestock. Surprisingly, the FAO claims that using cereals as animal feed does not contribute to hunger and undernutrition. Given current agricultural economics, if these cereals were not used as animal feed, they would probably not be produced at all and thus would not be available as food. According to this view, if everyone became a vegetarian, the lack of demand for cereals for livestock production would simply lead to lower crop production, not more food for the hungry.

The rapid proliferation of large-scale animal confinement operations raises a number of social and environmental questions. Up to 10,000 hogs enclosed in a giant barn, or 100,000 cattle in a feedlot complex, or a million chickens in a single facility (fig. 7.11) can cause serious local air and water pollution. Animal waste often is stored in enormous open lagoons, which can leak or rupture, contaminating both surface waters

(a) Brazilian zebu cattle

(b) Brazil soy area

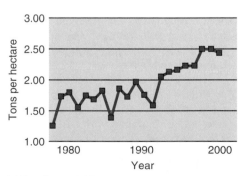

(c) Brazil soy yields

FIGURE 7.10 Beef and soy production in tropical countries, such as Brazil, could help satisfy the growing demand for meat in the developing world. Beef production and soy yields in Brazil have doubled in the past 40 years, while the area devoted to soybeans has increased more than 20-fold. This rapid expansion has had severe biological and social costs, however.
Source: (b) and (c) Data from USDA, 2005.

FIGURE 7.11 Most livestock grown in the United States are raised in large-scale, concentrated animal-feeding operations. Up to a million animals can be held in a single facility. High population densities require heavy use of antibiotics and can cause severe local air and water pollution.

and groundwater supplies. When Hurricane Floyd dumped torrential rains on North Carolina in 1999, an estimated 10 million m³ (2.5 billion gal) of hog and poultry waste overflowed from storage lagoons into local rivers (see related story "Flood of Pigs" at www.mhhe.com/cunningham4e). The high density of animals in these facilities and the rich diet they're fed to speed weight gain require a constant use of antibiotics and growth hormones. More than half of all antibiotics used in the United States are administered to livestock, mostly in confinement facilities. A big part of the rapid rise in antibiotic-resistant pathogens is due to this massive and constant dosing with drugs.

APPLICATIONS

Mapping Your Food Supply

Make a list of all the food you ate on a recent day. Insofar as you are able, map where each of those food items came from. For raw materials, you may have to make an educated guess. If you live in North America, for instance, the wheat in baked goods probably came from the upper Great Plains or Prairie Provinces; corn syrup in prepared foods probably came from the corn belt; cane sugar is mostly from Florida; beet sugar is mostly from the Red River Valley of the north. What was the average distance traveled by each food item from its source to your table that day? What do you know about the growing conditions or social circumstances at the source of your food? Are there pesticide controls, for instance, where your food originated? How many calories did you eat that day?

Emergent diseases (chapter 8) are having significant impacts on our diet. In 2004 mad cow disease, avian flu, rabies, and other infectious diseases blocked an estimated one-third of all global meat exports. The FAO estimated losses of $10 billion (U.S.). The impacts on many farmers were catastrophic. It's estimated that one-quarter of all U.S. cattle have the deadly bacterium *E. coli* 0157-H7 in their gut. As few as two or three of these cells can kill you.

Seafood is another protein source

The 140 million metric tons of seafood we eat every year is an important part of our diet. Seafood provides about 15 percent of all animal protein eaten by humans, and it is the main animal protein source for about 1 billion people in developing countries. Unfortunately, overharvesting and habitat destruction threaten most of the world's wild fisheries (see related stories "Saving the Reefs of Apo Island" in chapter 1 and "Shark Finning" at www.mhhe.com/cunningham4e). Annual catches of ocean fish rose by about 4 percent annually between 1950 and 1988. Since 1989, however, 13 of 17 major marine fisheries have declined dramatically or become commercially unsustainable. According to the United Nations, three-quarters of the world's edible ocean fish, crustaceans, and mollusks are declining and in urgent need of managed conservation.

The problem is too many boats using efficient but destructive technology to exploit a dwindling resource base. Boats as big as ocean liners travel thousands of kilometers and drag nets large enough to scoop up a dozen jumbo jets, sweeping a large patch of ocean clean of fish in a few hours. Long-line fishing boats set cables up to 10 km long with hooks every 2 meters that catch birds, turtles, and other unwanted "by-catch" along with targeted species. Trawlers drag heavy nets across the bottom, scooping up everything indiscriminately and reducing broad swaths of habitat to rubble. One marine biologist compared the technique to harvesting forest mushrooms with a bulldozer. In some operations, up to 15 kg of dead and dying by-catch are dumped back into the ocean for every kilogram of marketable food. The FAO estimates that operating costs for the 4 million boats now harvesting wild fish exceed sales by $50 billion (U.S.) per year. Countries subsidize fishing fleets to preserve jobs and to ensure access to this valuable resource.

Aquaculture is providing an increasing share of the world's seafood. Fish can be grown in farm ponds that take relatively little space but are highly productive. Cultivation of high-value carnivorous species, however, such as salmon, sea bass, and tuna, threaten wild stocks exploited to stock captive operations or to provide fish food. Building coastal fish-rearing ponds causes destruction of hundreds of thousands of hectares of mangrove forests and wetlands, which serve as irreplaceable nurseries for marine species. Net pens anchored in nearshore areas allow spread of diseases, escape of exotic species, and release of feces, uneaten food, antibiotics, and other pollutants into surrounding ecosystems (fig. 7.12).

FIGURE 7.12 Pens for fish-rearing in Thailand.

Polyculture systems of mixed species of herbivores or filter feeders can alleviate many aquaculture problems. Raising species in enclosed, land-based ponds or warehouses can eliminate the pollution problems associated with net pens in lakes or oceans. In China, for example, most fish are raised in ponds or rice paddies. One ecologically balanced system uses four carp species that feed at different levels of the food chain. The grass carp, as its name implies, feeds largely on vegetation, while the common carp is a bottom feeder, living on detritus that settles to the bottom. Silver carp and bighead carp are filter feeders that feed on phytoplankton and zooplankton, respectively. Agricultural wastes, such as manure, dead silkworms, and rice straw, fertilize ponds and encourage phytoplankton growth. All these carp species are considered dangerous invasive species in North America. Still, these integrated polyculture systems typically boost fish yields per hectare by 50 percent or more, compared with monoculture farming. Of the top ten seafood species recommended by the Monterey Aquarium as ecologically acceptable, six are farmed (see related story "Are Shrimp Safe to Eat?" at www.mhhe.com/cunningham4e).

SOIL: A RENEWABLE RESOURCE

Growing the food and fiber needed to support human life is a complex enterprise that requires knowledge from many different fields and cooperation from many different groups of people. In this section, we will survey some of the principles of soil science and look at some of the inputs necessary for continued agricultural production.

Of all the earth's crustal resources, the one we take most for granted is soil. We are terrestrial animals and depend on soil for life, yet most of us think of it only in negative terms. English is unique in using *soil* as an interchangeable word for earth and excrement. *Dirty* has a moral connotation of corruption and impurity. Perhaps these uses of the word enhance our tendency to abuse soil without scruples; after all, it's only dirt.

The truth is that **soil** is a marvelous substance, a living resource of astonishing beauty, complexity, and frailty. It is a complex mixture of weathered mineral materials from rocks, partially decomposed organic molecules, and a host of living organisms. It can be considered an ecosystem by itself. Soil is an essential component of the biosphere, and it can be used sustainably, or even enhanced, under careful management.

There are at least 20,000 different soil types in the United States and many thousands more worldwide. They vary because of the influences of parent material, time, topography, climate, and organisms on soil formation. There are young soils that, because they have not weathered much, are rich in soluble nutrients. There are old soils, such as the red soils of the tropics, from which rainwater has washed away most of the soluble minerals and organic matter, leaving behind clay and rust-colored oxides.

To understand the potential for feeding the world on a sustainable basis, we need to know how soil forms, how it is lost, and what we can do to protect and rebuild good agricultural soil. With careful husbandry, soil can be replenished and renewed indefinitely. Many farming techniques deplete soil nutrients, however, and expose the soil to the erosive forces of wind and moving water. As a result, in many places we are essentially "mining" this resource and using it much faster than it is being replaced.

Building good soil is a slow process. Under the best circumstances, good **topsoil** accumulates at a rate of about 10 tons per ha (2.5 acres) per year—enough soil to make a layer about 1 mm deep when spread over a hectare. Under poor conditions, it can take thousands of years to build that much soil. Perhaps one-third to one-half of the world's current croplands are losing topsoil faster than it is being replaced. In some of the worst spots, erosion carries away about 2.5 cm (1 in.) of topsoil per year. With losses like that, agricultural production has already begun to fall in many areas.

Living organisms create unique properties of soil

Without soil organisms, the earth would be covered with sterile mineral particles far different from the rich, living soil ecosystems on which we depend for most of our food. The activity of the myriad organisms living in the soil helps create structure, fertility, and tilth (condition suitable for tilling or cultivation).

Soil organisms usually stay close to the surface, but that thin living layer can contain thousands of species and billions of individual organisms per hectare. Algae live on the surface, while bacteria and fungi flourish in the top few centimeters of soil (fig. 7.13). A single gram of soil (about one-half teaspoon) can contain hundreds of millions of these microscopic cells. Worms and nematodes process plant roots and litter. Bacteria and fungi decompose organic detritus and recycle nutrients that plants can use for additional growth. The sweet aroma of freshly turned soil

FIGURE 7.13 Soil ecosystems include numerous consumer organisms, as depicted here: (1) snail, (2) termite, (3) nematode and nematode-killing constricting fungus, (4) earthworm, (5) wood roach, (6) centipede, (7) carabid (ground) beetle, (8) slug, (9) soil fungus, (10) wireworm (click beetle larva), (11) soil protozoan, (12) sow bug, (13) ant, (14) mite, (15) springtail, (16) pseudoscorpion, and (17) cicada nymph.

is caused by actinomycetes, bacteria that grow in fungus-like strands and give us the antibiotics streptomycin and tetracycline.

There are many different soil types

Most soils are stratified into horizontal layers called **soil horizons,** which reveal much about the soil's history and usefulness. The thickness, color, texture, and composition of each horizon are used to classify the soil. A cross-sectional view of the horizons in a soil is called a soil profile. Figure 7.14 shows the series of horizons generally seen in a soil profile. Soil scientists give each horizon a letter or descriptive name. Soils usually have one to seven or more horizons with different properties, depending on the soil type and history of a specific area.

The soil surface is often covered with a layer of leaf litter, crop residues, or other fresh or partially decomposed organic material (O horizon). Under this organic layer is the surface horizon, usually an A horizon or topsoil, composed of mineral particles mixed with organic material. The A horizon ranges from a thickness of several meters under virgin prairie to zero in some deserts. The surface horizon contains most of the living organisms and organic material in the soil, and it is in this layer that

most plants spread their roots to absorb water and nutrients. The surface horizon often blends into another horizon (the E horizon), which is subject to leaching (removal of soluble nutrients) by water percolating through it. This zone of leaching may have a very different appearance and composition from horizons above or below it.

Beneath the surface horizons, the subsurface horizons, or **subsoil,** usually have a lower organic content and higher concentrations of mineral particles. Under the subsoil is the parent material, or C horizon, made of weathered rock fragments with very little organic material. Weathering of this layer produces new soil particles and allows downward expansion of the horizons above. About 70 percent of all the parent horizon material in the United States was transported to its present site by geologic forces (glaciers, wind, and water) and is not directly related to the bedrock below it.

Soils are classified according to their structure and composition into orders, suborders, great groups, subgroups, families, and series. The richest farming soils are the mollisols (formed under grasslands) and alfisols (formed under moist, deciduous forests). North America is fortunate to have extensive areas of these fertile soils.

SOIL HORIZONS

O — Surface litter

A — Topsoil
Organic matter (humus), living organisms, inorganic minerals

E — Zone of leaching
Dissolved or suspended materials move downward

B — Subsoil
Accumulation of iron, aluminum, humic compounds, and clay leached down from the A and E horizons

C — Weathered parent material
Partially broken down inorganic minerals

Bedrock

FIGURE 7.14 Soil profile showing possible soil horizons. The actual number, composition, and thickness of these layers vary in different soil types.

WAYS WE USE AND ABUSE SOIL

Only about 11 percent of the earth's land area (14.66 million km^2 out of a total of 132.4 million km^2) is currently in agricultural production. Perhaps four times as much land could be converted to cropland, but much of this land serves as a refuge for cultural or biological diversity or suffers from constraints, such as steep slopes, shallow soils, poor drainage, tillage problems, low nutrient levels, metal toxicity, or excess soluble salts or acidity, that limit the types of crops that can be grown there (fig. 7.15).

Arable land is unevenly divided

In parts of Canada and the United States, temperate climates, abundant water, and high soil fertility produce high crop yields that contribute to high standards of living. Other countries, although rich in land area, lack suitable soil, topography, water, or climate to sustain our levels of productivity.

FIGURE 7.15 In many areas, soil or climate constraints limit agricultural production. These hungry goats in Sudan feed on a solitary Acacia shrub.

Arable land shrank from 0.38 ha (0.94 acre) per person in 1970 to 0.23 ha (0.56 acre) in 2000. If current population projections are correct, the amount of cropland per person will decline to 0.15 ha (0.37 acre) by 2050. In Asia, cropland will be even more scarce—0.09 ha (0.22 acre) per person—in 50 years. If you live on a typical quarter-acre suburban lot, look at your yard and imagine feeding yourself for a year on what you could produce there.

In the developed countries, 95 percent of agricultural growth in the twentieth century came from improved crop varieties or increased fertilization, irrigation, and pesticide use, rather than from bringing new land into production. In fact, less land is being cultivated now than 100 years ago in North America, or 600 years ago in Europe. As more effective use of labor, fertilizer, and water and improved seed varieties have increased in the more-developed countries, productivity per unit of land has increased, and marginal land has been retired, mostly to forests and grazing lands. In many developing countries, land continues to be cheaper than other resources, and new land is still being brought under cultivation, mostly at the expense of forests and grazing lands. Still, at least two-thirds of recent production gains have come from new crop varieties and more intense cropping, rather than expansion into new lands.

As the opening Case Study for this chapter shows, the largest increases in cropland over the past 30 years have occurred in South America and Oceania, where forests and grazing lands are rapidly being converted to farms. Many developing countries are reaching the limit of lands that can be exploited for agriculture without unacceptable social and environmental costs, but others still have considerable potential for opening new agricultural lands. East Asia, for instance, already uses about three-quarters of its potentially arable land. Most of its remaining land has severe restrictions for agricultural use. Further increases in crop production will probably have to come from higher yields per hectare. Latin America, by contrast, uses only about one-fifth of its potential land, and Africa uses only about one-fourth of the land that theoretically could grow crops. However, there would be serious ecological trade-offs in putting much of this land into agricultural production.

While land surveys tell us that much more land in the world could be cultivated, not all of that land necessarily should be farmed. Much of it is more valuable in its natural state. The soils over much of tropical Asia, Africa, and South America are old, weathered, and generally infertile. Most of the nutrients are in the standing plants, not in the soil. In many cases, clearing land for agriculture in the tropics has resulted in tragic losses of biodiversity and the valuable ecological services that it provides. Ultimately, much of this land is turned into useless scrub or semidesert.

On the other hand, there are large areas of rich, subtropical grassland and forest that are well watered, have good soil, and could become productive farmland without unduly reducing the world's biological diversity. Argentina, for instance, has pampas grasslands about twice the size of Texas that closely resemble the American Midwest a century ago in climate and potential for agricultural growth. Some of this land could probably be farmed with relatively little ecological damage if it were done carefully.

Land degradation reduces crop yields

Agriculture both causes and suffers from environmental degradation. The International Soil Reference and Information Centre in the Netherlands estimates that, every year, 3 million ha (7.4 million acres) of cropland are ruined by erosion, 4 million ha are turned into deserts, and 8 million ha are converted to nonagricultural uses, such as homes, highways, shopping centers, factories, and reservoirs. Over the past 50 years, some 1.9 billion ha of agricultural land (an area greater than that now in production) have been degraded to some extent. About 300 million ha of this land are strongly degraded (meaning that the soil has deep gullies, severe nutrient depletion, or poor crop growth or that restoration is difficult and expensive). Some 910 million ha—about the size of China—are moderately degraded. Nearly 9 million ha of former croplands are so degraded that they no longer support any crops at all. The causes of this extreme degradation vary: in Ethiopia, it is water erosion; in Somalia, it is wind; and in Uzbekistan, salt and toxic chemicals are responsible.

Definitions of degradation are based on both biological productivity and our expectations about what the land should be like. Often this is a subjective judgment, and it is difficult to distinguish between human-caused deterioration and natural conditions, such as drought. We generally consider the land degraded when the soil is impoverished or eroded, water runs off or is contaminated more than normal, vegetation is diminished, biomass production is decreased, or wildlife diversity diminishes. On farmlands, this results in lower crop yields. On ranchlands, it means that fewer livestock can be supported per unit area. On nature reserves, it means lower biological diversity.

The amount and degree of land degradation vary by region and country. About 20 percent of land in Africa and Asia is degraded, but most is in either the light or the moderate category. In Central America and Mexico, by contrast, 25 percent of all vegetated land suffers moderate to extreme degradation. Figure 7.16 shows some areas of greatest concern for soil degradation, as well as the major mechanisms for this problem in the United States. Water and wind erosion provide the motive force for the vast majority of all soil degradation worldwide. Chemical degradation includes nutrient depletion, salt accumulation, acidification, and pollution. Physical degradation includes compaction by heavy machinery or trampling by cattle, water accumulation from excess irrigation and poor drainage, and laterization (solidification of iron and aluminum-rich tropical soil when exposed to sun and rain).

Soil erosion is widespread

Erosion is an important natural process, resulting in the redistribution of the products of geologic weathering, and it is part of both soil formation and soil loss. The world's landscapes have been sculpted by erosion. When the results are spectacular enough, we enshrine them in national parks, as we did with the Grand Canyon.

Where erosion has worn down mountains and spread soil over the plains or deposited rich alluvial silt in river bottoms, we farm it. Erosion is a disaster only when it occurs in the wrong place at the wrong time.

In some places, erosion occurs so rapidly that anyone can see it happen. Deep gullies are created where water scours away the soil, leaving fenceposts and trees sitting on tall pedestals as the land erodes away around them. In most places, however, erosion is more subtle. It is a creeping disaster that occurs in small increments. A thin layer of topsoil washes off of fields year after year until, eventually, nothing is left but poor-quality subsoil that requires more and more fertilizer and water to produce any crop at all.

It's estimated that 25 billion metric tons of soil are lost from croplands every year due to wind and water erosion. The net effect, worldwide, of this widespread topsoil erosion is a reduction in crop production equivalent to removing about 1 percent of the world's cropland each year. Many farmers are able to compensate for this loss by applying more fertilizer and by bringing new land into cultivation. Continuation of current erosion rates, however, could reduce agricultural production by 25 percent in Central America and Africa and by 20 percent in South America by 2020. The total annual soil loss from croplands is thought to be 25 billion metric tons. About twice that much soil is lost from rangelands, forests, and urban construction sites each year.

In addition to reduced land fertility, erosion results in sediment loading of rivers and lakes, siltation of reservoirs and harbors, smothering of wetlands and coral reefs, creation of "dead zones" in coastal regions, and clogging of water intakes and waterpower turbines.

Wind and water move soil

Wind and water are the main agents that move soil around. Water flowing across a gently sloping, bare field removes a thin, uniform layer of soil in what is known as **sheet erosion.** When little rivulets of running water gather together and cut small channels in the soil, the process is called **rill erosion** (fig. 7.17a). If rills enlarge to form bigger channels or ravines that are too large to be removed by normal tillage operations, we call the process

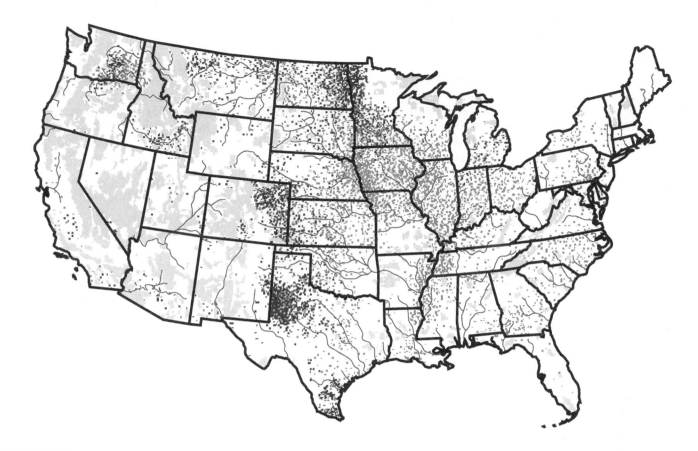

FIGURE 7.16 More than 43.7 million ha (108 million acres) in the United States are subjected to excess wind (red) and water (blue) erosion each year, removing 1.9 billion tons of soil from fields and pastures every year.
Source: Data from USDA.

gully erosion (fig. 7.17b). Streambank erosion is the washing away of soil from the banks of established streams, creeks, or rivers, often as a result of removing trees and brush along streambanks and of cattle damaging the banks.

Most soil erosion on agricultural land is rill erosion. Large amounts of soil can be transported a little bit at a time without being very noticeable. A farm field can lose 20 metric tons of soil per hectare during winter and spring runoff in rills so small that

(a) Sheet and rill erosion

(b) Gullying

(c) Wind erosion and desertification

FIGURE 7.17 Land degradation affects more than 1 billion ha yearly, or about two-thirds of all global cropland. Water erosion (a and b) account for about half that total. Wind erosion affects a nearly equal area (c).

they are erased by the first spring cultivation. That represents a loss of only a few millimeters of soil over the whole surface of the field, hardly apparent to any but the most discerning eye. But it doesn't take much mathematical skill to see that, if you lose soil twice as fast as it is being replaced, it eventually will run out.

Wind can equal or exceed water in erosive force, especially in a dry climate and on relatively flat land. When plant cover and surface litter are removed from the land by agriculture or grazing, wind lifts loose soil particles and sweeps them away. In extreme conditions, windblown dunes encroach on useful land and cover roads and buildings (fig. 7.17c). Over the past 30 years, China has lost 93,000 km^2 (about the size of Indiana) to **desertification,** or conversion of productive land to desert. Advancing dunes from the Gobi desert are now only 160 km (100 mi) from Beijing. Every year more than 1 million tons of sand and dust blow from Chinese drylands, often traveling across the Pacific Ocean to the West Coast of North America. Since 1985, China has planted more than 40 billion trees to try to stabilize the soil and hold back deserts.

Some areas of the United States and Canada have very high erosion rates. The U.S. Department of Agriculture reports that 69 million ha (170 million acres) of U.S. farmland and range are eroding at rates that reduce long-term productivity. Eleven metric tons per hectare (5 tons per acre) is generally considered the maximum tolerable rate of soil loss because that is generally the highest rate at which soil forms under optimum conditions. Some farms lose soil at twice that rate or more.

Intensive farming practices are largely responsible for this situation. Row crops, such as corn and soybeans, leave soil exposed for much of the growing season. Deep plowing and heavy herbicide applications create weed-free fields that look neat but are subject to erosion. Because big machines cannot easily follow contours, they often go straight up and down the hills, creating ready-made gullies for water to follow. Farmers sometimes plow through grass-lined watercourses (low areas where water runs off after a rain) and pull out windbreaks and fencerows to accommodate the large machines and to get every last square meter into production. Consequently, wind and water carry away the topsoil.

OTHER AGRICULTURAL RESOURCES

Soil is only part of the agricultural resource picture. Agriculture is also dependent upon water, nutrients, favorable climates to grow crops, productive crop varieties, and the mechanical energy to tend and harvest the crops.

All life needs water

Agriculture accounts for the largest single share of global water use. At least two-thirds of all fresh water withdrawn from rivers, lakes, and groundwater supplies is used for irrigation (chapter 10). Irrigation increases yields of most crops by 100 to 400 percent. Although estimates vary widely (as do definitions of irrigated land), about 15 percent of all cropland, worldwide, is irrigated.

Some countries are water rich and can readily afford to irrigate farmland, while other countries are water poor and must use water

FIGURE 7.18 Downward-facing sprinklers on this center-pivot irrigation system deliver water more efficiently than upward-facing sprinklers.

very carefully. The efficiency of irrigation water use varies greatly. High evaporative and seepage losses from unlined and uncovered canals in some places can mean that as much as 80 percent of water withdrawn for irrigation never reaches its intended destination. Poor farmers may overirrigate because they lack the technology to meter water and distribute just the amount needed. In wealthier countries, farmers can afford water-saving technologies such as drip irrigation or downward-facing sprinklers (fig. 7.18).

Excessive use not only wastes water but also often results in **waterlogging.** Waterlogged soil is saturated with water, and plant roots die from lack of oxygen. **Salinization,** in which mineral salts accumulate in the soil, is often a problem when irrigation water dissolves and mobilizes salts in the soil. As the water evaporates, it leaves behind a salty crust on the soil surface that is lethal to most plants. Flushing with excess water can wash away this salt accumulation, but the result is even more saline water for downstream users.

The FAO estimates that 20 percent of all irrigated land is damaged to some extent by waterlogging or salinity. Water conservation techniques can greatly reduce problems arising from excess water use. Conservation also makes more water available for other uses or for expanded crop production where water is in short supply (chapter 10).

Plants also need fertilizer

In addition to water, sunshine, and carbon dioxide, plants need small amounts of inorganic nutrients for growth. The major elements required by most plants are nitrogen, potassium, phosphorus, calcium, magnesium, and sulfur. Calcium and magnesium often are limited in areas of high rainfall and must be supplied in the form of lime. Lack of nitrogen, potassium, and phosphorus often limits plant growth even more. Adding these elements in fertilizer usually stimulates growth and greatly increases crop yields. A good deal of the doubling in worldwide crop production since 1950 has come from increased inorganic fertilizer use. In 1950 the average amount of fertilizer used was 20 kg per ha. In 2000 this had increased to an average of 90 kg per ha worldwide.

Farmers may overfertilize because they are unaware of the specific nutrient content of their soils or the needs of their crops. While European farmers use more than twice as much fertilizer per hectare as do North American farmers, their yields are not proportionally higher. Phosphates and nitrates from farm fields and cattle feedlots are a major cause of aquatic ecosystem pollution. Nitrate levels in groundwater have risen to dangerous levels in many areas where intensive farming is practiced. Young children are especially sensitive to the presence of nitrates. Using nitrate-contaminated water to mix infant formula can be fatal for newborns.

What are some alternative ways to fertilize crops? Manure and green manure (crops grown specifically to add nutrients to the soil) are important natural sources of soil nutrients. Nitrogen-fixing bacteria living symbiotically in root nodules of legumes are valuable for making nitrogen available as a plant nutrient (chapter 2). Interplanting and rotating beans or some other leguminous crop with such crops as corn and wheat are traditional ways of increasing nitrogen availability.

There is considerable potential for increasing the world's food supply by increasing fertilizer use in low-production countries if ways can be found to apply fertilizer more effectively and to reduce pollution. Africa, for instance, uses an average of only 19 kg of fertilizer per ha (17 lbs per acre), or about one-fourth of the world average. It has been estimated that the developing world could at least triple its crop production by raising fertilizer use to the world average.

Energy is a key resource

Farming as it is generally practiced in the industrialized countries is highly energy-intensive. Fossil fuels supply almost all of this energy. Between 1920 and 1980, direct energy use on farms rose as gasoline and diesel fuels were consumed by increasing mechanization of agriculture. Indirect energy use, in the form of synthetic fertilizers, pesticides, and other agricultural chemicals, increased even more, especially after World War II. On intensively fertilized farms, indirect energy use may be five times that of direct energy. The energy price shocks of the 1970s encouraged energy conservation that has reduced farm energy use, even though total food production has continued to rise.

After crops leave the farm, additional energy is used in food processing, distribution, storage, and cooking. It has been estimated that the average food item in the American diet travels 2,000 km (1,250 mi) between the farm that grew it and the person who consumes it. The energy required for this complex processing and distribution system may be five times as much as is used directly in farming. As a whole, the food system in the United States consumes about 16 percent of the total energy we use. Most of our foods require more energy to produce, process, and get to market than they yield when we eat them.

Farmers could assist in moving to a renewable energy future by growing energy crops that can be converted into biofuels. Many Midwestern states have encouraged construction of corn- or soy-based ethanol factories in recent years. Mixing ethanol with gasoline helps reduce air pollution, and having a new market

FIGURE 7.19 Biofuel crops, such as these sunflowers, can provide a good cash crop for farmers and contribute to national energy supplies.

helps struggling farmers. There has been much debate about whether corn-based ethanol produces more energy than is required to grow and process the crop. State-of-the-art facilities do, in fact, appear to have a 150 percent net energy yield. Plant oils from sunflowers, soybeans, corn, and other oil seed crops can be burned directly in most diesel engines, and may be closer than ethanol to a net energy gain (fig. 7.19). Fast-growing trees, kenaf, switch grass, cattails, and other energy crops may be grown on marginal land and burned in power plants.

Brazil now makes about 16 billion liters of ethanol per year, mostly from sugarcane waste. Ethanol now accounts for 40 percent of the fuel sold in Brazil. The United States plans to produce 38 billion liters (10 billion gal) of ethanol by 2008. Europe expects to manufacture 8 million tons (2 billion gal) of biofuels—half ethanol, half vegetable oils—in 2006. A German minister said that farmers could be the oil sheiks of tomorrow. Many American farmers are finding that their most lucrative crop may be the wind (chapter 12).

Pest control saves crops

Biological pests reduce crop yields and spoil as much as half of the crops harvested every year in some areas. Modern agriculture largely depends on toxic chemicals to kill or drive away these pests, but there are many serious concerns about the types and amounts of pesticides now in use.

There are many traditional methods for protecting crops from pests, but our war against unwanted organisms entered a new phase in the twentieth century with the invention of synthetic organic chemicals, such as DDT (dichlorodiphenyltrichloroethane). These chemicals have been an important part of our increased crop production and have helped control many disease-causing organisms. It's estimated that up to half our current crop yields might be lost if we had no pesticides to protect them. Indiscriminate and profligate pesticide use, however, also has caused many problems, such as killing nontarget species, creating new pests of organisms that

FIGURE 7.20 Spraying pesticides by air is quick and cheap, but toxins often drift to nearby fields.

previously were not a problem, and causing widespread pesticide resistance among pest species (fig. 7.20). Often highly persistent and mobile in the environment, many pesticides have moved through air, water, and soil and bioaccumulated or bioconcentrated in food chains, nearly exterminating several top predators, such as peregrine falcons.

Your exposure to pesticides is probably higher than you suspect. A study by the U.S. Department of Agriculture found that 73 percent of conventionally grown foods (out of 94,000 samples assayed) had residues from at least one pesticide. By contrast, only 23 percent of the organic samples of the same food groups had any residues. It is now clear that humans suffer adverse health effects from persistent organic pollutants (POPs), such as pesticides. In fact, most of the "dirty dozen" POPs recently banned globally are pesticides. See chapter 8 for more on health effects of POPs.

In 2002 Swiss researchers published results of a two-decades-long comparison of organic and conventional farming. Crops grown without synthetic fertilizers or pesticides produced, on average, about a 20 percent lower yield than similar crops grown by conventional methods, but they more than made up the difference with lower operating costs, premium crop prices, less ecological damage, and healthier farm families.

Some alternatives for reducing our dependence on dangerous chemical pesticides include management changes, such as using cover crops and mechanical cultivation, and planting mixed polycultures rather than vast monoculture fields. Consumers may have to learn to accept less than perfect fruits and vegetables if we want to avoid toxic chemicals in our diet. Biological controls, such as insect predators, pathogens, or natural poisons specific for a particular pest, can help reduce chemical use. Genetic breeding and biotechnology can produce pest-resistant crop and livestock strains as well. Integrated pest management (IPM) is a relatively new science that combines all of these alternative methods, together with judicious use of synthetic pesticides under precisely controlled conditions. There are ways that you can minimize pesticides in your diet (see What Can You Do? on this page).

What Can You Do?

Reducing the Pesticides in Your Food

If you want to reduce the amount of pesticide residues and other toxic chemicals in your diet, follow these simple rules.

- Wash and scrub all fresh fruits and vegetables thoroughly under running water.
- Peel fruits and vegetables when possible. Throw away the outer leaves of leafy vegetables, such as lettuce and cabbage.
- Store food carefully so it doesn't get moldy or pick up contaminants from other foods. Use food as soon as possible to ensure freshness.
- Cook foods that you suspect have been treated with pesticides to break down chemical residues.
- Trim the fat from meat, chicken, and fish. Eat lower on the food chain when possible to reduce bioaccumulated chemicals.
- Don't pick and eat berries or other wild foods that grow on the edges of roadsides where pesticides may have been sprayed.
- Grow your own fruits and vegetables without using pesticides or with minimal use of dangerous chemicals.
- Ask for organically grown food at your local grocery store, or shop at a farmer's market or co-op where you can get such food.

NEW CROPS AND GENETIC ENGINEERING

Although at least 3,000 species of plants have been used for food at one time or another, most of the world's food now comes from only 16 widely grown crops. Many new or unconventional varieties might be valuable human food supplies, however, especially in areas where conventional crops are limited by climate, soil, pests, or other problems. Among the plants now being investigated as potential additions to our crop roster is the winged bean, a perennial plant that grows well in hot climates where other legumes will not grow. It is totally edible (pods, mature seeds, shoots, flowers, leaves, and tuberous roots), resistant to diseases, and enriches the soil. Another promising crop is triticale, a hybrid between wheat (*Triticum*) and rye (*Secale*) that grows in light, sandy, infertile soil. It is drought-resistant, has nutritious seeds, and is being tested for salt tolerance for growth in saline soils or irrigation with seawater. Some traditional crop varieties grown by Native Americans, such as tepary beans, amaranth, and Sonoran panic grass, are being collected by seed conservator Gary Nabhan, both as a form of cultural revival for native people and as possible food crops for harsh environments.

The green revolution has increased yields

So far, the major improvements in farm production have come from technological advances and modification of a few well-known species. Yield increases often have been spectacular. A century ago,

FIGURE 7.21 Semidwarf wheat (*right*), bred by Norman Borlaug, has shorter, stiffer stems and is less likely to lodge (fall over) when wet than its conventional cousin (*left*). This "miracle" wheat responds better to water and fertilizer and has played a vital role in feeding a growing human population.

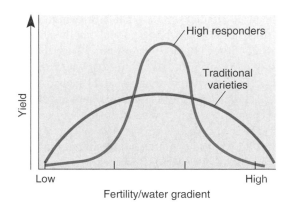

FIGURE 7.22 Green revolution miracle crops are really high responders, meaning that they have excellent yields under optimum conditions. For poor farmers who can't afford the fertilizer and water needed by high responders, traditional varieties may produce better yields.

when all maize (corn) in the United States was open pollinated, average yields were about 25 bushels per acre (bu/acre). In 2000 the average yield from rainfed fields in Iowa was 138 bu/acre, and irrigated maize in Arizona averaged 208 bu/acre. The highest yield ever recorded in field production was 370 bu/acre on an Illinois farm, but theoretical calculations suggest that 500 bu/acre (32 metric tons per hectare) could be possible. Most of this gain was accomplished by use of synthetic fertilizers along with conventional plant breeding: geneticists laboriously hand-pollinating plants and looking for desired characteristics in the progeny.

Starting about 50 years ago, agricultural research stations began to breed tropical wheat and rice varieties that would provide food for growing populations in developing countries. The first of the "miracle" varieties was a dwarf, high-yielding wheat developed by Norman Borlaug (who received a Nobel Peace Prize for his work) at a research center in Mexico (fig. 7.21). At about the same time, the International Rice Institute in the Philippines developed dwarf rice strains with three or four times the production of varieties in use at the time. The spread of these new high-yield varieties around the world has been called the **green revolution.** It is one of the main reasons that world food supplies have more than kept pace with the growing human population over the past few decades.

Most green revolution breeds really are "high responders," meaning that they yield more than other varieties if given optimum levels of fertilizer, water, and protection from pests and diseases (fig. 7.22). Under suboptimum conditions, on the other hand, high responders may not produce as well as traditional varieties. Poor farmers who can't afford the expensive seed, fertilizer, and water required to become part of this movement usually are left out of the green revolution. In fact, they may be driven out of farming altogether as rising land values and falling commodity prices squeeze them from both sides.

Genetic engineering could have benefits and costs

Genetic engineering, splicing a gene from one organism into the chromosome of another, has the potential to greatly increase both the quantity and the quality of our food supply. It is now possible to build entirely new genes, and even new organisms, often called "transgenic" organisms or **genetically modified organisms (GMOs),** by taking a bit of DNA from here, a bit from there, and even synthesizing artificial sequences to create desired characteristics in engineered organisms (fig. 7.23).

Proponents predict dramatic benefits from this new technology. Research is now underway to improve yields and create crops that resist drought, frost, or diseases. Other strains are being developed to tolerate salty, waterlogged, or low-nutrient

FIGURE 7.23 A researcher measures growth of genetically engineered rice plants. Superior growth and yield, pest resistance, and other useful traits can be moved from one species to another by molecular biotechnology.

soils, allowing degraded or marginal farmland to become productive. All of these could be important for reducing hunger in developing countries. Plants that produce their own pesticides might reduce the need for toxic chemicals, while engineering for improved protein or vitamin content could make our food more nutritious. Attempts to remove specific toxins or allergens from crops also could make our food safer. Crops such as bananas and tomatoes have been altered to contain oral vaccines that can be grown in developing countries where refrigeration and sterile needles are unavailable. Plants have been engineered to make industrial oils and plastics. Animals, too, are being genetically modified to grow faster, gain weight on less food, and produce pharmaceuticals, such as insulin, in their milk. Pigs are now being engineered to produce omega-3 fatty acids for a "heart healthy" diet. It may even be possible to create animals with human cell-recognition factors that could serve as organ donors.

Opponents worry that moving genes willy-nilly could create a host of problems, some of which we can't even imagine. GMOs themselves might escape and become pests or they might interbreed wild relatives. In either case, we may create superweeds or reduce native biodiversity. Constant presence of pesticides in plants could accelerate pesticide resistance in insects or leave toxic residues in soil or our food. Genes for toxicity or allergies could be transferred along with desirable genes, or novel toxins could be created as genes are mixed together. This technology may be available only to the richest countries or the wealthiest corporations, making family farms uncompetitive and driving developing countries even further into poverty.

Both the number of GMO crops and the acreage devoted to growing them is increasing rapidly. Between 1987 and 2005, more than 10,500 field tests for some 750 different crop varieties were carried out in the United States (fig. 7.24). Hawaii is the most popular site for GMO testing, with about 1,500 field releases over the past 15 years. Illinois, Iowa, California, and Puerto Rico each had more than 1,000 tests during that time. Worldwide, 72 million ha (about 25 percent of all cropland) were planted with GMO crops in 2005. The United States accounted for 63 percent of that acreage, followed by Argentina with 21 percent. Canada, Brazil, and China together make up 22 percent of all transgenic cropland.

Some transgenic crops have reached mainstream status. About 82 percent of all soybeans, one-quarter of all maize (corn), and 71 percent of all cotton grown in the United States are now GMOs. You've already probably eaten some genetically modified food. Estimates are that at least 60 percent of all processed food in America contains GMO ingredients. Since the United States, Argentina, and Brazil account for 90 percent of international trade in maize and soybeans (in which GMOs often are mixed with other grains), a large fraction of the world most likely has been exposed to some GMO products.

Most GMOs have been engineered for pest resistance or weed control

Biotechnologists have created plants with genes for endogenous insecticides. *Bacillus thuringiensis* (Bt), a bacterium, makes toxins lethal to Lepidoptera (butterfly family) and Coleoptera (beetle family). The genes for some of these toxins have been transferred into crops such as maize (to protect against European cutworms), potatoes (to fight potato beetles), and cotton (to protect against boll weevils). This allows farmers to reduce insecticide spraying. Arizona cotton farmers, for example, report reducing their use of chemical insecticides by 75 percent.

Most European nations have not approved Bt-containing varieties. Fear of transgenic products has been a sticking point in agricultural trade between the United States and the European Union.

In 2001 Starlink, a Bt-containing maize variety that had not yet been approved for human consumption, was found mixed into a wide variety of consumer products. By the time food was made from Starlink, no Bt could be detected, only the genes that code for it. There is no evidence that Bt is harmful to humans, but the form in Starlink may be allergenic. In fact, Bt is one of the few insecticides approved for organic farmers. Nevertheless, a public outcry ensued and seed companies, farmers, shippers, and food processors spent more than $1 billion trying to remove it from the food chain. What do you think: was the fear of Starlink justified?

Entomologists worry that Bt plants churn out toxin throughout the growing season, regardless of the level of infestation, creating perfect conditions for selection of Bt resistance in pests. Already 500 species of insect, mite, or tick are resistant to one or more pesticides. Having Bt constantly available can only aggravate this dilemma. One solution is to plant at least a part of every field in non-Bt crops that will act as a refuge for nonresistant pests. The hope is that interbreeding between these "wild-type" bugs and those exposed to Bt will dilute out recessive resistance genes. Deliberately harboring pests and letting them munch freely on crops is something that many farmers find hard to do. In addition, devoting a significant part of their land to nonproductive crops lowers the total yield and counteracts the profitability of engineered seed.

The other major group of transgenic crops is engineered to tolerate high doses of herbicides. The two dominant products in this category are Monsanto's "Roundup Ready" crops—so-called

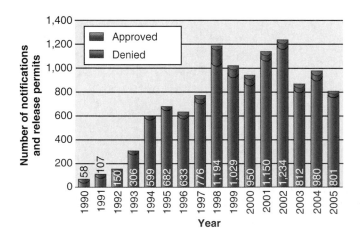

FIGURE 7.24 Transgenic crop field releases in the United States 1990 to 2005.
Source: Data from Information Systems for Biotechnology, Virginia Tech University.

because they can withstand treatment with Monsanto's best-selling herbicide, Roundup (glyphosate)—and AgrEvo's "Liberty Link" crops, which resist that company's Liberty (glufosinate) herbicide. Because crops with these genes can grow in spite of high herbicide doses, farmers can spray fields heavily to exterminate weeds. This allows for conservation tillage and leaving more crop residue on fields to protect topsoil from erosion—both good ideas—but it also can mean using more herbicide in higher doses than might otherwise be done.

Is genetic engineering safe?

Ever since scientists discovered how to move genes from one organism to another, critics have worried about irresponsible use of this technology or unforeseen consequences arising from novel combinations of genetic material. Greenpeace and several other environmental groups have campaigned against transgenic organisms, labeling them "Frankenfoods." Industry groups, on the other hand, describe their critics as Neo-Luddites, who blindly oppose any new technology. How can we decide what to believe in this welter of claims and counterclaims?

At the base of some people's unease with genetic modification is a feeling that it simply isn't right to mess with nature. It doesn't seem proper to recombine genes any way we want. Doing so raises specters of gruesome monsters stitched together from spare parts, like Mary Shelley's Frankenstein. Is this merely a fear of science, or is it a valid ethical issue?

The U.S. Food and Drug Administration declined to require labeling of foods containing GMOs, saying that these new varieties are "substantially equivalent" to related varieties bred via traditional practices. After all, proponents say, we have been moving genes around for centuries through plant and animal breeding. All domesticated organisms should be classified as genetically modified, some people argue. Biotechnology may be a more precise way of creating novel organisms than normal breeding procedures, its supporters suggest. We're moving only a few genes—or even part of one gene—at a time with biotechnology, rather than hundreds of unknown genes through classical techniques.

But what if GMOs escape and crossbreed with native species? In a ten-year study of genetically modified crops, a group from Imperial College, London, concluded that GMOs tested so far do not survive well in the wild and are no more likely to invade other habitats than their unmodified counterparts.

On the other hand, there is already evidence that some GMO crops can spread their genes to nearby fields. Normal rape seed (canola oil) varieties in Canada were found to contain genes from genetically modified varieties in nearby fields. It isn't clear, however, if the genes involved will have any adverse affect or whether they will persist for many generations. In 2001 researchers reported finding traces of genetically modified corn in wild relatives in Mexico, which supposedly has banned planting of GMOs, especially in regions where species were thought to have originated. This report was later withdrawn due to criticism of the techniques used to detect genetic markers, but it gave an interesting insight into the politics and sociology of science.

The first genetically modified animal designed to be eaten by humans is an Atlantic salmon (*Salmo salar*) containing extra growth hormone genes from an oceanic pout (*Macrozoarces americanus*). The greatest worry from this fish is not that it will introduce extra hormones into our diet—that's already being done by chickens and beef that get extra growth hormone via injections or their diet—but, rather, the ecological effects if the fish escape from captivity. The transgenic fish grow seven times faster and are more attractive to the opposite sex than a normal salmon. If they escape from captivity, they may outcompete already endangered wild relatives for food, mates, and habitat. Fish farmers say they will grow only sterile females and will keep them in secure net pens. Opponents point out that salmon frequently escape from aquaculture operations and that, if just a few fertile transgenic fish break out, it could be catastrophic for wild stocks.

Many scientists argue that we should take a better-safe-than-sorry "precautionary approach" that errs—if at all—on the side of safety. The 5,000-member Society of Toxicology recently concluded that, "based on available tests, there's no reason to suspect that transgenic plants differ in any substantive way from traditional varieties." On the other hand, a panel of biologists and agricultural scientists convened by the U.S. National Academy of Sciences urged the government to more carefully, and more publicly, review the potential environmental impacts of genetically modified plants and animals before approving them for commercial use.

Finally, there are social and economic implications of GMOs. Will they help feed the world, or will they lead to a greater consolidation of corporate power and economic disparity? Might higher yields and fewer losses to pests and diseases allow poor farmers in developing countries to stop using marginal land and avoid cutting down forests to create farmland? Is this simply a technological fix, or could it help promote agricultural sustainability? Critics suggest that there are simpler and cheaper ways other than high-tech crop varieties to provide vitamin A to children in developing countries or to increase the income of poor rural families. Adding a cow or a fishpond or training people in water harvesting or regenerative farming techniques (as we'll discuss in the next section) may have a longer-lasting impact than selling them expensive new seeds.

On the other hand, if we hope to reduce malnutrition and feed eight billion people in 50 years, maybe we need all the tools we can get. Where do you stand in this debate? What additional information would you need to reach a reasoned judgment about the risks and benefits of this new technology?

SUSTAINABLE AGRICULTURE

How, then, shall we feed the world? Can we make agriculture compatible with sustainable ecological and social systems? Having discussed some of the problems that beset modern agriculture, we now consider some suggested ways to overcome problems and make farming and food production just and lasting enterprises. This goal is usually termed **sustainable agriculture**, or **regenerative farming,** both of which aim to produce food and fiber on a sustainable basis and repair the damage caused by destructive practices. Some

What Do You Think?

Shade-Grown Coffee and Cocoa

Has it ever occurred to you that your purchases of coffee and chocolate may be contributing to the protection or destruction of tropical forests? Both coffee and cocoa are examples of food products grown exclusively in developing countries but consumed almost entirely in the wealthier, developed nations (vanilla and bananas are some other examples). Coffee grows in cool, mountain areas of the tropics, while cocoa is native to the warm, moist lowlands. Both are small trees of the forest understory, adapted to low light levels.

Until a few decades ago, most of the world's coffee and cocoa were **shade-grown** under a canopy of large forest trees. Recently, however, new varieties of both crops have been developed that can be grown in full sun. Because more coffee or cocoa trees can be crowded into these fields, and they get more solar energy than in a shaded plantation, yields for sun-grown crops are higher.

There are costs, however, in this new technology. Sun-grown trees die earlier from the stress and diseases common in these fields. Furthermore, ornithologists have found that the number of bird species can be cut in half in full-sun plantations, and the number of individual birds may be reduced by 90 percent. Shade-grown coffee and cocoa generally require fewer pesticides (or sometimes none) because the birds and insects residing in the forest canopy eat many of the pests. Shade-grown plantations also need less chemical fertilizer because many of the plants in these complex forests add nutrients to the soil. In addition, shade-grown crops rarely need to be irrigated because heavy leaf fall protects the soil, while forest cover reduces evaporation.

Currently, about 40 percent of the world's coffee and cocoa plantations have been converted to full-sun varieties and another 25 percent are in process. Traditional techniques for coffee and cocoa production are worth preserving. Thirteen of the world's 25 biodiversity hot spots occur in coffee or cocoa regions. If all the 20 million ha (49 million acres) of coffee and cocoa plantations in these areas are converted to monocultures, an incalculable number of species will be lost.

The Brazilian state of Bahia is a good example of both the ecological importance of these crops and how they might help preserve forest species. At one time, Brazil produced much of the world's cocoa, but in the early 1900s, the crop was introduced into West Africa. Now Côte d'Ivoire alone grows more than 40 percent of the world total, and the value of Brazil's harvest has dropped by 90 percent. Côte d'Ivoire is aided in this

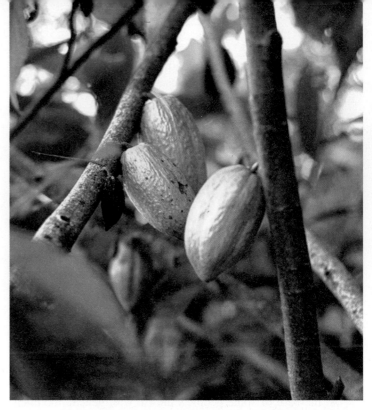

Cocoa pods grow directly on the trunk and large branches of cocoa trees.

competition by a labor system that reportedly includes widespread child slavery. Even adult workers in Côte d'Ivoire get only about $165 (U.S.) per year (if they get paid at all), compared with a minimum wage of $850 (U.S.) per year in Brazil. As African cocoa production ratchets up, Brazilian landowners are converting their plantations to pastures or other crops.

The area of Bahia where cocoa was once king is part of Brazil's Atlantic Forest, one of the most threatened forest biomes in the world. Only 8 percent of this forest remains undisturbed. Although cocoa plantations don't represent the full diversity of intact forests, they protect a surprisingly large sample of what once was there. And shade-grown cocoa can provide an economic rationale for preserving that biodiversity. Brazilian cocoa will probably never compete with that from other areas for lowest cost. There is room in the market, however, for specialty products. If consumers were willing to pay a small premium for organic, fair-trade, shade-grown chocolate and coffee, it might provide the incentive needed to preserve biodiversity. Wouldn't you like to know that your chocolate or coffee wasn't grown with child slavery and is helping protect plants and animal species that might otherwise go extinct?

---

alternative methods are developed through scientific research; others are discovered in traditional cultures and practices nearly forgotten in our mechanization and industrialization of agriculture.

Soil conservation is essential

With careful husbandry, soil is a renewable resource that can be replenished and renewed indefinitely. Since agriculture is the area in which soil is most essential and most often lost through erosion, agriculture offers the greatest potential for soil conservation and rebuilding. Some rice paddies in Southeast Asia, for instance, have been farmed continuously for a thousand years without any

apparent loss of fertility. The rice-growing cultures that depend on these fields have developed management practices that return organic material to the paddies and carefully nurture the soil's ability to sustain life.

While American agriculture hasn't reached that level of sustainability, there is evidence that soil conservation programs are having a positive effect. In one Wisconsin study, erosion rates in one small watershed were 90 percent less in 1975–1993 than they were in the 1930s. Among the most important elements in soil conservation are topography management, ground cover, and reduced tillage.

Water runs downhill. The faster it runs, the more soil it carries off the fields. Comparisons of erosion rates in Africa have

FIGURE 7.25 Countour plowing and strip cropping help prevent soil erosion on hilly terrain, as well as create a beautiful landscape.

FIGURE 7.26 Terracing, as in these Balinese rice paddies, can control erosion and make steep hillsides productive.

Planting **perennial species** (plants that grow for more than two years) is the only suitable use for some lands and some soil types. Establishing forest, grassland, or crops such as tea, coffee, or other crops that do not have to be cultivated every year may be necessary to protect certain unstable soils on sloping sites or watercourses. The $180 billion U.S. Farm Bill includes $17 billion (over ten years) to encourage soil, water, and wildlife conservation.

Groundcover protects the soil

Annual row crops, such as corn or beans, generally cause the highest erosion rates because they leave soil bare for much of the year (table 7.2). Often the easiest way to provide cover that protects soil from erosion is to leave crop residues on the land after harvest. They not only cover the surface to break the erosive effects of wind and water but also reduce evaporation and soil

shown that a 5 percent slope in a plowed field has three times the water runoff volume and eight times the soil erosion rate of a comparable field with a 1 percent slope. Water runoff can be reduced by grass strips in waterways and by **contour plowing**—that is, plowing across the hill rather than up and down. Contour plowing is often combined with **strip-farming,** the planting of different kinds of crops in alternating strips along the land contours (fig. 7.25). When one crop is harvested, the other is still present to protect the soil and keep water from running straight downhill. The ridges created by cultivation make little dams that trap water and allow it to seep into the soil, rather than running off. In areas where rainfall is very heavy, tied ridges are often useful. This method involves a series of ridges running at right angles to each other, so that water runoff is blocked in all directions and is encouraged to soak into the soil.

Terracing involves shaping the land to create level shelves of earth to hold water and soil (fig. 7.26). The edges of the terrace are planted with soil-anchoring plant species. This is an expensive procedure, requiring either much hand labor or expensive machinery, but it makes farming of very steep hillsides possible. The rice terraces in the Chico River Valley in the Philippines rise as much as 300 m (1,000 ft) above the valley floor. They are considered one of the wonders of the world. Some of these terraces have been farmed for 1,000 years without any apparent fertility loss. Will we be able to say the same thing a millenium from now?

TABLE 7.2

Soil Cover and Soil Erosion

Cropping System	Average Annual Soil Loss (tons/hectare)	Percent Rainfall Runoff
Bare soil (no crop)	41.0	30
Continuous corn	19.7	29
Continuous wheat	10.1	23
Rotation: corn, wheat, clover	2.7	14
Continuous bluegrass	0.3	12

Source: Based on 14 years of data from Missouri Experiment Station, Columbia, Missouri.

temperature in hot climates and protect ground organisms that help aerate and rebuild soil. In some experiments, 1 ton of crop residue per acre (0.4 ha) increased water infiltration 99 percent, reduced runoff 99 percent, and reduced erosion 98 percent. Leaving crop residues on the field also can increase disease and pest problems, however, and may require increased use of pesticides and herbicides.

Where crop residues are not adequate to protect the soil or are inappropriate for subsequent crops or farming methods, such **cover crops** as rye, alfalfa, and clover can be planted immediately after harvest to hold and protect the soil. These cover crops can be plowed under at planting time to provide green manure. Another method is to flatten cover crops with a roller and drill seeds through the residue to provide a continuous protective cover during early stages of crop growth.

In some cases, interplanting of two different crops in the same field not only protects the soil but also is a more efficient use of the land, providing double harvests. Native Americans and pioneer farmers, for instance, planted beans or pumpkins between the corn rows. The beans provided nitrogen needed by the corn, the pumpkins crowded out weeds, and both crops provided foods that nutritionally balance corn. Traditional swidden (slash-and-burn) cultivators in Africa and South America often plant as many as 20 different crops together in small plots. The crops mature at different times, so that there is always something to eat, and the soil is never exposed to erosion for very long. Shade-grown coffee and cocoa can play important roles in conserving biodiversity (see What Do You Think? p. 172).

Mulch is a general term for a protective groundcover that includes manure, wood chips, straw, seaweed, leaves, and other natural products. For some high-value crops, such as tomatoes, pineapples, and cucumbers, it is cost-effective to cover the ground with heavy paper or plastic sheets to protect the soil, save water, and prevent weed growth. Israel uses millions of square meters of plastic mulch to grow crops in the Negev Desert.

Reduced tillage can have many benefits

Farmers have traditionally used a moldboard plow to till the soil, digging a deep trench and turning the topsoil upside down. In the 1800s it was shown that tilling a field fully—until it was "clean"—increased crop production. It helped control weeds and pests, reducing competition; it brought fresh nutrients to the surface, providing a good seedbed; and it improved surface drainage and aerated the soil. This is still true for many crops and many soil types, but it is not always the best way to grow crops. Less plowing and cultivation often improves water management, preserves soil, saves energy, and increases crop yields.

There are three major **reduced tillage systems.** *Minimum till* involves reducing the number of times a farmer disturbs the soil by plowing, cultivating, and so on. This often involves a disc or chisel plow rather than a traditional moldboard plow. A chisel plow is a curved, chisel-like blade that doesn't turn the soil over but creates ridges on which seeds can be planted.

FIGURE 7.27 In ridge-tilling, a chisel plow is used to create ridges on which crops are planted and shallow troughs filled with crop residue. Less energy is used on plowing and cultivation, and weeds are suppressed and moisture is retained by the groundcover, or crop residue, left on the field.

It leaves up to 75 percent of plant debris on the surface between the rows, preventing erosion (fig. 7.27). *Conserv-till* farming uses a coulter (a sharp, disc like a pizza cutter), which slices through the soil, opening up a furrow, or slot, just wide enough to insert seeds. This disturbs the soil very little and leaves almost all plant debris on the surface. *No-till* planting is accomplished by drilling seeds into the ground directly through mulch and groundcover. This allows a cover crop to be interseeded with a subsequent crop.

Farmers who use these conservation tillage techniques often must depend on pesticides (insecticides, fungicides, and herbicides) to control insects and weeds. Increased use of toxic agricultural chemicals is a matter of great concern. Massive use of pesticides is not, however, a necessary corollary of soil conservation. It is possible to combat pests and diseases with integrated pest management that combines crop rotation, trap crops, natural repellents, and biological controls.

Low-input sustainable agriculture can benefit farmers, consumers, and the environment

In contrast to the trend toward industrialization and dependence on chemical fertilizers, pesticides, antibiotics, and artificial growth factors common in conventional agriculture, some farmers are going back to a more natural, agroecological farming style. Finding that they can't—or don't want to—compete with factory farms, these folks are making money and staying in farming by returning to small-scale, low-input agriculture. The Minar family, for instance, operates a highly successful 150-cow dairy

Principles of Environmental Science

FIGURE 7.28 On the Minar family's 230-acre dairy farm near New Prague, Minnesota, cows and calves spend the winter outdoors in the snow, bedding down on hay. Dave Minar is part of a growing counterculture that is seeking to keep farmers on the land and bring prosperity to rural areas. © 2004 Star Tribune/Minneapolis-St. Paul.

FIGURE 7.29 Your local farmers' market is a good source of locally grown organic produce.

operation on 97 ha (230 acres) near New Prague, Minnesota. No synthetic chemicals are used on their farm. Cows are rotated every day between 45 pastures or paddocks to reduce erosion and maintain healthy grass. Even in the winter, livestock remain outdoors to avoid the spread of diseases common in confinement (fig. 7.28). Antibiotics are used only to fight diseases. Milk and meat from this operation are marketed through co-ops and a community-supported agriculture (CSA) program. Sand Creek, which flows across the Minar land, has been shown to be cleaner when leaving the farm than when it enters.

Similarly, the Franzen family, who raise livestock on their organic farm near Alta Vista, Iowa, allow their pigs to roam in lush pastures, where they can supplement their diet of corn and soybeans with grasses and legumes. Housing for these happy hogs is in spacious, open-ended hoop structures. As fresh layers of straw are added to the bedding, layers of manure beneath are composted, breaking down into odorless organic fertilizer.

Low-input farms such as these typically don't turn out the quantity of meat or milk that their intensive agriculture neighbors do, but their production costs are lower, and they get higher prices for their crops, so that the all-important net gain is often higher. The Franzens, for example, calculate that they pay 30 percent less for animal feed, 70 percent less for veterinary bills, and half as much for buildings and equipment as their neighboring confinement operations. And on the Minar's farm, erosion after an especially heavy rain was measured to be 400 times lower than on a conventional farm nearby.

Preserving small-scale family farms also helps preserve rural culture. As Marty Strange of the Center for Rural Affairs in Nebraska asks, "Which is better for the enrollment in rural schools, the membership of rural churches, and the fellowship of rural communities—two farms milking 1,000 cows each or twenty farms milking 100 cows each?" Family farms help keep rural towns alive by purchasing machinery at the local implement

dealer, gasoline at the neighborhood filling station, and groceries at the mom-and-pop grocery store.

One way to support local agriculture is to shop at a farmers' market (fig. 7.29). The produce is fresh, and profits go directly to the farmer who grows the crop. A local food co-op or owner-operated grocery store also is likely to buy from local farmers and to feature pesticide-free foods. Many co-ops and buyers' associations sign contracts directly with producers to grow the types of food they want to eat. This benefits both parties. Producers are guaranteed a local market for organic food or specialty items. Consumers can be assured of quality and can even be involved in production.

Many colleges have adopted a policy to buy as much locally grown food as possible. If your school doesn't have such a policy, why not talk to administrators to start one? You could have healthy food while also helping preserve your local economy and environment.

Agroecology, or sustainable farming using ecological knowledge, also can be applied effectively in poorer countries. Brian Haliweil, of the Worldwatch Institute, reports that farmers in the Guatemalan Highlands raised their crop yields ten-fold without the use of chemical fertilizers or pesticides. Encouraged by World Neighbors, a nongovernmental development organization, to find innovative, local solutions to their problems, the farmers came up with low-cost techniques, such as planting cover crops and grass strips to reduce erosion and rotating corn with beans, peas, and other legumes to add nitrogen to the soil. Corn harvests increased from 0.4 tons to 4.5 tons per hectare, yields comparable with those of many richer countries.

One of the largest experiments in low-input farming is currently taking place in Cuba, where loss of aid from the former Soviet Union coupled with a trade embargo imposed by the United States have forced farmers to turn to organic, nonmechanized agriculture (see related story "Organic Farming in Cuba" at www.mhhe.com/cunningham4e).

REVIEW QUESTIONS

1. What is Brazil's Cerrado, and how is agriculture affecting it?

2. Explain how soybeans grown in Brazil are improving diets in China.

3. What does it mean to be chronically undernourished? How many people in the world currently suffer from this condition?

4. Why do nutritionists worry about food security? Who is most likely to suffer from food insecurity?

5. Describe the conditions that constitute a famine. Why does Amartya Sen say that famines are caused more by politics and economics than by natural disasters?

6. Define *malnutrition* and *obesity*. How many Americans are now considered obese?

7. How much does Dr. Alfred Sommer estimate that it would cost to treat a child with Vitamin A deficiency? How many childhood deaths and cases of blindness could be prevented every year by this treatment?

8. Summarize the diet recommendations of the Willett/Stampfer (Harvard) diet. How does this food pyramid differ from the one devised by the USDA. What criticisms have been made of the USDA recommendations?

9. What three crops provide most human caloric intake?

10. What are concentrated animal feeding operations, and why are they controversial?

11. What is *soil*? Why are soil organisms so important?

12. How much soil erodes from croplands every year worldwide? What are some effects of this loss?

13. How might farms contribute to sustainable energy?

14. What do we mean by the *green revolution*?

15. What is *genetic engineering,* or *biotechnology,* and how might it help or hurt agriculture?

16. What is *sustainable agriculture*?

17. How could your choices of coffee or cocoa help preserve forests, biodiversity, and local economies in tropical countries?

18. What is *perennial polyculture* and how might it protect soil?

19. What is *reduced tillage* or *minimum tillage* with annual crops?

20. What are the economic advantages of low-input farming?

CRITICAL THINKING AND DISCUSSION

Apply the principles you have learned in this chapter to discuss these questions with other students.

1. Suppose that you were engaged in biotechnology, or genetic engineering; what environmental safeguards would you impose on your own research? Are there experiments that would be ethically off-limits for you?

2. Debate the claim that famines are caused more by human actions (or inactions) than by environmental forces. What scientific evidence would you need to have to settle this question? What hypotheses could you test to help resolve the debate?

3. Should farmers be forced to use ecologically sound techniques that serve farmers' best interests in the long run, regardless of short-term consequences? How could we mitigate hardships brought about by such policies?

4. How would you provide assurance to consumers, government officials, and other interested parties that new genetically engineered products were environmentally and socially safe? What experiments would you set up to test for unknown unknowns?

5. Former U.S. President Jimmy Carter said, "Responsible biotechnology is not the enemy; starvation is." What did he mean? Do you agree?

SUGGESTED READINGS

Brown, Lester. 2005. *Outgrowing the Earth: The Food Security Challenge in an Age of Falling Water Tables and Rising Temperatures.* Earth Policy Institute.

Food and Agriculture Organization (UN). 2004. *The State of Food Insecurity in the World: Monitoring Progress Towards the World Food Summit and the Millennium Development Goals.* United Nations Publications.

Halweil, Brian. 2004. *Eat Here: Homegrown Pleasures in a Global Supermarket.* W.W. Norton Co.

Hites, R. A., et al. 2004. Global assessment of organic contaminants in farmed salmon. *Science* 303:226–29.

Kimbrell, Andrew, ed. 2002. *Fatal Harvest: The Tragedy of Industrial Agriculture.* Island Press.

Marris, Emma. 2005. The forgotten ecosystem. *Nature* 437: 944–45.

Mas, A. H., and T. V. Dietsch. 2004. Linking shade coffee certification to biodiversity conservation: Butterflies and birds in Chiapas, Mexico. *Ecological Applications* 14(3):642–54.

McKibben, Bill. 2005. The Cuba diet. *Harper's* 310(1859): 61–70.

Runge, C. F., et al. 2003. *Ending Hunger in Our Lifetime: Food Security and Globalization.* International Food Policy Research Institute.

Willett, Walter C., and Meir J. Stampfer. 2003. Rebuilding the food pyramid. *Scientific American* 288(1):64–71.

DATA analysis   Using Relative Values

The graph in this box shows changes in abundance of marine species in six oceanic regions between 1970 and 2000. Look at the scales on the axes. Do you see anything unusual about them? The X-axis is simply five-year intervals. Nothing unusual there. But what are the units for the Y-axis? What is a species index, and why do all the curves start at 100? Essentially, the species index is a percent change from the abundance of species in each region in 1970. Why use an index rather than absolute numbers? We do this because the abundance values vary so much between regions. It's difficult to fit the full range of real numbers of species on the Y-axis scale, and if we did so, the curves would be so far apart that comparison would be difficult. What we're mostly interested in here are the comparative changes and the trends in different areas. A relative scale is useful because it makes it easy to see that species abundance in most areas has declined sharply over this 30-year period as well as where the losses have been greatest.

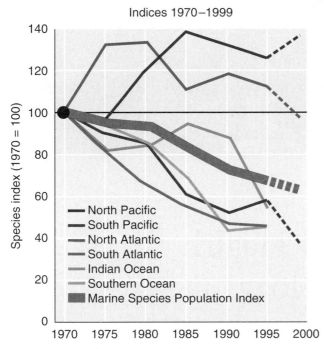

Indices 1970–1999

Legend:
- North Pacific
- South Pacific
- North Atlantic
- South Atlantic
- Indian Ocean
- Southern Ocean
- Marine Species Population Index

Changes in abundance of 217 species of marine fish, birds, mammals, and reptiles, from 1970 to 1999, by continent.
Source: World Wide Fund for Nature (WWF), 2000.

- In which regions have species abundance remained relatively constant?

- What do you think may be responsible for the sharp declines shown here?

- Can you think of any reason that these regions have experienced less decline than other places?

- How much has marine abundance declined in the South Atlantic during the period shown?

- What is the significance of the thick gray line?

- Why are some of the curves dotted at their ends?

- How would you describe the marine biodiversity history of the North Atlantic over the time shown?

- How do you think the agenda of the WWF, which created this graph, might be furthered by the way it's presented?

- Might some other organization be likely to present the data in another way?

For Additional Help in Studying This Chapter, please visit our website at www.mhhe.com/cunningham4e. You will find practice quizzes, a chapter summary, key terms, answers to review questions, additional case studies, regional examples, and an extensive reading list, all of which will help you understand the material in this chapter.

A new strain of avian flu caused by the H5N1 virus was first identified in Hong Kong in 1997. Despite heroic efforts to contain this highly lethal disease, it continues to spread around the world. Could this be the next big pandemic?

Environmental Health and Toxicology

It's not a question of if a pandemic occurs, but when.

–Dr. David Nabarro, United Nations–

OBJECTIVES

Studying this chapter will help you answer the following questions:

- What is environmental health?
- Epidemiologists say, "The next pandemic is just plane ride away." Do you agree?
- What health risks should worry us most?
- Emergent diseases seem to be more frequent now. What human factors may be involved in this?

- What are toxins, and how do they affect us?
- When Paracelcus said, "The dose makes the poison," what did he mean?
- What makes some chemicals dangerous and others harmless?
- How much risk is acceptable, and to whom?

Case Study The Next Pandemic?

Every year one or more new flu virus strains emerge somewhere in the world and quickly spread around the globe. For most people, in most years, flu merely means a few days of feeling terrible. Every couple of decades, however, an especially virulent strain appears. The worst flu **pandemic** (a widespread, deadly epidemic) in history occurred in 1918 when somewhere between 50 and 100 million people are thought to have died.

Now, health officials warn us, a highly contagious form of flu is moving through domestic and wild bird populations in Asia and Europe. First identified in 1997 in Hong Kong, the highly pathogenic H5N1 bird flu has spread to more than 50 countries in Asia, Europe, and Africa. More than 150 million domestic ducks and chickens were culled or died from the disease in its first two years. Of the 175 people known to have contracted this flu strain, at least 105 died, making it a highly deadly disease. As we'll discuss later in this chapter, wild geese and ducks are now spreading this dangerous pathogen between continents and are already infecting poultry.

Dr. David Nabarro, who coordinates the United Nations response to avian influenza, warned that if this virus starts a pandemic, it might kill 150 million people. There are now three times as many people in the world than in 1918, he pointed out; people travel farther and faster, and the H5N1 virus has killed nearly 60 percent of the people it has infected so far, compared with only 1 to 2 percent lethality in 1918. This grim estimate was quickly downplayed by the World Health Organization, which fears spreading panic, but many epidemiologists agree with Dr. Nabarro that the question is not *if* but *when* the next pandemic will occur.

A major breakthrough in understanding this disease occurred in 2005 when a team of U.S. scientists successfully sequenced the entire genome of the 1918 flu virus. They did this using tissue samples from people who died in the epidemic. One of these was a woman who died in a remote Alaskan village, and whose body was preserved for 75 years in the permafrost. They discovered that it was an avian flu in the same genetic family as the virus now circulating around the world. Until this genetic analysis, scientists thought a bird flu virus would first have to mix with a human flu virus for it to infect humans. But now we know that the 1918 pandemic was caused by a pure bird flu virus that jumped to humans directly. No transition phase was necessary.

There is some good news in this grim scenario. In the first place, we know much more about influenza and how it spreads than medical

Domestic livestock represents a huge reservoir of pathogens that can spread to humans. Since 2004, China has killed 150 million ducks and chickens in an effort to contain a deadly new strain of bird flu. Will this be the next great pandemic?

officials did in 1918. We now have new antiviral drugs that can be mass produced and stockpiled to fight a pandemic. There is hope that having the genome sequence of H5N1 will help create an effective vaccine. World leaders are being more proactive in addressing the current epidemic. And China is being remarkably open and cooperative in dealing with the current bird flu. This is a marked contrast to the situation in 2003 when Severe Acute Respiratory Syndrome (SARS) emerged in China. In that outbreak, the Chinese government denied the existence and severity of the disease for months while it spread to 31 countries around the world and infected at least 8,500 people. Stung by international criticism of their handling of that epidemic, the Chinese are being much more forthcoming now. Instead of imposing an information blackout, the government has been quick to report bird flu developments.

In 2005, China announced plans to vaccinate 14 billion birds, and to ship 45 tons of vaccine to Vietnam. Some veterinary scientists doubt whether this plan can be effective. They point out that inoculation will prevent birds from dying, but won't prevent them from being infected and acting as a reservoir for the virus. Furthermore, most poultry in China is raised in small flocks in the backyards of rural homes. Reaching all those birds may be impossible, especially when poor farmers are suspicious of the government and would rather eat dead birds than report them to officials. Some experts call for a shutdown of all backyard poultry production and elimination of live-bird markets where animals are killed under very unsanitary conditions. This would be a severe economic and cultural blow to billions of people.

Meanwhile, the close cohabitation of huge numbers of people, poultry, and pigs throughout Asia continues to provide ideal conditions for viruses to mutate into new, virulent forms and continue to spread throughout the world. As some experts warn, the next deadly pandemic could be only a plane ride away. The emergence of these new infectious diseases reminds us of how our environment and health are interconnected in the modern world. In this chapter, we'll look at some of the principles of environmental health and toxicology to help you understand some of the risks we face and what we might do about them.

ENVIRONMENTAL HEALTH

What is health? The World Health Organization (WHO) defines **health** as a state of complete physical, mental, and social well-being, not merely the absence of disease or infirmity. By that definition, we all are ill to some extent. Likewise, we all can improve our health to live happier, longer, more productive, and more satisfying lives if we think about what we do.

What is disease? A **disease** is an abnormal change in the body's condition that impairs important physical or psychological

FIGURE 8.1 Major sources of environmental health risks.

functions. Diet and nutrition, infectious agents, toxic substances, genetics, trauma, and stress all play roles in **morbidity** (illness) and **mortality** (death). **Environmental health** focuses on factors that cause disease, including elements of the natural, social, cultural, and technological worlds in which we live. Figure 8.1 shows some major environmental disease agents, as well as the media through which we encounter them. Ever since the publication of Rachel Carson's *Silent Spring* in 1962, the discharge, movement, fate, and effects of synthetic chemical toxins have been a special focus of environmental health, but, as the opening Case Study shows, infectious diseases remain a grave threat. In this chapter, we'll study these topics in detail. First, however, let's look at some of the major causes of illness worldwide.

Global disease burden is changing

In the past, health organizations have focused on the leading causes of death as the best summary of world health. Mortality data, however, fail to capture the impacts of nonfatal outcomes of disease and injury, such as dementia or blindness, on human well-being. When people are ill, work isn't done, crops aren't planted or harvested, meals aren't cooked, and children can't study and learn. Health agencies now calculate **disability-adjusted life years (DALYs)** as a measure of disease burden. DALYs combine premature deaths and loss of a healthy life resulting from illness or disability. This is an attempt to evaluate the total cost of disease, not simply how many people die. Clearly, many more years of expected life are lost when a child dies of neonatal tetanus than when an 80-year-old dies

of pneumonia. Similarly, a teenager permanently paralyzed by a traffic accident will have many more years of suffering and lost potential than will a senior citizen who has a stroke. According to the WHO, chronic diseases now account for nearly 60 percent of the 56.5 million total deaths worldwide each year and about half of the global disease burden.

The world is now undergoing a dramatic epidemiological transition. Chronic conditions, such as cardiovascular disease and cancer, no longer afflict only wealthy people. Marvelous progress in eliminating communicable diseases, such as smallpox, polio, and malaria, is allowing people nearly everywhere to live longer. As chapter 6 points out, over the past century the average life expectancy worldwide has risen by about two-thirds. In some poorer countries, such as India, life expectancies nearly tripled in the twentieth century. Although the traditional killers in developing countries—infections, maternal and perinatal (birth) complications, and nutritional deficiencies—still take a terrible toll, diseases such as depression and heart attacks that once were thought to occur only in rich countries are rapidly becoming the leading causes of disability and premature death everywhere.

In 2020, the WHO predicts that heart disease, which was fifth in the list of causes of global disease burden a decade ago, will be the leading source of disability and deaths worldwide (table 8.1). Most of that increase will be in the poorer parts of the world where people are rapidly adopting the lifestyles and diet of the richer countries. Similarly, global cancer rates will increase by 50 percent. By 2020 it's expected that 15 million people will have cancer and 9 million will die from it.

TABLE 8.1

Leading Causes of Global Disease Burden

Rank	1990	Rank	2020
1	Pneumonia	1	Heart disease
2	Diarrhea	2	Depression
3	Perinatal conditions	3	Traffic accidents
4	Depression	4	Stroke
5	Heart disease	5	Chronic lung disease
6	Stroke	6	Pneumonia
7	Tuberculosis	7	Tuberculosis
8	Measles	8	War
9	Traffic accidents	9	Diarrhea
10	Birth defects	10	HIV/AIDS
11	Chronic lung disease	11	Perinatal conditions
12	Malaria	12	Violence
13	Falls	13	Birth defects
14	Iron anemia	14	Self-inflicted injuries
15	Malnutrition	15	Respiratory cancer

Source: Data from World Health Organization, 2002.

Principles of Environmental Science

Taking disability as well as death into account in our assessment of disease burden reveals the increasing role of mental health as a worldwide problem. WHO projections suggest that psychiatric and neurological conditions could increase their share of the global burden from 10 percent currently to 15 percent of the total load by 2020. Again, this isn't just a problem of the developed world. Depression is expected to be the second largest cause of all years lived with disability worldwide, as well as the cause of 1.4 percent of all deaths. For women in both developing and developed regions, depression is the leading cause of disease burden, while suicide, which often is the result of untreated depression, is the fourth largest cause of female deaths.

Notice in table 8.1 that diarrhea, which was the second leading cause of disease burden in 1990, is expected to be ninth on the list in 2020, while measles and malaria are expected to drop out of the top 15 causes of disability. Tuberculosis, which is becoming resistant to antibiotics and is spreading rapidly in many areas (especially in Russia), is the only infectious disease whose ranking is not expected to change over the next 20 years. Traffic accidents are now soaring as more people drive. War, violence, and self-inflicted injuries similarly are becoming much more important health risks than ever before.

Chronic obstructive lung diseases (e.g., emphysema, asthma, and lung cancer) are expected to increase from eleventh to fifth in disease burden by 2020. A large part of the increase is due to rising use of tobacco in developing countries, sometimes called "the tobacco epidemic." Every day about 100,000 young people—most of them in poorer countries—become addicted to tobacco. At least 1.1 billion people now smoke, and this number is expected to increase at least 50 percent by 2020. If current patterns persist, about 500 million people alive today will eventually be killed by tobacco. This is expected to be the biggest single cause of death worldwide (because illnesses such as heart attack and depression are triggered by multiple factors). In 2003 the World Health Assembly adopted a historic tobacco-control convention that requires countries to impose restrictions on tobacco advertising, establish clean indoor air controls, and clamp down on tobacco smuggling. Dr. Gro Harlem Brundtland, former director-general of the WHO, predicted that the convention, if ratified by enough nations, could save billions of lives.

As chapter 7 points out, the world is now experiencing an epidemic of obesity. Poor diet and lack of exercise are now the second leading underlying cause of death in America, causing at least 400,000 deaths per year. Obesity is expected to overtake tobacco soon as the largest single health risk in many countries.

Emergent and infectious diseases still kill millions of people

Although the ills of modern life have become the leading killers almost everywhere in the world, communicable diseases still are responsible for about one-third of all disease-related mortality. Diarrhea, acute respiratory illnesses, malaria, measles, tetanus,

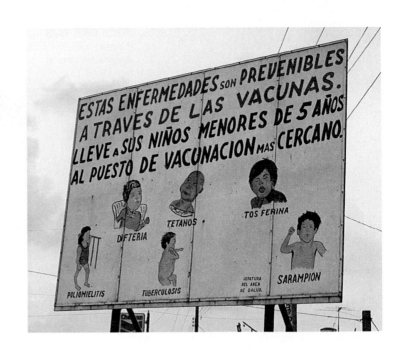

FIGURE 8.2 Millions of children die each year from easily prevented childhood diseases. This Guatemalan billboard urges that children be vaccinated against polio, diphtheria, TB, tetanus, pertussis (whooping cough), and scarlet fever (*left to right*).

and a few other infectious diseases kill about 11 million children under age five every year in the developing world. Better nutrition, clean water, improved sanitation, and inexpensive inoculations could eliminate most of those deaths (fig. 8.2).

A wide variety of **pathogens** (disease-causing organisms) afflict humans, including viruses, bacteria, protozoans (single-celled animals), parasitic worms, and flukes (fig. 8.3). The greatest loss of life from an individual disease in a single year was the great influenza pandemic of 1918. Epidemiologists now estimate that at least one-third of all humans living at the time were infected, and that between 50 to 100 million died. Businesses, schools, churches, and sports or entertainment events were shut down for months. If a similar pandemic were to occur now, it could sicken 2 billion people, kill up to 150 million, and bring the world economy to a standstill. Influenza is caused by a family of viruses (fig. 8.3a) that mutate rapidly and move from wild and domestic animals to humans, making control of this disease very difficult.

Every year there are 76 million cases of foodborne illnesses in the United States, resulting in 300,000 hospitalizations and 5,000 deaths. *Giardia,* a parasitic intestinal protozoan, is thought to be the largest single cause of diarrhea in the United States (fig. 8.3b). It is spread from human feces through food and water. You may think of this as a wilderness disease, but researchers report that the best place to find the pathogens is day care centers and nursery schools. At any given time, around 2 billion people—one-third of the world population—suffer from worms, flukes, and other internal parasites. Guinea worms (*Dracunculus medinesis*) are round worms transmitted via

FIGURE 8.3 (a) A group of influenza viruses magnified about 300,000 times. (b) *Giardia,* a parasitic intestinal protozoan, magnified about 10,000 times. (c) A guinea worm emerges from the foot of an infected person.

(a) Influenza viruses

(b) *Giardia*

(c) Guinea worm

contaminated water. After a year of migrating through the body, the adult, which can be 1 meter long, emerges to lay eggs (fig. 8.3c). A campaign to eradicate this scourge has eliminated it from most countries. (See related story "Fighting the Fiery Serpent" at www.mhhe.com/cunningham4e.)

Malaria is one of the most prevalent remaining infectious diseases. Every year about 300 million new cases of this disease occur, and about 1 million people die from it. The territory infected by this disease is expanding as global climate change allows mosquito vectors to move into new territory. Simply providing insecticide-treated bed nets and a few dollars worth of chloroquine pills could prevent tens of millions of cases of this debilitating disease every year. Tragically, some of the countries where malaria is most widespread tax both bed nets and medicine as luxuries, placing them out of reach for ordinary people.

Emergent diseases are those not previously known or that have been absent for at least 20 years. The new strain of bird flu described in the opening Case Study is a good example. There have been at least 30 outbreaks of emergent diseases over the past two decades, including the extremely deadly Ebola and Marburg fevers, which have afflicted Central Africa in at least six different locations in the past decade. Similarly, cholera, which had been absent from South American for more than a century, reemerged in Peru in 1991. Some other examples include a new drug-resistant form of tuberculosis, now spreading in Russia; dengue fever, which is spreading through Southeast Asia and the Caribbean; malaria, which is reappearing in places where it had been nearly eradicated; and a new human lymphotropic virus (HTLV), which is thought to have jumped from monkeys into people in Cameroon who handled or ate bushmeat. Theses HTLV strains are now thought to infect 25 million people.

Health experts point out that West Nile virus is the most virulent and rapidly moving disease to hit the United States in recent

years. West Nile belongs to a family of mosquito-transmitted viruses that cause encephalitis (brain inflammation). Although recognized in Africa in 1937, the West Nile virus was absent from North America until 1999, when it apparently was introduced by a bird or mosquito from the Middle East. The disease spread from New York, where it was first reported, throughout the eastern United States in only two years (fig. 8.4). The virus infects 230 species of animals, including 130 bird species. In 2002 more than 4,000 people contracted the disease and 284 died in the United States. There's a worry that, if the disease reaches Hawaii, it

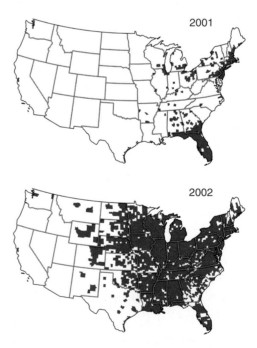

2001

2002

FIGURE 8.4 The spread of West Nile virus in birds, 2001–2002.
Source: Data from U.S. Centers for Disease Control and U.S. Geological Survey.

could completely eradicate rare species, such as the Hawaiian crow and several types of honeycreeper.

The largest recent human death toll from an emergent disease is due to HIV/AIDS. Although it was first recognized in the early 1980s, acquired immune deficiency syndrome has now become the fifth greatest cause of contagious deaths. The WHO estimates that 60 million people are now infected with the human immune-deficiency virus and that 3 million die every year from AIDS complications. Although two-thirds of all current HIV infections are now in sub-Saharan Africa, the disease is spreading rapidly in South and East Asia. Over the next 20 years, there could be an additional 65 million AIDS deaths. In Botswana, health officials estimate about 40 percent of all adults are HIV-positive and that two-thirds of all current 15-year-olds will die of AIDS before age 50. Without AIDS, the life expectancy in Botswana would be expected to be 69.7 years. With AIDS, Botswana's average life expectancy is now 36.1 years. Worldwide, more than 14 million children—the equivalent of every child under age five in America—have lost one or both parents to AIDS. The economic costs of treating patients and lost productivity from premature deaths resulting from this disease are estimated to be at least $35 billion (U.S.) per year, or about one-tenth of the total GDP of sub-Saharan Africa.

Conservation medicine attempts to combine ecology and medicine

Humans aren't the only ones to suffer from new and devastating diseases. Domestic animals and wildlife also experience sudden and widespread epidemics, which are sometimes called **ecological diseases.** Bird flu, for example, isn't just a threat to humans. It also is spreading through wild bird populations, and could push some rare species into extinction. Ornithologists have identified endangered species, such as the aquatic warbler, the dalmatian pelican, the marbled teal, the slender-billed curlew, and the spoon-billed sandpiper, all of which share summer or winter habitat and migration routes with species known to harbor the avian flu virus.

Chronic wasting disease (CWD) is spreading through deer and elk populations in North America (fig. 8.5). Caused by a strange protein called a prion, CWD is one of a family of irreversible, degenerative neurological diseases known as transmissible spongiform encephalopathies (TSE) that include mad cow disease in cattle, scrapie in sheep, and Creutzfelt-Jacob disease in humans. CWD probably started when elk ranchers fed contaminated animal by-products to their herds. Infected animals were sold to other ranches, and now the disease has spread to wild populations. First recognized in 1967 in Saskatchewan, CWD has been identified in wild deer populations and ranch operations in at least eight American states.

The Canadian government is estimated to have spent $65 million in an attempt to stop the spread of CWD. Wisconsin has encouraged hunters to kill some 20,000 deer in an area near

FIGURE 8.5 Wild elk and deer are widely affected by chronic wasting disease, which probably originated with domestic herds.

Madison in an effort to contain the disease. No humans are known to have contracted TSE from deer or elk, but there is a concern that we might see something like the mad cow disaster that inflicted Europe in the 1990s. At least 100 people died, and nearly 5 million European cattle and sheep were slaughtered in an effort to contain that disease. Although a few cattle in the United States have been found to have TSE, so far, there have been no known human deaths from eating domestically raised animals.

One thing that emergent diseases in humans and ecological diseases in natural communities have in common is environmental change that stresses biological systems and upsets normal ecological relationships. We cut down forests and drain wetlands, destroying habitat for native species. Invasive organisms and diseases are accidentally or intentionally introduced into new areas where they can grow explosively. Increasing incursion into former wilderness is spurred by human population growth and ecotourism. In 1950, only about 3 million people a year flew on commercial jets; by 2000, more than 300 million did. Diseases can spread around the globe in mere days as people pass through international travel hubs. In the case of severe acute respiratory syndrome (SARS), which emerged in southern China in 2003, one flight attendant is thought to have spread the virus to 160 people in seven countries before he was diagnosed with the disease (see related story "The Cough Heard Round the World" at www.mhhe.com/cunningham4e).

As the opening Case Study of this chapter shows, wild animals are capable of spreading diseases widely. Wild birds have been shown to carry the H5N1 avian flu virus over long-distance migration routes. The first evidence of this came from Qinghai Lake in western China where waterfowl congregate on their way from Southeast Asia to Mongolia and Siberia (fig. 8.6). At least 100,000 birds—mostly barheaded geese—died at Qinghai Lake in the spring of 2005 from the H5N1 flu. Ornithologists warned that in the fall, birds exposed to the virus would be traveling from

FIGURE 8.6 Flyways for water birds link South and Southeast Asia with Russia and central Africa. Barheaded geese, in particular, have been shown to carry bird flu virus from Southeast Asia to Russia, and from there to Europe, where domestic flocks have been infected. As birds spread this highly lethal disease into central Africa, health experts warn, it could unleash a catastrophic worldwide pandemic. Shaded countries have reported cases of bird flu.
Source: Wetlands International.

Siberia not just back to Southeast Asia, but also through Central Asia to the Black and Caspian Seas and, eventually to Europe and Africa. The predictions proved correct. By 2006, infected birds had been found in 50 countries in Asia, Africa, Europe, and the Middle East. Spread of this virus to Africa is especially worrisome. Most African countries lack the money, infrastructure, and political organization to monitor and treat human flu cases, much less the health status of wild birds. Rural communities around the lakes of the Rift Valley region depend heavily on poultry to survive, and live in close contact with both domestic and migratory birds, conditions that increase the potential for a worldwide pandemic.

Wild birds occasionally cross the Bering Strait from Siberia to Alaska. U.S. and Canadian wildlife researchers are testing migratory ducks and geese for evidence of H5N1 infection. This is considered the most likely route of bird flu into North America. The virus has been shown to survive for a month in bird droppings in cold weather, expanding the window for infection.

Increased consumption of wild species, whether as bushmeat in Africa, civet cats and raccoon dogs in China, or fresh duck blood in Vietnam, allows diseases to leap species barriers. We used to believe that most viruses can't cross species barriers, but

canine distemper virus decimated lions in the Serengeti and almost wiped out black-footed ferrets in the American West. Similarly, Ebola fever was recently shown to be endemic to fruit bats, which are commonly eaten by people in Central Africa. The SARS virus, also, is thought to have passed from horseshoe bats to palm civets, raccoon dogs, and ferret badgers, all of which are considered delicacies in Chinese cuisine. And HIV, the virus that causes AIDS, almost certainly jumped from apes and monkeys to people who killed and ate them. By some estimates, three-quarters of all emerging diseases are "zoonotic," shared by animals and humans.

Climate change also facilitates—or forces—expansion of species into new territories. In 2001, a congressionally mandated assessment of climate change predicted greater incidence of malaria, yellow fever, and other tropical diseases in the United States, as mosquitoes, rodents, and other animals expand their range. This phenomenon is occurring worldwide. In the 1970s, only nine countries suffered from dengue fever. Today, dengue has spread to over 100 countries. Other diseases linked to environmental changes include lyme disease, schistosomiasis, tuberculosis, bubonic plague, and cholera.

We are coming to recognize that the delicate ecological balances that we value so highly—and disrupt so frequently—are

Principles of Environmental Science http://www.mhhe.com/cunningham4e

important to our own health. **Conservation medicine** is an emerging discipline that attempts to understand how our environmental changes threaten our own health as well as that of the natural communities on which we depend for ecological services. While still small, this new field is gaining recognition from mainstream funding sources such as the World Bank, the World Health Organization, and the U.S. National Institutes of Health.

Resistance to antibiotics and pesticides is increasing

Malaria, the most deadly of all insectborne diseases, is an example of the return of a disease that once was thought to be nearly vanquished. Malaria now claims about a million lives every year—90 percent in Africa, and most of them children. With the advent of modern medicines and pesticides, malaria had nearly been wiped out in many places but recently has come roaring back. The protozoan parasite that causes the disease is now resistant to most antibiotics, while the mosquitoes that transmit it have developed resistance to many insecticides. Spraying of DDT in India and Sri Lanka, for instance, reduced malaria from millions of infections per year to only a few thousand in the 1950s and 1960s. Now South Asia is back near its pre-DDT levels with about half a million new cases of malaria every year. Other places that never had malaria cases now have them as a result of climate change and habitat alteration.

Why have vectors, such as mosquitoes, and pathogens, such as the malaria parasite, become resistant to pesticides and antibiotics? Part of the answer is natural selection and the ability of many organisms to evolve rapidly. Another factor is the human tendency to use control measures carelessly. When we discovered that DDT and other insecticides could control mosquito populations, we spread them everywhere. This not only harmed wildlife and beneficial insects but also created perfect conditions for natural selection. Many pests and pathogens were exposed only minimally to control measures, allowing those with natural resistance to survive and spread their genes through the population (fig. 8.7). After repeated cycles of exposure and selection, many microorganisms and their vectors have become insensitive to almost all our weapons against them.

Raising huge numbers of cattle, hogs, and poultry in densely packed barns and feedlots is another reason for widespread antibiotic resistance in pathogens. Confined animals are dosed constantly with antibiotics and steroid hormones to keep them disease-free and to make them gain weight faster. More than half of all antibiotics used in the United States each year are fed to livestock. A significant amount of these antibiotics and hormones are excreted in urine and feces, which are spread, untreated, on the land or discharged into surface water, where they contribute further to the evolution of supervirulent pathogens.

At least half of the 100 million antibiotic doses prescribed for humans every year in the United States are unnecessary or are the wrong ones. Furthermore, many people who start a course of antibiotic treatment fail to carry it out for the time prescribed. For your own health and that of the people around you, if you are taking an antibiotic, follow your doctor's orders. Finish your prescribed doses and don't stop taking the medicine as soon as you start feeling better.

Who should pay for health care?

The heaviest burden of illness is borne by the poorest people, who can afford neither a healthy environment nor adequate health care. Women in sub-Saharan Africa, for example, suffer six times the disease burden per 1,000 population of women in most European countries. The WHO estimates that 90 percent of all disease burden occurs in developing countries, where less than one-tenth of all health care dollars is spent. The group Medecins Sans Frontieres (MSF, or Doctors without Borders) calls this the 10/90 gap. While wealthy nations pursue drugs to treat baldness and obesity, depression in dogs, and erectile dysfunction, billions of people are sick or dying from treatable infections and parasitic diseases to which little attention is paid. Worldwide, only 2 percent of the people with AIDS have access to modern medicines. Every year, some 600,000 infants acquire

(a) Mutation and selection create drug-resistant strains

Mutation

Antibiotic

Antibiotic

Bacterial colony   Resistant bacterium   Resistant colony

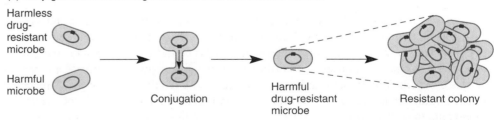

(b) Conjugation transfers drug resistance from one strain to another

Harmless drug-resistant microbe

Harmful microbe

Conjugation

Harmful drug-resistant microbe

Resistant colony

FIGURE 8.7 How microbes acquire antibiotic resistance. (a) Random mutations make a few cells resistant. When challenged by antibiotics, only those cells survive to give rise to a resistant colony. (b) Sexual reproduction (conjugation) or plasmid transfer moves genes from one strain or species to another. **Source:** © 1994 Time Inc. Reprinted by permission.

HIV—almost all of them through mother-to-child transmission during birth or breast-feeding. Antiretroviral therapy costing only a few dollars can prevent most of this transmission. The Bill and Melinda Gates Foundation has pledged $200 million for medical aid to developing countries to help fight AIDS, TB, and malaria.

Dr. Jeffrey Sachs of the Columbia University Earth Institute says that disease is as much a cause as a consequence of poverty and political unrest, yet the world's richest countries now spend just $1 per person per year on global health. He predicts that raising our commitment to about $25 billion annually (about 0.1 percent of the annual GDP of the 20 richest countries) would not only save about 8 million lives each year but also would boost the world economy by billions of dollars. There also would be huge social benefits for the rich countries in not living in a world endangered by mass social instability; the spread of pathogens across borders; the spread of other ills, such as terrorism; and drug trafficking caused by social problems. Sachs also argues that reducing disease burden would help reduce population growth. When parents believe their offspring will survive, they have fewer children and invest more in food, health, and education for smaller families.

The United States is among the least generous of the world's rich countries, donating only about 12 cents per $100 of GDP on international development aid. Could this country do better? During this time of fear of terrorism and rising anti-American feelings around the globe, it's difficult to interest legislators in international aid, yet helping reduce disease might win the United States more friends and make the nation safer than buying more bombs and bullets. Improved health care in poorer countries may also help prevent the spread of emergent diseases, such as SARS, in a globally interconnected world.

Epidemiologists note that almost all of the 2.2 billion people expected to be added to the world population in the next 30 years will live in megacities of the developing world. The economic and environmental conditions in those cities will have a profound impact on global disease burden. Many world leaders urge us to address the "lethal disease of poverty." More discussion of urban areas and their problems is presented in chapter 14.

TOXICOLOGY

Toxicology is the study of **toxins** (poisons) and their effects, particularly on living systems. Because many substances are known to be poisonous to life (whether plant, animal, or microbial), toxicology is a broad field, drawing from biochemistry, histology, pharmacology, pathology, and many other disciplines. Toxins damage or kill living organisms because they react with cellular components to disrupt metabolic functions. Because of this reactivity, toxins often are harmful even in extremely dilute concentrations. In some cases, billionths, or even trillionths, of a gram can cause irreversible damage.

What Can You Do?

Tips for Staying Healthy

- Eat a balanced diet with plenty of fresh fruits, vegetables, legumes, and whole grains. Wash fruits and vegetables carefully; they may have come from a country where pesticide and sanitation laws are lax.
- Use unsaturated oils, such as olive or canola, rather than hydrogenated or semisolid fats, such as margarine.
- Cook meats and other foods at temperatures high enough to kill pathogens; clean utensils and cutting surfaces; store food properly.
- Wash your hands frequently. You transfer more germs from hand to mouth than any other means of transmission.
- When you have a cold or flu, don't demand antibiotics from your doctor—they aren't effective against viruses.
- If you're taking antibiotics, continue for the entire time prescribed—quitting as soon as you feel well is an ideal way to select for antibiotic-resistant germs.
- Practice safe sex.
- Don't smoke; avoid smoky places.
- If you drink, do so in moderation. Never drive when your reflexes or judgment are impaired.
- Exercise regularly: walk, swim, jog, dance, garden. Do something you enjoy that burns calories and maintains flexibility.
- Get enough sleep. Practice meditation, prayer, or some other form of stress reduction. Get a pet.
- Make a list of friends and family who make you feel more alive and happy. Spend time with one of them at least once a week.

All toxins are hazardous, but not all hazardous materials are toxic. Some substances, for example, are dangerous because they're flammable, explosive, acidic, caustic, irritants, or sensitizers. Many of these materials must be handled carefully in large doses or high concentrations, but they can be rendered relatively innocuous by dilution, neutralization, or other physical treatment. They don't react with cellular components in ways that make them poisonous at low concentrations.

Environmental toxicology, or ecotoxicology, specifically deals with the interactions, transformation, fate, and effects of natural and synthetic chemicals in the biosphere, including individual organisms, populations, and whole ecosystems. In aquatic systems the fate of the pollutants is primarily studied in relation to mechanisms and processes at interfaces of the ecosystem components. Special attention is devoted to the sediment/water, water/organisms, and water/air interfaces. In terrestrial environments, the emphasis tends to be on effects of metals on soil fauna community and population characteristics.

Table 8.2 is a list of the top 20 toxins compiled by the U.S. Environmental Protection Agency from the 275 substances

TABLE 8.2

Top 20 Toxic and Hazardous Substances

Material	Major Sources
1. Arsenic	Treated lumber
2. Lead	Paint, gasoline
3. Mercury	Coal combustion
4. Vinyl chloride	Plastics industrial uses
5. Polychlorinated biphenyls (PCBs)	Electric insulation
6. Benzene	Gasoline, industrial use
7. Cadmium	Batteries
8. Benzo(a)pyrene	Waste incineration
9. Polycyclic aromatic hydrocarbons	Combustion
10. Benzo(b)fluoranthene	Fuels
11. Chloroform	Water purification, industry
12. DDT	Pesticide use
13. Aroclor 1254	Plastics
14. Aroclor 1260	Plastics
15. Trichloroethylene	Solvents
16. Dibenz(a,h)anthracene	Incineration
17. Dieldrin	Pesticides
18. Chromium, hexavalent	Paints, coatings, welding, anticorrosion agents
19. Chlordane	Pesticides
20. Hexachlorobutadiene	Pesticides

Source: Data from U.S. Environmental Protection Agency, 2003.

regulated by the Comprehensive Environmental Response, Compensation, and Liability Act (CERCLA), commonly known as the Superfund Act. These materials are listed in order of assessed importance in terms of human and environmental health.

How do toxins affect us?

Allergens are substances that activate the immune system. Some allergens act directly as **antigens;** that is, they are recognized as foreign by white blood cells and stimulate the production of specific antibodies (proteins that recognize and bind to foreign cells or chemicals). Other allergens act indirectly by binding to and changing the chemistry of foreign materials so they become antigenic and cause an immune response.

Formaldehyde is a good example of a widely used chemical that is a powerful sensitizer of the immune system. It is directly allergenic and can trigger reactions to other substances. Widely used in plastics, wood products, insulation, glue, and fabrics, formaldehyde concentrations in indoor air can be thousands of times higher than in normal outdoor air. Some people suffer from what is called **sick building syndrome:** headaches, allergies, and chronic fatigue caused by poorly vented indoor air contaminated by molds, carbon monoxide, nitrogen oxides, formaldehyde, and other toxic chemicals released by carpets, insulation, plastics, building materials, and other sources (fig. 8.8). The Environmental Protection Agency estimates that poor indoor air quality may cost the nation $60 billion a year in absenteeism and reduced productivity.

Neurotoxins are a special class of metabolic poisons that specifically attack nerve cells (neurons). The nervous system is so important in regulating body activities that disruption of its activities is especially fast-acting and devastating. Different types of neurotoxins act in different ways. Heavy metals, such as lead and mercury, kill nerve cells and cause permanent neurological damage. Anesthetics (ether, chloroform, halothane, etc.) and chlorinated hydrocarbons (DDT, Dieldrin, Aldrin) disrupt nerve cell membranes necessary for nerve action. Organophosphates (Malathion, Parathion) and carbamates (carbaryl, zeneb, maneb) inhibit acetylcholinesterase, an enzyme that regulates signal transmission between nerve cells and the tissues or organs they innervate (for example, muscle). Most neurotoxins are both acute and extremely toxic. More than 850 compounds are now recognized as neurotoxins.

Mutagens are agents, such as chemicals and radiation, that damage or alter genetic material (DNA) in cells. This can lead to birth defects if the damage occurs during embryonic or fetal growth. Later in life, genetic damage may trigger neoplastic (tumor) growth. When damage occurs in reproductive cells, the results can be passed on to future generations. Cells have repair mechanisms to detect and restore damaged genetic material, but some changes may be hidden, and the repair process itself can be flawed. It is generally accepted that there is no "safe" threshold for exposure to mutagens. Any exposure has some possibility of causing damage.

Teratogens are chemicals or other factors that specifically cause abnormalities during embryonic growth and development. Some compounds that are not otherwise harmful can cause tragic problems in these sensitive stages of life. Perhaps the most prevalent teratogen in the world is alcohol. Drinking during pregnancy can lead to **fetal alcohol syndrome**—a cluster of symptoms including craniofacial abnormalities, developmental delays, behavioral problems, and mental defects, that last throughout a child's life. Even one alcoholic drink a day during pregnancy has been associated with decreased birth weight.

Carcinogens are substances that cause **cancer**—invasive, out-of-control cell growth that results in malignant tumors. Cancer rates rose in most industrialized countries during the twentieth century, and cancer is now the second leading cause of death in the United States, killing more than half a million people in 2000. Twenty-three of the 28 compounds listed by the U.S. EPA as greatest risk to human health are probable or possible human carcinogens. More than 200 million people live in areas where the

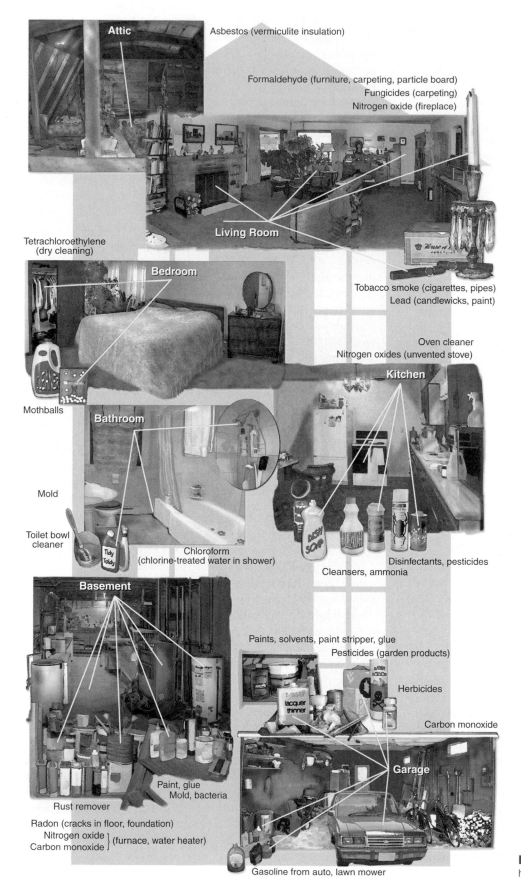

Asbestos (vermiculite insulation)

Attic

Formaldehyde (furniture, carpeting, particle board)
Fungicides (carpeting)
Nitrogen oxide (fireplace)

Living Room

Tetrachloroethylene
(dry cleaning)

Bedroom

Tobacco smoke (cigarettes, pipes)
Lead (candlewicks, paint)

Oven cleaner
Nitrogen oxides (unvented stove)

Kitchen

Mothballs

Bathroom

Mold

Toilet bowl
cleaner

Chloroform
(chlorine-treated water in shower)

Disinfectants, pesticides
Cleansers, ammonia

Basement

Paints, solvents, paint stripper, glue
Pesticides (garden products)

Herbicides

Carbon monoxide

Garage

Paint, glue
Mold, bacteria

Rust remover

Radon (cracks in floor, foundation)
Nitrogen oxide ⎫
              ⎬ (furnace, water heater)
Carbon monoxide ⎭

Gasoline from auto, lawn mower

FIGURE 8.8 Some sources of toxic and hazardous substances in a typical home.

combined upper limit lifetime cancer risk from these carcinogens exceeds 10 in 1 million, or 10 times the risk normally considered acceptable.

The American Cancer Society calculates that one in two males and one in three females in the United States will have some form of cancer in their lifetime. Some authors blame this cancer increase on toxic synthetic chemicals in our environment and diet. Others argue that it is attributable mainly to lifestyle (smoking, sunbathing, drinking alcohol) or simply living longer.

Endocrine hormone disrupters are of special concern

One of the most recently recognized environmental health threats are **endocrine hormone disrupters,** chemicals that interrupt the normal endocrine hormone functions. Hormones are chemicals released into the bloodstream by glands in one part of the body to regulate the development and function of tissues and organs elsewhere in the body (fig. 8.9). You undoubtedly have heard about sex hormones and their powerful effects on how we look and behave, but these are only one example of the many regulatory hormones that rule our lives.

We now know that some of the most insidious effects of persistent chemicals, such as DDT and PCBs, are that they interfere with normal growth, development, and physiology of a variety of animals—presumably including humans—at very low doses. In some cases, picogram concentrations (trillionths of a gram per liter) may be enough to cause developmental abnormalities in sensitive organisms. These chemicals are sometimes called environmental estrogens or androgens, because they often cause sexual dysfunction (reproductive health problems in females or feminization of males, for example). They are just as likely, however, to disrupt thyroxin functions or those of other important regulatory molecules as they are to obstruct sex hormones.

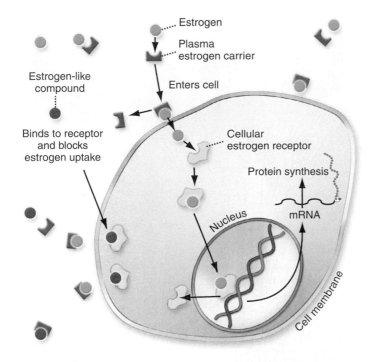

FIGURE 8.9 Steroid hormone action. Plasma hormone carriers deliver regulatory molecules to the cell surface, where they cross the cell membrane. Intracellular carriers deliver hormones to the nucleus, where they bind to and regulate expression of DNA.

Labels in figure:
- Estrogen
- Plasma estrogen carrier
- Enters cell
- Estrogen-like compound
- Binds to receptor and blocks estrogen uptake
- Cellular estrogen receptor
- Protein synthesis
- mRNA
- Nucleus
- Cell membrane

MOVEMENT, DISTRIBUTION, AND FATE OF TOXINS

There are many sources of toxic and hazardous chemicals in the environment and many factors related to each chemical itself, its route or method of exposure, and its persistence in the environment, as well as characteristics of the target organism (table 8.3), that determine the danger of the chemical. We can think of both individuals and an ecosystem as sets of interacting compartments between which chemicals move, based on molecular size, solubility, stability, and reactivity (fig. 8.10). The dose (amount), route of entry, timing of exposure, and sensitivity of the organism all play important roles in determining toxicity. In this section, we will consider some of these characteristics and how they affect environmental health.

Solubility and mobility determine when and where chemicals move

Solubility is one of the most important characteristics in determining how, where, and when a toxic material will move through the environment or through the body to its site of action. Chemicals can be divided into two major groups: those that dissolve more readily in water and those that dissolve more readily in oil. Water-soluble compounds move rapidly and widely through the environment because water is ubiquitous. They also tend to have ready access to most cells in the body because aqueous solutions bathe all our cells.

Molecules that are oil- or fat-soluble (usually organic molecules) generally need a carrier to move through the environment and into or within the body. Once inside the body, however, oil-soluble toxins penetrate readily into tissues and cells because the membranes that enclose cells are themselves made of similar oil-soluble chemicals. Once inside cells, oil-soluble materials are likely to accumulate and to be stored in lipid deposits, where they may be protected from metabolic breakdown and persist for many years.

Exposure and susceptibility determine how we respond

Just as there are many sources of toxins in our environment, there are many routes for entry of dangerous substances into our bodies (fig. 8.11). Airborne toxins generally cause more ill health than any other exposure source. We breathe far more air every day than the volume of food we eat or water we drink. Furthermore, the lining of our lungs, which is designed to exchange gases very efficiently, also absorbs toxins very well. Still, food, water, and skin contact also can expose us to a wide variety of toxins. The largest exposures for many toxins are found in industrial settings, where workers may encounter doses thousands of times higher than would be found anywhere else. The European Agency for Safety and Health at Work warns that 32 million people (20 percent of

FIGURE 8.10 Movement and fate of chemicals in the environment. Processes that modify, remove, or sequester compounds are shown below each compartment. Toxins also move directly from a source to soil and sediment.

FIGURE 8.11 Routes of exposure to toxic and hazardous environmental factors.

all employees) in the European Union are exposed to unacceptable levels of carcinogens and other toxins in their workplace.

Condition of the organism and timing of exposure also have strong influences on toxicity. Healthy adults, for example, may be relatively insensitive to doses that would be very dangerous for young children or for someone already weakened by disease (What Do You Think? p. 191). Similarly, exposure to a toxin may be very dangerous at certain stages of developmental or metabolic cycles but may be innocuous at other times. A single dose of the notorious teratogen thalidomide, for example, taken in the third week of pregnancy (a time when many women aren't aware they're pregnant) can cause severe abnormalities in fetal limb development. A complication in measuring toxicity is that great differences in sensitivity exist among species. Thalidomide was tested on a number of laboratory animals without any deleterious effects. Unfortunately, however, it is a powerful teratogen in humans.

Bioaccumulation and biomagnification increase chemical concentrations

Cells have mechanisms for **bioaccumulation,** the selective absorption and storage of a great variety of molecules. This allows them to accumulate nutrients and essential minerals, but at the same time, they also may absorb and store harmful substances through the same mechanisms. Toxins that are rather dilute in the environment can reach dangerous levels inside cells and tissues through this process of bioaccumulation.

The effects of toxins also are magnified in the environment through food webs. **Biomagnification** occurs when the toxic burden of a large number of organisms at a lower trophic level is accumulated and concentrated by a predator in a higher trophic level. Phytoplankton and bacteria in aquatic ecosystems, for instance, take up heavy metals or toxic organic molecules from water or sediments (fig. 8.12). Their predators—zooplankton and small fish—collect and retain the toxins from many prey organisms, building up higher concentrations of toxins. The top carnivores in the food chain—game fish, fish-eating birds, and humans—can accumulate such high toxin levels that they suffer adverse health effects.

One of the first known examples of bioaccumulation and biomagnification was DDT, which accumulated through food chains, so that by the 1960s it was shown to be interfering with reproduction of peregrine falcons, brown pelicans, and other predatory birds at the top of their food chains.

What Do You Think?

Children's Health

Increasing evidence shows that children are much more vulnerable than adults to environmental toxins. Pound for pound, children drink more water, eat more food, and breathe more air than do adults. Constantly putting fingers, toys, and other objects into their mouths increases children's exposure to toxins in dust or soil. Compared with adults, children generally have less developed protective mechanisms, such as immune systems or metabolic processes to degrade or excrete toxins. Rapid growth and development in childhood makes their metabolism and maturation highly susceptible to disruption. And because children have more years of life remaining than do most adults, they have more time to develop chronic diseases, such as cancer, that may require decades to be expressed.

Childhood cancer incidence has risen 1 percent per year since the early 1970s. Asthma also is up sharply. Some 6.3 million children now suffer from this disease, a four-fold increase over the past 30 years. Asthmatic attacks, which cause blockage of the airways and result in coughing, wheezing, and difficulty breathing, are now the fourth-ranking cause of childhood hospitalization. Inner-city children seem to be particularly at risk for this disease. One study found that 25 percent of all children in New York City's central Harlem suffer from asthma. Scientists believe genetic predisposition plays an important role in this disease but that environmental exposure to factors such as dust, cockroaches and mites, cigarette smoke, mold and mildew, and diesel engine fumes also act as a trigger.

Currently, about 430,000 American children between the ages of 1 and 5 have elevated blood lead levels that can cause irreversible disabilities, such as lower IQ and neurological damage. Autism and attention-deficit hyperactivity disorder (ADHD) are reported to have increased sharply in the past 20 years, although increasing awareness and diagnosis of these conditions may be partly responsible for this increase. A growing percentage of boys with reproductive tract defects also has been reported.

A recently recognized environmental hazard for children is the chromated copper arsenate (CCA)-treated lumber often used to build rot-resistant decks, sidewalks, and playground equipment. All three metals are toxic. Arsenic, the worst of the three, can leach out of weathered wood and has been shown to cause muscle damage as well as lung, bladder, and skin cancer. CCA-treated wood is being phased out in the United States, but not fast enough, according to some groups.

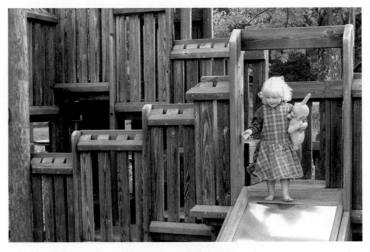
Should all arsenic-treated lumber be removed from places where children play?

In 2003 the U.S. EPA issued its first guidelines to define the greater risks that children face from environmental toxins. According to these proposed rules, regulators should assume that children have ten times the exposure risk of adults to cancer-causing chemicals. Some health groups criticized these guidelines as not being strong enough, pointing out that certain chemicals, such as vinyl chloride (used in PVC plastics), diethylnitrosamine (from tobacco smoke), and polychlorinated biphenyls (PCBs, in electrical transformers), can be up to 65 times more potent in infants and toddlers than in adults.

To study connections between environmental exposures and potential health effects issues, the U.S. government has started a landmark National Children's Study, which will follow 100,000 children from birth to age 18. In establishing this study, former EPA Administrator Christie Whitman said, "Children represent 25 percent of our population, but 100 percent of the future."

What do you think? What would be the best way to safeguard children from environmental hazards? Would you ban CCA-treated lumber completely, or just remove it from playgrounds and other places in which children might be exposed? How would you distinguish between the effects of poverty and environmental hazards for inner-city children? Might it be more effective to focus on providing them with better housing, nutrition, and education than on protecting them from environmental chemicals? What research would you request if you were an environmental regulator?

Persistence makes some materials a greater threat

Some chemical compounds are very unstable and degrade rapidly under most environmental conditions, so that their concentrations decline quickly after release. Most modern herbicides and pesticides, for instance, quickly lose their toxicity. Other substances are more persistent and last for years or even centuries in the environment. PVC plastics, chlorinated hydrocarbon pesticides, and asbestos are valuable because they are resistant to degradation, among other properties. This stability, however, also causes problems because these materials persist in the environment and have unexpected effects far from the sites of their original use.

Some **persistent organic pollutants (POPs)** have become extremely widespread, being found now from the tropics to the Arctic. They often accumulate in food webs and reach toxic concentrations in long-living top predators, such as humans, sharks, raptors, swordfish, and bears. Some of greatest current concerns are

- Polybrominated diphenyl ethers (PBDE) are widely used as flame-retardants in textiles, foam in upholstery, and plastic in appliances and computers. This compound was first reported accumulating in women's breast milk in Sweden in the 1990s. It was subsequently found in humans and other species everywhere from Canada to Israel. Nearly 150 million metric tons (330 million lbs) of PBDEs are used

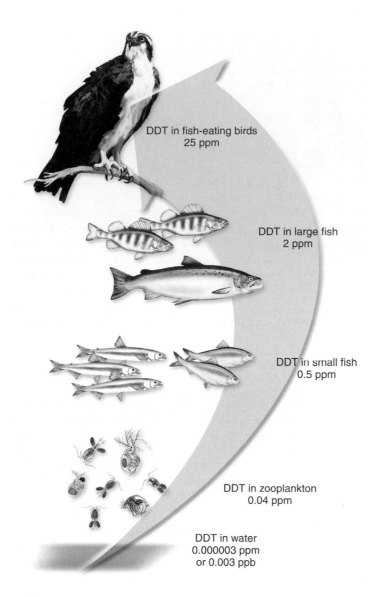

DDT in fish-eating birds
25 ppm

DDT in large fish
2 ppm

DDT in small fish
0.5 ppm

DDT in zooplankton
0.04 ppm

DDT in water
0.000003 ppm
or 0.003 ppb

FIGURE 8.12 Bioaccumulation and biomagnification. Organisms lower on the food chain take up and store toxins from the environment. They are eaten by larger predators, who are eaten, in turn, by even larger predators. The highest members of the food chain can accumulate very high levels of the toxin.

every year worldwide. The toxicity and environmental persistence of PBDE are much like those of PCBs, to which it is closely related chemically. The dust at ground zero in New York City after September 11 was heavily laden with PBDE. The European Union has already banned this compound.

- Perfluorooctane sulfonate (PFOS) and perfluorooctanoic acid (PFOA, also known as C8) are members of a chemical family used to make nonstick, waterproof, and stain-resistant products such as Teflon, Gortex, Scotchguard, and Stainmaster. Industry makes use of their slippery, heat-stable properties to manufacture everything from airplanes and computers to cosmetics and household cleaners. Now these chemicals—which are reported to be infinitely persistent in the environment—are found throughout the world, even the most remote and seemingly pristine sites. Almost all Americans have one

or more perfluorinated compounds in their blood. Heating some nonstick cooking pans above 260°C (500°F) can release enough PFOA to kill pet birds. This chemical family has been shown to cause liver damage as well as various cancers and reproductive and developmental problems in rats. Exposure may be especially dangerous to women and girls, who may be 100 times more sensitive than men to these chemicals.

- Perchlorate is a waterborne contaminant left over from propellants and rocket fuels. About 12,000 sites in the United States were used by the military for live munition testing and are contaminated with perchlorate. Polluted water used to irrigate crops such as alfalfa and lettuce has introduced the chemical into the human food chain. Tests of cow's milk and human breast milk detected perchlorate in nearly every sample from throughout the United States. Perchlorate can interfere with iodine uptake in the thyroid gland, disrupting adult metabolism and childhood development.

- Phthalates (pronounced *thalates*) are found in cosmetics, deodorants, and many plastics (such as soft polyvinyl chloride, or PVC) used for food packaging, children's toys, and medical devices. Some members of this chemical family are known to be toxic to laboratory animals, causing kidney and liver damage and possibly some cancers. In addition, many phthalates act as endocrine hormone disrupters and have been linked to reproductive abnormalities and decreased fertility. A correlation has been found between phthalate levels in urine and low sperm numbers and decreased sperm motility in men. Nearly everyone in the United States has phthalates in his or her body at levels reported to cause these problems. While not yet conclusive, these results could help explain a 50-year decline in semen quality in most industrialized countries.

- Bisphenol A (BPA), a prime ingredient in polycarbonate plastic (commonly used for products ranging from water bottles to tooth-protecting sealants), has been widely found in humans, even those who have no known exposure to these chemicals. So far, there is little direct evidence linking BPA exposure to human health risks, but studies in animals have found that the chemical can cause abnormal chromosome numbers, a condition called aneuploidy, which is the leading cause of miscarriages and several forms of mental retardation. It also is an environmental estrogen and may alter sexual development in both males and females.

- Atrazine is the most widely used herbicide in America. More than 60 million pounds of this compound are applied per year, mainly on corn and cereal grains, but also on golf courses, sugar cane, and Christmas trees. It has long been known to disrupt endocrine hormone functions in mammals, resulting in spontaneous abortions, low birth weights, and neurological disorders. Studies of families in corn-producing areas in the American Midwest have found higher rates of developmental defects among infants, and certain cancers in families with elevated atrazine levels in their drinking water. University of California professor Tyrone Hayes has shown that atrazine levels as low as 0.1 ppb

(30 times less than the EPA maximum contaminant level) caused severe reproductive effects in amphibians, including abnormal gonadal development and hermaphroditism. Atrazine now is found in rain and surface waters nearly everywhere in the United States at levels that could cause abnormal development in frogs. In 2003, the European Union withdrew regulatory approval for this herbicide, and several countries banned its use altogether. Some toxicologists have suggested a similar rule in the United States.

Chemical interactions can increase toxicity

Some materials produce *antagonistic* reactions. That is, they interfere with the effects or stimulate the breakdown of other chemicals. For instance, vitamins E and A can reduce the response to some carcinogens. Other materials are *additive* when they occur together in exposures. Rats exposed to both lead and arsenic show twice the toxicity of only one of these elements. Perhaps the greatest concern is synergistic effects. **Synergism** is an interaction in which one substance exacerbates the effects of another. For example, occupational asbestos exposure increases lung cancer rates 20-fold. Smoking increases lung cancer rates by the same amount. Asbestos workers who also smoke, however, have a 400-fold increase in cancer rates. How many other toxic chemicals are we exposed to that are below threshold limits individually but combine to give toxic results?

MECHANISMS FOR MINIMIZING TOXIC EFFECTS

A fundamental concept in toxicology is that every material can be poisonous under some conditions, but most chemicals have a safe level or threshold below which their effects are undetectable or insignificant. Each of us consumes lethal doses of many chemicals over the course of a lifetime. One hundred cups of strong coffee, for instance, contain a lethal dose of caffeine. Similarly, 100 aspirin tablets, 10 kg (22 lbs) of spinach or rhubarb, or a liter of alcohol would be deadly if consumed all at once. Taken in small doses, however, most toxins can be broken down or excreted before they do much harm. Furthermore, the damage they cause can be repaired. Sometimes, however, mechanisms that protect us from one type of toxin or at one stage in the life cycle become deleterious with another substance or in another stage of development. Let's look at how these processes help protect us from harmful substances, as well as how they can go awry.

Metabolic degradation and excretion eliminate toxins

Most organisms have enzymes that process waste products and environmental poisons to reduce their toxicity. In mammals, most of these enzymes are located in the liver, the primary site of detoxification of both natural wastes and introduced poisons. Sometimes, however, these reactions work to our disadvantage. Compounds such as benzepyrene, for example, that are not toxic in their original form are processed by the same liver enzymes into cancer-causing carcinogens. Why would we have a system that makes a chemical more dangerous? Evolution and natural selection are expressed through reproductive success or failure. Defense mechanisms that protect us from toxins and hazards early in life are "selected for" by evolution. Factors or conditions that affect postreproductive ages (such as cancer or premature senility) usually don't affect reproductive success or exert "selective pressure."

We also reduce the effects of waste products and environmental toxins by eliminating them from the body through excretion. Volatile molecules, such as carbon dioxide, hydrogen cyanide, and ketones, are excreted via breathing. Some excess salts and other substances are excreted in sweat. Primarily, however, excretion is a function of the kidneys, which can eliminate significant amounts of soluble materials through urine formation. Toxin accumulation in the urine can damage this vital system, however, and the kidneys and bladder often are subjected to harmful levels of toxic compounds. In the same way, the stomach, intestine, and colon often suffer damage from materials concentrated in the digestive system and may be afflicted by diseases and tumors.

Repair mechanisms mend damage

In the same way that individual cells have enzymes to repair damage to DNA and protein at the molecular level, tissues and organs that are exposed regularly to physical wear-and-tear or to toxic or hazardous materials often have mechanisms for damage repair. Our skin and the epithelial linings of the gastrointestinal tract, blood vessels, lungs, and urogenital system have high cellular reproduction rates to replace injured cells. With each reproduction cycle, however, there is a chance that some cells will lose normal growth controls and run amok, creating a tumor. Thus, any agent, such as smoking or drinking, that irritates tissues is likely to be carcinogenic. And tissues with high cell-replacement rates are among the most likely to develop cancers.

MEASURING TOXICITY

In 1540 the Swiss scientist Paracelsus said, "The dose makes the poison," by which he meant that almost everything is toxic at some level. This remains the most basic principle of toxicology. Sodium chloride (table salt), for instance, is essential for human life in small doses. If you were forced to eat a kilogram of salt all at once, however, it would make you very sick. A similar amount injected into your bloodstream would be lethal. How a material is delivered—at what rate, through which route of entry, and in what medium—plays a vital role in determining toxicity.

This does not mean that all toxins are identical, however. Some are so poisonous that a single drop on your skin can kill you. Others require massive amounts injected directly into the blood to be lethal. Measuring and comparing the toxicity of various materials are difficult because species differ in sensitivity, and individuals within a species respond differently to a given exposure. In this section, we will look at methods of toxicity testing and at how results are analyzed and reported.

Assessing Toxins

The earliest studies of human toxicology came from experiments in which volunteers (usually students or prisoners) were given measured doses of suspected toxins. Today it is considered neither ethical nor humane to deliberately expose individuals to danger, even if they volunteer. Toxicology is now done in either retrospective or prospective studies. In a **retrospective study,** you identify a group of people who have been exposed to a suspected risk factor and then compare their health with that of a control group who are as nearly identical as possible to the experimental group, except for exposure to that particular factor. Unfortunately, people often can't remember where they were or what they were doing many years ago. In a **prospective study,** you identify a study group and a control group and then keep track of everything they do and how it affects their health. Then you watch and wait for years to see if a response appears in the study group but not in the control group. This kind of study is more accurate, but expensive because you may need a very large group to study a rare effect, and it is still difficult to distinguish among many simultaneous variables.

Suppose that you and your classmates have been chosen to be part of a prospective study of the health risks of a particular soft drink.

1. The researchers can't afford to keep records of everything that you do or are exposed to over the next 20 or 30 years. What do you think would be the most important factors and/or effects to monitor?

2. In a study group of a hundred students, how many would have to get sick to convince you that the soft drink was a risk factor? (Remember that there are many potential variables in the health of your study group.)

We usually test toxins on lab animals

The most commonly used and widely accepted toxicity test is to expose a population of laboratory animals to measured doses of a specific substance under controlled conditions. This procedure is expensive, time-consuming, and often painful and debilitating to the animals being tested. It commonly takes hundreds—or even thousands—of animals, several years of hard work, and hundreds of thousands of dollars to thoroughly test the effects of a toxin at very low doses. More humane toxicity tests using computer simulations of model reactions, cell cultures, and other substitutes for whole living animals are being developed. However, conventional large-scale animal testing is the method in which scientists have the most confidence and on which most public policies about pollution and environmental or occupational health hazards are based.

In addition to humanitarian concerns, several other problems in laboratory animal testing trouble both toxicologists and policymakers. One problem is differences in toxin sensitivity among the members of a specific population. Figure 8.13 shows a typical

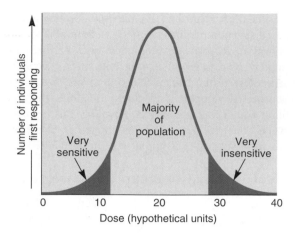

FIGURE 8.13 Probable variations in sensitivity to a toxin within a population. Some members of a population may be very sensitive to a given toxin, while others are much less sensitive. The majority of the population falls somewhere between the two extremes.

dose/response curve for exposure to a hypothetical toxin. Some individuals are very sensitive to the toxin, while others are insensitive. Most, however, fall in a middle category, forming a bell-shaped curve. The question for regulators and politicians is whether we should set pollution levels that will protect everyone, including the most sensitive people, or only aim to protect the average person. It might cost billions of extra dollars to protect a very small number of individuals at the extreme end of the curve. Is that a good use of resources?

Dose/response curves are not always symmetrical, making it difficult to compare toxicity of unlike chemicals or different species of organisms. A convenient way to describe toxicity of a chemical is to determine the dose to which 50 percent of the test population is sensitive. In the case of a lethal dose (LD), this is called the **LD50** (fig. 8.14).

Unrelated species can react very differently to the same toxin, not only because body sizes vary but also because physiology

FIGURE 8.14 Cumulative population response to increasing doses of a toxin. The LD50 is the dose that is lethal to half the population.

and metabolism differ. Even closely related species can have very dissimilar reactions to a particular toxin. Hamsters, for instance, are nearly 5,000 times less sensitive to some dioxins than are guinea pigs. Of 226 chemicals found to be carcinogenic in either rats or mice, 95 cause cancer in one species but not the other. These variations make it difficult to estimate the risks for humans, since we don't consider it ethical to perform controlled experiments in which we deliberately expose people to toxins.

There is a wide range of toxicity

It is useful to group materials according to their relative toxicity. A moderate toxin takes about 1 g per kg of body weight (about 2 oz for an average human) to make a lethal dose. Very toxic materials take about one-tenth that amount, while extremely toxic substances take one-hundredth as much (only a few drops) to kill most people. Supertoxic chemicals are extremely potent; for some, a few micrograms (millionths of a gram—an amount invisible to the naked eye) make a lethal dose. These materials are not all synthetic. One of the most toxic chemicals known, for instance, is ricin, a protein found in castor bean seeds. It is so toxic that 0.3 billionths of a gram given intravenously will kill a mouse. If aspirin were this toxic for humans, a single tablet, divided evenly, could kill 1 million people.

Many carcinogens, mutagens, and teratogens are dangerous at levels far below their direct toxic effect because abnormal cell growth exerts a kind of biological amplification. A single cell, perhaps altered by a single molecular event, can multiply into millions of tumor cells or an entire organism. Just as there are different levels of direct toxicity, however, there are different degrees of carcinogenicity, mutagenicity, and teratogenicity. Methanesulfonic acid, for instance, is highly carcinogenic, while the sweetener saccharin is a possible carcinogen whose effects may be vanishingly small.

Acute versus chronic doses and effects

Most of the toxic effects that we have discussed so far have been **acute effects.** That is, they are caused by a single exposure to the toxin and result in an immediate health crisis. Often, if the individual experiencing an acute reaction survives this immediate crisis, the effects are reversible. **Chronic effects,** on the other hand, are long-lasting, perhaps even permanent. A chronic effect can result from a single dose of a very toxic substance, or it can be the result of a continuous or repeated sublethal exposure.

We also describe long-lasting exposures as chronic, although their effects may or may not persist after the toxin is removed. It usually is difficult to assess the specific health risks of chronic exposures because other factors, such as aging or normal diseases, act simultaneously with the factor under study. It often requires very large populations of experimental animals to obtain statistically significant results for low-level chronic exposures. Toxicologists talk about "megarat" experiments in which it might take a million rats to determine the health risks of some supertoxic chemicals at very low doses. Such an experiment would be terribly

expensive for even a single chemical, let alone for the thousands of chemicals and factors suspected of being dangerous.

An alternative to enormous studies involving millions of animals is to give massive amounts—usually the maximum tolerable dose—of a toxin being studied to a smaller number of individuals and then to extrapolate what the effects of lower doses might have been. This is a controversial approach because it is not clear that responses to toxins are linear or uniform across a wide range of doses.

Figure 8.15 shows three possible results from low doses of a toxin. Curve (a) shows a baseline level of response in the population, even at zero dose of the toxin. This suggests that some other factor in the environment also causes this response. Curve (b) shows a straight-line relationship from the highest doses to zero exposure. Many carcinogens and mutagens show this kind of response. Any exposure to such agents, no matter how small, carries some risks. Curve (c) shows a threshold for the response where some minimal dose is necessary before any effect can be observed. This generally suggests the presence of a defense mechanism that prevents the toxin from reaching its target in an active form or repairs the damage that the toxin causes. Low levels of exposure to the toxin in question may have no deleterious effects, and it might not be necessary to try to keep exposures to zero.

Which, if any, environmental health hazards have thresholds is an important but difficult question. The 1958 Delaney Clause to the U.S. Food and Drug Act forbids the addition of any amount of known carcinogens to food and drugs, based on the assumption that any exposure to these substances represents unacceptable risks. This standard was replaced in 1996 by a "no reasonable harm" requirement, defined as less than one cancer for every million people exposed over a lifetime. This change was supported by a report from the National Academy of Sciences concluding that synthetic chemicals in our diet are unlikely to represent an appreciable cancer risk. We will discuss risk analysis in the section "Risk Assessment and Acceptance."

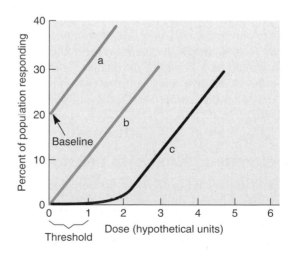

FIGURE 8.15 Three possible dose-response curves at low doses. (a) Some individuals respond, even at zero dose, indicating that some other factor must be involved. (b) Response is linear down to the lowest possible dose. (c) Threshold must be passed before any response is seen.

Detectable levels aren't always dangerous

You may have seen or heard dire warnings about toxic materials detected in samples of air, water, or food. A typical headline announced recently that 23 pesticides were found in 16 food samples. What does that mean? The implication seems to be that any amount of dangerous materials is unacceptable and that counting the numbers of compounds detected is a reliable way to establish danger. We have seen, however, that the dose makes the poison. It matters not only what is there but also how much, where it is located, how accessible it is, and who is exposed. At some level, the mere presence of a substance is insignificant.

Toxins and pollutants may seem to be more widespread now than in the past, and this is surely a valid perception for many substances. The daily reports we hear of new materials found in new places, however, are also due, in part, to our more sensitive measuring techniques. Twenty years ago, parts per million were generally the limits of detection for most chemicals. Anything below that amount was often reported as "zero" or "absent," rather than more accurately as "undetected." A decade ago, new machines and techniques were developed to measure parts per billion. Suddenly, chemicals were found where none had been suspected. Now we can detect parts per trillion or even parts per quadrillion in some cases. Increasingly sophisticated measuring capabilities may lead us to believe that toxic materials have become more prevalent. In fact, our environment may be no more dangerous; we are just better at finding trace amounts.

RISK ASSESSMENT AND ACCEPTANCE

Even if we know with some certainty how toxic a specific chemical is in laboratory tests, it still is difficult to determine **risk** (the probability of harm times the probability of exposure) if that chemical is released into the environment. As we have seen, many factors complicate the movement and fate of chemicals both around us and within our bodies. Furthermore, public perception of relative dangers from environmental hazards can be skewed so that some risks seem much more important than others.

Our perception of risks isn't always rational

A number of factors influence how we perceive relative risks associated with different situations.

- People with social, political, or economic interests—including environmentalists—tend to downplay certain risks and emphasize others that suit their own agendas. We do this individually as well, building up the dangers of things that don't benefit us, while diminishing or ignoring the negative aspects of activities we enjoy or profit from.

- Most people have difficulty understanding and believing probabilities. We feel that there must be patterns and connections in events, even though statistical theory says otherwise. If the coin turned up heads last time, we feel certain that it will turn up tails next time. In the same way, it is difficult to understand the meaning of a 1-in-10,000 risk of being poisoned by a chemical.

- Our personal experiences often are misleading. When we have not personally experienced a bad outcome, we feel it is more rare and unlikely to occur than it actually may be. Furthermore, the anxieties generated by life's gambles make us want to deny uncertainty and to misjudge many risks (fig. 8.16).

FIGURE 8.16 How dangerous are motorcycles? Many parents regard them as extremely risky, while many students—especially males—believe the risks (which are about the same as your chances of dying from surgery or other medical care) are acceptable. Perhaps the more important question is whether the benefits outweigh the risks.

- We have an exaggerated view of our own abilities to control our fate. We generally consider ourselves above-average drivers, safer than most when using appliances or power tools, and less likely than others to suffer medical problems, such as heart attacks. People often feel they can avoid hazards because they are wiser or luckier than others.

- News media give us a biased perspective on the frequency of certain kinds of health hazards, overreporting some accidents or diseases, while downplaying or underreporting others. Sensational, gory, or especially frightful causes of death, such as murders, plane crashes, fires, or terrible accidents, receive a disproportionate amount of attention in the public media. Heart disease, cancer, and stroke kill nearly 15 times as many people in the United States as do accidents and 75 times as many people as do homicides, but the emphasis placed by the media on accidents and homicides is nearly inversely proportional to their relative frequency, compared with either cardiovascular disease or cancer. This gives us an inaccurate picture of the real risks to which we are exposed.

- We tend to have an irrational fear or distrust of certain technologies or activities that leads us to overestimate their dangers. Nuclear power, for instance, is viewed as very risky, while coal-burning power plants seem to be familiar and relatively benign; in fact, coal mining, shipping, and combustion cause an estimated 10,000 deaths each year in the United States, compared with none known so far for nuclear power generation. An old, familiar technology seems safer and more acceptable than does a new, unknown one.

- Alarmist myths and fallacies spread through society, often fueled by xenophobia, politics, or religion. For example, the World Health Organization campaign to eradicate polio worldwide has been thwarted by religious leaders in northern Nigeria—the last country where the disease remains widespread—who claim that oral vaccination is a U.S. plot to spread AIDS or infertility among Muslims.

How much risk is acceptable?

How much is it worth to minimize and avoid exposure to certain risks? Most people will tolerate a higher probability of occurrence of an event if the harm caused by that event is low. Conversely, harm of greater severity is acceptable only at low levels of frequency. A 1-in-10,000 chance of being killed might be of more concern to you than a 1-in-100 chance of being injured. For most people, a 1-in-100,000 chance of dying from some event or some factor is a threshold for changing what they do. That is, if the chance of death is less than 1 in 100,000, we are not likely to be worried enough to change our ways. If the risk is greater, we will probably do something about it. The Environmental Protection Agency generally assumes that a risk of 1 in 1 million is acceptable for most environmental hazards. Critics of this policy ask, acceptable to whom?

For activities that we enjoy or find profitable, we are often willing to accept far greater risks than this general threshold. Conversely, for risks that benefit someone else, we demand far higher protection. For instance, your chance of dying in a motor vehicle accident in any given year are about 1 in 5,000, but that doesn't deter many people from riding in automobiles. Your chances of dying from lung cancer if you smoke one pack of cigarettes per day are about 1 in 1,000. By comparison, the risk from drinking water with the EPA limit of trichloroethylene is about 2 in 1 billion. Strangely, many people demand water with zero levels of trichloroethylene while continuing to smoke cigarettes.

More than 1 million Americans are diagnosed with skin cancer each year. Some of these cancers are lethal, and most are disfiguring, yet only one-third of teenagers routinely use sunscreen. Tanning beds more than double your chances of cancer, especially if you're young, but about 10 percent of all teenagers admit regularly using these devices.

Table 8.4 lists lifetime odds of dying from some leading causes. These are statistical averages, of course, and there clearly are differences in where one lives and how one behaves that affect the danger level of these activities. Although the average lifetime chance of dying in an automobile accident is 1 in 100, there are clearly things you can do—such as wearing a seat belt, driving defensively, and avoiding risky situations—to improve your odds. Still, it is interesting how we readily accept some risks while shunning others.

TABLE 8.4

Lifetime Chances of Dying in the United States

Source	Odds (1 in x)
Heart disease	2
Cancer	3
Smoking	4
Lung disease	15
Pneumonia	30
Automobile accident	100
Suicide	100
Falls	200
Firearms	200
Fires	1,000
Airplane accident	5,000
Jumping from high places	6,000
Drowning	10,000
Lightning	56,000
Hornets, wasps, bees	76,000
Dog bite	230,000
Poisonous snakes, spiders	700,000
Botulism	1 million
Falling space debris	5 million
Drinking water with EPA limit of trichloroethylene	10 million

Source: Data from U.S. National Safety Council, 2003.

Our perception of relative risks is strongly affected by whether risks are known or unknown, whether we feel in control of the outcome, and how dreadful the results are. Risks that are unknown or unpredictable and results that are particularly gruesome or disgusting seem far worse than those that are familiar and socially acceptable.

Studies of public risk perception show that most people react more to emotion than to statistics. We go to great lengths to avoid some dangers while gladly accepting others. Factors that are involuntary, unfamiliar, undetectable to those exposed or catastrophic; those that have delayed effects; and those that are a threat to future generations are especially feared. Factors that are voluntary, familiar, detectable, or immediate cause less anxiety. Even though the actual number of deaths from automobile accidents, smoking, or alcohol, for instance, is thousands of times greater than those from pesticides, nuclear energy, or genetic engineering, the latter group preoccupies us far more than the former.

ESTABLISHING PUBLIC POLICY

Risk management combines principles of environmental health and toxicology with regulatory decisions based on socioeconomic, technical, and political considerations (fig. 8.17). The biggest problem in making regulatory decisions is that we are usually exposed to many sources of harm, often unknowingly. It is difficult to separate the effects of all these different hazards and to evaluate their risks accurately, especially when the exposures are near the threshold of measurement and response. In spite of often vague and contradictory data, public policymakers must make decisions.

The case of the sweetener saccharin is a good example of the complexities and uncertainties of risk assessment in public health. Studies in the 1970s at the University of Wisconsin and the Canadian Health Protection Branch suggested a link between saccharin and bladder cancer in male rats. Critics of these studies pointed out that humans would have to drink 800 cans of diet soda *per day* to get a saccharin dose equivalent to that given to the rats. Furthermore, they argued this response may be unique to male rats. In 2000 the U.S. Department of Health concluded a study that found no association between saccharin and cancer in humans. Congress ordered that all warnings be removed from saccharin-containing products. Still, some groups, such as the Center for Science in the Public Interest, consider this sweetener dangerous and urge us to avoid it if possible.

In setting standards for environmental toxins, we need to consider (1) combined effects of exposure to many different sources of damage, (2) different sensitivities of members of the population, and (3) effects of chronic as well as acute exposures. Some people argue that pollution levels should be set at the highest amount that does *not* cause measurable effects. Others demand that pollution be reduced to zero if possible, or as low as is technologically feasible. It may not be reasonable to demand that we be protected from every potentially harmful contaminant in our environment, no matter how small the risk. As we have seen, our bodies have mechanisms that enable us to avoid or repair many kinds of damage, so that most of us can withstand a minimal level of exposure without harm (fig. 8.18).

On the other hand, each challenge to our cells by toxic substances represents stress on our bodies. Although each individual stress may not be life-threatening, the cumulative effects of all the environmental stresses, both natural and human-caused, to which we are exposed may seriously shorten or restrict our lives. Furthermore, some individuals in any population are more susceptible to those stresses than others. Should we set pollution standards so that no one is adversely affected, even the most sensitive individuals, or should the acceptable level of risk be based on the average member of the population?

FIGURE 8.17 Risk assessment organizes and analyzes data to determine relative risk. Risk management sets priorities and evaluates relevant factors to make regulatory decisions.
Source: Data from D. E. Patton, "USEPA's Framework for Ecological Risk Assessment" in *Human Ecological Risk Assessment*, Vol. 1, No. 4.

FIGURE 8.18 "Do you want to stop reading those ingredients while we're trying to eat?"
Source: Reprinted with permission of the *Star-Tribune*, Minneapolis-St. Paul.

Finally, policy decisions about hazardous and toxic materials also need to be based on information about how such materials affect the plants, animals, and other organisms that define and maintain our environment. In some cases, pollution can harm or destroy whole ecosystems with devastating effects on the life-supporting cycles on which we depend. In other cases, only the most sensitive species are threatened. Table 8.5 shows the Environmental Protection Agency's assessment of relative risks to human welfare. This ranking reflects a concern that our exclusive focus on reducing pollution to protect human health has neglected risks to natural ecological systems. While there have been many benefits from a case-by-case approach in which we evaluate the health risks of individual chemicals, we have often missed broader ecological problems that may be of greater ultimate importance.

REVIEW QUESTIONS

1. What is bird flu, and how is it spreading around the world?
2. How many people might die if bird flu jumps from poultry to people?
3. Define the terms *health* and *disease*.
4. What are *disability-adjusted life years (DALYs)*, and why use them rather than death rates to evaluate health?
5. Name the five leading causes of global disease burden expected by 2020.
6. Define *emergent diseases* and give some recent examples.
7. What is *conservation medicine*?
8. How has conservation medicine been useful in combating bird flu?
9. How do bacteria acquire antibiotic resistance?
10. What percentage of medical expenditures go to the major diseases that affect most of the world population? How much does the United States donate for health care in the developing world?
11. What is the difference between toxic and hazardous? Give some examples of materials in each category.
12. What are *endocrine disrupters*, and why are they of concern?
13. How do the physical and chemical characteristics of materials affect their toxicity?
14. What are *bioaccumulation* and *biomagnification*?
15. Why is atrazine a concern?
16. What is an *LD50*?
17. Distinguish between acute and chronic toxicity.
18. Why aren't our perceptions of risk always rational?
19. What level of risk does the EPA regard as acceptable?
20. What does the EPA regard as the four greatest risks to human welfare?

CRITICAL THINKING AND DISCUSSION

Apply the principles you have learned in this chapter to discuss these questions with other students.

1. Is it ever possible to be completely healthy?
2. How much would be appropriate for wealthy countries to contribute to global health? Why should we do more than we do now? What's in it for us?
3. Why do we spend more money on heart diseases or cancer than childhood diseases?
4. Why do we tend to assume that natural chemicals are safe while industrial chemicals are evil? Is this correct?
5. In the list of reasons why people have a flawed perception of certain risks, do you see things that you or someone you know sometimes do?
6. Do you agree that 1 in 1 million risk of death is an acceptable risk? Notice that almost everything in table 8.4 carries a greater risk than this. Does this make you want to change your habits?

SUGGESTED READINGS

Aldhous, Peter. 2005. Infectious disease: Vietnam's war on flu. *Nature* 433:102–4.

Chen, H., et al. 2005. Avian flu: H5N1 virus outbreak in migratory waterfowl. *Nature* 436:191.

Gold, L. S., T. H. Slone, and B. N. Ames. 2001. Natural and synthetic chemicals in the diet: A critical analysis of possible cancer hazards. In *Issues in Environmental Science and Technology: Food Safety and Food Quality*. R. E. Hester and R. M. Harrison, eds. Royal Society of Chemistry.

Goldstein, Inge F., and Martin Goldstein. 2002. *How Much Risk? A Guide to Understanding Environmental Health Hazards*. Oxford University Press.

Hayes, T. B, et al. 2002. Hermaphroditic, demasculinized frogs after exposure to the herbicide, atrazine, at low ecologically relevant doses. *Proceedings of the National Academy of Sciences (U.S.)* 99:5476–80.

Hufnagel, L., et al. 2004. Forecast and control of epidemics in a globalized world. *Proceedings of the National Academy of Sciences (U.S.)* 101(42):15124–29.

Taubenberger, J. K., et al. 2005. Characterization of the 1918 influenza virus polymerase genes. *Nature* 437: 889–93.

World Health Organization. 2006. *World Health Report: Working for Health*. Oxford University Press.

DATA analysis Graphing Multiple Variables

Is it possible to show relationships between two dependent variables on the same graph? Sometimes that's desirable when you want to make comparisons between them. The graph on this page does just that. It's a description of how people perceive different risks. We judge the severity of risks based on how familiar they are and how much control we have over our exposure. The Y-axis represents how mysterious, unknown or delayed the risk seems to be. Things that are unobservable, unknown to those exposed, delayed in their effects, and unfamiliar or unknown to science tend to be more greatly feared than those that are observable, known, immediate, familiar, and known to science. The X-axis represents a measure of dread, which combines how much control we feel we have over the risk, how terrible the results could potentially be, and how equitably the risks are distributed. The size of the symbol for each risk indicates the combined effect of these two variables. Notice that things such as DNA technology or nuclear waste, which have high levels of both mystery and dread, tend to be regarded with the greatest fear, while familiar, voluntary, personally rewarding behaviors such as riding in automobiles or on bicycles, or drinking alcohol are thought to be relatively minor risks. Actuarial experts (statisticians who gather mortality data) would tell you that automobiles, bicycles, and alcohol have killed far more people (so far) than DNA technology or radioactive waste. But this isn't just a question of data. It's a reflection of how much we fear various risks. Notice that this is a kind of scatter plot mapping categories of data that have no temporal sequence. Still, you can draw some useful inferences from this sort of graphic presentation.

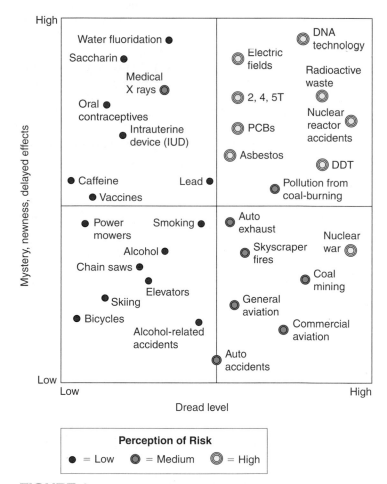

FIGURE 1 Public perception of risk depending on the familiarity, apparent potential for harm, and personal control over the risk.
Source: Data from Slovic, Paul. 1987. Perception of Risk, *Science,* 236 (4799): 286–290.

For Additional Help in Studying This Chapter, please visit our website at www.mhhe.com/cunningham4e. You will find practice quizzes, a chapter summary, key terms, answers to review questions, additional case studies, regional examples, and an extensive reading list, all of which will help you understand the material in this chapter.

Planting trees is a good way to combat global warming while also beautifying your city or neighborhood.

C H A P T E R

Air

9

Climate and Pollution

Climate is an angry beast, and we are poking it with sticks.

–Wallace Broecker–

OBJECTIVES

Studying this chapter will help you answer the following questions:

- What or where is the stratosphere, and why do we care?
- Explain the greenhouse effect and how it is changing our climate.
- How do jet streams, monsoon winds, and convection currents shape our weather and climate?
- What is the ENSO cycle, and how does it affect weather patterns?
- Is it too late to do anything about global climate change?

- Why has the United States refused to ratify the Kyoto Protocol?
- Why has stratospheric ozone been disappearing? Should we worry about it?
- What are the main sources and effects of air pollution?
- Has world air quality been getting better or worse?
- What are the new source review and the clear air interstate rule?

Case Study Stabilizing Our Climate

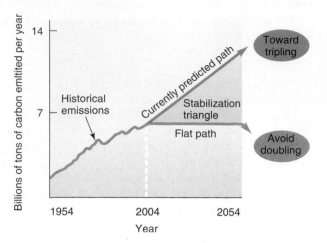

What would it take to stabilize our carbon emissions? Some authors argue that we already have the science and technology to do so. All we need is the political will.

Data from Socolow, et al. Environment vol. 46, no. 10 p. 11 2004.

An ever-increasing body of evidence suggests that human actions are changing our global climate. Worldwide, average annual temperatures have increased by about 0.6°C (1°F) over the past century, but polar regions are warming much faster—as much as 4°C (7°F)—than the rest of the globe. Alpine glaciers are melting rapidly. Arctic sea ice is about half as thick and one-third less extensive than it was a century ago. Huge sections are breaking off of antarctic ice shelves. Sea levels are rising. Fall is significantly later and spring earlier than ever in recorded history. Many plant and animal species are shifting their ranges into new territory or changing their behavior to accommodate to changing climates. Nineteen of the 20 warmest years in the past 150 have occurred since 1980, with 2005 being the hottest year since thermometers have been available. Droughts are becoming more frequent and widespread, and storms are increasing in severity. The 2005 Atlantic storm season was the most severe on record, with 26 named tropical storms, 13 of which were classified as hurricanes. The ocean conveyor, which carries huge amounts of warm water into the north Atlantic and keeps Europe far warmer than it would ordinarily be, is getting weaker.

The forcing factor behind these changes is our release of gases—in particular carbon dioxide (CO_2)—that trap heat and warm the atmosphere. At the beginning of the industrial age, 250 years ago, the atmosphere had about 280 ppm of CO_2. The level has risen to about 380 ppm now. If current trends continue, we could triple the preindustrial CO_2 levels by the end of this century. This would be higher than any time in millions of years. Predictions about the timing and consequences of such a drastic change in our atmosphere vary, but most of them are grim.

Is the situation hopeless? Can we do anything to reverse the path we're on? No single step can reverse our release of greenhouse gases, but many simple policy changes could help stabilize our climate. One of the most encouraging analyses in this area is that of Robert Socolow and his colleagues at Princeton University. They say we already have the scientific, technical, and industrial know-how to stabilize our CO_2 emissions over the next half century, while still meeting the world's energy needs. In several recent publications, this group examines a portfolio of 14 policy options for reducing our CO_2 output. No single step can do the entire job by itself, they conclude, but the potential benefits of the entire portfolio are large enough that not every option must be used.

Professor Socolow and his colleagues estimate that to avoid a doubling of atmospheric CO_2 we need to reduce our annual carbon emissions by about 7 billion tons by 2054. To make this easier to understand, the Princeton group divides the "stabilization triangle" between our current path and the flat emissions that we need in order to avoid doubling CO_2 levels into seven "wedges." Each wedge represents 1 GT (1 billion tons) of carbon emissions avoided in 2054, or a cumulative total of 25 billion tons over the next 50 years, compared to a "business as usual" scenario.

Because most of our CO_2 emissions come from fossil fuel combustion, energy conservation and a switch to renewable fuels probably are the first places we should look to reduce our climate problem. For example, 1.5 wedges (1.5 GT reduction in carbon emissions) could be accomplished by doubling our average fuel economy from the expected 30 miles per gallon in 2054 to 60 mpg, and by halving the number of miles driven each year by switching to walking, biking, or using mass transit. Simply installing the most efficient lighting and appliances available, along with improved insulation in buildings, could save another 2 GT of carbon emissions. Similarly, improving power plant efficiency and reducing the energy consumption of industrial processes could reduce carbon output one wedge (1 GT) by 2054. Capturing and storing CO_2 released by power plants, methane wells, and other large sources could save another billion tons of carbon. Much of this CO_2 could be injected into oil wells to improve crude oil production. Building 2 million wind turbines (of the current one-megawatt size) or doubling our current global nuclear generating capacity would each reduce carbon emissions by about 1 GT in 50 years. This would, of course, require solving the energy storage problems with wind power as well as public worries about the safety of nuclear reactors and storing radioactive waste. Increasing production of biofuels, such as ethanol and vegetable oil, 50-fold, could replace 1 GT of carbon from fossil fuels. Halting tropical forest destruction and replanting 250 million ha of forest (or 400 million ha in temperate regions) would avoid almost 1 GT of carbon emissions, while practices such as conservation tillage and planting of cover crops could reduce carbon releases by another GT.

As you can see, altogether these policy changes add up to considerably more than the 7 GT per year reduction we need to make by 2054 to avoid doubling atmospheric CO_2 concentrations. You'll find much more about fossil fuels and renewable energy in chapter 12. In this chapter, we'll look in more detail about the evidence that climate change is happening as well as more about our options for controlling it. First, however, we'll look at the factors that shape our normal weather and climate.

THE ATMOSPHERE AND CLIMATE

We live at the bottom of a virtual ocean of air that extends upward about 500 km (300 mi). In the layer closest to the earth's surface, known as the troposphere, the air moves ceaselessly, flowing, swirling, and continually redistributing heat and moisture from one part of the globe to another. The composition and behavior of the troposphere and other layers control our **weather** (daily temperature and moisture conditions in a place) and our **climate** (long-term weather patterns).

The earth's earliest atmosphere probably consisted mainly of hydrogen and helium. Over billions of years, most of that hydrogen and helium diffused into space. Volcanic emissions added carbon, nitrogen, oxygen, sulfur, and other elements to the atmosphere. Virtually all of the molecular oxygen (O_2) we breathe was probably produced by photosynthesis in blue-green bacteria, algae, and green plants.

Clean, dry air is 78 percent nitrogen and almost 21 percent oxygen, with the remaining 1 percent composed of argon, carbon dioxide (CO_2), and a variety of trace gases. Water vapor concentrations vary from near 0 to 4 percent, depending on air temperature and available moisture. Minute particles and liquid droplets—collectively called **aerosols**—also are suspended in the air. Atmospheric aerosols play important roles in the earth's energy budget and in rain production.

The atmosphere has four distinct zones of contrasting temperature, due to differences in absorption of solar energy (fig. 9.1). The layer of air immediately adjacent to the earth's surface is called the **troposphere** (*tropein* means to turn or change, in Greek). Within the troposphere, air circulates in great vertical and horizontal **convection currents,** constantly redistributing heat and moisture around the globe. The troposphere ranges in depth from about 18 km (11 mi) over the equator to about 8 km (5 mi) over the poles, where air is cold and dense. Because gravity holds most air molecules close to the earth's surface, the troposphere is much denser than the other layers: it contains about 75 percent of the total mass of the atmosphere. Air temperature drops rapidly with increasing altitude in this layer, reaching about –60°C (–76°F) at the top of the troposphere. A sudden reversal of this temperature gradient creates a sharp boundary called the tropopause, which limits mixing between the troposphere and upper zones.

The **stratosphere** extends from the tropopause up to about 50 km (31 mi). It is vastly more dilute than the troposphere, but it has a similar composition—except that it has almost no water vapor and nearly 1,000 times more **ozone** (O_3). This ozone absorbs some wavelengths of ultraviolet solar radiation, known as UV-B (290–330 nm, see fig. 2.11). This absorbed energy makes the atmosphere warmer toward the top of the stratosphere. Since UV radiation damages living tissues, this UV absorption in the stratosphere also protects life on the surface. Recently discovered depletion of stratospheric ozone, especially over Antarctica, is allowing increased amounts of UV radiation to reach the earth's surface. If observed trends continue, this radiation could cause higher rates of skin cancer, genetic mutations, crop failures,

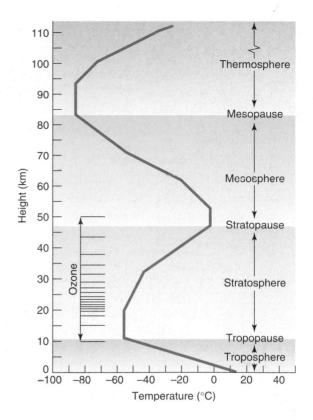

FIGURE 9.1 Temperatures change drastically in the four layers of the atmosphere. Bars in the ozone graph represent relative concentrations of stratospheric ozone with altitude.
Source: Courtesy of Dr. William Culver, St. Petersburg Junior College.

and disruption of important biological communities, as you will see later in this chapter.

Unlike the troposphere, the stratosphere is relatively calm. There is so little mixing in the stratosphere that volcanic ash and human-caused contaminants can remain in suspension there for many years.

Above the stratosphere, the temperature diminishes again, creating the mesosphere, or middle layer. The thermosphere (heated layer) begins at about 50 km. This is a region of highly ionized (electrically charged) gases, heated by a steady flow of high-energy solar and cosmic radiation. In the lower part of the thermosphere, intense pulses of high-energy radiation cause electrically charged particles (ions) to glow. This phenomenon is what we know as the *aurora borealis* and *aurora australis,* or northern and southern lights.

No sharp boundary marks the end of the atmosphere. Pressure and density decrease with distance from the earth until they become indistinguishable from the near vacuum of interstellar space.

The sun warms our world

The sun supplies the earth with an enormous amount of energy, but that energy is not evenly distributed over the globe. Incoming solar radiation (insolation) is much stronger near the equator than

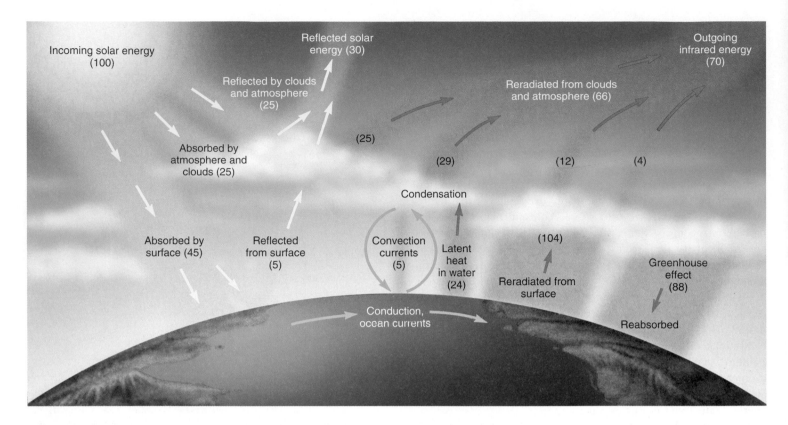

FIGURE 9.2 Energy balance between incoming and outgoing radiation. The atmosphere absorbs or reflects about half of the solar energy reaching the earth. Most of the energy reemitted from the earth's surface is long-wave, infrared energy. Most of this infrared energy is absorbed by aerosols and gases in the atmosphere and is re-radiated toward the planet, keeping the surface much warmer than it would otherwise be. This is known as the greenhouse effect.

at high latitudes. Of the solar energy that reaches the outer atmosphere, about one-quarter is reflected by clouds and atmospheric gases, and another quarter is absorbed by carbon dioxide, water vapor, ozone, methane, and a few other gases (fig. 9.2). This energy absorption warms the atmosphere slightly. About half of incoming solar radiation (insolation) reaches the earth's surface. Most of this energy is in the form of light or infrared (heat) energy (see fig. 2.11). Some of this energy is reflected by bright surfaces, such as snow, ice, and sand. The rest is absorbed by the earth's surface and by water. Surfaces that *reflect* energy have a high **albedo** (reflectivity). Fresh snow and dense clouds, for instance, can reflect as much as 85 to 90 percent of the light falling on them. Surfaces that *absorb* energy have a low albedo and generally appear dark. Black soil, asphalt pavement, and water, for example, have low albedos (reflectivity as low as 3 to 5 percent).

Absorbed energy heats materials (such as an asphalt parking lot in summer), evaporates water, and provides the energy for photosynthesis in plants. Following the second law of thermodynamics, absorbed energy is gradually reemitted as lower-quality heat energy. A brick building, for example, absorbs energy in the form of light and reemits that energy in the form of heat.

The change in energy quality is very important because the atmosphere selectively absorbs longer wavelengths. Most solar energy comes in the form of intense, high-energy light or near-infrared wavelengths. This short-wavelength energy passes relatively easily through the atmosphere to reach the earth's surface. Energy re-released from the earth's warmed surface is lower-intensity, longer-wavelength energy in the far-infrared part of the spectrum. Atmospheric gases, especially carbon dioxide and water vapor, absorb much of this long-wavelength energy, re-releasing it in the lower atmosphere and letting it leak out to space only slowly. This re-irradiated energy provides most of the heat in the lower atmosphere. If the atmosphere were as transparent to infrared radiation as it is to visible light, the earth's average surface temperature would be about 20°C (36°F) colder than it is now.

This phenomenon is called the **greenhouse effect** because the atmosphere, loosely comparable to the glass of a greenhouse, transmits sunlight while trapping heat inside. The greenhouse effect is a natural atmospheric process that is necessary for life as we know it. However, too much greenhouse effect, caused by burning of fossil fuels and deforestation, may cause harmful environmental change (for a more detailed discussion of this phenomenon, see our web page www.mhhe.com/cunningham4e).

Water stores heat and winds redistribute it

Much of the incoming solar energy is used to evaporate water. Every gram of evaporating water absorbs 580 calories of energy as it transforms from liquid to gas. Globally, water vapor contains a huge amount of stored energy, known as **latent heat.** When water vapor condenses, returning from a gas to a liquid form, the 580 calories of heat energy are released. Imagine the sun shining on the Gulf of Mexico in the winter. Warm sunshine and plenty of water allow continuous evaporation that converts an immense amount of solar (light) energy into latent heat stored in evaporated water. Now imagine a wind blowing the humid air north from the Gulf toward Canada. The air cools as it moves north (especially if it encounters cold air moving south). Cooling causes the water vapor to condense. Rain (or snow) falls as a consequence. Note that it is not only water that has moved from the Gulf to the Midwest: 580 calories of heat have also moved with every gram of moisture. The heat and water have moved from a place with strong incoming solar energy to a place with much less solar energy and much less water. The redistribution of heat and water around the globe is essential to life on earth.

Uneven heating, with warm air close to the equator and colder air at high latitudes, also produces pressure differences that cause wind, rain, storms, and everything else we know as weather. As the sun warms the earth's surface, the air nearest the surface warms and expands, becoming less dense than the air above it. The warm air must then rise above the denser air. Vertical convection currents result, which circulate air from warm latitudes to cool latitudes and vice versa. These convection currents can be as small and as localized as a narrow column of hot air rising over a sun-heated rock, or they can cover huge regions of the earth. At the largest scale, the convection cells are described by a simplified model known as Hadley cells, which redistribute heat globally (fig. 9.3).

Where air rises in convection currents, air pressure at the surface is low. Where air is sinking, or subsiding, air pressure is high. On a weather map these high and low pressure centers, or rising and sinking currents of air, move across continents. In most of North America, they generally move from west to

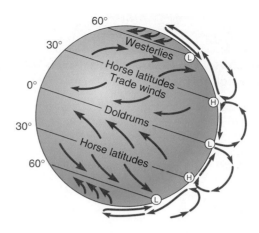

FIGURE 9.3 General circulation patterns redistribute heat and moisture around the globe. The approximate locations of vertical convection currents, generally referred to as Hadley cells, are noted on the right side. Low-pressure belts (L) occur at latitudes where air rises. High-pressure belts (H) occur where air sinks. Dominant winds, such as trade winds and westerlies, also occur in latitudinal bands.

east. Rising air tends to cool with altitude, releasing latent heat, which causes further rising. Very warm and humid air can rise very vigorously, especially if it is rising over a mass of very cold air. Storms associated with low pressure and rising air are known as cyclonic storms. These include some of the most violent storms we know: hurricanes, tornadoes, and intense rain and hail (fig. 9.4).

Pressure differences are an important cause of wind. There is always someplace with sinking (high-pressure) air and someplace with rising (low-pressure) air. Air moves from high-pressure centers toward low-pressure areas, and we call this movement wind.

(a) Hurricane Floyd 1999

(b) A tornado touches down

(c) Gulf Shores, Alabama 2005

FIGURE 9.4 (a) Hurricane Floyd was hundreds of kilometers wide as it approached Florida in 1999. Note the hole, or eye, in the center of the storm. (b) Tornadoes are much smaller than hurricanes, but can have stronger local winds. (c) Destruction caused by Hurricane Katrina in 2005. More than 230,000 km^2 (90,000 mi^2) of coastal areas were devastated by this massive storm and many cities were almost completely demolished.

Why does it rain?

To understand why it rains, remember two things: water condenses as air cools, and air cools as it rises. Anytime air is rising, clouds, rain, or snow might form. Cooling occurs because of changes in pressure with altitude: air cools as it rises (as pressure decreases); air warms as it sinks (as pressure increases). Air rises in convection currents where solar heating is intense, such as over the equator. Moving masses of air also rise over each other and cool. Air also rises when it encounters mountains. If the air is moist (if it has recently come from over an ocean or an evaporating forest region, for example), condensation and rainfall are likely as the air is lifted (fig. 9.5). Regions with intense solar heating, frequently colliding air masses, or mountains tend to receive a great deal of precipitation.

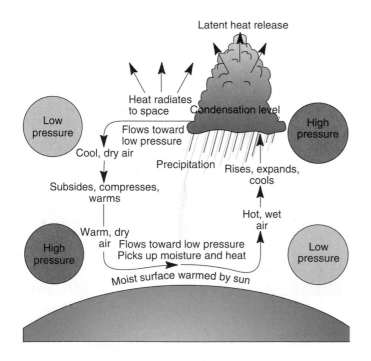

FIGURE 9.5 Convection currents and latent energy cause atmospheric circulation and redistribution of heat and water around the globe.

Where air is sinking, on the other hand, it tends to warm because of increasing pressure. As it warms, available moisture evaporates. Rainfall occurs relatively rarely in areas of high pressure. High pressure and clear, dry conditions occur where convection currents are sinking. High pressure also occurs where air sinks after flowing over mountains. Figure 9.3 shows sinking, dry air at about 30° north and south latitudes. If you look at a world map, you will see a band of deserts at approximately these latitudes.

Another ingredient is usually necessary to initiate condensation of water vapor: condensation nuclei. Tiny particles of smoke, dust, sea salts, spores, and volcanic ash all act as condensation nuclei. These particles form a surface on which water molecules can begin to coalesce. Without them even supercooled vapor can remain in gaseous form. Even apparently clear air can contain large numbers of these particles, which are generally too small to be seen by the naked eye.

Large-scale winds don't move in a straight line

Large-scale winds tend not to move in a straight line across the earth's surface. In the Northern Hemisphere, they generally bend clockwise (right), and in the Southern Hemisphere, they bend counterclockwise (left). This curving pattern results from the fact that the earth rotates in an eastward direction as the winds move above the surface. The apparent curvature of the winds is known as the **Coriolis effect.** On a global scale, this effect produces steady, reliable wind patterns, such as the trade winds and the midlatitude westerlies (see fig. 9.3). Ocean currents similarly curve clockwise in the Northern Hemisphere and counterclockwise in the south (see appendix 3, p. 382). On a regional scale, the Coriolis effect produces cyclonic winds, or wind movements controlled by the earth's spin. Cyclonic winds spiral clockwise out of an area of high pressure in the Northern Hemisphere and counterclockwise into a low-pressure zone. If you look at a weather map in the newspaper, can you find this counterclockwise spiral pattern?

Why does this curving or spiraling motion occur? Imagine you were looking down on the North Pole of the rotating earth. Now imagine that the earth is a merry-go-round in a playground, with the North Pole at its center and the equator around the edge. As it spins counterclockwise (eastward), the spinning edge moves very fast (a full rotation, 39,800 km, every 24 hours for the real earth, or more than 1,600 km/hour). Near the center, though, there is very little eastward velocity. If you threw a ball from the edge toward the center, it would be traveling faster (edge speed) than the middle. It would appear, to someone standing on the merry-go-round, to curve toward the right. If you threw the ball from the center toward the edge, it would start out with no eastward velocity, but the surface below it would spin eastward, making the ball end up, to a person on the merry-go-round, west of its starting point. Winds move above the earth's surface much as the ball does. If you were looking down at the South Pole, you would see the earth spinning clockwise, and winds—or thrown balls—would appear to bend left. Incidentally, this effect does not apply to drains in your house. Their movement is far too small to be affected by the spinning of the earth. And, contrary to popular myth, there is no significant difference in sink draining between the Northern and Southern Hemispheres.

At the top of the troposphere are **jet streams,** hurricane-force winds that circle the earth. These powerful winds follow an undulating path approximately where the vertical convection currents known as the Hadley and Ferrell cells meet. Although we can't perceive jet streams on the ground, they are important to us because they greatly affect weather patterns. Sometimes jet streams dip down near the top of the world's highest mountains, exposing mountain climbers to violent, brutally cold winds.

Ocean currents modify our climate

Warm and cold ocean currents strongly influence climate conditions on land. Surface ocean currents result from wind pushing on the ocean surface. As surface water moves, deep water wells up to replace it, creating deeper ocean currents. Differences in water density—depending on the temperature and saltiness of the water—also drive ocean circulation. Huge cycling currents called gyres carry water north and south, redistributing heat from low latitudes to high latitudes (see appendix 3, p. 382, global climate map). The Alaska current, flowing from Alaska southward to California, keeps San Francisco cool and foggy during the summer. The Gulf Stream, one of the best known currents, carries warm Caribbean water north past Canada's maritime provinces to northern Europe. This current is immense, some 800 times the volume of the Amazon, the world's largest river. The heat transported from the Gulf keeps Europe much warmer than it should be for its latitude. Stockholm, Sweden, for example, where temperatures rarely fall much below freezing, is at the same latitude as Churchill, Manitoba, which is famous as one of the best places in the world to see polar bears. As the warm Gulf Stream passes Scandinavia and swirls around Iceland, the water cools and evaporates, becomes dense and salty, and plunges downward, creating a strong, deep, southward current.

Together, this surface- and deep-water circulation system is called the **thermohaline ocean conveyor.** Dr. Wallace Broecker of the Lamont Doherty Earth Observatory, who first described this great conveyor system, also showed how it can shut down suddenly. About 11,000 years ago, as the earth was gradually warming at the end of the Pleistocene ice age, a huge body of meltwater, called Lake Agassiz, collected along the south margin of the North American ice sheet. At its peak, Lake Agassiz contained more water than all the current freshwater lakes in the world. Drainage of the lake was blocked by ice laying over what is now the Great Lakes. When that ice dam suddenly gave way, it's estimated that 163,000 km^3 of fresh water roared down the St. Lawrence Seaway and out into the North Atlantic, where it layered on top of the ocean and prevented sinking of deep, cold, dense seawater. This stopped the oceanic conveyor and plunged the whole planet back into an ice age (called the Younger Dryas after a small tundra flower that became more common in colder conditions) that lasted for another 1,300 years. Cessation of the ocean conveyor may have occurred in years, maybe even in a few months.

Could this happen again? Certainly not as suddenly as 11,000 years ago. There's no meltwater buildup today similar to that of Lake Agassiz. But increased river flow, rain, and melting ice thought to be linked to global warming are now causing fresh water to accumulate on top of the Arctic Ocean. Changing ocean salinity seems to be slowing the ocean conveyor. British oceanographers recently reported that the deep-ocean return flow has slowed about 30 percent in the past 50 years. Much of the water that once traveled to the North Atlantic now seems trapped in a subtropical loop. If this situation persists, it could give Northern Europe a climate more like that of Siberia, not a pleasant prospect for many Europeans.

Seasonal winds and monsoons have powerful effects

While figure 9.3 shows regular global wind patterns, large parts of the world, especially the tropics, receive seasonal winds and rainy seasons that are essential for sustaining both ecosystems and human life. Sometimes these seasonal rains are extreme, causing disastrous flooding. (See related story "Floods in Mozambique" at www.mhhe.com/cunningham4e.)

Sometimes the rains fail, causing crop failures and famine. The most regular seasonal winds and rains are known as **monsoons.** In India and Bangladesh, monsoon rains come when seasonal winds blow hot, humid air from the Indian Ocean. Strong convection currents lift this air, causing heavy rain across the subcontinent. When the rising air reaches the Himalayas, it rises even farther, creating some of the heaviest rainfall in the world. During the five-month rainy season of 1970, a weather station in the foothills of the Himalayas recorded 25 m (82 ft) of rain!

Tropical and subtropical regions around the world have seasonal rainy and dry seasons (see the discussion of tropical biomes, chapter 5). The main reason for this variable climate is that the region of most intense solar heating and evaporation shifts through the year. Remember that the earth's axis of rotation is at an angle. Sometimes the sun hits the surface just south of the equator, and sometimes it hits just north of the equator. Wherever the sun shines most directly, evaporation and convection currents—and rainfall and thunderstorms—are very strong. As the earth orbits the sun, different regions fall below the sun, and different regions receive seasonal rain and winds. These seasonal rains support seasonal tropical forests, and they fill some of the world's greatest rivers, including the Ganges and the Amazon. As the year shifts from summer to winter, solar heating weakens, the rainy season ends, and little rain may fall for months.

During the 1970s and 1980s, rains failed repeatedly in parts of the Sahel in northern Africa. Although rains have often been irregular in this region, increasing populations and civil disorder made these droughts result in widespread starvation and death (fig. 9.6).

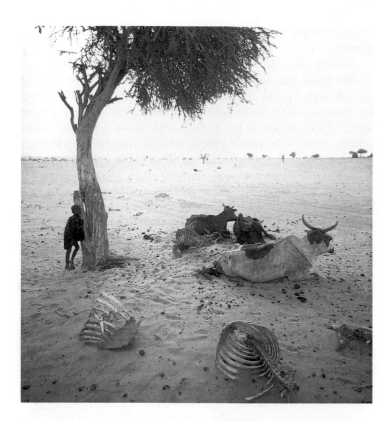

FIGURE 9.6 Failure of monsoon rains brings drought, starvation, and death to both livestock and people in the Sahel desert margin of Africa. Although drought is a fact of life in Africa, many governments fail to plan for it, and human suffering is much worse than it needs to be.

CLIMATE CHANGE

Both global and local climate can be highly variable, often over relatively short time scales. Figure 9.7 shows global temperature estimates and atmospheric CO_2 concentrations over the past 400,000 years as determined from a 3,100 m long ice core drilled in the Antarctic ice cap. Note the close correlation between climate and atmospheric chemistry. It isn't yet clear, however, which is cause and which is effect. These abrupt climate changes can be disastrous for biological organisms. A sudden cooling 65 million years ago, for example, is thought to have ended the age of dinosaurs, along with 75 percent of the species existing at the time. There may have been a dozen such mass extinctions (see table 5.3). On a shorter time scale, several ice ages, each lasting hundreds of thousands of years, have come and gone in the past 2 million years. Even shorter climate shifts occur, such as the "little ice age," which began in the 1300s and caused crop failures throughout Europe for the next two centuries. On a scale of decades there seem to be approximately 20-year drought cycles in North America.

What causes climate shifts at these different time scales? One explanation is periodic changes in sunlight intensity resulting from **Milankovitch cycles.** These cycles, named for the Serbian scientist who first described them in the 1920s, include (1) the shape of the earth's orbit around the sun elongates and shortens in a 100,000-year cycle, (2) the axis of rotation changes its angle of tilt in a 40,000-year cycle, and (3) over a 26,000-year period, the axis wobbles like an out-of-balance spinning top

FIGURE 9.7 Atmospheric CO_2 concentrations and global mean temperatures estimated from the antarctic Vostok ice core. Note the relatively rapid changes and close correlation in both climate and atmospheric chemistry.
Source: Data from United Nations Environment Programme.

Principles of Environmental Science http://www.mhhe.com/cunningham4e

FIGURE 9.8 Milankovitch cycles that may affect long-term climate conditions: (a) changes in the occupancy of the earth's orbit, (b) shifting tilt of the axis, and (c) wobble of the earth.

FIGURE 9.9 The El Niño/La Niña/Southern Oscillation cycle. Every three to five years, surface trade winds that normally push warm water westward toward Indonesia weaken and allow this pool of water to flow eastward toward South America.

(fig. 9.8). Worldwide expansion of glaciers every 100,000 years or so could result from the orbital shifts. Decadal droughts, meanwhile, may be explained by 11-year sunspot cycles, shifts in stormy activity on the sun's surface.

Climate doesn't necessarily change gradually. Evidence from ice cores suggests that, during the last major interglacial period 135,000 to 115,000 years ago, temperatures flipped suddenly over a period of years or decades, rather than centuries. Meteor impacts, volcanic activity, and sudden shifts in ocean currents, could have also caused rapid climate changes. Some climatologists suspect that rapid releases of methane trapped beneath the ocean floor, a "big burp," may have produced a sudden warming that caused the mass extinction 183 million years ago. This theory raises questions about the stability of huge methane deposits currently stored beneath the seafloor.

Two important shorter-term climate changes, El Niño and anthropogenic global warming, are discussed in the next sections.

The El Niño/Southern Oscillation can have far-reaching effects

El Niño, **La Niña**, and the **Southern Oscillation** are all terms referring to a major ocean-current/climate connection that affects weather throughout the Pacific—and possibly throughout the world. The core of this system is a huge pool of warm surface water in the Pacific Ocean that sloshes slowly back and forth between Indonesia and South America like water in a giant bathtub. Most years, this pool is held in the western Pacific by steady equatorial trade winds that push ocean surface currents westward (fig. 9.9). From Southeast Asia to Australia, this

concentration of warm equatorial water provides latent heat (water vapor) that drives strong upward convection (low pressure) in the atmosphere. Heavy rain results, supporting dense tropical forests. On the other side of the Pacific, the westward-moving surface waters are replaced by cold water welling up along the South American coast. Cold, nutrient-rich waters support dense schools of anchovies and other fish. While the trade winds blow westward on the ocean's surface, returning winds high in the troposphere flow back from Indonesia to Chile and to Mexico and southern California. There the returning air sinks, creating dry, desert conditions.

Every three to five years, for reasons that we don't fully understand, the Indonesian low-pressure system collapses, and the mass of warm surface water surges back east across the Pacific. One theory is that the high cirrus clouds block enough sunshine to cool the ocean surface in Asia. Convection would then weaken, and trade winds—and ocean currents—would reverse, flowing eastward instead of westward. Another theory is that eastward-flowing deep currents called baroclinic waves periodically interfere with coastal upwelling, warming the sea surface off South America and eliminating the temperature gradient across the Pacific (see the Data Analysis box at the end of this chapter).

Fishermen in Peru were the first to notice irregular cycles of rising ocean temperatures because the fish disappeared when the water warmed. They named this event El Niño (Spanish for the Christ child) because they often occur around Christmastime. The counterpart to El Niño, when the eastern tropical Pacific cools, has come to be called La Niña (little girl). Together, these cycles are called the El Niño Southern Oscillation (ENSO).

ENSO cycles have far-reaching effects. During an El Niño year, the northern jet stream—which is normally over Canada—splits and is drawn south over the United States. This pulls moist air from the Pacific and Gulf of Mexico inland, bringing intense storms and heavy rains from California across the midwestern states. The intervening La Niña years bring hot, dry weather to the same areas. Oregon, Washington, and British Columbia, on the other hand, tend to have warm, sunny weather in El Niño years rather than their usual rain. Droughts in Australia and Indonesia during El Niño episodes cause disastrous crop failures and forest fires, including one in Borneo in 1983 that burned 3.3 million ha (8 million acres).

Some climatologists believe that ENSO events are becoming stronger or more frequent because of global climate change. There are signs that warm ocean-surface temperatures are spreading, which could contribute to El Niño strength or frequency. On the other hand, increased cloud cover over warmer oceans could raise global albedo, and strong convection currents generated by these storms could pump heat into the stratosphere. This might have an overall cooling effect and act as a safety valve for global warming.

Our actions are now causing global climate change

Most scientists now regard anthropogenic (human-caused) global climate change to be the most important environmental issue of our times. The possibility that humans might alter world climate is not a new idea. In 1895 Svante Arrhenius, who subsequently received a Nobel Prize for his work in chemistry, predicted that CO_2 released by coal burning could cause global warming. Most of Arrhenius' contemporaries dismissed his calculations as impossible, but his work seems remarkably prescient now.

The first evidence that human activities are increasing atmospheric CO_2 came from an observatory on top of the Mauna Loa volcano in Hawaii. The observatory was established in 1957 as part of an International Geophysical Year and was intended to provide data on air chemistry in a remote, pristine environment. Surprisingly, measurements showed CO_2 levels increasing about 0.5 percent per year, rising from 315 ppm in 1958 to 383 ppm in 2006 (fig. 9.10). This increase isn't a straight line, however. Because a majority of the world's land and vegetation are in the Northern Hemisphere, northern seasons dominate the signal. Every May, CO_2 levels drop slightly as plant growth on northern continents uses CO_2 in photosynthesis. During the northern winter, levels rise again as respiration releases CO_2.

In 1988 the United Nations Environment Programme and World Meteorological Organization formed the **Intergovernmental Panel on Climate Change (IPCC)**. This panel brings together scientists from many nations and a wide variety of fields to assess the current state of knowledge about climate change. The IPCC's First Assessment Report played an important role in adoption of the UN Framework Convention on Climate Change at the Rio Earth Summit in 1992. In 2001 the IPCC released its third climate report, which said "with a high degree of confidence" that "recent changes in the world's climate have had discernible

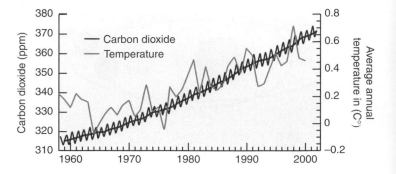

FIGURE 9.10 Carbon dioxide concentrations on Mauna Loa (monthly and annual averages) and global annual average temperature variations from 1960 to 2000.
Data from NASA Earth Observatory, 2002.

impacts on physical and biological systems." For this report, more than 700 scientists representing 100 countries reviewed results from some 3,000 scientific studies, which showed changes in 420 different physical and biological systems.

Noting that the earth's surface temperature has risen by about 0.6°C (1.1°F) over the past century, with most of that warming occurring in the previous two decades, the IPCC concluded that human activities must be at least partially responsible. Separating the impact of human activity from natural climate variation is extremely difficult. Nonetheless, the IPCC claims, the observed global warming is unlikely to be the result of natural variability alone. "We have altered the chemical composition of the atmosphere through the buildup of greenhouse gases—primarily carbon dioxide, methane, and nitrous oxide," the report declared. "The heat-trapping property of these gases is undisputed although uncertainties exist about exactly how earth's climate responds to them."

Greenhouse gases have many sources

Since preindustrial times, atmospheric concentrations of CO_2, CH_4, and N_2O have climbed by over 31 percent, 151 percent, and 17 percent, respectively. Carbon dioxide is by far the most important cause of anthropogenic climate change (fig. 9.11a). Burning fossil fuels, making cement, burning forests and grasslands, and other human activities release nearly 30 billion tons of CO_2 every year, on average, containing some 8 billion tons of carbon (fig. 9.11b). About 3 billion tons of this excess carbon is taken up by terrestrial ecosystems, and around 2 billion tons are absorbed by the oceans, leaving an annual atmospheric increase of some 3 billion tons per year. If current trends continue, CO_2 concentrations could reach about 500 ppm (approaching twice the preindustrial level of 280 ppm) by the end of the twenty-first century.

Although rarer than CO_2, methane absorbs 20 to 30 times as much infrared and is accumulating in the atmosphere about twice as fast as CO_2. Methane is released by ruminant animals, wet-rice paddies, coal mines, landfills, and pipeline leaks. Chlorofluorocarbons (CFCs) also are powerful infrared absorbers. CFC

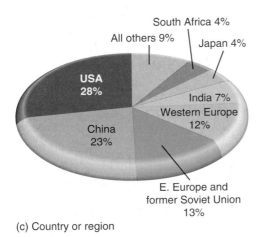

FIGURE 9.11 Percentage contributions to global warming by (a) gas type, (b) source activity, and (c) country or region.
Data from World Resources Institute.

per person. Eastern Europe and the former Soviet Union have many old, inefficient factories and coal-buring power plants that emit high CO_2 levels with relatively little economic return.

Evidence of climate change is becoming overwhelming

As the opening Case Study for this chapter notes, there is little doubt that our planet is warming. The American Geophysical Union, one of the nation's largest and most respected scientific organizations, has said, "As best as can be determined, the world is now warmer than it has been at any point in the last two millennia, and, if current trends continue, by the end of the century it will likely be hotter than at any point in the last two million years." Some of the evidence for this warming includes the following:

- Over the last century the average global temperature has climbed about 0.6°C (1°F). Nineteen of the 20 warmest years in the past 150 have occurred since 1980. The hottest year since the instrumental temperature record began was 1998, but the next four hottest were all since 2000.

- Polar regions have warmed much faster than the rest of the world. In Alaska, western Canada, and eastern Russia, average temperatures have increased as much as 4°C (7°F) over the past 50 years. Permafrost is melting; houses, roads, pipelines, sewage systems, and transmission lines are being damaged as the ground sinks beneath them. Trees are tipping over, and a beetle infestation in Alaska (made possible by warmer winters) killed 25 million spruce trees on the Kenai Peninsula. Coastal villages, such as Shishmaref, Point Hope, and Barrow, are having to relocate inland as ice breakup and increasingly severe storms erode the arctic shoreline.

- Arctic sea ice is 40 percent thinner now, and covers one-third less area than it did a century ago. As sea ice forms later in the fall and melts earlier in the spring, polar bears have a shorter hunting season. Hudson Bay polar bears now weigh as much as 100 kg (220 lbs) less than in the 1960s. An indigenous way of life based on hunting marine mammals along the edge of the pack ice is also disappearing.

- Ice shelves on the Antarctic Peninsula are breaking up and disappearing rapidly. A recent survey found that 90 percent of the glaciers on the peninsula are now retreating an average of 50 m per year. The Greenland ice cap also is reported to be thinning by as much as 1 m per year. Because ice shelves are floating, they don't affect sea level when they melt. Greenland's massive ice cap, however, holds enough water to raise sea level by about 7 m (about 23 ft) if it melts. (See related story "Collapse of the Larsen B" at www.mhhe.com/cunningham4e.)

- Alpine glaciers everywhere are retreating rapidly (fig. 9.12). Mt. Kilimanjaro has lost 85 percent of its famous ice cap since 1915. By 2015, all the permanent ice on the mountaintop is expected to be gone. In 1972, Venezuela had six glaciers; now it has only two. When Montana's Glacier

releases have declined since many of their uses were banned, but the CFCs already in the atmosphere will persist for many years. Nitrous oxide is produced by burning organic material and by soil denitrification. As figure 9.11a shows, CFCs and N_2O together are thought to account for only about 17 percent of human-caused global warming.

The United States, with less than 5 percent of the world's population, produces 28 percent of all anthropogenic CO_2 (fig. 9.11c). China, with 1.3 billion people, is second in total CO_2 emissions, but fourteenth in per capita production. Japan and Western Europe, which by most measures have standards of living at least equal to the United States, produce only half as much CO_2

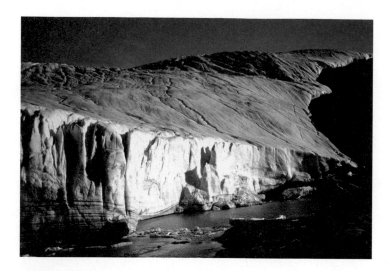

FIGURE 9.12 Alpine glaciers are retreating rapidly nearly everywhere in the world. For the billions of people whose drinking water comes from glacial-fed rivers, this could be a serious threat.

National Park was created in 1910, it held some 150 glaciers. Now, fewer than 30, greatly shrunken glaciers remain. If current trends continue, all will have melted by 2030.

- So far, the oceans have been buffering the effects of our greenhouse emissions both by absorbing CO_2 directly and by storing heat. This slows current warming, but also means that even if we reduce our greenhouse gas emissions today, it will take centuries to dissipate that stored heat. The higher levels of CO_2 being absorbed are acidifying the oceans, and could have adverse effects on sea life. Mollusks and corals, for example, have a more difficult time making calcium carbonate shells and skeletons at the lower pH.

- Sea level has risen worldwide approximately 15–20 cm (6–8 in.) in the past century. About one-quarter of this increase is ascribed to melting glaciers; roughly half is due to thermal expansion of seawater. Melting of the Greenland ice cap could raise sea levels by about 7 m. Far more water is locked up in Antarctica. If all of it were to melt, the resulting rise in sea level could raise sea levels by several hundred meters.

- Satellite images and surface measurements show that growing seasons are now as much as three weeks longer in a band across northern Eurasia and North America than they were 30 years ago (see fig. 1.13). New plants and animals, never before seen in the far north, are now extending their territories, while native species, such as musk ox, caribou, walrus, and seal populations are declining as climate changes.

- Droughts are becoming more frequent and widespread. In Africa, for example, droughts have increased about 30 percent since 1970. In 2005, the United Nations reported that 60 million people in 36 countries needed emergency food aid. Drought was the single largest cause of famine, although wars, economics, and internal politics always exacerbate these emergencies.

- Biologists report that many animals are breeding earlier or extending their range into new territory as the climate changes. In Europe and North America, for example, 57 butterfly species have either died out at the southern end of their range, or extended the northern limits, or both. Plants, also, are moving into new territories. Given enough time and a route for migration, many species may adapt to new conditions, but we now are forcing them to move much faster than many achieved at the end of the last ice age (fig. 9.13).

- The disappearance of amphibians, such as the beautiful golden toads from the cloud forests of Costa Rica or western toads from Oregon's Cascade Range, are thought to have been caused at least in part by changing weather patterns. Emperor and Adélie penguin populations have declined by half over the past 50 years as the ice shelves on which they depend for feeding and breeding disappear.

- Coral reefs worldwide are "bleaching," losing key algae and resident organisms, as water temperatures rise above 30°C (85°F). With reefs nearly everywhere threatened by pollution, overfishing, and other stressors, scientists worry that rapid climate change could be the final blow for many species in these complex, biologically rich ecosystems.

- Storms are becoming stronger and more damaging. The 2005 Atlantic storm season was the most severe on record, with 26 named tropical storms, twice as many as the average over the past 30 years. Some of this increased frequency could be a natural cycle, but the greater strength of these storms (an unprecedented 7 hurricanes reached level 5, the highest category) undoubtedly reflects higher sea surface temperatures. Hurricane Katrina, which demolished

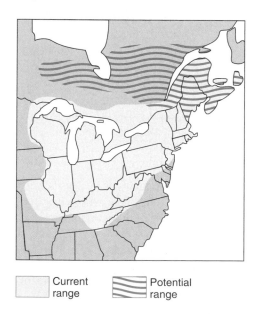

Current range	Potential range

FIGURE 9.13 Change in suitable range for sugar maple trees, according to the climate change model.
Source: Data from Margaret Davis and Catherine Zabriskie in *Global Warming and Biological Diversity,* ed. By Peters and Lovejoy, 1991, Yale University Press.

Principles of Environmental Science

FIGURE 9.14 More than 1 million people had to be evacuated from New Orleans in 2005 when a storm surge created by Hurricane Katrina breached the levees and allowed Lake Pontchartrain to flood into the city.

a 230,000 km² (90,000 mi²) area of the U.S. Gulf Coast (an area about the size of Great Britain) and flooded New Orleans (fig. 9.14) caused at least $200 billion in damage, the greatest storm loss in American history.

Who wins and who loses?

Local climate changes could well have severe effects on human societies, agriculture, and natural ecosystems. In many cases, both organisms and human infrastructure will not be able to move or adapt quickly enough to accommodate the rates at which climate seems to be changing.

Does anyone win in these scenarios? Residents of northern Canada, Siberia, and Alaska probably would enjoy warmer temperatures, longer growing seasons, and longer ice-free shipping seasons from their ports (fig. 9.15). Stronger monsoons bringing more moisture to parts of Central Africa and South Asia could

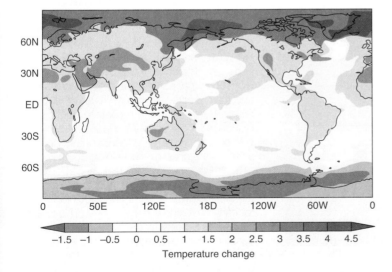

FIGURE 9.15 Predicted warming for carbon dioxide doubling in 2100. Note general warming of 1–2°C with much greater temperature rise in polar regions.
Source: Data from Dr. Tim Barnet, Scripps Institute of Oceanography, 2002.

increase crop yields. But soils in some areas may not be suitable for crops no matter how agreeable the climate. Carbon dioxide is a fertilizer for plants, and higher CO_2 levels bring faster growth with less water for many plant species. Sherwood Idso of the U.S. Department of Agriculture says that increased CO_2 concentrations could trigger lush plant growth that would make crops abundant. "See Exploring Science p. 214".

Other areas, such as the North American Great Plains, central Siberia, and the Pampas of Argentina, are projected to be much drier than now. This could be a blow to biodiversity. The prairie potholes of the American midwest and Canadian prairie provinces produce more than half of all ducks, geese, and other North American water birds. The vast wetlands of Siberia play a similar role for Eurasia. If these myriad ponds and wetlands dry up, bird populations will plummet.

Enough water is stored in glacial ice caps in Greenland and Antarctica to raise sea levels around 100 meters (300 feet) if they all melt. About one-third of the world's population now live in areas that would be flooded if that happened. Even the 75 cm (30 in.) sea-level rise expected by 2050 will flood much of south Florida, Bangladesh, Pakistan, and many other low-lying coastal areas. Most of the world's largest urban areas are on coastlines. Wealthy cities, such as New York or London, can probably afford to build dikes to keep out rising seas, but poorer cities, such as Jakarta, Calcutta, or Manila, might simply be abandoned as residents flee to higher ground. Several small island countries, such as the Maldives, the Bahamas, Kiribati, and the Marshall Islands, could become uninhabitable if sea levels rose a meter or more. In 2002 the South Pacific nation of Tuvalu announced that it was abandoning its island homeland. All 11,000 residents will move to New Zealand, perhaps the first of many climate change refugees.

Insurance companies worry that the $2 trillion in insured property along U.S. coastlines could be at risk from a combination of high seas and catastrophic storms. Some 87,000 homes in the United States within 150 m (500 ft) of a shoreline are in danger of coastal erosion or flooding in the next 50 years. Accountants state that loss of land and structures to flooding and coastal erosion together with damage to fishing stocks, agriculture, and water supplies could raise worldwide insurance claims from about $50 billion per year in the 1970s to a projected $150 billion per year in 2010. Some of this increase in insurance claims is because more people are living in dangerous places, but extra-severe storms and rising sea levels only exacerbate this problem.

Infectious diseases are likely to increase as the insects, rodents, and ticks that carry them spread to new areas. Already we have seen diseases such as malaria, dengue fever, and West Nile virus appear in parts of North America where they had never before been reported. Coupled with the movement of hundreds of millions of environmental refugees and greater crowding as people are forced out of areas made uninhabitable by rising seas and changing climates, the spread of epidemic diseases will also increase.

Permafrost is already beginning to melt in the Arctic, causing roads to buckle and houses to sink into the mud. An ominous possibility from this melting is the release of vast stores of methane hydrate now locked in frozen ground and in sediments

Among the greenhouse gases, none has been studied more than carbon dioxide (CO_2). As you've learned already in this chapter, CO_2 has been analyzed in 400,000-year-old air bubbles taken from glacial ice, measured atop Hawaii's Mauna Loa volcano since 1957, and implicated in a whole host of current climate changes. No one disputes that CO_2 levels in the troposphere increased from 280 to 380 ppm between 1750 and 2005. Few argue that it may exceed 500 ppm—nearly twice its preindustrial concentration—before the twenty-first century ends, but we're not sure what the effects of these high CO_2 levels may be on ecological processes.

Plants need CO_2 to carry out photosynthesis. We know that many plant species flourish in elevated CO_2 levels. That's why greenhouses often pipe in extra CO_2 to accelerate growth. Some horticulturists predict that doubling CO_2 atmospheric concentrations will produce such lush vegetation that the earth will return to a verdant Eden. Others contend that how plants behave in the ideal conditions of a greenhouse or growth chamber is very different from what happens in the real world. How can we know what to expect if current trends continue?

In 1997, Peter Reich and his colleagues began experiments at the University of Minnesota's Cedar Creek Natural History Area that attempt to mimic natural conditions more faithfully than is possible in a growth chamber. At Cedar Creek, the researchers erected several large rings of towers from which CO_2 is released (fig. 1). Sensors track wind speeds and CO_2 concentrations within the ring, while a computer regulates gas flows to maintain 560 ppm throughout the experimental plot. Even when winds are swirling around the research site, the computer can adjust gas flows to keep CO_2 levels constant. Because the CO_2 isn't contained, these experiments are called Free Air Carbon Enrichment (FACE). Control plots have exactly the same design as the experimental ones except that the control towers blow plain air rather than CO_2-enriched air. The Cedar Creek Long-Term Ecological Research is an important center for the study of the relationship between biodiversity and prairie ecosystem stability, so the FACE research also included several different levels of native prairie

FIGURE 1 Student researchers collect samples for analysis at Cedar Creek's Free Air Carbon Enrichment study. The ring of striped poles blow CO_2 over the prairie plants growing in experimental plots.

species richness, as well as different fertilizer treatments, in experimental plots.

Dr. Reich and colleagues discovered that the response of plants to elevated CO_2 and nitrogen fertilization depended on how many species were in a test plot. In a paper published in 2001 in *Nature,* they showed that the amount of plant material (biomass) produced each year rose when CO_2 and nitrogen were added to the test plots. But test plots containing 9 or 16 plant species benefited much more than test plots with a single plant species (fig. 2). In a later paper in the *Proceedings of the National Academy of Sciences,* Dr. Reich and his team also demonstrated that, if test

FIGURE 2 The Cedar Creek BioCON study demonstrates strength in numbers. Single species plots did not respond as well to CO_2 increases as multi-species plots.
Data from Peter Reich.

plots contained ecologically different plants (e.g., a legume and a summer-blooming grass), more biomass developed than in test plots with a single one of those functional plant groups. In other words, vegetation that contained a single plant species responded positively to rising carbon dioxide, but several plant species with different ecological functions responded much better, even without fertilization. The different species and functional groups compensated for each other in the face of a varying climate and environment.

Similar research is being done on forest species and agricultural crops. Peter Curtis, an Ohio State University researcher, writing in the journal *Phytologist,* summarized the results of dozens of FACE experiments on a number of species. In grain crop plants the total seed weight—and therefore food yield—increased by up to 25 percent with a doubling of CO_2 concentration, but in all species except legumes the nutritional value of those plants fell. In one study of tall fescue (*Festuca alatior*), one of the most important pasture grasses in United States, the digestible protein that grazers could extract from the CO_2-stimulated grass actually fell below levels in existing forage, despite fertilization. So, while we may have luxuriant plant growth under elevated CO_2, it may be less valuable for certain uses than what we have today.

In the end, raising CO_2 through human activities represents a huge global experiment not unlike the FACE experiments, but with an important difference. If CO_2 reaches 500 ppm by 2100, we'll be stuck with those conditions—and their effects—for many years. What basic research tells us in the meantime about plant responses, including our important crop plants, will be important in understanding how to adapt to the challenges we face.

Dr. Reich says, "The idea that carbon dioxide will uniformly increase plant growth and therefore provide a large and predictable sink for carbon dioxide is naive at best. This type of model fails to consider the real complexity of the carbon dioxide fertilization effect. Species composition, community diversity, and the amount of nitrogen in the system interact to produce varying responses. In particular, reduced biodiversity and the widespread limited amount of nitrogen in the soil will reduce the CO_2 fertilization effect in terrestrial plant biomass."

in the ocean floor. Together with the increased oxidation of high-latitude peat lands potentially caused by warmer and dryer conditions, release of these carbon stores could add as much CO_2 to the atmosphere as all the fossil fuels ever burned. We could trigger a disastrous positive feedback loop in which the effects of warming cause even more warming.

The Kyoto Protocol attempts to slow climate change

One of the centerpieces of the 1992 United Nations Earth Summit meeting in Rio de Janeiro was the Framework Convention on Climate Change, which set an objective of stabilizing greenhouse gas emissions to reduce the threats of global warming. At a follow-up conference in Kyoto, Japan, in 1997, 160 nations agreed to roll back CO_2, methane, and nitrous oxide emissions about 5 percent below 1990 levels by 2012. Three other greenhouse gases, hydrofluorocarbons, perfluorocarbons, and sulfur hexafluoride, would also be reduced, although from what level was not decided. Known as the **Kyoto Protocol,** this treaty sets different limits for individual nations, depending on their output before 1990. Poorer nations, such as China and India, were exempted from emission limits to allow development to increase their standard of living. Wealthy countries created the problem, the poorer nations argue, and the wealthy should deal with it.

Although the United States took a lead role in negotiating a compromise at Kyoto that other countries could accept, President George W. Bush refused to honor U.S. commitments. Claiming that reducing carbon emissions would be too costly for the U.S. economy, he said, "We're going to put the interests of our own country first and foremost." The United States, which depends solely on voluntary efforts to limit greenhouse gas emissions, had a 2 percent rise in those releases in 2005. At this rate, the U.S. will be 25 percent above 1990 emissions by 2012.

Meanwhile, 126 countries have ratified the Kyoto Protocol. With the addition of Russia to this list in 2004, the provisions became legally binding on all participants. If the U.S. continues to refuse to comply, it could have severe ramifications on U.S. corporations engaged in international business. Thousands of American businesses—including most of the largest ones—fall into this category. Having to modify their products and practices for overseas markets while not doing so for domestic sales would be expensive and would put them at a disadvantage with competitors not subject to these costs. Moreover, because Kyoto is based on a global cap-and-trade program, companies that get in on it early will have an advantage—they can buy cheap emissions credits before the price gets bid up. At a follow-up to the Kyoto Conference in Montreal in 2005, the United States continued its opposition not only to any new targets for cutting greenhouse gas emissions but also to any discussions that might even touch on the subject. Although some participants took an optimistic attitude toward the results of the conference, particularly its resolve to work on reducing the effects of global warming in poor countries, others took a harsher view. In an unusually direct rebuke, Canadian Prime Minister Paul Martin said, "To the reticent nations, including the United States, I say this: There is such a thing as a global conscience, and now is the time to listen to it." Meanwhile, representatives of the polar Inuit people filed a complaint against the United States at the Inter-American Commission on Human Rights, claiming that emissions of greenhouse gases are threatening their traditional way of life.

Even if the United States joins the rest of the world in accepting the Kyoto provisions, however, many climatologists argue that much more will have to be done to combat global warming. According to some calculations, we will have to reduce CO_2 emissions by 60 percent or more to stabilize atmospheric concentrations. In the next section, we'll look at some of the ways we might accomplish this.

There are many ways we can control greenhouse emissions

As the opening Case Study in this chapter shows, there are many ways to slow or reverse global warming. Progress already is being made in many places toward reducing greenhouse emissions. The United Kingdom, for example, had already rolled CO_2 emissions back to 1990 levels by 2000 and vowed to reduce them 60 percent by 2050. Britain already has started to substitute natural gas for coal, promote energy efficiency in homes and industry, and raise its already high gasoline tax. Plans are to "decarbonize" British society and to decouple GNP growth from CO_2 emissions. A revenue-neutral carbon levy is expected to lower CO_2 releases and trigger a transition to renewable energy over the next five decades.

Germany, also, has reduced its CO_2 emissions at least 10 percent by switching from coal to gas and by encouraging energy efficiency throughout society. Atmospheric scientist Steve Schneider calls this a "no regrets" policy; even if we don't need to stabilize our climate, many of these steps save money, conserve resources, and have other environmental benefits. Nuclear power also is being promoted as an alternative to fossil fuels. It's true that nuclear reactions don't produce greenhouse gases, but security worries and unresolved problems of how to store wastes safely make this option unacceptable to many people.

Many people believe renewable energy sources offer a better solution to climate problems. Chapter 12 discusses options for conserving energy and switching to renewable sources, such as solar, wind, geothermal, biomass, and fuel cells. Denmark, the world's leader in wind power, now gets 20 percent of its electricity from windmills. Plans are to generate half of the nation's electricity from offshore wind farms by 2030. Even China reduced its CO_2 emissions 20 percent between 1997 and 2005 through greater efficiency in coal burning and industrial energy use.

In addition to reducing its output, there are options for capturing and storing CO_2. Dr. Klaus S. Lackner of Columbia University has proposed building synthetic "trees" to capture CO_2 from the air (fig. 9.16). A calcium hydroxide solution trickling across the "leaves" would absorb CO_2. A structure with 2,500 m^2 of collecting area could remove 90,000 tons of CO_2 per year from

FIGURE 9.16 Synthetic "trees" could capture CO_2 from the air. Each of these structures would be about 100 m (300 ft) tall and would have an array of metal slats 50 × 50 m across. Each of these trees could remove 90,000 tons of CO_2 per year from the air, an amount equal to the CO_2 produced by 15,000 cars. It would take a quarter of a million of them to remove all the CO_2 produced annually by humans.

the air (1,000 times more than a natural tree), an amount equal to the CO_2 produced by 15,000 cars. About 250,000 synthetic trees worldwide would be needed to soak up the 22 billion tons of CO_2 produced annually by burning fossil fuels. Critics claim that synthesizing the hydroxide needed for this reaction and disposing of the vast amount of calcium carbonate it would create may take more energy than the fuel that produced the CO_2 released. The first of these "trees" is now being built in Arizona. Planting real trees can also be effective if they're allowed to mature to old-growth status or are made into products, such as window frames and doors, that will last for many years before they're burned or recycled. Farmland also can serve as a carbon sink if farmers change their crop mixture and practice minimum till cultivation that keeps carbon in the soil.

In 1988 John H. Martin suggested that phytoplankton growth in the ocean may be limited by iron deficiency. This theory was tested in 2000, when researchers spread about 3.5 tons of dissolved iron over 75 km^2 of the South Pacific. Monitored by satellite, a ten-fold rise in chlorophyll concentration eventually spread over about 1,700 km^2 and was calculated to have pulled several thousand tons of CO_2 out of the air. Still, some ecologists warn about such large-scale tinkering. We don't know whether the carbon fixed by phytoplankton growth will be stored in sediments or simply eaten by predators and returned immediately to the atmosphere. It's possible that dying algae might create an anoxic zone that would devastate oceanic food webs.

Another way to store CO_2 is to inject it into underground strata or deep ocean waters. Since 1996, Norway's Statoil has been pumping more than 1 million metric tons of CO_2 per year into an aquifer 1,000 m below the seafloor at a North Sea gas well. It is economical to do so because otherwise the company would have to pay a $50 per ton carbon tax on its emissions. Around the world, deep, briny aquifers could store a century worth of CO_2 output at current fossil fuel consumption rates. It might also be possible to

pump liquid CO_2 into deep ocean trenches, where it would form lakes contained by enormous water pressures. There are worries, however, about what this might do to deep ocean fauna and what might happen if earthquakes or landslides caused a sudden release of this CO_2. In Canada, an Alberta power plant is injecting CO_2 into a coal seam too deep to be mined. The CO_2 releases natural gas, which is burned to produce electricity and more CO_2. Proponents of **carbon management,** as these various projects are called, argue that it may be cheaper to clean up fossil fuel effluents than to switch to renewable energy sources.

Most attention is focused on CO_2 because it lasts in the atmosphere, on average, for about 120 years. Methane and other greenhouse gases are much more powerful infrared absorbers but remain in the air for a much shorter time. NASA's Dr. James Hansen made a controversial suggestion, however, that the best short-term attack on warming might come by focusing not on carbon dioxide but on methane. Reducing gas pipeline leaks would conserve this valuable resource as well as help the environment.

Methane from landfills, oil wells, and coal mines that once would have been simply vented into the air is now being collected and used to generate electricity. Rice paddies are a major methane source. Changing flooding schedules and fertilization techniques can reduce anoxia (oxygen depletion) that produces marsh gas. Finally, ruminant animals (such as cows, camels, and buffalo) create large amounts of digestive system gas. Modified diets can reduce methane releases significantly.

Soot, while not mentioned in the Kyoto Protocol, might also be important in global warming. Dark, airborne particles absorb both ultraviolet and visible light, converting them to heat energy. According to some calculations, reducing soot from diesel engines, coal-fired generators, forest fires, and wood stoves might reduce net global warming by 40 percent within three to five years. Curbing soot emissions would also have beneficial health effects.

Some individual cities and states have announced their own plans to combat global warming. Among the first of these were Toronto, Copenhagen, and Helsinki, which pledged to reduce CO_2 emissions 20 percent from 1990 levels by 2010. In the United States, 12 states and 25 cities have adopted policies to reduce greenhouse emissions. New York and California passed regulations limiting CO_2 emissions from automobiles and trucks, but industry has vowed to fight these measures. Some corporations are following suit. British Petroleum has set a goal of cutting CO_2 releases from all its facilities by 10 percent before 2010. Each of us can make a contribution in this effort. As Professor Socolow and his colleagues point out, simply driving less and buying high-mileage vehicles could save about 1.5 billion tons of carbon emissions by 2054 (fig. 9.17).

In the midst of all the debate about how serious the consequences of global climate change may or may not be, we need to remember that many of the proposed solutions are advantageous in their own right. Even if climate change turns out not to be as much of a threat as we think now, they have other positive benefits. Moving from fossil fuels to renewable energy sources such as solar or wind power, for example, would free us from dependence on foreign oil and improve air quality. Planting trees makes cities more pleasant places to live and provides habitat for

Principles of Environmental Science

FIGURE 9.17 Burning fossil fuels produces about half our greenhouse gas emissions, and transportation accounts for about half our fossil fuel consumption. Driving less, choosing efficient vehicles, carpooling, and other conservation measures are among our most important personal choices in the effort to control global warming.

TABLE 9.1
Reductions in U.S. Pollution Emissions, 1970–1997

Pollutant	Percent Change in Annual Emissions
Carbon monoxide	−32
Lead[1]	−98
Nitrogen oxides[2]	+11
Ozone	−37
Particulate matter	−75
Sulfur dioxide	−35

[1]Reduction due mainly to elimination of leaded gasoline.
[2]Increase mainly from more vehicle emissions.
Source: Courtesy of Dr. William Culver, St. Petersburg Junior College.

wildlife. Making buildings more energy efficient and buying high-mileage vehicles saves money in the long run. Walking, biking, and climbing stairs are good for your health as well as reducing traffic congestion and energy consumption. Reducing waste, recycling, and other forms of sustainable living improve our environment in many ways in addition to helping fight climate change. It's important to focus on these positive effects rather than to look only at the gloom-and-doom scenarios for global climate catastrophes. Which of the steps suggested for controlling climate change do you think should be our highest priorities for individual and collective action? Chapter 12 discusses ways we can cut our individual energy consumption, the principal cause of greenhouse gas emissions (see p. 294).

AIR POLLUTION

As the global warming debate shows, human activities can influence climate processes. In this section, we will discuss ways that human activities affect air quality and environmental processes on a more local scale.

According to the Environmental Protection Agency (EPA), Americans release some 147 million metric tons of air pollution (not counting carbon dioxide or wind-blown soil) each year. Worldwide emissions of these pollutants are around 2 billion metric tons per year. Even remote, pristine wilderness areas are now affected. Over the past 20 years, however, air quality has improved in most cities in Western Europe, North America, and Japan. Many young people might be surprised to learn that, a generation ago, most American cities were much dirtier than they are today. The EPA estimates that, since 1990, when regulation of the most hazardous materials began, air toxics emissions have been reduced more than 1 million tons per year. This is almost ten times the reductions achieved in the previous 20 years. Since the 1970s, the levels of major pollutants monitored by the EPA have decreased in the United States, despite population growth of more than 30 percent (table 9.1). Pollution reductions have resulted mainly from greater efficiency and pollution-control technologies in factories,

power plants, and automobiles. Our success in controlling some of the most serious air pollutants gives us hope for similar progress in other environmental problems.

While developed countries have been making progress, however, air quality in the developing world has been getting much worse. Especially in the burgeoning megacities of rapidly industrializing countries (chapter 14), air pollution often exceeds World Health Organization standards by large margins. In Lahore, Pakistan, and Xi'an, China, for instance, airborne dust, smoke, and dirt often are ten times higher than levels considered safe for human health (fig. 9.18).

Studies of air pollutants over southern Asia reveal that a 3 km (2 mi) thick toxic cloud of ash, acids, aerosols, dust, and

FIGURE 9.18 While air quality is improving in many industrialized countries, newly developing countries have growing pollution problems. Xi'an, China, often has particulate levels above 300 μg/m³.

FIGURE 9.19 Primary pollutants are released directly from a source into the air. A point source is a specific location of highly concentrated discharge, such as this smokestack.

photochemical reactants covers the entire Indian subcontinent for much of the year. Nobel laureate Paul Crutzen estimates that up to 2 million people in India alone die each year from atmospheric pollution. Produced by forest fires, the burning of agricultural wastes, and dramatic increases in the use of fossil fuels, the Asian smog layer cuts the amount of solar energy reaching the earth's surface beneath it by up to 15 percent. Meteorologists suggest that the cloud—80 percent of which is human-made—could disrupt monsoon weather patterns and cut rainfall over northern Pakistan, Afghanistan, western China, and central Asia by up to 40 percent. Shifting monsoon flows may also have contributed to the catastrophic floods in Nepal, Bangladesh, and eastern India in 2002 that killed at least 1,000 people and left more than 25 million homeless.

When this "Asian Brown Cloud" drifts out over the Indian Ocean at the end of the monsoon season, it cools sea temperatures and may be changing El Niño/Southern Oscillation patterns in the Pacific Ocean as well. As UN Environment Programme Executive Director Klaus Töpfer said, "There are global implications because a pollution parcel like this, which stretches 3 km high, can travel half way round the globe in a week."

We have different ways to describe pollutants

In the United States, principal pollutants are identified and regulated by the Clean Air Act of 1970. A **point source** is a smokestack or some other concentrated pollution origin (fig. 9.19). **Primary pollutants** are released in a harmful form. Secondary pollutants, by contrast, become hazardous after reactions in the air. **Photochemical oxidants** (compounds created by reactions driven by solar energy) and atmospheric acids are probably the most important secondary pollutants. **Fugitive,** or **nonpoint-source, emissions** are those that do not go through a smokestack. By far the most massive example of this category is dust from soil erosion,

strip mining, rock crushing, and building construction (and destruction). Leaking valves and pipe joints contribute as much as 90 percent of the hydrocarbons and volatile organic chemicals emitted from oil refineries and chemical plants.

Conventional, or **criteria, pollutants** are a group of seven major pollutants (sulfur dioxide, carbon monoxide, particulates, volatile organic compounds, nitrogen oxides, ozone, and lead) that contribute the largest volume of air-quality degradation and are considered the most serious threat of all air pollutants to human health and welfare. Figure 9.20 shows the main sources of six of these criteria pollutants. Since 1970 the Clean Air Act has mandated that the EPA set allowable limits for concentrations of these pollutants in **ambient air** (the air around us), especially in cities.

The EPA also monitors **unconventional pollutants,** compounds that are produced in less volume than conventional pollutants but that are especially toxic or hazardous. Among these are asbestos, benzene, beryllium, mercury, polychlorinated biphenyls (PCBs), and vinyl chloride. Most of these materials have no natural source in the environment (to any great extent) and are, therefore, only anthropogenic in origin.

Aesthetic degradation is another important form of pollution. Noise, odors, and light pollution may not be life threatening, but they reduce the quality of our lives. In most urban areas, it is difficult or impossible to see stars in the sky at night because of dust in the air and stray light from buildings, outdoor advertising, and streetlights.

Sources and problems of major pollutants

Most conventional pollutants are produced primarily from burning fossil fuels, especially in coal-powered electric plants and in cars and trucks, as well as in processing natural gas and oil. Others, especially sulfur and metals, are by-products of mining and manufacturing processes. Of the 188 air toxics listed in the Clean Air Act, about two-thirds are volatile organic compounds (VOCs), and most of the rest are metal compounds. In this section we will discuss the characteristics and origin of the major pollutants.

Sulfur dioxide (SO_2) is a colorless, corrosive gas that damages both plants and animals. Once in the atmosphere, it can be further oxidized to sulfur trioxide (SO_3), which reacts with water vapor or dissolves in water droplets to form sulfuric acid (H_2SO_4), a major component of acid rain. Sulfur dioxide and sulfate ions are probably second only to smoking as causes of air-pollution–related health damage. Sulfate particles and droplets also reduce visibility in the United States by as much as 80 percent.

Nitrogen oxides (NO_x) are highly reactive gases formed when combustion initiates reactions between atmospheric nitrogen and oxygen. The initial product, nitric oxide (NO), oxidizes further in the atmosphere to nitrogen dioxide (NO_2), a reddish-brown gas that gives photochemical smog its distinctive color. Because these gases convert readily from one form to the other, the general term NO_x is used to describe these gases. Nitrogen oxides combine with water to form nitric acid (HNO_3), which is

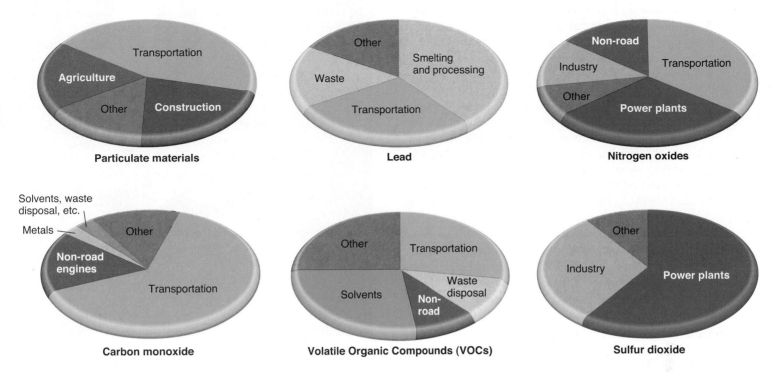

FIGURE 9.20 Anthropogenic sources of six of the primary "criteria" air pollutants in the United States. **Source:** UNEP, 1999.

also a major component of acid precipitation. Excess nitrogen in water is causing eutrophication of inland waters and coastal seas. It may also encourage the growth of weedy species that crowd out native plants.

Carbon monoxide (CO) is less common than the principal form of atmospheric carbon, carbon dioxide (CO_2), but more dangerous. CO is a colorless, odorless, but highly toxic gas produced mainly by incomplete combustion of fuel (coal, oil, charcoal, wood, or gas). CO inhibits respiration in animals by binding irreversibly to hemoglobin. In the United States, two-thirds of the CO emissions are created by internal combustion engines in transportation. Land-clearing fires and cooking fires also are major sources. About 90 percent of the CO in the air is consumed in photochemical reactions that produce ozone.

Particulate material includes dust, ash, soot, lint, smoke, pollen, spores, algal cells, and many other suspended materials. Aerosols, or extremely minute particles or liquid droplets suspended in the air, are included in this class. Particulates often are the most apparent form of air pollution, since they reduce visibility and leave dirty deposits on windows, painted surfaces, and textiles. Breathable particles smaller than 2.5 μm are among the most dangerous of this group because they can damage lung tissues. Asbestos fibers and cigarette smoke are among the most dangerous respirable particles in urban and indoor air because they are carcinogenic.

Volatile organic compounds (VOCs) are organic gases. Plants, bogs, and termites are the largest sources of VOCs, especially isoprenes (C_5H_8), terpenes ($C_{10}H_{15}$), and methane (CH_4).

These volatile hydrocarbons are generally oxidized to CO and CO_2 in the atmosphere.

More dangerous synthetic organic chemicals, such as benzene, toluene, formaldehyde, vinyl chloride, phenols, chloroform, and trichloroethylene, are released by human activities. Principal sources are incompletely burned fuels from vehicles, power plants, chemical plants, and petroleum refineries. These chemicals play an important role in the formation of photochemical oxidants.

Photochemical oxidants are products of secondary atmospheric reactions driven by solar energy (table 9.2). One of the most important of these reactions involves formation of singlet (atomic) oxygen by splitting nitrogen dioxide (NO_2). This atomic oxygen then reacts with another molecule of O_2 to make ozone

TABLE 9.2

Photochemical Oxidant Production

Steps	Photochemical Products
1. NO + VOC →	NO_2 (nitrogen dioxide)
2. NO_2 + UV sunlight →	NO + O (nitric oxide + atomic oxygen)
3. O + O_2 →	O_3 (ozone)
4. NO_2 + VOC →	PAN (peroxyacetyl nitrate)

(O_3). Although ozone is important in the stratosphere, in ambient air it is highly reactive and damages vegetation, animal tissues, and building materials. Ozone's acrid, biting odor is a distinctive characteristic of photochemical smog.

Lead and other toxic elements

Toxic metals and halogens are chemical elements that are toxic when concentrated and released in the environment. Principal metals of concern are lead, mercury, arsenic, nickel, beryllium, cadmium, thallium, uranium, cesium, and plutonium. Halogens (fluorine, chlorine, bromine, and iodine) are highly reactive toxic elements. Most of these materials are mined and used in manufacturing. Metals commonly occur as trace elements in fuels, especially coal.

Lead and mercury are widespread neurotoxins that damage the nervous system. By some estimates, 20 percent of all inner-city children suffer some degree of developmental retardation from high environmental lead levels. Long-range transport of lead and mercury through the air is causing bioaccumulation in remote aquatic ecosystems, such as arctic lakes and seas. Chlorine is a toxic halogen widely used in bleach, plastics, and other products. Methyl bromide (a powerful fungicide used in agriculture) and **chlorofluorocarbons** (propellants and refrigerants) are also implicated in ozone depletion.

Indoor air can be more dangerous than outdoor air

We have spent a considerable amount of effort and money to control the major outdoor air pollutants, but we have only recently become aware of the dangers of indoor air pollutants. The U.S. EPA has found that indoor concentrations of toxic air pollutants are often higher than outdoors. Furthermore, people generally spend more time inside than out and therefore are exposed to higher doses of these pollutants. In some cases, indoor air in homes has chemical concentrations that would be illegal outside or in the workplace. Under some circumstances, compounds such as chloroform, benzene, carbon tetrachloride, formaldehyde, and styrene can be 70 times higher in indoor air than in outdoor air. Molds, pathogens, and other biohazards also represent serious indoor pollutants.

Cigarette smoke is without doubt the most important air contaminant in developed countries in terms of human health. The U.S. surgeon general has estimated that 400,000 people die each year in the United States from emphysema, heart attacks, strokes, lung cancer, or other diseases caused by smoking. These diseases are responsible for 20 percent of all mortality in the United States, or four times as much as infectious agents. Total costs for early deaths and smoking-related illnesses are estimated to be $100 billion per year. Eliminating smoking probably would save more lives than any other pollution-control measure.

In the less-developed countries of Africa, Asia, and Latin America, where such organic fuels as firewood, charcoal, dried dung, and agricultural wastes make up the majority of household energy, smoky, poorly ventilated heating and cooking fires represent the greatest source of indoor air pollution (fig. 9.21).

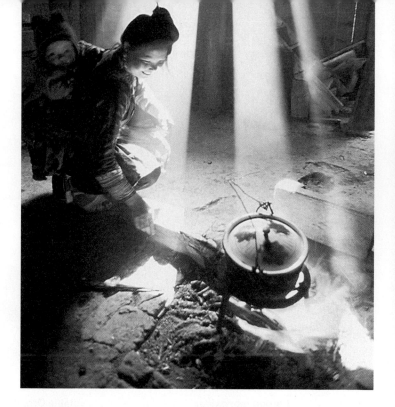

FIGURE 9.21 Some 2.5 billion people, mainly women and children, spend hours each day in poorly ventilated kitchens and living spaces where carbon monoxide, particulates, and cancer-causing hydrocarbons often reach dangerous levels.

The World Health Organization (WHO) estimates that 2.5 billion people—more than one-third of the world's population—are adversely affected by pollution from this source. In particular, women and small children spend long hours each day around open fires or unventilated stoves in enclosed spaces.

INTERACTIONS BETWEEN CLIMATE PROCESSES AND AIR POLLUTION

Physical processes in the atmosphere transport, concentrate, and disperse air pollutants. To comprehend the global effects of air pollution, it is necessary to understand how climate processes interact with pollutants. Global warming in which pollutants are altering the earth's energy budget, is the best-known case of interaction between anthropogenic pollutants and the atmosphere. In this section we will survey other important climate-pollution interactions.

Air pollutants can travel far

Dust and fine aerosols can be carried great distances by the wind. Pollution from the industrial belt between the Great Lakes and the Ohio River Valley regularly contaminates the Canadian Maritime Provinces and sometimes can be traced as far as Ireland. Similarly, dust storms from China's Gobi and Takla Makan Deserts routinely close schools, factories, and airports in Japan and Korea, and often reach western North America. In one particularly severe dust storm in 1998, chemical analysis showed

FIGURE 9.22 A massive dust storm extends more than 1,600 km (1,000 mi) from the coast of western Sahara and Morocco. Storms such as this can easily reach the Americas, and they have been linked both to the decline of coral reefs in the Caribbean and to the frequency and intensity of hurricanes formed in the eastern Atlantic Ocean.

that 75 percent of the particulate pollution in Seattle, Washington, air came from China. Similarly, dust from North Africa regularly crosses the Atlantic and contaminates the air in Florida and the Caribbean Islands (fig. 9.22). This dust can carry pathogens and is thought to be the source of diseases attacking Caribbean corals. Soil scientists estimate that 3 billion tons of sand and dust are blown around the world every year.

Increasingly sensitive monitoring equipment has begun to reveal industrial contaminants in places usually considered among the cleanest in the world. Samoa, Greenland, and even Antarctica and the North Pole all have heavy metals, pesticides, and radioactive elements in their air. Since the 1950s, pilots flying in the high Arctic have reported dense layers of reddish-brown haze clouding the arctic atmosphere. Aerosols of sulfates, soot, dust, and toxic heavy metals, such as vanadium, manganese, and lead, travel to the pole from the industrialized parts of Europe and Russia.

In a process called "grasshopper" transport, or atmosphere distillation, volatile compounds evaporate from warm areas, travel through the atmosphere, then condense and precipitate in cooler regions. Over several years, contaminants accumulate in the coldest places, generally at high latitudes, where they bioaccumulate in food chains. Whales, polar bears, sharks, and other top carnivores in polar regions have been shown to have dangerously high levels of pesticides, metals, and other hazardous air pollutants in their bodies. The Inuit people of Broughton Island, well above the Arctic Circle, have higher levels of polychlorinated biphenyls (PCBs) in their blood than any other known population, except victims of industrial accidents. Far from any source of this industrial by-product, these people accumulate PCBs from the flesh of fish, caribou, and other animals they eat.

Stratospheric ozone is declining

Long-range pollution transport and the chemical reactions of atmospheric gases and pollution produce the phenomenon known as the ozone hole (fig. 9.23). The ozone "hole," really a thinning of ozone concentrations in the stratosphere, was discovered in 1985 but has probably been developing since at least the 1960s. Chlorine-based aerosols, especially chlorofluorocarbons (CFCs) and other halon gases, are the principal agents of ozone depletion. Nontoxic, nonflammable, chemically inert, and cheaply produced, CFCs were extremely useful as industrial gases and in refrigerators, air conditioners, styrofoam inflation, and aerosol spray cans for many years. From the 1930s until the 1980s, CFCs were used all over the world and widely dispersed through the atmosphere.

Although ozone is a pollutant in the ambient air, ozone in the stratosphere is valuable because it absorbs much of the ultraviolet (UV) radiation entering the atmosphere. UV radiation harms plant and animal tissues, including the eyes and the skin. A 1 percent loss of ozone could result in about a million extra human skin cancers per year worldwide if no protective measures are taken. Excessive UV exposure could reduce agricultural production and disrupt ecosystems. Scientists worry, for example, that high UV levels in Antarctica could reduce populations of plankton, the tiny floating organisms that form the base of a food chain that includes fish, seals, penguins, and whales in Antarctic seas.

Antarctica's exceptionally cold winter temperatures (−85° to −90°C) help break down ozone. During the long, dark, winter

FIGURE 9.23 In 2003 stratospheric ozone was depleted over an area (*dark, irregular circle*) that covered 28.5 million square kilometers, or more than the entire Antarctic continent. Although CFC production is declining, this was the second highest area ever recorded.

TABLE 9.3

Stratospheric Ozone Destruction by Chlorine Atoms and UV Radiation

Step	Products
1. $CFCl_3$ (chlorofluorocarbon) + UV energy	$CFCl_2$ + Cl
2. $Cl + O_3$	$ClO + O_2$
3. O_2 + UV energy	2O
4. $ClO + 2O$	O_2 + Cl
5. Return to step 2	

months, strong winds known as the circumpolar vortex isolate antarctic air and allow stratospheric temperatures to drop low enough to create ice crystals at high altitudes—something that rarely happens elsewhere in the world. Ozone and chlorine-containing molecules are absorbed on the surfaces of these ice particles. When the sun returns in the spring, it provides energy to liberate chlorine ions, which readily bond with ozone, breaking it down to molecular oxygen (table 9.3). It is only during the antarctic spring (September through December) that conditions are ideal for rapid ozone destruction. During that season, temperatures are still cold enough for high-altitude ice crystals, but the sun gradually becomes strong enough to drive photochemical reactions.

As the antarctic summer arrives, temperatures moderate somewhat, the circumpolar vortex breaks down, and air from warmer latitudes mixes with antarctic air, replenishing ozone concentrations in the ozone hole. Slight decreases worldwide result from this mixing, however. Ozone re-forms naturally, but not nearly as fast as it is destroyed. Since the chlorine atoms are not themselves consumed in reactions with ozone, they continue to destroy ozone for years, until they finally precipitate or are washed out of the air. Almost every year since it was discovered, the antarctic ozone hole has grown. In 2000 the region of ozone depletion covered 29.8 million km^2 (about the size of North America). In 2005 the hole was nearly as large.

Ominously, this phenomenon is now spreading to other parts of the world as well. About 10 percent of all stratospheric ozone worldwide has been destroyed in recent years, and levels over the Arctic have averaged 40 percent below normal. Ozone depletion has been observed over the North Pole as well, although it is not as concentrated as that in the south.

There are signs of progress

The discovery of stratospheric ozone losses brought about a remarkably quick international response. In 1987 an international meeting in Montreal, Canada, produced the Montreal Protocol, the first of several major international agreements on phasing out most use of CFCs by 2000. As evidence accumulated, showing that losses were larger and more widespread than previously

thought, the deadline for the elimination of all CFCs (halons, carbon tetrachloride, and methyl chloroform) was moved up to 1996, and a $500 million fund was established to assist poorer countries in switching to non-CFC technologies. Fortunately, alternatives to CFCs for most uses already exist. The first substitutes are hydrochlorofluorocarbons (HCFCs), which release much less chlorine per molecule. Eventually, scientists hope to develop halogen-free molecules that work just as well and are no more expensive than CFCs.

There is some evidence that the CFC ban is already having an effect. CFC production in most industrialized countries has fallen sharply since 1989 (fig. 9.24), and CFCs are now being removed from the atmosphere more rapidly than they are being added. In 50 years or so, stratospheric ozone levels are expected to be back to normal. Unfortunately, there may be a downside to stopping ozone destruction. Ozone is a potent greenhouse gas. Some models suggest that lower ozone levels have been offsetting the effects of increased CO_2. When ozone is restored, global warming may be accelerated. As so often is the case, when we disturb one environmental factor, we affect others as well.

Cities create dust domes, smog, and heat islands

Temperature inversions can concentrate dangerous levels of pollutants within cities. Normally, air temperatures decrease with elevation above the earth's surface. An inversion reverses these conditions, with cool, dense air lying below a warmer, lighter layer. In these very stable conditions, convection currents cannot develop, and there is no mechanism for dispersing pollutants. The most stable inversion conditions are usually created by rapid nighttime cooling in a valley or basin where air movement is restricted. Los Angeles is a classic example of an area with conditions that create temperature inversions and photochemical smog. Mountains surround the city on three sides, reducing wind

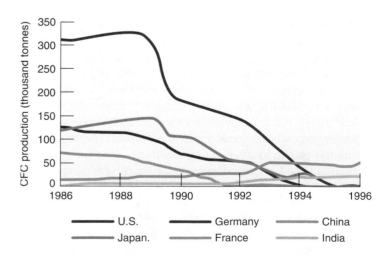

FIGURE 9.24 Production of chlorofluorocarbons has fallen sharply in many countries since the Montreal Protocol was passed in 1987.
Source: Data from United Nations Environment Programme.

Principles of Environmental Science

movement; the climate is sunny; and heavy traffic creates high pollution levels. Skies are generally clear at night, allowing rapid radiant heat loss, and the ground cools quickly. Surface air layers are cooled by conduction, while upper layers remain relatively warm. Density differences retard vertical mixing. During the night, cool, humid, onshore breezes slide in under the contaminated air, squeezing it up against the cap of warmer air above and concentrating the pollutants accumulated during the day.

Morning sunlight initiates photochemical oxidation in the concentrated aerosols and gaseous chemicals in the inversion layer. A brown haze of hazardous chemicals, especially ozone and nitrogen dioxide, quickly develops. Although recent air quality regulations have helped tremendously, on summer days, ozone concentrations in the Los Angeles basin still can reach unhealthy levels.

Heat islands and **dust domes** occur in cities even without inversion conditions. With their low albedo, concrete and brick surfaces in cities absorb large amounts of solar energy. A lack of vegetation or water results in very slight evaporation (latent heat production); instead, available solar energy is turned into heat. As a result, temperatures in cities are frequently 3° to 5°C (5° to 9°F) warmer than in the surrounding countryside, a condition known as an urban heat island. Tall buildings create convective updrafts that sweep pollutants into the air. Stable air masses created by this heat island over the city concentrate pollutants in a dust dome.

EFFECTS OF AIR POLLUTION

Air pollution is equally serious for ecosystem health and for human health. In this section we will review the most important effects of air pollution.

Polluted air is unhealthy

Consequences of breathing dirty air include increased probability of heart attacks, respiratory diseases, and lung cancer. This can mean as much as a five- to ten-year decrease in life expectancy if you live in the worst parts of Los Angeles or Baltimore, for example. Of course, the intensity and duration of your exposure, as well as your age and general health, are extremely important: you are much more likely to be at risk if you are very young, very old, or already suffering from some respiratory or cardiovascular disease. Bronchitis and emphysema are common chronic conditions resulting from air pollution. The U.S. Office of Technology Assessment estimates that 250,000 people suffer from pollution-related bronchitis and emphysema in the United States, and some 50,000 excess deaths each year are attributable to complications of these diseases, which are probably second only to heart attack as a cause of death.

Conditions are often much worse in developing countries. The United Nations estimates that at least 1.3 billion people around the world live in areas where the air is dangerously polluted. In the "black triangle" region of Poland, Hungary, the Czech Republic, and Slovakia, for example, respiratory ailments, cardiovascular diseases, lung cancer, infant mortality, and miscarriages are as much as 50 percent higher than in cleaner parts

FIGURE 9.25 Sulfur dioxide emissions and acid precipitation from the International Nickel Company copper-nickel smelter (*background*) killed all vegetation over a large area near Sudbury, Ontario, in the mid-twentieth century. The scrubbers have dramatically reduced sulfur emissions and vegetation now has recovered most of this area.

of those countries. And in China, city dwellers are four to six times more likely than country folk to die of lung cancer. The World Health Organization estimates that 4 million people die each year from diseases exacerbated by air pollution.

How does air pollution cause these health effects? Because they are strong oxidizing agents, sulfates, SO_2, NO_x, and O_3 irritate and damage delicate tissues in the eyes and lungs. Fine, suspended particulate materials penetrate deep into the lungs, causing irritation, scarring, and even tumor growth. Heart stress results from impaired lung functions. Carbon monoxide binds to hemoglobin, reducing oxygen flow to the brain. Headaches, dizziness, and heart stress result. Lead also binds to hemoglobin, damaging critical neurons in the brain and resulting in mental and physical impairment and developmental retardation.

Plants are sensitive to pollutants

In the early days of industrialization, fumes from furnaces, smelters, refineries, and chemical plants often destroyed vegetation and created desolate, barren landscapes around mining and manufacturing centers. The copper-nickel smelter at Sudbury, Ontario, is a spectacular example. Starting in 1886, open-bed roasting was used to purify sulfide ores of nickel and copper. The resulting sulfur dioxide and sulfuric acid destroyed almost all plant life within about 30 km (18.6 mi) of the smelter. Rains washed away the exposed soil, leaving a barren moonscape of blackened bedrock (fig. 9.25). Recently, emission controls have been introduced, and the environment is beginning to recover, although exposed rock is still black, and the forest cover is mostly small and rather spindly.

There are two probable ways that air pollutants damage plants. They can damage sensitive cell membranes, much as irritants do in human lungs. Toxic levels of oxidants produce discoloration (destruction of chlorophyll) and then necrotic (dead) spots. Often, however, the symptoms are vague and difficult to separate from diseases or insect damage. Pollutants can also act as hormones, disrupting plant metabolism, growth, and development. Ethylene, a volatile organic compound released in automobile exhaust and from petroleum refineries and chemical plants, often injures plants around roads and factories.

Certain combinations of environmental factors have **synergistic effects** in which the injury caused by exposure to two factors together is more than the sum of exposure to each factor individually. For instance, white pine seedlings exposed to subthreshold concentrations of ozone and sulfur dioxide individually do not suffer any visible injury. If the same concentrations of pollutants are given together, however, visible damage occurs.

Pollutant levels too low to produce visible symptoms of damage may still have important effects. Field studies show that yields in some crops, such as soybeans, may be reduced as much as 50 percent by currently existing levels of oxidants in ambient air. Some plant pathologists suggest that ozone and photochemical oxidants are responsible for as much as 90 percent of agricultural, ornamental, and forest losses from air pollution. The total costs of this damage may be as much as $10 billion per year in North America alone.

Smog and haze reduce visibility

We have only recently realized that pollution affects rural areas as well as cities. Even supposedly pristine places such as our national parks are suffering from air pollution. Grand Canyon National Park, where maximum visibility used to be 300 km (185 mi), is now so smoggy on some winter days that visitors can't see the opposite rim only 20 km (12.5 mi) across the canyon. Mining operations, smelters, and power plants (some of which were moved to the desert to improve air quality in cities such as Los Angeles) are the main culprits. Huge regions are affected by pollution. A gigantic "haze blob" as much as 3,000 km (about 2,000 mi) across covers much of the eastern United States in the summer, cutting visibility as much as 80 percent. People become accustomed to these conditions and don't realize that the air once was clear. Studies indicate, however, that, if all human-made sources of air pollution were shut down, the air would clear up in a few days, and there would be about 150-km (90-mi) visibility nearly everywhere, rather than the 15 km to which we have become accustomed.

Acid deposition has many effects

Acid precipitation, the deposition of wet, acidic solutions or dry, acidic particles from the air, became widely recognized as a pollution problem only in the last 20 years. But the concept has been recognized since the 1850s. We describe acidity in terms of pH, with substances below pH 7 being acidic (chapter 2). Normal, unpolluted rain generally has a pH of about 5.6 due to carbonic acid created when rainwater reacts with CO_2 in the air. Downwind of industrial areas, rainfall acidity can reach levels below pH 4.3, more than ten times as acidic as normal rain. Acid fog,

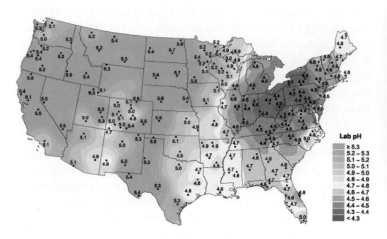

FIGURE 9.26 Acid precipitation over the United States, 2000.
Source: Data from National Acid Deposition Program, 2001.

snow, mist, and dew can deposit damaging acids on plants, in water systems, and on buildings. Furthermore, fallout of dry sulfate and nitrate particles can account for as much as half of the acidic deposition in some areas.

A vigorous program of pollution control has been undertaken by Canada, the United States, and several European countries since the widespread recognition of acid rain. SO_2 and NO_x emissions from power plants have decreased dramatically over the past three decades over much of Europe and eastern North America as a result of pollution-control measures. However, rain falling in these areas remains acidic (fig. 9.26). Apparently, alkaline dust that would once have neutralized acids in air has been depleted by years of acid rain and is now no longer effective.

Aquatic effects of acid deposition

Aquatic ecosystems in Scandinavia were among the first discovered to be damaged by acid precipitation. Prevailing winds from Germany, Poland, and other parts of Europe deliver acids generated by industrial and automobile emissions—principally H_2SO_4 and HNO_3. The thin, acidic soils and nutrient-poor lakes and streams in the mountains of southern Norway and Sweden have been severely affected by this acid deposition. Most noticeable is the reduction of trout, salmon, and other game fish, whose eggs and fry die below pH 5. Aquatic plants, insects, and invertebrates also suffer. Many lakes in Sweden are now so acidic that they will no longer support game fish or other sensitive aquatic organisms. Large parts of Europe and eastern North America have also been damaged by acid precipitation.

Forest damage

Since the early 1980s observers have documented forest declines apparently related to acid damage in both Europe and North America. A detailed 1980 ecosystem inventory on Camel's Hump Mountain in Vermont found that seedling production, tree density,

Principles of Environmental Science

FIGURE 9.27 A Fraser fir forest on Mount Mitchell, North Carolina, killed by acid rain, insect pests, and other stressors.

FIGURE 9.28 Atmospheric acids, especially sulfuric and nitric acids, have almost completely eaten away the face of this medieval statue. Each year, the total losses from air pollution damage to buildings and materials amount to billions of dollars.

and viability of spruce-fir forests at high elevations had declined about 50 percent in 15 years. On Mount Mitchell in North Carolina, nearly all the trees above 2,000 m (6,000 ft) are losing needles, and about half are dead (fig. 9.27). Damage has been reported throughout Europe, from the Netherlands to Switzerland, as well as in China and the states of the former Soviet Union. In 1985 West German foresters estimated that about half the total forest area in West Germany (more than 4 million ha) was declining. The loss to the forest industry is estimated to be about 1 billion DM (deutsche marks) per year.

High-elevation forests are most severely affected. Mountain tops often have thin, acidic soils under normal conditions, with little ability to buffer, or neutralize, acid precipitation. Acidic fog and mist frequently rest on mountaintops, lengthening plants' exposure to acidity. Mountaintops also catch more rain and snow than lowlands, so they are more exposed to acidic precipitation.

The most visible mechanism of forest decline is direct damage to plant tissues and seedlings. Nutrient availability in forest soils is also depleted, however. Acids can also dissolve and mobilize toxic metals, such as aluminum, and weakened trees become susceptible to diseases and insect pests.

Buildings and monuments

In cities throughout the world, air pollution is destroying some of the oldest and most glorious buildings and works of art. Smoke and soot coat buildings, paintings, and textiles. Acids dissolve limestone and marble, destroying features and structures of historic buildings (fig. 9.28). The Parthenon in Athens, the Taj Mahal in Agra, the Coliseum in Rome, medieval cathedrals in Europe, and the Washington Monument in Washington, D.C., are slowly dissolving and flaking away because of acidic fumes in the air. On a more mundane level, air pollution and acid precipitation corrode steel in reinforced concrete, weakening buildings, roads, and bridges. Limestone, marble, and some kinds of sandstone

flake and crumble. The Council on Environmental Quality estimates that U.S. economic losses from architectural damage caused by air pollution amount to about $4.8 billion in direct costs and $5.2 billion in property-value losses each year.

AIR POLLUTION CONTROL

"Dilution is the solution to pollution" was one of the early approaches to air pollution control. Tall smokestacks were built to send emissions far from the source, where they became unidentifiable and largely untraceable. But dispersed and diluted pollutants are now the source of some of our most serious pollution problems. We are finding that there is no "away" to which we can throw our waste products.

The most effective pollution-control strategy is to minimize production

Since most air pollution in the developed world is associated with transportation and energy production, the most effective strategy would be conservation: reducing electricity consumption, insulating homes and offices, and developing better public transportation could all greatly reduce air pollution in the United States, Canada, and Europe. Alternative energy sources, such as wind and solar power, produce energy with little or no pollution, and these and other technologies are becoming economically competitive (chapter 12). In addition to conservation, pollution can be controlled by technological innovation.

FIGURE 9.29 An electrostatic precipitator traps particulate material on electrically charged plates as effluent makes its way to the smokestack.

Labels in figure: Cleaned gas, Electrodes, Dust discharge, Dirty gas

Particulate removal involves filtering air emissions. Filters trap particulates in a mesh of cotton cloth, spun glass fibers, or asbestos-cellulose. Industrial air filters are generally giant bags 10 to 15 m long and 2 to 3 m wide. Effluent gas is blown through the bag, much like the bag on a vacuum cleaner. Every few days or weeks, the bags are opened to remove the dust cake. Electrostatic precipitators are the most common particulate controls in power plants. Ash particles pick up an electrostatic surface charge as they pass between large electrodes in the effluent stream (fig. 9.29). Charged particles then collect on an oppositely charged collecting plate. These precipitators consume a large amount of electricity, but maintenance is relatively simple, and collection efficiency can be as high as 99 percent. The ash collected by both of these techniques is a solid waste (often hazardous due to the heavy metals and other trace components of coal or other ash source) and must be buried in landfills or other solid waste disposal sites.

Sulfur removal is important because sulfur oxides are among the most damaging of all air pollutants in terms of human health and ecosystem viability. Switching from soft coal with a high sulfur content to low-sulfur coal is the surest way to reduce sulfur emissions. High-sulfur coal is frequently politically or economically expedient, however. In the United States, Appalachia, a region of chronic economic depression, produces mostly high-sulfur coal. In China, much domestic coal is rich in sulfur. Switching to cleaner oil or gas would eliminate metal effluents as well as sulfur. Cleaning fuels is an alternative to switching. Coal can be crushed, washed, and gasified to remove sulfur and metals before combustion. This improves heat content and firing properties but may replace air pollution with solid waste and water pollution problems; furthermore, these steps are expensive.

Sulfur can also be removed to yield a usable product instead of simply a waste disposal problem. Elemental sulfur, sulfuric acid, and ammonium sulfate can all be produced using catalytic converters to oxidize or reduce sulfur. Markets have to be reasonably close and fly ash contamination must be reduced as much as possible for this procedure to be economically feasible.

Nitrogen oxides (NO_x) can be reduced in both internal combustion engines and industrial boilers by as much as 50 percent by carefully controlling the flow of air and fuel. Staged burners, for example, control burning temperatures and oxygen flow to prevent formation of NO_x. The catalytic converter on your car uses platinum-palladium and rhodium catalysts to remove up to 90 percent of NO_x, hydrocarbons, and carbon monoxide at the same time.

Hydrocarbon controls mainly involve complete combustion or the control of evaporation. Hydrocarbons and volatile organic compounds are produced by incomplete combustion of fuels or by solvent evaporation from chemical factories, paints, dry cleaning, plastic manufacturing, printing, and other industrial processes. Closed systems that prevent escape of fugitive gases can reduce many of these emissions. In automobiles, for instance, positive crankcase ventilation (PCV) systems collect oil that escapes from around the pistons and unburned fuel and channels them back to the engine for combustion. Controls on fugitive losses from industrial valves, pipes, and storage tanks can have a significant impact on air quality. Afterburners are often the best method for destroying volatile organic chemicals in industrial exhaust stacks.

CLEAN AIR LEGISLATION

Throughout history, countless ordinances have prohibited the emission of objectionable smoke, odors, and noise. Air pollution traditionally has been treated as a local problem, however. The Clean Air Act of 1963 was the first national legislation in the United States aimed at air pollution control. The act provided federal grants to states to combat pollution but was careful to preserve states' rights to set and enforce air quality regulations. It soon became obvious that some pollution problems cannot be solved on a local basis.

In 1970 an extensive set of amendments essentially rewrote the Clean Air Act. These amendments identified the criteria pollutants discussed earlier in this chapter and established primary and secondary standards for ambient air quality. Primary standards (table 9.4) are intended to protect human health, while secondary standards are set to protect materials, crops, climate, visibility, and personal comfort.

Since 1970 the Clean Air Act has been modified, updated, and amended. The most significant amendments were in the 1990 update. Amendments have involved acrimonious debate, with bills sometimes languishing in Congress from one session to the next because of disputes over burdens of responsibility and cost and definitions of risk. A 2002 report concluded that, simply by enforcing existing clean air legislation, the United States could save at least 6,000 lives per year and prevent 140,000 asthma attacks.

TABLE 9.4

National Ambient Air Quality Standards (NAAQS)

Pollutant	Primary (health-based) Averaging Time	Standard Concentration
TSP[1]	Annual geometric mean[2]	50 μg/m^3
	24 hours	150 μg/m^3
SO$_2$	Annual arithmetic mean[3]	80 μg/m^3 (0.03 ppm)
	24 hours	120 μg/m^3 (0.14 ppm)
CO	8 hours	10 mg/m^3 (9 ppm)
	1 hour	40 mg/m^3 (35 ppm)
NO$_2$	Annual arithmetic mean	80 μg/m^3 (0.05 ppm)
O$_3$	Daily max 8 hours avg.	157 μg/m^3 (0.08 ppm)
Lead	Maximum quarterly avg.	1.5 μg/m^3

[1]Total suspended particulate material.

[2]The geometric mean is obtained by taking the nth root of the product of n numbers. This tends to reduce the impact of a few very large numbers in a set.

[3]An arithmetic mean is the average determined by dividing the sum of a group of data points by the number of points.

Air regulations are controversial

In 2002 President Bush announced his "Clear Skies" plan, calling it the "most aggressive initiative in American history to cut power plant emissions." Supporters applauded his declared intent to eliminate "outdated and obstructive features" of the Clean Air Act and to establish a market-based approach for controlling three key air pollutants (sulfur dioxide, nitrogen oxide, and mercury). Opponents, however, criticized the plan as "the biggest rollback of the Clean Air Act in history."

The most controversial aspect of Bush's plan is elimination of the "new source review," which was established in 1977. This provision was originally adopted because industry argued that it would be intolerably expensive to install new pollution-control equipment on old power plants and factories that were about to close down, anyway. Congress agreed to "grandfather" or exempt existing equipment from new pollution limits with the stipulation that, when they were upgraded or replaced, more stringent rules would apply. The result was that owners kept old facilities operating precisely because they were exempted from pollution control. In fact, corporations poured millions into aging power plants and factories, expanding their capacity rather than building new ones. A quarter of a century later, most of those grandfathered plants are still going strong and continue to be among the biggest contributors to smog and acid rain.

The Clinton administration tried to force utilities to install modern pollution control on old plants when they replaced or repaired equipment. President Bush, however, said that determining which facilities are new, and which are not, represents a cumbersome and unreasonable imposition on industries. The EPA subsequently announced it would abandon new source reviews, depending instead on voluntary emissions controls and a trading program for air pollution allowances. Several of the utilities that had already agreed to install new pollution controls immediately put their plans on hold.

Environmental groups generally agree that cap-and-trade (which sets maximum amounts for pollutants and then lets facilities facing costly cleanup bills pay others with lower costs to reduce emissions on their behalf) has worked fairly well for sulfur dioxide. When trading began in 1990, economists estimated that eliminating 10 million tons of sulfur dioxide would cost $15 billion per year. Left to find the most economical ways to reduce emissions, however, utilities have been able to reach clean air goals for one-tenth that price. A serious shortcoming of this approach is that, while trading has resulted in overall pollution reduction, some local "hot spots" remain where owners have found it cheaper to pay someone else to reduce pollution than to do it themselves. Knowing that the average person is enjoying cleaner air isn't much comfort if you're living in one of the persistently dirty areas.

After several failed congressional attempts to pass his "Clear Skies" initiative, which critics said would have done more for the industries that pollute the air (and helped write the legislation) than for the people who breathe it, President Bush announced an alternative plan in 2005. Called the Clean Air Interstate Rule, or CAIR, this new policy essentially eliminates the new source review, but does set limits for emissions of sulfur dioxide, nitrogen oxides, and particulate. Pollution reduction won't come immediately; it will be phased-in over a decade. By 2018, however, SO$_2$, NO$_x$, and particulate emissions are expected to decline about 70 percent. As we mentioned earlier in this chapter, scrubbing out these pollutants is expected to reduce mercury emissions as well. To meet these goals, many plants will have to install new filters and other emissions-capturing equipment. The plan still depends largely on "cap-and-trade" provisions that allow plants that can't meet standards to buy credits from those that are ahead of schedule. CAIR is an interstate rule because many eastern states can't meet air quality standards when pollution drifts across their borders from upwind sources. Under CAIR, those states are allowed flexibility and discretion in how they meet the new standards.

Meanwhile, a new power plant in West Virginia is showing that pollution controls can work. Once ranked second in the United States in toxic mercury emissions, the giant 1,600-megawatt Mountain Storm plant installed sulfur and nitrogen oxide scrubbers in 2004. This system also captures mercury at no extra cost.

Many environmentalists argue that carbon dioxide should have been classified as a pollutant and regulated under this rule because of its role in global warming. Some also complain that market mechanisms allow industry to postpone installing pollution controls and forces residents of many states to continue to breathe dirty air for far longer than in necessary. Still, several environmental groups applaud CAIR as a positive step and much better than the Clear Skies amendment. And industry contends that "command and control" mechanisms aren't effective because they don't provide an incentive to continue to search for new, more efficient means of pollution control. What do you think? Which of these regulatory approaches would you favor? Does the kind of pollutant or its effects influence your answer?

CURRENT CONDITIONS AND FUTURE PROSPECTS

Although the United states has not yet achieved the Clean Air Act goals in many parts of the country, air quality has improved dramatically in the past decade in terms of the major large-volume pollutants. For 23 of the largest U.S. cities, the number of days each year in which air quality reached the hazardous level is down 93 percent from a decade ago. Of 97 metropolitan areas that failed to meet clean air standards in the 1980s, 41 were in compliance in 1991–1992. For many cities, this was the first time they met air quality goals in 20 years. Still, the EPA estimates that some 86 million Americans breathe unhealthy air at least part of the time.

The EPA estimates that, between 1970 and 1994, lead fell 98 percent, SO_2 declined 35 percent, and CO shrank 32 percent (fig. 9.30). Filters, scrubbers, and precipitators on power plants and other large stationary sources are responsible for most of the particulate and SO_2 reductions. Catalytic converters on automobiles are responsible for most of the CO and O_3 reductions.

The only conventional criteria pollutants that have not dropped significantly are particulates and NO_x. Because automobiles are the main source of NO_x, cities such as Nashville, Tennessee, and Atlanta, Georgia, where pollution comes largely from traffic, still have serious air-quality problems. Particulate

matter (mostly dust and soot) is produced by agriculture, fuel combustion, metal smelting, concrete manufacturing, and other activities. Industrial cities, such as Baltimore, Maryland, and Baton Rouge, Louisiana, also have continuing problems. In 1999, for the first time in many years, Houston passed Los Angeles as having the worst air quality in the country. Eighty-five other urban areas are still considered nonattainment regions. In spite of these local failures, however, 80 percent of the United States now meets the National Ambient Air Quality Standards. This improvement in air quality is perhaps the greatest environmental success story in our history.

Air pollution remains a problem in developing countries

The outlook is not as encouraging in other parts of the world. The major metropolitan areas of many developing countries are growing at explosive rates to incredible sizes (chapter 14), and environmental quality is abysmal in many of them. Mexico City remains notorious for bad air. Pollution levels exceed WHO health standards 350 days per year, and more than half of all city children have lead levels in their blood high enough to lower intelligence and retard development. Mexico City's 131,000 industries and 2.5 million vehicles spew out more than 5,500 tons of air pollutants daily. Santiago, Chile, averages 299 days per year on which suspended particulates exceed WHO standards of 90 mg/m^3.

While there are few statistics on China's pollution situation, it is known that many of China's 400,000 factories have no air pollution controls. Experts estimate that home coal burners and factories emit 10 million tons of soot and 15 million tons of sulfur dioxide annually and that emissions have increased rapidly over the past 20 years. Sheyang, an industrial city in northern China, is thought to have the world's worst continuing particulate problem, with peak winter concentrations over 700 mg/m^3 (nine times U.S. maximum standards). Airborne particulates in Sheyang exceed WHO standards on 347 days per year. Beijing, Xi'an, and Guangzhou are nearly as bad (see fig. 9.18). The high incidence of cancer in Shanghai is thought to be linked to air pollution.

As political walls came down across Eastern Europe and the Soviet Union at the end of the 1980s, horrifying environmental conditions in these centrally planned economies were revealed. Inept industrial managers, a rigid bureaucracy, and lack of democracy have created ecological disasters. Where governments own, operate, and regulate industry, there are few checks and balances or incentives to clean up pollution. Much of the Eastern bloc depends heavily on soft, brown coal for its energy, and pollution controls are absent or highly inadequate.

Southern Poland, the northern Czech Republic, and Slovakia were covered most of the time by a cloud of smog from factories and power plants. Acid rain ate away historic buildings and damaged already inadequate infrastructures. Home gardening has been banned in Katowice, Poland, because vegetables raised there have unsafe levels of lead and cadmium. For miles around

FIGURE 9.30 Air pollution trends in the United States, 1970 to 1998. Although population and economic activity increased during this period, emissions of all criteria air pollutants, except for nitrogen oxides and particulate matter, decreased significantly.
Source: Data from U.S. Environmental Protection Agency, 2002.

Principles of Environmental Science

the infamous Romanian "black town" of Copsa Mica, the countryside is so stained by soot that it looks as if someone poured black ink over everything. Birth defects afflict ten percent of infants in northern Bohemia. Workers in factories there get extra hazard pay—burial money, they call it. Life expectancy in these industrial towns is as much as ten years less than the national average.

There are signs of hope

Not all is pessimistic, however. There have been some spectacular successes in air pollution control. Sweden and West Germany (countries affected by forest losses due to acid precipitation) cut their sulfur emissions by two-thirds between 1970 and 1985. Austria and Switzerland have gone even further, regulating even motorcycle emissions. The Global Environmental Monitoring System (GEMS) reports declines in particulate levels in 26 of 37 cities worldwide. Sulfur dioxide and sulfate particles, which cause acid rain and respiratory disease, have declined in 20 of these cities.

Twenty years ago, Cubatao, Brazil, was described as the "Valley of Death," one of the most dangerously polluted places in the world. Steel mills, oil refineries, and fertilizer and chemical factories churned out thousands of tons of air pollutants every year (fig. 9.31). Trees died on the surrounding hills. Birth defects and respiratory diseases were alarmingly high. Since then, however, the citizens of Cubatao have made remarkable progress in cleaning up their environment. The end of military rule and the restoration of democracy allowed residents to publicize their complaints. The environment became an important political issue. The state of São Paulo invested about $100 million and the private sector spent

FIGURE 9.31 Cubatao, Brazil, was once considered one of the most polluted cities in the world. Better environmental regulations and enforcement along with massive investments in pollution control equipment have improved air quality significantly.

twice as much to clean up most pollution sources in the valley. Particulate pollution was reduced 75 percent, ammonia emissions were reduced 97 percent, hydrocarbons that cause ozone and smog were cut 86 percent, and sulfur dioxide production fell 84 percent. Fish are returning to the rivers, and forests are regrowing on the mountains. Progress is possible! We hope that similar success stories will be obtainable elsewhere.

REVIEW QUESTIONS

1. What are the "stabilization wedges" suggested by Professor Socolow and his colleagues at Princeton University? How many wedges do we need to accomplish to flatten our CO_2 emissions?

2. List some things we could do to reverse global climate change.

3. What or where is the *stratosphere*, and why do we care?

4. What is the *greenhouse effect*, and how does it work?

5. Why are we worried about greenhouse gases?

6. How does the Coriolis effect work?

7. What are *monsoons*, and why do they occur?

8. What is the *thermohaline ocean conveyor* and what is happening to it?

9. Describe the El Niño/Southern Oscillation.

10. What gas, action, and country make the largest contribution to global warming?

11. List ten lines of evidence that global warming is already occurring.

12. What did Canadian Prime Minister Paul Martin say to the United States at the 2005 Montreal Conference on Climate Change?

13. What has been the greatest air pollution control success in the United States since 1970?

14. Define *primary air pollutant, secondary air pollutant, photochemical oxidant, point source,* and *fugitive emissions.*

15. What is destroying stratospheric ozone, and where does this happen?

16. What happened to the forest at Sudbury, Ontario?

17. What are the "new source review" and the "Clean Air Interstate Rule"?

18. Why is Cubatao, Brazil, famous? What happened there?

CRITICAL THINKING AND DISCUSSION

Apply the principles you have learned in this chapter to discuss these questions with other students.

1. El Niño is a natural climate process, but some scientists suspect it is changing because of global warming. What sort of evidence would you look for to decide if El Niño has gotten stronger recently?

2. Explain the process of photochemical oxidation. How does it work? What are important examples?

3. One of the problems with the Kyoto Protocol and with the Clean Air Act is that economists and scientists define problems differently and have contrasting priorities. How would an economist and an ecologist explain disputes over the Kyoto Protocol differently?

4. Economists and scientists often have difficulty reaching common terms for defining and solving issues such as the Clean Air Act renewal. How might their conflicting definitions be reshaped to make the discussion more successful?

5. Why do you think controlling greenhouse gases is such a difficult problem? List some of the technological, economic, political, emotional, and other factors involved. Whose responsibility is it to reduce our impacts on climate?

6. Air pollution often originates in one state or country but causes health and crop damage in other areas. For example, mercury from Midwestern power plants is harming plants, water, and health in eastern states. How should states, or countries, negotiate the costs of controlling these pollutants?

7. Global energy resources, including coal-bed methane, tar sands, and coal, are still abundant, but environmental costs of extracting and using them are increasing. What are some of these costs? Is there a way to account for lost environmental resources in setting energy prices?

8. What are some of the reasons for the rising economic costs of hurricanes and other storms? How much of these costs should be borne by local communities in hurricane-prone areas? By individuals? By society as a whole? How much does your answer depend on where you live?

9. Explain how ocean evaporation at the equator helps move heat and winds around the globe. How might higher temperatures influence evaporation, winds, and storms? (See p. 204–205.)

SUGGESTED READINGS

Bryden, H., et al. 2005. Ocean circulation is weakening. *Nature* 438:665–7.

Cryanoski, David. 2005. Climate change: The long-range forecast. *Nature* 438:275–6.

Emanuel, Kerry. 2005. Increasing destructiveness of tropical cyclones over the past 30 years. *Nature* 436:686–8.

Field, David B., et al. 2006. Planktonic foraminifera of the California Current reflect 20th-century warming. *Science* 311 (5757):63–66.

Hopkinson, C. S., and J. J. Vallino. 2005. Efficient export of carbon to the deep ocean through dissolved organic matter. *Nature* 433:142–5.

Martrat, Belen, et al. 2004. Abrupt temperature changes in the western Mediterranean over the past 250,000 years. *Science* 306(5702):1762–5.

Nunes, Flavia, and Richard D. Norris. 2006. Abrupt reversals in ocean overturning during the Palaeocene/Eocene warm period. *Nature* 439:60–63.

Patz, J. A., et al. 2005. Impact of regional climate change on human health. *Nature* 438:310–17.

Reich, Peter B., et al. 2001. Plant diversity enhances ecosystem responses to elevated CO_2 and nitrogen deposition. *Nature* 410:809–10.

Socolow, Robert, et al. 2004. Solving the climate problem: Technologies available to curb CO_2 emissions. *Environment* 46(10):8–19.

Stainforth, D. A., et al. 2005. Uncertainty in predictions of the climate response to rising levels of greenhouse gases. *Nature* 433:403–6.

Wu, Pei-Chih, et al., 2004. Increased levels of ambient fungal spores in Taiwan are associated with dust events from China. *Atmospheric Environment* 38(29):4879–86.

Before doing this exercise, review the El Niño discussion, p. 209. Then visit the website **www.ssmi.com/ssmi_monthly.html**.

Satellite images are one of the principal ways scientists map and monitor environmental conditions worldwide. This exercise uses satellite data gathered by the U.S. government for a variety of environmental monitoring purposes. This website is one of many excellent sources providing up-to-date environmental data. The monthly "static data" website provides graphic images of satellite data. If you are using a slow server, you may want to save a few of the images on your computer for answering the questions that follow.

For the following questions you will mostly be looking at *atmospheric water vapor,* or humidity. Keep this in mind: humidity is higher where evaporation is high, so *high humidity* indicates *warm water;* low humidity indicates cooler water.

The Special Sensor Microwave Imager (SSMI) monthly data page lets you select a month and year to view and provides four types of atmospheric information.

1. What are the four types of information shown? What units are used to measure these data types? (How fast is 1 wind-speed unit, compared with how fast you walk?)

2. Look at the Atmospheric Water Vapor map for the current month. Bright red and yellow colors indicate high humidity in the air—and high temperatures. Where is the greatest concentration of red and yellow?

3. Look at December 1998. Then change to December 1997. (Each time you change years or months, you must click on the "Update Display" button to redraw.) Sketch on a piece of paper the general location of the red concentration in the Indian and Pacific Oceans in 1998 and in 1997. How do the two years differ? December 2002 and December 2003 also make a good pair. Which was an El Niño?

4. Now look at the surface wind-speed maps for 1998 and 1997. Which year has the lowest wind speeds in the Pacific Ocean? In the Indian Ocean? Can you think of how wind speeds and ocean temperatures might be related?

5. Go backward, year by year, to see how the concentration of red changes over the years. For each year, mark an *X* on the following graph to show the approximate location of the biggest red concentration in the Pacific Ocean. During which years does the region of strong red/yellow expand farthest eastward?

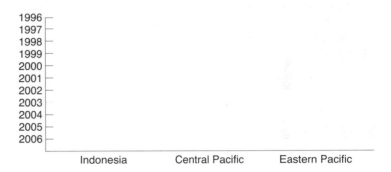

6. If you have a fast network connection, start with January 1997 (or an earlier year) and hit the "next month" button to watch conditions change through the years. During which months do the 1998 El Niño conditions really start to develop?

For Additional Help in Studying This Chapter, please visit our website at www.mhhe.com/cunningham4e. You will find practice quizzes, a chapter summary, key terms, answers to review questions, additional case studies, regional examples, and an extensive reading list, all of which will help you understand the material in this chapter.

Iguazu Falls on the Brazil/Argentina border has 275 individual falls
spread along a 4-km-long cliff.

C H A P T E R

Water

10

Resources and Pollution

If there is magic on this planet, it is in water.

–Loren Eisley–

OBJECTIVES

Studying this chapter will help you answer the following questions:

- Where does our water come from? How do we use it?

- Where and why do water shortages occur?

- How can we increase water supplies? What are some costs of
these methods?

- How can *you* conserve water?

- What is *water pollution*? What are its sources and effects?

- Why are sewage treatment and clean water important in
developing countries?

- How can we control water pollution?

Case Study   Sharing the Chattahoochee

Beginning from a small spring in northern Georgia's Blue Ridge Mountains, the Chattahoochee River provides drinking water, irrigation, hydropower, wildlife habitat, and recreation for millions of people. As it tumbles down from the mountains, the Chattahoochee is a clean, cool, swift-flowing stream, a popular destination for fishing, swimming, and boating. Far downstream, the Chattahoochee joins the Flint to form the Apalachicola River, which flows across the Florida Panhandle into Apalachicola Bay, in the Gulf of Mexico. Together, the Apalachicola, Chattahoochee, and Flint (ACF) watershed supports rich and diverse ecosystems, including valuable marine fisheries for oysters, shrimp, and finfish.

But between the mountains and the sea, the Chattahoochee River passes through Atlanta, where it undergoes dramatic changes. Atlanta is one of America's fastest-growing cities. Expanding suburbs and industries crowd the Chattahoochee's banks, producing runoff of silt, salts, and yard fertilizers that wash into the river. Growing numbers of households and industries withdraw water from the river, which becomes lower, slower, and warmer. Additional contaminants mix with remaining water—oils, metals, and dust from city streets and storm sewers, as well as chlorine and other contaminants from sewage treatment plants. The warm, shallow, turbid water holds less oxygen and supports fewer of the river's native plants and animals. The Environmental Protection Agency has named the 100 km of the Chattahoochee south of Atlanta one of the five most polluted river segments in the United States.

In an effort to force action, the EPA has levied millions of dollars in fines on Atlanta for polluting the river. Atlanta dilutes the contaminated waters by releasing more clean water from reservoirs upstream, but the city government says little can be done to prevent widespread runoff from the large and growing city. These diffuse sources of pollutants, running off of roads, yards, and parking lots, can be much harder to control than localized sources, such as factory pipe outlets.

Meanwhile, Alabama and Florida complain that Georgia is taking too much water. They say the 2.3 billion liters (600 million gal) of water

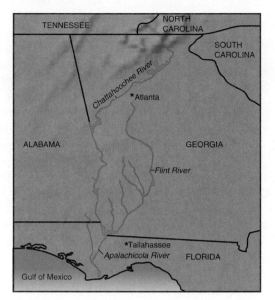

Rising in the mountains of northern Georgia, the Chattahoochee flows south through Atlanta before joining the Flint to form the Apalachicola River.

withdrawn each year in Georgia is needed downstream for homes, industries, and farms. Low river flows and contaminated water also threaten wildlife, shipping, and recreation. The Apalachicola Bay is particularly endangered. Pollution, reduced river flow, and increasing salinity in the estuary jeopardize the bay's multimillion-dollar-per-year fishing and tourism industries.

Atlanta recognized the looming problem of water supplies in the 1970s, and the city has been working with the Army Corps of Engineers to develop additional water supplies. In 2004 Georgia began working on a comprehensive state water plan, which is scheduled to be completed in four years. Ecologists worry that wildlife and biodiversity won't be adequately represented in this planning. Meanwhile, hydrologists point out that climate change could result in more erratic, less dependable rainfall in the future (chapter 9). If river flows become more unpredictable as a result, water management could become an even more urgent and difficult concern.

In the United States, water shortages have long been a western problem, with growing cities in California and other western states struggling to acquire enough water without destroying river ecosystems. But like Atlanta, many growing metropolitan areas in other regions are increasingly competing with other users for water. Around the world, cities are experiencing water shortages, while pollution makes the water we do have less useful. The United Nations warns that water supplies are likely to become one of the most pressing environmental issues of the twenty-first century. By 2025 two-thirds of all humans could be living in countries where water resources are inadequate. In this chapter, we'll look at where our fresh water comes from, what we do with it, and how we can protect its quality and extend its usefulness.

WATER RESOURCES

Water is a marvelous substance—flowing, swirling, seeping, constantly moving from sea to land and back again. It shapes the earth's surface and moderates our climate. Water is essential for life. It is the medium in which all living processes occur (chapter 2). Water dissolves nutrients and distributes them to cells, regulates body temperature, supports structures, and removes waste products. About 60 percent of your body is water. You could survive for weeks without food, but only a few days without water.

Water also is needed for agriculture, industry, transportation, and a host of other human uses. In short, clean freshwater is one of our most vital natural resources.

The hydrologic cycle constantly redistributes water

The water we use cycles endlessly through the environment. The total amount of water on our planet is immense—more than 1,404 million km³ (370 billion billion gal) (table 10.1). This water

TABLE 10.1

Units of Water Measurement

One cubic kilometer (km^3) equals 1 billion cubic meters (m^3), 1 trillion liters, or 264 billion gal.

One acre-foot is the amount of water required to cover an acre of ground 1 ft deep. This is equivalent to 325,851 gal, or 1.2 million liters, or 1,234 m^3, approximately the amount consumed annually by a family of four in the United States.

One cubic foot per second of river flow equals 28.3 liters per second, or 449 gal per minute.

evaporates from moist surfaces, falls as rain or snow, passes through living organisms, and returns to the ocean in a process known as the **hydrologic cycle** (see fig. 2.19). Every year, about 500,000 km^3, or a layer 1.4 m thick, evaporates from the oceans. More than 90 percent of that moisture falls back on the ocean. The 47,000 km^3 carried onshore joins some 72,000 km^3 that evaporate from lakes, rivers, soil, and plants to become our annual, renewable freshwater supply. Plants play a major role in the hydrologic cycle, absorbing groundwater and pumping it into the atmosphere by transpiration (transport plus evaporation). In tropical forests, as much as 75 percent of annual precipitation is returned to the atmosphere by plants.

Solar energy drives the hydrologic cycle by evaporating surface water, which becomes rain and snow. Because water and sunlight are unevenly distributed around the globe, water resources are very uneven. At Iquique in the Chilean desert, for instance, no rain has fallen in recorded history. At the other end of the scale, 22 m (72 ft) of rain was recorded in a single year at Cherrapunji in India. Figure 10.1 shows broad patterns of precipitation around the world. Most of the world's rainiest regions are tropical, where heavy rainy seasons occur, or in coastal mountain regions. Most of the driest areas are in the high-pressure bands of deserts (chapter 9). Deserts occur on every continent just outside the tropics (the Sahara, the Namib, the Gobi, the Sonoran, and many others). Rainfall is also slight at very high latitudes, another high-pressure region.

Mountains also influence moisture distribution. The windward sides of mountain ranges, including the Pacific Northwest and the flanks of the Himalayas, are typically wet and have large rivers; on the leeward sides of mountains, in areas known as the rain shadow, dry conditions dominate, and water can be very scarce. The windward side of Mount Waialeale on the island of Kauai, for example, is one of the wettest places on earth, with an annual rainfall around 1,200 cm (460 in.). The leeward side, only a few kilometers away, has an average yearly rainfall of only 46 cm (18 in.).

An important consideration in water availability is whether there is rainfall year-round, especially in the growing season. Another question is whether hot, dry weather evaporates available moisture. These factors help determine the amount and variety of biological activity in a place (chapter 5).

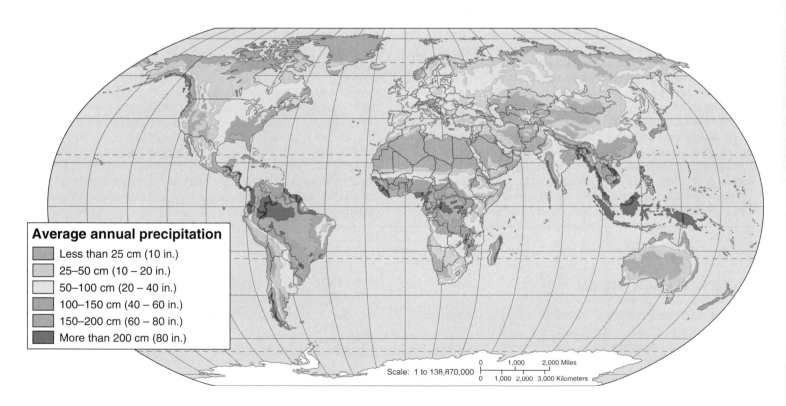

Average annual precipitation

- Less than 25 cm (10 in.)
- 25–50 cm (10 – 20 in.)
- 50–100 cm (20 – 40 in.)
- 100–150 cm (40 – 60 in.)
- 150–200 cm (60 – 80 in.)
- More than 200 cm (80 in.)

Scale: 1 to 138,870,000

FIGURE 10.1 Average annual precipitation. Note wet areas that support tropical rainforests occur along the equator, while the major world deserts occur in zones of dry, descending air between 20° and 40° north and south.

Principles of Environmental Science http://www.mhhe.com/cunningham4e

TABLE 10.2

Earth's Water Compartments

Compartment	Volume (1,000 km³)	Percent of Total Water	Average Residence Time
Total	1,386,000	100	2,800 years
Oceans	1,338,000	96.5	3,000 to 30,000 years*
Ice and snow	24,364	1.76	1 to 100,000 years*
Saline groundwater	12,870	0.93	Days to thousands of years*
Fresh groundwater	10,530	0.76	Days to thousands of years*
Fresh lakes	91	0.007	1 to 500 years*
Saline lakes	85	0.006	1 to 1,000 years*
Soil moisture	16.5	0.001	2 weeks to 1 year*
Atmosphere	12.9	0.001	1 week
Marshes, wetlands	11.5	0.001	Months to years
Rivers, streams	2.12	0.0002	1 week to 1 month
Living organisms	1.12	0.0001	1 week

Source: Data from UNEP, 2002.

*Depends on depth and other factors.

MAJOR WATER COMPARTMENTS

The distribution of water often is described in terms of interacting compartments in which water resides, sometimes briefly and sometimes for eons (table 10.2). The length of time water typically stays in a compartment is its **residence time.** On average, a water molecule stays in the ocean for about 3,000 years, for example, before it evaporates and starts through the hydrologic cycle again. Nearly all the world's water is in the oceans (fig. 10.2). Oceans play a crucial role in moderating the earth's temperature, and over 90 percent of the world's living biomass is contained in the oceans. What we mainly need, though, is fresh water. Of the 2.4 percent that is fresh, most is locked up in glaciers or in groundwater. Amazingly, only about .02 percent of the world's water is in a form accessible to us and to other organisms that rely on fresh water (fig. 10.3).

Groundwater stores most fresh, liquid water

Originating as precipitation that percolates into layers of soil and rock, groundwater makes up the largest compartment of liquid, fresh water. The groundwater within 1 km of the surface is more than 100 times the volume of all the freshwater lakes, rivers, and reservoirs combined.

Plants get moisture from a relatively shallow layer of soil containing both air and water, known as the **zone of aeration** (fig. 10.4). Depending on rainfall amount, soil type, and surface topography, the zone of aeration may be a few centimeters or many meters deep. Lower soil layers, where all soil pores are filled with water, make up the **zone of saturation,** the source of water in most wells; the top of this zone is the **water table.**

Geologic layers that contain water are known as **aquifers.** Aquifers may consist of porous layers of sand or gravel or of

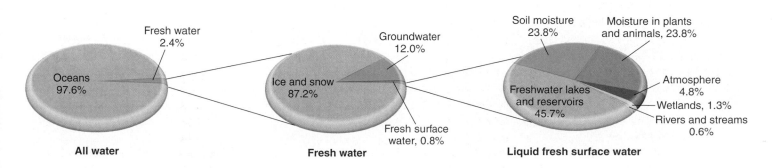

FIGURE 10.2 Less than 1 percent of fresh water, and less than 0.02 percent of all water, is fresh, liquid surface water on which terrestrial life depends. **Source:** U.S. Geological Survey.

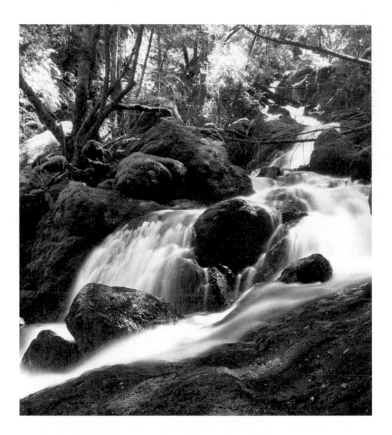

FIGURE 10.3 Water is essential for life, yet only about .02 percent of the world's supply is accessible, fresh, liquid water.

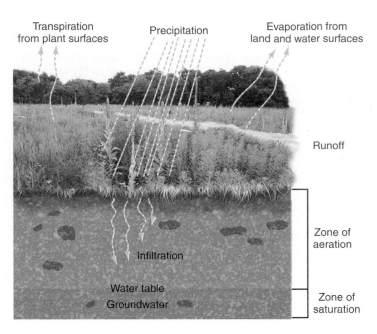

FIGURE 10.4 Precipitation that does not evaporate or run off over the surface percolates through the soil in a process called infiltration. The upper layers of soil hold droplets of moisture between air-filled spaces. Lower layers, where all spaces are filled with water, make up the zone of saturation, or groundwater.

cracked or porous rock. Below an aquifer, relatively impermeable layers of rock or clay keep water from seeping out at the bottom. Instead, water seeps more or less horizontally through the porous layer. Depending on geology, it can take from a few hours to several years for water to move a few hundred meters through an aquifer. If impermeable layers lie above an aquifer, pressure can develop within the water-bearing layer. Pressure in the aquifer can make a well flow freely at the surface. These free-flowing wells and springs are know as *artesian* wells or springs.

Areas where surface water filters into an aquifer are **recharge zones** (fig. 10.5). Most aquifers recharge extremely slowly, and road and house construction or water use at the surface can further slow recharge rates. Contaminants can also enter aquifers through recharge zones. Urban or agricultural runoff in recharge zones is often a serious problem. About 2 billion

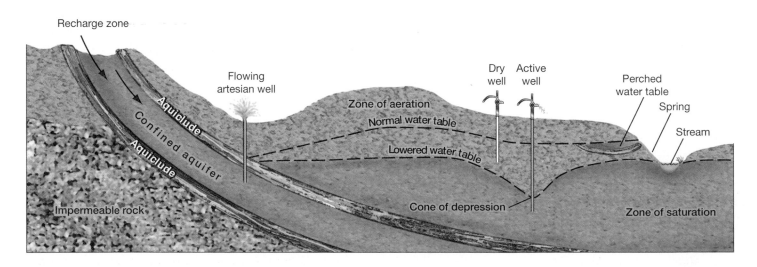

FIGURE 10.5 An aquifer is a porousor cracked layer of rock. Impervious rock layers (aquicludes) keep water within a confined aquifer. Pressure from uphill makes an artesian well flow freely. Pumping can create a cone of depression, which leaves shallower wells dry.

people—approximately one-third of the world's population—depend on groundwater for drinking and other uses. Every year 700 km^3 are withdrawn by humans, mostly from shallow, easily polluted aquifers.

Rivers, lakes, and wetlands cycle quickly

Fresh, flowing surface water is one of our most precious resources. Rivers contain a minute amount of water at any one time. Most rivers would begin to dry up in weeks or days if they were not constantly replenished by precipitation, snowmelt, or groundwater seepage.

We compare the size of rivers in terms of **discharge,** or the amount of water that passes a fixed point in a given amount of time. This is usually expressed as liters or cubic feet of water per second. The 16 largest rivers in the world carry nearly half of all surface runoff on the earth, and a large fraction of that occurs in a single river, the Amazon, which carries 10 times as much water as the Mississippi (table 10.3).

Lakes contain nearly 100 times as much water as all rivers and streams combined, but much of this water is in a few of the world's largest lakes. Lake Baikal in Siberia, the Great Lakes of North America, the Great Rift Lakes of Africa, and a few other lakes contain vast amounts of water, not all of it fresh. Worldwide, lakes are almost as important as rivers in terms of water supplies, food, transportation, and settlement.

Wetlands—bogs, swamps, wet meadows, and marshes—play a vital and often unappreciated role in the hydrologic cycle. Their lush plant growth stabilizes soil and holds back surface runoff, allowing time for infiltration into aquifers and producing even, year-long stream flow. When wetlands are disturbed, their natural water-absorbing capacity is reduced, and surface waters run

TABLE 10.3

World's Ten Largest Rivers

River	Location	Annual Discharge (m^3/second)*
Amazon	Brazil, Peru	175,000
Orinoco	Venezuela, Colombia	45,300
Congo	Congo	39,200
Yangtze	Tibet, China	28,000
Brahmaputra	South Asia	19,000
Mississippi	United States	18,400
Mekong	Southeast Asia	18,300
Paraná	Paraguay, Argentina	18,000
Yenisey	Russia	17,200
Lena	Russia	16,000

Source: Data from World Resource Institute.

*1 m^3 = 264 gal.

off quickly, resulting in floods and erosion during the rainy season and low stream flow the rest of the year.

The atmosphere is one of the smallest compartments

The atmosphere contains only 0.001 percent of the total water supply, but it is the most important mechanism for redistributing water around the world. An individual water molecule resides in the atmosphere for about ten days, on average. Some water evaporates and falls within hours. Water can also travel halfway around the world before it falls, replenishing streams and aquifers on land.

WATER AVAILABILITY AND USE

Clean, fresh water is essential for nearly every human endeavor (fig. 10.6). Collectively, we now appropriate more than half of all the freshwater in the world. Perhaps more than any other environmental factor, the availability of water determines the location and activities of humans on the earth. **Renewable water supplies** are resources that are replenished regularly—mainly surface water and shallow groundwater. Renewable water is most plentiful in the tropics, where rainfall is heavy, followed by midlatitudes, where rainfall is regular.

Water-poor countries have little rain and large populations

Water availability is usually measured in terms of renewable water per capita, so population density, as well as total water volumes, dictate renewable supplies for human use. The highest per

FIGURE 10.6 We depend on fresh water in many ways, but for billions of people, water shortages are a brutal fact of life.

Legend:
- ★ State capitols
- • Major cities
- — Major rivers
- ▨ Indian lands and native entities
- ☐ States

Water supply issue areas
- Unmet rural water needs
- Conflict potential — Moderate
- Conflict potential — Substantial
- Conflict potential — Highly likely

FIGURE 10.7 Rapidly growing populations in arid regions are straining available water supplies. By 2025, the Department of the Interior warns, shortages could cause conflicts in many areas.
Source: Data from U.S. Department of Interior.

capita water supplies generally occur in countries with moist climates and low population densities. Iceland, for example, has about 160 million gal per person per year. In contrast, Kuwait, where temperatures are extremely high and rain almost never falls, has less than 3,000 gal per person per year from renewable natural sources. Almost all of Kuwait's water comes from imports and desalinized seawater. In countries such as Libya and Israel, where water is one of the most crucial environmental resources, renewable supplies do not meet basic needs. These countries manage by "mining" groundwater, depleting sources that are probably unrenewable on a human time scale.

Much of the western United States has insufficient water to meet all the demands placed on this vital resource. The U.S. Department of the Interior warns that by 2025 many western states will face water crises (fig. 10.7). After five years of severe drought, flow in the Colorado River has decreased so much that Lake Powell has lost 60 percent of its volume and the lake surface has dropped more than 33 m (100 ft). Electric generation could cease in four years, if current conditions continue, and the reservoir may never refill. Serious suggestions have been made to remove the dam and let the river run free.

As you can see, interannual variability in rainfall is an important issue in water availability. In the African Sahel region, like the American southwest, abundant rainfall occurs some years but not others. Usually natural ecosystems can survive these changes, but human societies, and ecosystems greatly altered by grazing, farming, or urban development, can be badly destabilized by rainfall fluctuations. Many climatologists now worry that the greenhouse effect (chapter 9) will bring about more serious or frequent droughts in dry parts of the world.

Agriculture is our greatest water user

In contrast to energy resources, which usually are consumed when used, water can be used over and over if it is not too badly contaminated. Water **withdrawal** is the total amount of water taken from a water body. Much of this water could be returned to circulation in a reusable form. Water **consumption,** on the other hand, is loss of water due to evaporation, absorption, or contamination.

The natural cleansing and renewing functions of the hydrologic cycle replace the water we need if natural systems are not overloaded or damaged. Water is a renewable resource, but renewal takes time. The rate at which many of us now use water may make it necessary to conscientiously protect, conserve, and replenish our water supply.

Water use has been increasing about twice as fast as population growth over the past century. Water withdrawals are expected to continue to grow as more land is irrigated to feed an expanding population (fig. 10.8). Conflicts increase as different countries, economic sectors, and other stakeholders compete for the same, limited water supply. Water wars may well be the major source of hostilities in the twenty-first century.

(a) Flood irrigation

(b) Center pivot sprinkler with low sprayers

(c) Drip irrigation

FIGURE 10.8 Growth of water withdrawal and consumption, by sector, with projected levels to 2025.
Source: UNEP, 2002.

FIGURE 10.9 Agricultural irrigation consumes more water than any other use. Methods vary from flood and furrow (a), which use extravagant amounts of water but also flush salts from soils, to sprinklers (b), to highly efficient drip systems (c).

Worldwide, agriculture claims about 70 percent of total water withdrawal, ranging from 93 percent of all water used in India to only 4 percent in Kuwait, which cannot afford to spend its limited water on crops. In many developing countries and in parts of the United States, the most common type of irrigation is to simply flood the whole field or run water in rows between crops. As much as half the water can be lost through evaporation or seepage from unlined irrigation canals bringing water to fields. Sprinklers are more efficient in distributing water, but they are more costly and energy intensive (fig. 10.9). Water-efficient drip irrigation can save significant amounts of water but currently is used on only about 1 percent of the world's croplands.

Industry uses about one-fourth of water withdrawals worldwide. Some European countries use 70 percent of water for industry; less-industrialized countries use as little as 5 percent. Cooling water for power plants is by far the largest single industrial use of water, typically accounting for 50 to 75 percent of industrial withdrawal.

Domestic, or household, water use accounts for only about 6 percent of world water use. This includes water for drinking, cooking, and washing. The amount of water used per household varies enormously, however, depending on a country's wealth. The United Nations reports that people in developed countries consume on average, about ten times more water daily than those in developing nations. Poorer countries can't afford the infrastructure to obtain and deliver water to citizens. Inadequate water supplies, on the other hand, prevent agriculture, industry, sanitation, and other developments that reduce poverty.

FRESHWATER SHORTAGES

Clean drinking water and basic sanitation are necessary to prevent communicable diseases and to maintain a healthy life. For many of the world's poorest people, one of the greatest environmental threats to health remains the continued use of polluted water. In 2004 the United Nations estimated that at least 1.5 billion people lacked access to safe drinking water and 3 billion didn't have adequate sanitation. These deficiencies result in hundreds of millions of cases of water-related illness and more than 5 million deaths every year. As populations grow, more people move into cities, and agriculture and industry compete for increasingly scarce water supplies, water shortages are expected to become even worse.

By 2025 two-thirds of the world's people will be living in water-stressed countries—defined by the United Nations as consumption of more than 10 percent of renewable freshwater resources. One of the highest priorities announced at the UN World Summit in Johannesburg in 2002 was to reduce by one-half the proportion of people without reliable access to clean water and improved sanitation.

There have been many attempts to enhance local supplies and redistribute water. Towing icebergs from Antarctica has been proposed, and creating rain in dry regions has been accomplished, with mixed success, by cloud seeding—distributing condensation nuclei in humid air to help form raindrops. Desalination is locally important: in the arid Middle East, where energy and money are available but water is scarce, desalination is sometimes the principal source of water. Some American cities, such as San Diego, also depend partly on energy-intensive desalination.

Many people lack access to clean water

The World Health Organization considers an average of 1,000 m^3 (264,000 gal) per person per year to be a necessary amount of water for modern domestic, industrial, and agricultural uses. Some 45 countries, most of them in Africa or the Middle East, cannot meet the minimum essential water needs of all their citizens. In some countries, the problem is access to *clean* water. In Mali, for example, 88 percent of the population lacks clean water; in Ethiopia, it is 94 percent. Rural people often have less access to clean water than do city dwellers. Causes of water shortages include natural deficits, overconsumption by agriculture or industry, and inadequate funds for purifying and delivering good water.

More than two-thirds of the world's households have to fetch water from outside the home (fig. 10.10). This is heavy work, done mainly by women and children and sometimes taking several hours a day. Improved public systems bring many benefits to these poor families.

Availability does not always mean affordability. A typical poor family in Lima, Peru, for instance, uses one-sixth as much water as a middle-class American family but pays three times as much for it. If they followed government recommendations to boil all water to prevent cholera, up to one-third of the poor family's income could be used just in acquiring and purifying water.

Investments in rural development have brought significant improvements in recent years. Since 1990, nearly 800 million people—about 13 percent of the world's population—have gained access to clean water. The percentage of rural families with safe drinking water has risen from less than 10 percent to nearly 75 percent.

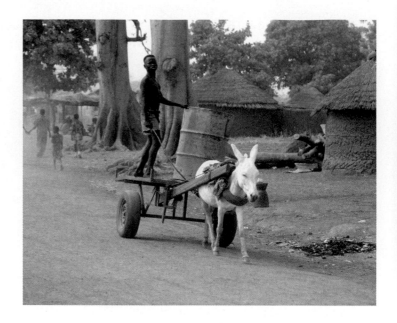

FIGURE 10.10 Village water supplies in Ghana.

Agriculture is reducing groundwater supplies

Groundwater provides nearly 40 percent of the fresh water for agricultural and domestic use in the United States. Nearly half of all Americans and about 95 percent of the rural population depend on groundwater for drinking and other domestic purposes. Overuse of these supplies dries up wells, natural springs, and even groundwater-fed wetlands, rivers, and lakes. Pollution of aquifers through dumping of contaminants on recharge zones, leaks through abandoned wells, or deliberate injection of toxic wastes can make this valuable resource unfit for use.

In many areas of the United States, groundwater is being withdrawn from aquifers faster than natural recharge can replace it. On a local level, this causes a cone of depression in the water table. On a broader scale, heavy pumping can deplete a whole aquifer. The Ogallala Aquifer underlies eight Great Plains states from Texas to North Dakota. This porous bed of sand, gravel, and sandstone once held more water than all the freshwater lakes, streams, and rivers on the earth. Excessive pumping for irrigation has removed so much water that wells have dried up in many places, and farms, ranches, even whole towns are being abandoned. Recharging many such aquifers will take thousands of years. Using "fossil" water like this is essentially water mining. For all practical purposes, these aquifers are nonrenewable resources.

Water withdrawal also allows aquifers to collapse. Subsidence, or sinking of the ground surface, follows. The San Joaquin Valley in California has sunk more than 10 m in the past 50 years because of excessive groundwater pumping. Where aquifers become compressed, recharge becomes impossible.

Another consequence of aquifer depletion is saltwater intrusion. Along coastlines and in areas where saltwater deposits are left from ancient oceans, overuse of freshwater reservoirs often allows saltwater to intrude into aquifers used for domestic and agricultural purposes.

Dams, reservoirs, and canals redistribute water

Dams and canals are a foundation of civilization because they store and redistribute water for farms and cities. Many great civilizations have been organized around large-scale canal systems, including ancient empires of Sumeria, Egypt, and the Inca. As modern dams and water diversion projects have grown in scale and number, though, their environmental costs have raised serious questions about efficiency, costs, and the loss of river ecosystems.

More than half of the world's 227 largest rivers have been dammed or diverted. Of the 50,000 large dams in the world, 90 percent were built in the twentieth century. Half of those are in China, and China continues to build and plan dams on its remaining rivers. Dams are justified in terms of flood control, water storage, and electricity production. However, the costs of relocating villages, as well as lost fishing, farming, and water losses to evaporation, are enormous. Economically speaking, at least one-third of the world's large dams should never have been built (see related story, "South Water North" at www.mhhe.com/cunningham4e).

One of the most dramatic water diversion histories is that of the Aral Sea. Situated in arid Central Asia, on the border of Kazakhstan and Uzbekistan, the Aral Sea is a shallow inland sea fed by rivers from distant mountains. Starting in the 1950s, the Soviet Union began diverting these rivers to water cotton fields and rice fields. Gradually, the Aral Sea has evaporated, leaving vast, toxic salt flats (fig. 10.11). The economic value of the cotton and rice has probably never met the cost of lost fisheries, villages, and health. Similar problems have beset the Colorado River delta, the Dead Sea, Africa's Lake Chad, and many other places.

Questions of justice often surround dam projects

While dams provide hydroelectric power and water to distant cities, local residents often suffer economic and cultural losses. In some cases, dam builders have been charged with using public money to increase the value of privately held farmlands (see What Do You Think? p. 242),

as well as encouraging inappropriate farming and urban growth in arid lands.

China's Three Gorges Dam, scheduled to be complete in 2009, is the world's largest and probably most controversial dam. This mile-wide dam on the Yangtze River inundates 550 km (350 mi) of the river, including three narrow gorges, famous for their dramatic scenery and rich archeological heritage, as well as scores of cities and towns. The project involves relocating between 1 and 2 million people—estimates vary. Many of those people are farmers, who are unlikely to be able to find new farmland and are unprepared for factory jobs (see related story "Three Gorges" at www.mhhe.com/cunningham4e). More recently, China has proposed a series of 13 dams on the Nu River, which has been called one of the last undeveloped rivers in Asia.

On India's Narmada River, a proposed series of 30 dams has incited bitter protest. Many of the 1 million villagers and tribal people being displaced by this project have engaged in civil disobedience and mass protests, which have gone on for the past 15 years or more. In neighboring Nepal, there are fears

1975

1987

1997

2005

FIGURE 10.11 For 30 years, rivers feeding the Aral Sea have been diverted to irrigate cotton and rice fields. The Aral Sea has lost more than 60 percent of its water. Dust storms from remaining salt flats now contaminate irrigated fields.

What Do You Think?

Water Wars on the Klamath

Until a century ago, the Klamath River was the third most productive salmon fishery south of the Canadian border. The 35 km (22 mi) long Upper Klamath Lake, the largest lake in the Pacific Northwest, once contained great schools of C'wam and Qapdo (Lost River and shortnose suckers). Millions of hectares of marshes, lakes, and dry steppe in the Klamath Basin teemed with waterfowl and wildlife. At least 80 percent of the birds following the Pacific Flyway stopped to feed in the area. Native American tribes, including the Klamath and Paiute in the upper basin, and the Yurok and Hupa downstream, depended on the abundant fish and wildlife for their survival. the United States' first waterfowl refuge, the Lower Klamath Wildlife Refuge, was established in 1908 to protect the area's critical habitat.

In 1905 the newly formed United States Bureau of Reclamation was directed to "reclaim the sunbaked prairies and worthless swamps" in the upper Klamath Basin. Spending $50 million, the bureau built 7 dams, 45 pumping stations, and more than 1,600 km (1,000 mi) of canals and ditches. The project drained three-quarters of the wetlands in the upper basin, and provided irrigation water to 90,000 ha (220,000 acres) of cropland. Promises of cheap land and subsidized water lured some 1,400 farmers to the valley to grow potatoes, alfalfa, sugar beets, mint, onions, and cattle. In the 1990s, irrigators in the Klamath Basin used almost 1 million acre-feet (325 billion gallons or 1.2 trillion liters) of surface water per year.

These water diversions left salmon stranded in dried-up streams, while oxygen-depleted lake water was contaminated with agricultural runoff and clogged with algae. Downstream, Native American tribes and commercial fishermen, who once had brought in about 500,000 kg (roughly a million pounds) of salmon per year, saw their catches decline by as much as 90 percent. More than 7,000 jobs were lost when the fisheries collapsed. In 1997, the C'wam and Qapdo were declared endangered, and Coho salmon were listed as threatened. A coalition of commercial and sports fishermen, environmentalists, and native people sued the government for damaging fish and wildlife resources. A federal judge agreed, and ordered the Bureau of Reclamation to reduce irrigation flow and to maintain minimum water levels in rivers and lakes.

A severe drought in 2001 precipitated a crisis. For the first time in its history, the bureau closed the gates on its irrigation canals and cut off water to the farmers. Outraged farmers broke open the headgate locks and released water into drying fields. News media flocked to the site to film the confrontation. Wildlife enthusiasts and hunters clamored for the return of the water to the river and the area's wetlands. Farmers fumed over lost profits. The irrigation headgates were closed again.

Then Republican National Committee strategist Karl Rove stepped in to this small, remote conflict. Rove and other conservative politicians

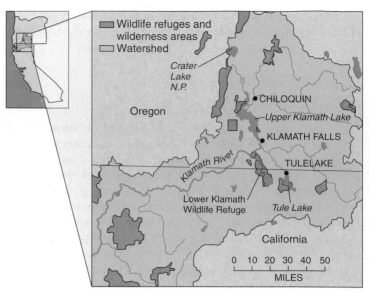

The Klamath River rises in the Cascade Mountains and flows about 400 km (250 mi) across Oregon and Washington to the Pacific Ocean.

publicized the debate around the property rights of farmers—a concern shared by farmers and conservatives across the nation. Following Rove's advice, Interior Secretary Gale Norton flew to the site and, with great publicity, reopened the irrigation headgates.

Norton suggested that the Klamath water dispute could be best resolved by the free market. She proposed a "water bank," with which farmers could voluntarily sell surplus water to Indian tribes or the Fish and Wildlife Service. The tribes and wildlife groups objected that this approach would only work when water was abundant. In critical drought years, wildlife and fish would perish.

Conservationists argue that irrigation practices should change. Two-thirds of the basin uses flood irrigation, which loses about 50 percent of water to evaporation. Farmers could install sprinkler systems or switch to less water-intensive crops.

Farmers counter that their survival depends on Klamath water. Denial of water rights constitutes unlawful "takings" that threaten their livelihoods. Property rights activists argue that water is a property right accorded to landowners, not to wildlife and fish.

Meanwhile, because of several poor spawning years, California's 2006 salmon fishing quotas were cut, costing thousands of fishery workers an estimated $100 million in fishing revenues.

Conflicting demands for dwindling resources often lead to complex disputes. What would you do if you were Secretary of the Interior? What arguments would you use to resolve this dispute? What goals are achieved and lost when the national spotlight turns on a local debate? How would you weigh the rights of farmers and fish, and of fisheries and landowners?

---

that earthquakes will cause dams to collapse, leading to catastrophic flooding.

Canada's James Bay project, built by Hydro-Quebec to sell electricity to New York markets, has diverted three major rivers flowing west into Hudson Bay and has flooded more than 10,000 km^2 (4,000 m^2) of forest and tundra. In 1984, 10,000

caribou drowned while trying to follow ancient migration routes across newly flooded land. Native Cree communities have been devastated by the loss of hunting and fishing sites. In addition, mercury leeching out of rocks in the newly submerged lands has entered the food chain, and many residents have shown signs of food poisoning.

Environmental costs of dams include evaporation and siltation

Dams ensure a year-round water supply, but they also waste tremendous amounts of water from evaporation and through seepage into porous rock beds. Some dams built in the western United States lose more water than they make available. Evaporative losses from Lake Mead and Lake Powell on the Colorado River are about 1 km^3 (264 billion gal) per year (fig. 10.12). The salts left behind by evaporation and agricultural runoff nearly double the salinity of the river. Mexico had to sue the United States to force construction of a $350 million desalination plant at Yuma, Arizona, to make the water partially usable again.

Dams also collect silt, decreasing the effectiveness of reservoirs and starving streambeds and sandbars downstream. As the turbulent Colorado River slows in the reservoirs created by Glen Canyon and Boulder Dams, it drops its load of suspended material. More than 10 million metric tons of silt per year collect behind these dams. Imagine a line of 20,000 dump trucks backed up to Lake Mead and Lake Powell every day, dumping dirt into the water. Within as little as 100 years, these reservoirs could be full of silt and useless for either water storage or hydroelectric generation (fig. 10.13).

In Egypt, the Aswân High Dam was built to irrigate thousands of hectares of farmland, but the dam loses much of the Nile's river to evaporation. Without the annual floods that carried rich silt to farmlands for thousands of years, many farming areas are becoming infertile, and the famous Nile Delta—and its rich fisheries—are disappearing.

Dams and river channelization also drown or destroy free-flowing rivers. One of the first and most divisive battles over this loss was in the Hetch Hetchy Valley in Yosemite National Park. In the early 1900s, San Francisco wanted to dam the Tuolumne River to produce hydroelectric power and provide water for the city water system. This project was supported by many prominent San Francisco citizens because it represented an opportunity for both clean water and municipal power. John Muir, founder of the Sierra Club and protector of Yosemite Park, said, "These temple destroyers, devotees of ravaging commercialism, seem to have perfect contempt for Nature, and, instead of lifting their eyes to the God of the mountains, lift them to the Almighty Dollar. Dam Hetch Hetchy! As well dam for water-tanks the people's cathedrals and churches, for no holier temple has ever been consecrated by the heart of man." After a prolonged and bitter fight, the developers won, and the dam was built.

Water pricing and policies encourage waste

Throughout most of U.S. history, water policies generally worked against conservation. In the well-watered eastern United States, water policy was based on riparian use rights: those who lived along a river bank had the right to use as much water as they liked, as long as they didn't interfere with its quality or availability to neighbors downstream. In the drier western regions, where water is often a limiting resource, water law is based primarily on the Spanish system of prior appropriation rights, or "first in time

FIGURE 10.12 Hoover Dam provides valuable electric power to Nevada and California but Lake Mead, behind the dam, loses about 1.3 billion m^3 of water per year to evaporation.

FIGURE 10.13 This tower is part of a dam structure, the field of mud was a reservoir, now filled with silt and sediment.

are first in right." But the appropriated water has to be put to "beneficial" use by being consumed. This creates a policy of "use it or lose it." Water left in a stream, even if essential for recreation, aesthetic enjoyment, or to sustain ecological communities, is not being appropriated or put to "beneficial" (that is, economic) use. Under this system, water rights can be bought and sold, but if water owners fail to use their allotment, they may lose it. This policy has frequently led to extravagant misuse of Western water.

Water has also traditionally been supplied at unrealistically low cost, especially for farming. In most federal "reclamation" projects, customers have been charged only for the immediate costs of water delivery. The costs of building dams and distribution systems have been subsidized, and the potential value of competing uses has been routinely ignored. Farmers in California's Central Valley, for instance, for many years paid only about one-tenth of what it cost the government to supply water to them. This didn't encourage conservation. Subsidies created by underpriced water amounted to as much as $500,000 per farm per year in some areas.

WATER MANAGEMENT AND CONSERVATION

Watershed management and conservation are often more economical and environmentally sound ways to prevent flood damage and store water for future use than building huge dams and reservoirs. A **watershed,** or catchment, is all the land drained by a stream or river. It has long been recognized that retaining vegetation and groundcover in a watershed helps hold back rainwater and lessens downstream floods. In 1998 Chinese officials acknowledged that unregulated timber cutting upstream on the Yangtze contributed to massive floods that killed 30,000 people. Similarly, after disastrous floods in the upper Mississippi Valley in 1993, it was suggested that, rather than allowing residential, commercial, or industrial development on floodplains, these areas should be reserved for water storage, aquifer recharge, wildlife habitat, and agriculture. Further discussion of flooding hazards can be found in chapter 11.

Sound farming and forestry practices can reduce runoff. Retaining crop residue on fields reduces flooding, and minimizing plowing and forest cutting on steep slopes protects watersheds. Effects of deforestation on weather and water supplies are discussed in chapter 6. Wetlands conservation preserves natural water storage capacity and aquifer recharge zones. A river fed by marshes and wet meadows tends to run consistently clear and steady, rather than in violent floods.

A series of small dams on tributary streams can hold back water before it becomes a great flood. Ponds formed by these dams provide useful wildlife habitat and stock-watering facilities. They also catch soil where it could be returned to the fields. Small dams can be built with simple equipment and local labor, eliminating the need for massive construction projects and huge dams.

In 1998 U.S. Forest Service chief, Mike Dombeck, announced a major shift in his agency's priorities. "Water," he said, "is the most valuable and least appreciated resource the national forests provide. More than 60 million people in 33 states obtain their drinking water from national forest lands. Protecting watersheds is far more economically important than logging or mining, and will be given the highest priority in forest planning."

Everyone can help conserve water

We could probably save as much as half of the water we now use for domestic purposes without great sacrifice or serious changes in our lifestyles. Simple steps, such as taking shorter showers, fixing leaks, and washing cars, dishes, and clothes as efficiently as possible, can go a long way toward forestalling the water shortages that many authorities predict. Isn't it better to adapt to more conservative uses now when we have a choice than to be forced to do it by scarcity in the future?

Conserving appliances, such as low-volume shower heads and efficient dishwashers, can reduce water consumption greatly (see What Can You Do? p. 245). If you live in an arid part of the country, you might consider whether you really need a lush, green lawn that requires constant watering, feeding, and care. Planting native groundcover in a "natural lawn" or developing a

FIGURE 10.14 By using native plants in a natural setting, residents of Phoenix save water and fit into the surrounding landscape.

rock garden or landscape in harmony with the surrounding ecosystem can be both ecologically sound and aesthetically pleasing (fig. 10.14). Cultivated lawns, golf courses, and parks in the United States use more water, fertilizer, and pesticides per hectare than any other kind of land.

Toilets are our greatest water user (fig. 10.15). Usually each flush uses several gallons of water to dispose of a few ounces of waste. Each person in the United States uses about 50,000 l (13,000 gal) of drinking-quality water annually to flush toilets. Low-flush toilets and composting toilets can drastically reduce this water use. Low-flow shower heads can reduce our second-largest household water use.

These steps are so important that a number of cities (including Los Angeles, Orlando, Austin, and Phoenix) ordered that water-saving toilets, showers, and faucets be installed in all new buildings. The motivation was two-fold: to relieve overburdened sewer systems and to conserve water.

Significant amounts of water also can be reclaimed and recycled. California uses more than 555 million m³ (450,000

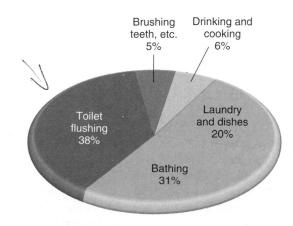

FIGURE 10.15 Typical household water use in the United States.
Source: Data from U.S. Environmental Protection Agency, 2004.

What Can You Do?

Saving Water and Preventing Pollution

Each of us can conserve much of the water we use and avoid water pollution in many simple ways.

- Don't flush every time you use the toilet. Take shorter showers, and shower instead of taking baths.
- Don't let the faucet run while brushing your teeth or washing dishes. Draw a basin of water for washing and another for rinsing dishes. Don't run the dishwasher when it's half full.
- Use water-conserving appliances: low-flow showers, low-flush toilets, and aerated faucets.
- Fix leaking faucets, tubs, and toilets. A leaky toilet can waste 50 gal per day. To check your toilet, add a few drops of dark food coloring to the tank and wait 15 minutes. If the tank is leaking, the water in the bowl will change color.
- Put a brick or full water bottle in your toilet tank to reduce the volume of water in each flush.
- Dispose of used motor oil, household hazardous waste, batteries, and so on responsibly. Don't dump anything down a storm sewer that you wouldn't want to drink.
- Avoid using toxic or hazardous chemicals for simple cleaning or plumbing jobs. A plunger or plumber's snake will often unclog a drain just as well as caustic acids or lye. Hot water and soap can accomplish most cleaning tasks.
- If you have a lawn, use water, fertilizer, and pesticides sparingly. Plant native, low-maintenance plants that have low water needs.
- If possible, use recycled (gray) water for lawns, house plants, and car washing.

acre-feet) of recycled water annually. That's equivalent to about two-thirds of the water consumed by Los Angeles every year.

Efficiency is reducing water use in many areas

Growing recognition that water is a precious and finite resource has changed policies and encouraged conservation across the United States. Despite a growing population, the United States is now saving some 144 million l (38 million gal) per day—a tenth the volume of Lake Erie—compared with per capita consumption rates of 20 years ago. With 37 million more people in the United States now than in 1980, we get by with 10 percent less water. New requirements for water-efficient fixtures and low-flush toilets in many cities help conserve water on the home front. More efficient irrigation methods on farms also are a major reason for the downward trend. New sprinkler systems have small spray heads just a foot or so above the plant tops and apply water much more directly. Even better is drip irrigation, which applies water

directly to plant roots (see fig. 10.9). California and Florida farmers currently water about 500,000 ha with this technique.

Charging a higher proportion of real costs to users of public water projects has helped encourage conservation, and so have water marketing policies that allow prospective users to bid on water rights. Both the United States and Australia have had effective water pricing and allocation policies that encourage the most socially beneficial uses and discourage wasteful water uses. It will be important, as water markets develop, to be sure that environmental, recreational, and wildlife values are not sacrificed to the lure of high-bidding industrial and domestic users.

WATER POLLUTION

Any physical, biological, or chemical change in water quality that adversely affects living organisms or makes water unsuitable for desired uses can be considered pollution. There are natural sources of water contamination, such as poison springs, oil seeps, and sedimentation from erosion, but here we will focus primarily on human-caused changes that affect water quality or usability.

Pollution includes point sources and nonpoint sources

Pollution control standards and regulations usually distinguish between point and nonpoint pollution sources. Factories, power plants, sewage treatment plants, underground coal mines, and oil wells are classified as **point sources** because they discharge pollution from specific locations, such as drain pipes, ditches, or sewer outfalls (fig. 10.16). These sources are discrete and identifiable, so they are relatively easy to monitor and regulate. It is generally possible to divert effluent from the waste streams of these sources and treat it before it enters the environment.

In contrast, **nonpoint sources** of water pollution are diffuse, having no specific location where they discharge into a particular body of water. They are much harder to monitor and regulate than point sources because their origins are hard to identify. Nonpoint sources include runoff from farm fields and feedlots, golf courses, lawns and gardens, construction sites, logging areas, roads, streets, and parking lots (fig. 10.17). While point sources may be fairly uniform and predictable throughout the year, nonpoint sources are often highly episodic. The first heavy rainfall after a dry period may flush high concentrations of gasoline, lead, oil, and rubber residues off city streets, for instance, while subsequent runoff may be much cleaner.

Perhaps the ultimate in diffuse, nonpoint pollution is atmospheric deposition of contaminants carried by air currents and precipitated into watersheds or directly onto surface waters as rain, snow, or dry particles. The Great Lakes, for example, have been found to be accumulating industrial chemicals, such as PCBs (polychlorinated biphenyls) and dioxins, as well as agricultural toxins, such as the insecticide toxaphene, that cannot be accounted for by local sources alone. The nearest sources for many of these chemicals are sometimes thousands of kilometers away.

FIGURE 10.16 Sewer outfalls, industrial effluent pipes, acid draining out of abandoned mines, and other point sources of pollution are generally easy to recognize. Pollution control laws have made this sight less common today than it once was.

Biological pollution includes pathogens and waste

Although the types, sources, and effects of water pollutants are often interrelated, it is convenient to divide them into major categories for discussion (table 10.4). Here, we look at some of the important sources and effects of different pollutants.

Pathogens

The most serious water pollutants in terms of human health worldwide are pathogenic (disease-causing) organisms (chapter 8). Among the most important waterborne diseases are typhoid, cholera, bacte-

FIGURE 10.17 This bucolic scene looks peaceful and idyllic, but allowing cows to trample stream banks is a major cause of bank erosion and water pollution. Nonpoint sources such as this have become the leading unresolved cause of stream and lake pollution in the United States.

rial and amoebic dysentery, enteritis, polio, infectious hepatitis, and schistosomiasis. Malaria, yellow fever, and filariasis are transmitted by insects that have aquatic larvae. Altogether, at least 25 million deaths each year are blamed on these water-related diseases. Nearly two-thirds of the mortalities of children under 5 years old are associated with waterborne diseases.

The main source of these pathogens is untreated or improperly treated human wastes. Animal wastes from feedlots or fields near

TABLE 10.4

Major Categories of Water Pollutants

Category	Examples	Sources
Cause Ecosystem Disruption		
1. Oxygen-demanding wastes	Animal manure, plant residues	Sewage, agricultural runoff, paper mills, food processing
2. Plant nutrients	Nitrates, phosphates, ammonium	Agricultural and urban fertilizers, sewage, manure
3. Sediment	Soil, silt	Land erosion
4. Thermal changes	Heat	Power plants, industrial cooling
Cause Health Problems		
1. Pathogens	Bacteria, viruses, parasites	Human and animal excreta
2. Inorganic chemicals	Salts, acids, caustics, metals	Industrial effluents, household cleansers, surface runoff
3. Organic chemicals	Pesticides, plastics, detergents, oil, gasoline	Industrial, household, and farm use
4. Radioactive materials	Uranium, thorium, cesium, iodine, radon production, natural sources	Mining and processing of ores, power plants, weapons

waterways and food processing factories with inadequate waste treatment facilities also are sources of disease-causing organisms.

In developed countries, sewage treatment plants and other pollution-control techniques have reduced or eliminated most of the worst sources of pathogens in inland surface waters. Furthermore, drinking water is generally disinfected by chlorination, so epidemics of waterborne diseases are rare in these countries. The United Nations estimates that 90 percent of the people in developed countries have adequate (safe) sewage disposal, and 95 percent have clean drinking water.

The situation is quite different in less-developed countries, where billions of people lack adequate sanitation and access to clean drinking water. Conditions are especially bad in remote, rural areas, where sewage treatment is usually primitive or nonexistent and purified water is either unavailable or too expensive to obtain. The World Health Organization estimates that 80 percent of all sickness and disease in less-developed countries can be attributed to waterborne infectious agents and inadequate sanitation.

Detecting specific pathogens in water is difficult, time-consuming, and costly, so water quality is usually described in terms of concentrations of **coliform bacteria**—any of the many types that live in the colon, or intestines, of humans and other animals. The most common of these is *Escherichia coli* (or *E. coli*). These bacteria occur naturally in our intestines. Other bacteria such as *Shigella, Salmonella,* or *Listeria,* can also cause serious, even fatal, illness. If any coliform bacteria are present in a water sample, infectious pathogens are usually assumed to be present also. Therefore, the Environmental Protection Agency (EPA) considers water with any coliform bacteria at all to be unsafe for drinking.

Biological oxygen demand

The amount of oxygen dissolved in water is a good indicator of water quality and of the kinds of life it will support. An oxygen content above 6 parts per million (ppm) will support game fish and other desirable forms of aquatic life. At oxygen levels below 2 ppm, water will support mainly worms, bacteria, fungi, and other detritus feeders and decomposers. Oxygen is added to water by diffusion from the air, especially when turbulence and mixing rates are high, and by photosynthesis of green plants, algae, and cyanobacteria. Turbulent, rapidly flowing water is constantly aerated, so it often recovers quickly from oxygen-depleting processes. Oxygen is removed from water by respiration and chemical processes that consume oxygen. Because oxygen is so important in water, **dissolved oxygen (DO)** levels are often measured to compare water quality in different places.

Adding organic materials, such as sewage or paper pulp, to water stimulates activity and oxygen consumption by decomposers. Consequently, **biochemical oxygen demand (BOD),** or the amount of dissolved oxygen consumed by aquatic microorganisms, is another standard measure of water contamination. Alternatively, chemical oxygen demand (COD) is a measure of all organic matter in water.

Downstream from a point source, such as a municipal sewage plant discharge, a characteristic decline and restoration of water quality can be detected either by measuring DO content or by observing the types of flora and fauna that live in successive sections of the river. The oxygen decline downstream is called the **oxygen sag** (fig. 10.18). Upstream from the pollution source,

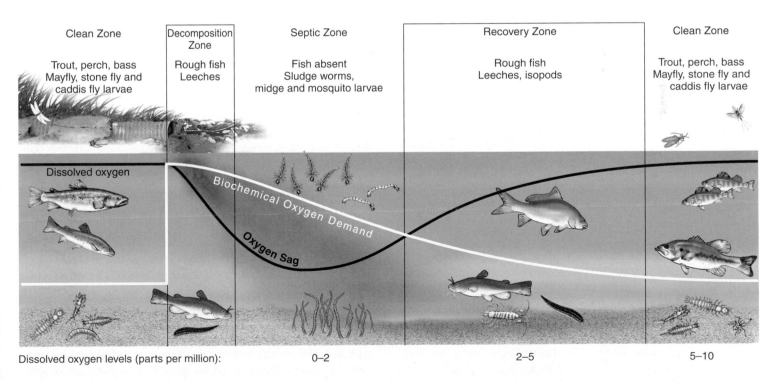

Clean Zone	Decomposition Zone	Septic Zone	Recovery Zone	Clean Zone
Trout, perch, bass Mayfly, stone fly and caddis fly larvae	Rough fish Leeches	Fish absent Sludge worms, midge and mosquito larvae	Rough fish Leeches, isopods	Trout, perch, bass Mayfly, stone fly and caddis fly larvae

Dissolved oxygen levels (parts per million): 0–2   2–5   5–10

FIGURE 10.18 Oxygen sag downstream of an organic source. A great deal of time and distance may be required for the stream and its inhabitants to recover.

oxygen levels support normal populations of clean-water organisms. Immediately below the source of pollution, oxygen levels begin to fall as decomposers metabolize waste materials. Rough fish, such as carp, bullheads, and gar, are able to survive in this oxygen-poor environment, where they eat both decomposer organisms and the waste itself.

Farther downstream, the water may become so oxygen depleted that only the most resistant microorganisms and invertebrates can survive. Eventually, most of the nutrients are used up, decomposer populations are smaller, and the water becomes oxygenated once again. Depending on the volumes and flow rates of the effluent plume and the river receiving it, normal communities may not appear for several miles downstream.

Plant nutrients and cultural eutrophication

Water clarity (transparency) is affected by sediments, chemicals, and the abundance of plankton organisms; clarity is a useful measure of water quality and water pollution. Rivers and lakes that have clear water and low biological productivity are said to be **oligotrophic** (*oligo* = little + *trophic* = nutrition). By contrast, **eutrophic** (*eu* + *trophic* = well-nourished) waters are rich in organisms and organic materials. Eutrophication, an increase in nutrient levels and biological productivity, often accompanies successional changes (chapter 5) in lakes. Tributary streams bring in sediments and nutrients that stimulate plant growth. Over time, ponds and lakes often fill in, becoming marshes or even terrestrial biomes. The rate of eutrophication depends on water chemistry and depth, volume of inflow, mineral content of the surrounding watershed, and biota of the lake itself.

Human activities can greatly accelerate eutrophication, an effect called **cultural eutrophication.** Cultural eutrophication is mainly caused by increased nutrient input into a water body. Increased productivity in an aquatic system sometimes can be beneficial. Fish and other desirable species may grow faster, providing a welcome food source. Often, however, eutrophication produces "blooms" of algae or thick growths of aquatic plants stimulated by elevated phosphorus or nitrogen levels (fig. 10.19). Bacterial populations then increase, fed by larger amounts of organic matter. The water often becomes cloudy, or turbid, and has unpleasant tastes and odors. Cultural eutrophication can accelerate the "aging" of a water body enormously over natural rates. Lakes and reservoirs that normally might exist for hundreds or thousands of years can be filled in a matter of decades.

Eutrophication also occurs in marine ecosystems, especially in nearshore waters and partially enclosed bays or estuaries. Partially enclosed seas, such as the Black, Baltic, and Mediterranean Seas, tend to be in especially critical condition. During the tourist season, the coastal population of the Mediterranean, for example, swells to 200 million people. Eighty-five percent of the effluents from large cities go untreated into the sea. Beach pollution, fish kills, and contaminated shellfish result. Extensive "dead zones" often form where rivers dump nutrients

FIGURE 10.19 Eutrophic lake. Nutrients from agriculture and domestic sources have stimulated growth of algae and aquatic plants. This reduces water quality, alters species composition, and lowers the lake's recreational and aesthetic values.

into estuaries and shallow seas. One of the largest in the world occurs during summer months in the Gulf of Mexico at the mouth of the Mississippi River. This hypoxic (oxygen-depleted) zone can cover 21,000 km^2, or about the area of New Hampshire. A federal study of the condition of U.S. coastal waters announced in 2004 that 28 percent of estuaries are impaired for aquatic life, and 80 percent of all coastal water is in fair to poor condition.

Marine animals in hypoxic zones die not only because of depleted oxygen, but also because of high concentrations of harmful organisms, including toxic algae, pathogenic fungi, and parasitic protists. Excessive nutrients support blooms of these deadly aquatic microorganisms in polluted nearshore waters. Red tides—and other colors, depending on the species involved—have become increasingly common where nutrients and wastes wash down rivers. (See related story "A Flood of Pigs" at www.mhhe.com/cunningham4e.)

Inorganic pollutants include metals, salts, and acids

Some toxic inorganic chemicals are naturally released into water from rocks by weathering processes (chapter 11). Humans accelerate the transfer rates in these cycles thousands of times above natural background levels by mining, processing, using, and discarding minerals.

Among the chemicals of greatest concern are heavy metals, such as mercury, lead, tin, and cadmium. Supertoxic elements, such as selenium and arsenic, also have reached hazardous levels in some waters. Other inorganic materials, such as acids, salts, nitrates, and chlorine, that are nontoxic at low concentrations may

Principles of Environmental Science

become concentrated enough to lower water quality and adversely affect biological communities.

Metals

Many metals, such as mercury, lead, cadmium, and nickel, are highly toxic in minute concentrations. Because metals are highly persistent, they accumulate in food chains and have a cumulative effect in humans.

Currently the most widespread toxic metal contamination in North America is mercury released from incinerators and coal-burning power plants. Transported through the air, mercury precipitates in water supplies, where it bioconcentrates in food webs to reach dangerous levels in top predators. As a general rule, Americans are warned not to eat more than one meal of fish per week. Top marine predators, such as shark, swordfish, bluefin tuna, and king mackerel, tend to have especially high mercury content. Pregnant women and small children should avoid these species entirely. Public health officials estimate that 600,000 American children now have mercury levels in their bodies high enough to cause mental and developmental problems, while one woman in six in the United States has blood-mercury concentrations that would endanger a fetus.

Mine drainage and leaching of mining wastes are serious sources of metal pollution in water. A survey of water quality in eastern Tennessee found that 43 percent of all surface streams and lakes and more than half of all groundwater used for drinking supplies were contaminated by acids and metals from mine drainage. In some cases, metal levels were 200 times higher than what is considered safe for drinking water.

Nonmetallic salts

Some soils contain high concentrations of soluble salts, including toxic selenium and arsenic (Exploring Science, p. 250). You have probably heard of poison springs and seeps in the desert, where percolating groundwater brings these compounds to the surface. Irrigation and drainage of desert soils can mobilize these materials on a larger scale and result in serious pollution problems, as in Kesterson Marsh in California, where selenium poisoning killed thousands of migratory birds in the 1980s.

Salts, such as sodium chloride (table salt), that are nontoxic at low concentrations also can be mobilized by irrigation and concentrated by evaporation, reaching levels that are toxic for plants and animals. Salinity levels in the Colorado River and surrounding farm fields have become so high in recent years that millions of hectares of valuable croplands have had to be abandoned. In northern states, millions of tons of sodium chloride and calcium chloride are used to melt road ice in the winter. Leaching of road salts into surface waters has a devastating effect on some aquatic ecosystems.

Acids and bases

Acids are released as by-products of industrial processes, such as leather tanning, metal smelting and plating, petroleum distillation, and organic chemical synthesis. Coal mining is an especially important source of acid water pollution. Sulfur compounds in coal react with oxygen and water to make sulfuric acid. Thousands of kilometers of streams in the United States have been acidified by acid mine drainage, some so severely that they are essentially lifeless.

Acid precipitation (chapter 9) also acidifies surface-water systems. In addition to damaging living organisms directly, these acids leach aluminum and other elements from soil and rock, further destabilizing ecosystems.

Organic chemicals include pesticides and industrial substances

Thousands of different natural and synthetic organic chemicals are used in the chemical industry to make pesticides, plastics, pharmaceuticals, pigments, and other products that we use in everyday life. Many of these chemicals are highly toxic (chapter 8). Exposure to very low concentrations (perhaps even parts per quadrillion, in the case of dioxins) can cause birth defects, genetic disorders, and cancer. Some can persist in the environment because they are resistant to degradation and toxic to organisms that ingest them.

The two principal sources of toxic organic chemicals in water are (1) improper disposal of industrial and household wastes and (2) pesticide runoff from farm fields, forests, roadsides, golf courses, and private lawns. The EPA estimates that about 500,000 metric tons of pesticides are used in the United States each year. Much of this material washes into the nearest waterway, where it passes through ecosystems and may accumulate in high levels in nontarget organisms. The bioaccumulation of DDT in aquatic ecosystems was one of the first of these pathways to be understood (chapter 8). Dioxins and other chlorinated hydrocarbons (hydrocarbon molecules that contain chlorine atoms) have been shown to accumulate to dangerous levels in the fat of salmon, fish-eating birds, and humans and to cause health problems similar to those resulting from toxic metal compounds.

Hundreds of millions of tons of hazardous organic wastes are thought to be stored in dumps, landfills, lagoons, and underground tanks in the United States (chapter 13). Many, perhaps most, of these sites have leaked toxic chemicals into surface waters, groundwater, or both. The EPA estimates that about 26,000 hazardous waste sites will require cleanup because they pose an imminent threat to public health, mostly through water pollution.

Sediment and heat also degrade water

Sediment is a natural and necessary part of river systems. Sediment fertilizes floodplains and creates fertile deltas. But human activities, chiefly farming and urbanization, greatly accelerate erosion and increase sediment loads in rivers. Silt and sediment are considered the largest source of water pollution in the United States, being responsible for 40 percent of the impaired river miles in EPA water quality surveys. Cropland erosion contributes

When we think of water pollution, we usually visualize sewage or industrial effluents pouring out of a discharge pipe, but there are natural toxins that threaten us as well. One of these is arsenic, a common contaminate in drinking water that may be poisoning millions of people around the world. Arsenic has been known since the fourth century B.C. to be a potent poison. It has been used for centuries as a rodenticide, insecticide, and weed killer, as well as a way of assassinating enemies. Because it isn't metabolized or excreted from the body, arsenic accumulates in hair and fingernails, where it can be detected long after death. Napoleon Bonaparte was recently found to have high enough levels of arsenic in his body to suggest he was poisoned.

Perhaps the largest population to be threatened by naturally occurring groundwater contamination by arsenic is in West Bengal, India, and adjacent areas of Bangladesh. Arsenic occurs naturally in the sediments that make up the Ganges River delta (see map). Rapid population growth, industrialization, and intensification of agricultural irrigation, however, have put increasing stresses on the limited surface-water supplies. Most surface water is too contaminated to drink, so groundwater has all but replaced other water sources for most people in this region.

In the 1960s, thousands of deep tube wells were sunk throughout the region to improve water supplies. Much of this humanitarian effort was financed by loans from the World Bank. At first, villagers were suspicious of well water, regarding it as unnatural and possibly evil. But as surface-water supplies diminished and populations grew, Bengal and Bangladesh became more and more dependent on this new source of supposedly fresh, clean water. By the late 1980s, health workers had become aware of widespread signs of chronic arsenic poisoning among villagers. Symptoms include watery and inflamed eyes, gastrointestinal cramps, gradual loss of strength, scaly skin and skin tumors, anemia, confusion, and eventually death.

Why is arsenic poisoning appearing now? Part of the reason is increased dependence on well water, but some villages have had wells for centuries with no problem. One theory is that excessive withdrawals now lower the water table during the dry season, exposing arsenic-bearing minerals to air, which converts normally insoluble salts to soluble oxides. When aquifers are refilled during the next rainy season, dissolved arsenic can be pumped out. Between 30 and 90 million people depend on these contaminated water sources.

Discovering and understanding arsenic poisoning in Bangladesh required a combination of scientific insights from medicine and geology. Chemists and development experts are working to find solutions. A variety of inexpensive filters are being developed, using waste coal ash, sand, or clay fragments coated with chemicals that bind to arsenic, such as iron oxides or iron sulfates. These innovative strategies may save millions of lives and reduce the economic burdens of chronic illness in poor villages.

Arsenic-tainted water occurs around the world. In the United States it has been the

West Bengal and adjoining areas of Bangladesh have hundreds of millions of people who may be exposed to dangerous arsenic levels in well water.

subject of long and bitter legal disputes, because many city water supplies, especially in the American Southwest, rely on arsenic-tainted groundwater. In 1942, the U.S. government set the acceptable level of arsenic at 50 ppb. A 1999 study by the National Academy of Sciences found that this level produced a 1 in 100 risk for cancer—10,000 times the normally accepted risk level for cancer. The World Health Organization's standard limit is much lower, at 10 ppb. For years, policy makers debated whether it was fair to force communities to install filters in municipal water supplies, or whether the public health costs of increased cancer and chronic illness justified the installation of filters. In 2002, public outrage over water quality, together with the economic cost of health care, convinced the federal government to lower arsenic limits to meet international standards.

---

about 25 billion metric tons of soil, sediment, and suspended solids to world surface waters each year. Forest disturbance, road building, urban construction sites, and other sources add at least 50 billion additional tons.

This sediment fills lakes and reservoirs, obstructs shipping channels, clogs hydroelectric turbines, and makes purification of drinking water more costly. Sediments smother gravel beds in which insects take refuge and fish lay their eggs. Sunlight is blocked, so that plants cannot carry out photosynthesis, and oxygen levels decline. Murky, cloudy water also is less attractive for swimming, boating, fishing, and other recreational uses (fig. 10.20). Sediment washed into the ocean clogs estuaries and coral reefs.

Thermal pollution, usually effluent from cooling systems of power plants or other industries, alters water temperature. Raising or lowering water temperatures from normal levels can adversely affect water quality and aquatic life. Water temperatures are usually much more stable than air temperatures, so aquatic organisms tend to be poorly adapted to rapid temperature changes. Lowering the temperature of tropical oceans by even 1° can be lethal to some corals and other reef species. Raising water temperatures can have similar devastating effects on sensitive organisms. Oxygen solubility in water decreases as temperatures increase, so species requiring high oxygen levels are adversely affected by warming water.

FIGURE 10.20 Sediment and industrial waste flow from this drainage canal into Lake Erie.

Humans also cause thermal pollution by altering vegetation cover and runoff patterns. Reducing water flow, clearing streamside trees, and adding sediment all make water warmer and alter the ecosystems in a lake or stream.

Warm-water plumes from power plants often attract fish and birds, which find food and refuge there, especially in cold weather. This artificial environment can be a fatal trap, however. Florida's manatees, an endangered mammal, are attracted to the abundant food supply and warm water in power plant thermal plumes. Often they are enticed into spending the winter much farther north than they normally would. On several occasions, a midwinter power plant breakdown has exposed a dozen or more of these rare animals to a sudden, deadly thermal shock.

WATER QUALITY TODAY

Surface-water pollution is often both highly visible and one of the most common threats to environmental quality. In more developed countries, reducing water pollution has been a high priority over the past few decades. Billions of dollars have been spent on control programs, and considerable progress has been made. Still, much remains to be done.

The 1972 Clean Water Act
protects our water

Like most developed countries, the United States and Canada have made encouraging progress in protecting and restoring water quality in rivers and lakes over the past 40 years. In 1948 only about one-third of Americans were served by municipal sewage systems, and most of those systems discharged sewage without any treatment or with only primary treatment (the bigger lumps of waste are removed). Most people depended on cesspools and septic systems to dispose of domestic wastes.

Areas of progress

The 1972 Clean Water Act established a National Pollution Discharge Elimination System (NPDES), which requires an easily revoked permit for any industry, municipality, or other entity dumping wastes in surface waters. The permit requires disclosure of what is being dumped and gives regulators valuable data and evidence for litigation. As a consequence, only about 10 percent of our water pollution now comes from industrial and municipal point sources. One of the biggest improvements has been in sewage treatment.

Since the Clean Water Act was passed in 1972, the United States has spent more than $180 billion in public funds and perhaps ten times as much in private investments on water pollution control. Most of that effort has been aimed at point sources, especially to build or upgrade thousands of municipal sewage treatment plants. As a result, nearly everyone in urban areas is now served by municipal sewage systems, and no major city discharges raw sewage into a river or lake except as overflow during heavy rainstorms.

This campaign has led to significant improvements in surface-water quality in many places. Fish and aquatic insects have returned to waters that formerly were depleted of life-giving oxygen. Swimming and other water-contact sports are again permitted in rivers, in lakes, and at ocean beaches that once were closed by health officials.

The Clean Water Act goal of making all U.S. surface waters "fishable and swimmable" has not been fully met, but in 2003 the EPA reported that 91 percent of all monitored river miles and 88 percent of all assessed lake acres are suitable for their designated uses. This sounds good, but you have to remember that not all water bodies are monitored. Furthermore, the designated goal for some rivers and lakes is merely to be "boatable." Water quality doesn't have to be very high to allow boating. Even in "fishable" rivers and lakes, there isn't a guarantee that you can catch anything other than rough fish, such as carp or bullheads, nor can you be sure that what you catch is safe to eat. Even with billions of dollars of investment in sewage treatment plants, elimination of much of the industrial dumping and other gross sources of pollutants, and a general improvement in water quality, the EPA reports that 21,000 water bodies still do not meet their designated uses. According to the EPA, an overwhelming majority of the American people—almost 218 million—live within 16 km (10 mi) of an impaired water body.

In 1998 a new regulatory approach to water quality assurance was instituted by the EPA. Rather than issue standards on a river by river approach or factory-by-factory permit discharge, the focus was changed to watershed-level monitoring and protection. Some 4,000 watersheds are now monitored for water

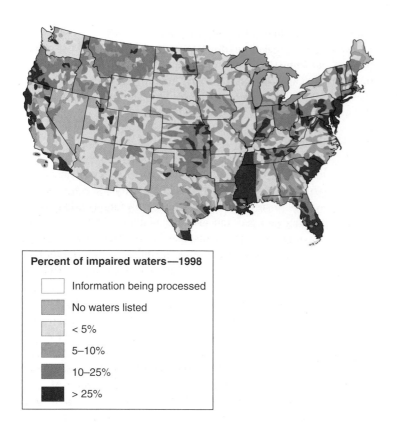

Percent of impaired waters—1998

☐	Information being processed
	No waters listed
	< 5%
	5–10%
	10–25%
	> 25%

FIGURE 10.21 Percent of impaired U.S. rivers in the contiguous 48 states by watershed in 1998.
Source: Data from U.S. Environmental Protection Agency, 1999.

quality (fig. 10.21). You can find information about your watershed at www.epa.gov/owow/tmdl/. The intention of this program is to give the public more and better information about the health of their watersheds. In addition, states can have greater flexibility as they identify impaired water bodies and set priorities, and new tools can be used to achieve goals. States are required to identify waters not meeting water quality goals and to develop **total maximum daily loads (TMDL)** for each pollutant and each listed water body. A TMDL is the amount of a particular pollutant that a water body can receive from both point and nonpoint sources. It considers seasonal variation and includes a margin of safety.

By 1999, all 56 U.S. states and territories had submitted TMDL lists, and the EPA had approved most of them. Of the 5.6 million km of rivers monitored, only 480,000 km fail to meet their clean water goals. Similarly, of 40 million lake ha, only 12.5 percent (in about 20,000 lakes) fail to meet their goal. To give states more flexibility in planning, the EPA has proposed new rules that include allowances for reasonably foreseeable increases in pollutant loadings to encourage "Smart Growth." In the future, TMDLs also will include load allocations from all nonpoint sources, including air deposition and natural background levels.

An encouraging example of improved water quality is seen in Lake Erie. Although widely regarded as "dead" in the 1960s, the lake today is promoted as the "walleye capital of the world." Bacteria counts and algae blooms have decreased more than 90 percent since 1962. Water that once was murky brown is now clear. Interestingly, part of the improved water quality is due to immense numbers of exotic zebra mussels, which filter the lake water very efficiently. Swimming is now officially safe along 96 percent of the lake's shoreline. Nearly 40,000 nesting pairs of double-crested cormorants nest in the Great Lakes region, up from only about 100 in the 1970s.

Canada's 1970 Water Act has produced comparable results. Seventy percent of all Canadians in towns over 1,000 population are now served by some form of municipal sewage treatment. In Ontario, the vast majority of those systems include tertiary treatment. After ten years of controls, phosphorus levels in the Bay of Quinte in the northeast corner of Lake Ontario have dropped nearly by half, and algal blooms that once turned waters green are less frequent and less intense than they once were. Elimination of mercury discharges from a pulp and paper mill on the Wabigoon-English River system in western Ontario has resulted in a dramatic decrease in mercury contamination. Twenty years ago this mercury contamination was causing developmental retardation in local residents. Extensive flooding associated with hydropower projects has raised mercury levels in fish to dangerous levels elsewhere, however.

Remaining problems

The greatest impediments to achieving national goals in water quality in both the United States and Canada are sediment, nutrients, and pathogens, especially from nonpoint discharges of pollutants. These sources are harder to identify and to reduce or treat than are specific point sources. About three-fourths of the water pollution in the United States comes from soil erosion, fallout of air pollutants, and surface runoff from urban areas, farm fields, and feedlots. In the United States, as much as 25 percent of the 46,800,000 metric tons (52 million tons) of fertilizer spread on farmland each year is carried away by runoff (fig. 10.22).

Cattle in feedlots produce some 129,600,000 metric tons (144 million tons) of manure each year, and the runoff from these sites is rich in viruses, bacteria, nitrates, phosphates, and other contaminants. A single cow produces about 30 kg (66 lbs) of manure per day. Some feedlots have 100,000 animals with no provision for capturing or treating runoff water. Imagine drawing your drinking water downstream from such a facility. Pets also can be a problem. It is estimated that the wastes from about a half million dogs in New York City are disposed of primarily through storm sewers and therefore do not go through sewage treatment.

Loading of both nitrates and phosphates in surface water have decreased from point sources but have increased about four-fold since 1972 from nonpoint sources. Fossil fuel combustion has become a major source of nitrates, sulfates, arsenic, cadmium, mercury, and other toxic pollutants that find their way into water. Carried to remote areas by atmospheric transport, these combustion

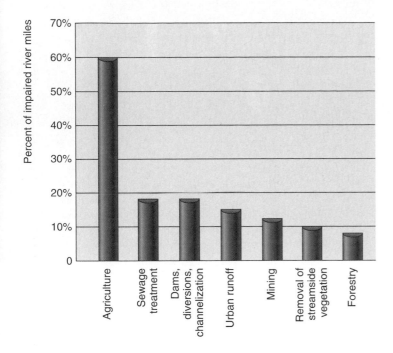

FIGURE 10.22 Percentage of impaired river miles in the United States by source of damage. Totals add up to more than 100 percent because one river can be affected by many sources.

Source: Data from USDA and Natural Resources Conservation Service, *America's Private Land: A Geography of Hope,* USDA.

products now are found nearly everywhere in the world. Toxic organic compounds, such as DDT, PCBs, and dioxins, also are transported long distances by wind currents.

Waters are improving in wealthier countries but not in poorer ones

Japan, Australia, and most of Western Europe also have improved surface-water quality in recent years. Sewage treatment in the wealthier countries of Europe generally equals or surpasses that in the United States. Sweden, for instance, serves 98 percent of its population with at least secondary sewage treatment (compared with 70 percent in the United States), and the other 2 percent have primary treatment. Poorer countries have much less to spend on sanitation. Spain serves only 18 percent of its population with even primary sewage treatment. In Ireland, it is only 11 percent, and in Greece, less than 1 percent of the people have even primary treatment. Most of the sewage, both domestic and industrial, is dumped directly into the ocean.

The fall of the "iron curtain" in 1989 revealed appalling environmental conditions in much of the former Soviet Union and its satellite states in Eastern and central Europe. The countries closest geographically and socially to Western Europe, the Czech Republic, Hungary, former East Germany, and Poland, have made massive investments and encouraging progress toward cleaning up environmental problems. Parts of Russia itself, however, along with former socialist states in the Balkans and central Asia,

remain some of the most polluted places on earth. In Russia, for example, only about half the tap water is fit to drink. In cities such as St. Petersburg, even boiling and filtering aren't enough to make municipal water safe. About one-third of all Russians live in regions where air pollution levels are ten times higher than World Health Organization safety standards. Life expectancies for Russian men have plummeted from about 72 years in 1980 to 59 years in 2003. Deaths now exceed births in Russia by about 1 million per year. Only about one-quarter of all Russian children are considered healthy. In heavily industrialized cities, such as Magnitogorsk, a steel-manufacturing center in the Ural Mountains, nine out of ten children suffer from pollution-related illnesses and birth defects.

High levels of radioactivity from the Chernobyl nuclear power plant accident in 1986 remain in many former Eastern Bloc countries. The Danube River, which originates in Bavaria and Austria and flows through Slovakia, Poland, Hungary, Serbia, Romania, and Bulgaria before emptying into the Black Sea, illustrates some of the problems besetting the area. Draining a landscape with a long history of unregulated mining and heavy industry, the Danube carries chrome, copper, mercury, lead, zinc, and oil to the Black Sea at 20 times the levels these materials flow into the North Sea. Just one city, Bratislava, the capital of Slovakia, dumps 73 million m³ (about 2 billion gal) of industrial and municipal wastes into the river each year. With a population of about 100 million people in its catchment area, the beautiful blue Danube isn't blue anymore. Bombing in Serbia during the Kosovo War in 1999 further exacerbated the situation by releasing oil, industrial chemicals, agricultural fertilizers, pesticides, and other toxic materials into the Danube.

However, there are some encouraging pollution-control stories. In 1997 Minamata Bay in Japan, long synonymous with mercury poisoning, was declared officially clean again. Another important success is found in Europe, where one of its most important rivers has been cleaned up significantly through international cooperation. The Rhine, which starts in the rugged Swiss Alps and winds 1,320 km through five countries before emptying through a Dutch delta into the North Sea, has long been a major commercial artery into the heart of Europe. More than 50 million people live in its catchment basin, and nearly 20 million get their drinking water from the river or its tributaries. By the 1970s, the Rhine had become so polluted that dozens of fish species disappeared and swimming was discouraged along most of its length.

Efforts to clean up this historic and economically important waterway began in the 1950s, but a disastrous fire at a chemical warehouse near Basel, Switzerland, in 1986 provided the impetus for major changes. Through a long and sometimes painful series of international conventions and compromises, land-use practices, waste disposal, urban runoff, and industrial dumping have been changed and water quality has significantly improved. Oxygen concentrations have gone up five-fold since 1970 (from less than 2 mg/l to nearly 10 mg/l, or about 90 percent of saturation) in long stretches of the river. Chemical oxygen demand has fallen five-fold during the same period, and organochlorine levels have decreased as much as ten-fold. Many species of fish and aquatic invertebrates have returned to the river. In 1992, for the first time in decades, mature salmon were caught in the Rhine.

The less-developed countries of South America, Africa, and Asia have even worse water quality than do the poorer countries of Europe (fig. 10.23). Sewage treatment is usually either totally lacking or woefully inadequate. In urban areas, 95 percent of all sewage is discharged untreated into rivers, lakes, or the ocean. Low technological capabilities and little money for pollution control are made even worse by burgeoning populations, rapid urbanization, and the shift of much heavy industry (especially the dirtier ones) from developed countries where pollution laws are strict to less-developed countries where regulations are more lenient.

Appalling environmental conditions often result from these combined factors (fig. 10.24). Two-thirds of India's surface waters are contaminated sufficiently to be considered dangerous to human health. The Yamuna River in New Delhi has 7,500 coliform bacteria per 100 ml (37 times the level considered safe for swimming in the United States) *before* entering the city. The coliform count increases to an incredible 24 *million* cells per 100 ml as the river leaves the city! At the same time, the river picks up some 20 million liters of industrial effluents every day from New Delhi. It's no wonder that disease rates are high and life expectancy is low in this area. Only 1 percent of India's towns and cities have any sewage treatment, and only eight cities have anything beyond primary treatment.

In Malaysia, 42 of 50 major rivers are reported to be "ecological disasters." Residues from palm oil and rubber manufacturing, along with heavy erosion from logging of tropical rainforests, have destroyed all higher forms of life in most of these rivers. In the Philippines, domestic sewage makes up 60 to 70 percent of the total volume of Manila's Pasig River. Thousands of people

FIGURE 10.24 Ditches in this Haitian slum serve as open sewers, into which all manner of refuse and waste are dumped. The health risks of living under these conditions are severe.

use the river not only for bathing and washing clothes but also as their source of drinking and cooking water. China treats only 2 percent of its sewage. Of 78 monitored rivers in China, 54 are reported to be seriously polluted. Of 44 major cities in China, 41 use "contaminated" water supplies, and few do more than rudimentary treatment before it is delivered to the public.

Groundwater is especially hard to clean up

About half the people in the United States, including 95 percent of those in rural areas, depend on underground aquifers for their drinking water. This vital resource is threatened in many areas by overuse and pollution and by a wide variety of industrial, agricultural, and domestic contaminants. For decades it was widely assumed that groundwater was impervious to pollution because soil would bind chemicals and cleanse water as it percolated through. Springwater or artesian well water was considered to be the definitive standard of water purity, but that is no longer true in many areas.

One of the serious sources of groundwater pollution throughout the United States is MTBE (methyl tertiary butyl ether), a suspected carcinogen added to gasoline to reduce carbon monoxide and ozone in urban air. Aquifers across the United States have been contaminated—mainly from leaking underground storage tanks at gas stations. In one U.S. Geological Survey (USGS) study, 27 percent of shallow urban wells tested contained MTBE. The additive is being phased out, but plumes of tainted water will continue to move through aquifers for decades to come. Liability for this contamination is a highly contentious issue.

The EPA estimates that every day some 4.5 trillion l (1.2 trillion gal) of contaminated water seep into the ground in the United States from septic tanks, cesspools, municipal and industrial landfills and waste disposal sites, surface impoundments, agricultural fields, forests, and wells (fig. 10.25). The most toxic of these are

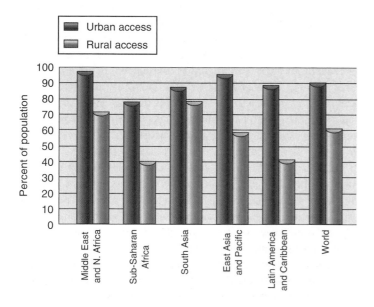

FIGURE 10.23 Proportion of people in developing regions with access to safe drinking water.
Source: UNESCO, 2002.

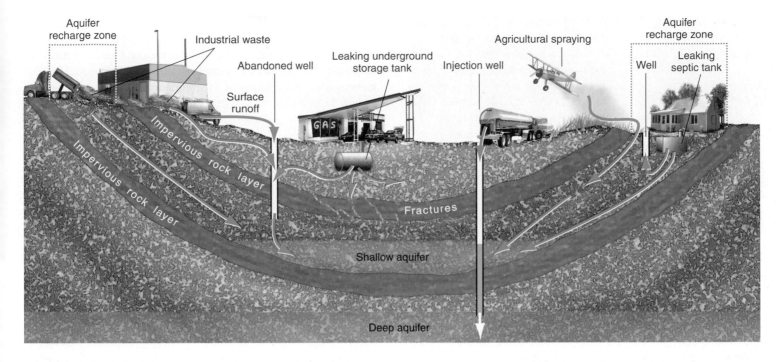

FIGURE 10.25 Sources of groundwater pollution. Septic systems, landfills, and industrial activities on aquifer recharge zones leach contaminants into aquifers. Wells provide a direct route for injection of pollutants into aquifers.

probably waste disposal sites. Agricultural chemicals and wastes are responsible for the largest total volume of pollutants and area affected. Because deep underground aquifers often have residence times of thousands of years, many contaminants are extremely stable once underground. It is possible, but expensive, to pump water out of aquifers, clean it, and then pump it back.

In farm country, especially in the Midwest's corn belt, fertilizers and pesticides commonly contaminate aquifers and wells. Herbicides such as atrazine and alachlor are widely used on corn and soybeans and show up in about half of all wells in Iowa, for example. Nitrates from fertilizers often exceed safety standards in rural drinking water. These high nitrate levels are dangerous to infants (nitrates combine with hemoglobin in the blood and result in "blue-baby" syndrome).

Every year, epidemiologists estimate that around 1.5 million Americans fall ill from infections caused by fecal contamination. In 1993, for instance, a pathogen called cryptosporidium got into the Milwaukee public water system, making 400,000 people sick and killing at least 100 people. The total costs of these diseases amount to billions of dollars per year. Preventative measures, such as protecting water sources and aquifer recharge zones and updating treatment and distribution systems, would cost far less.

Ocean pollution has few controls

Although we don't use ocean waters directly, ocean pollution is serious and one of the fastest-growing water pollution problems. Coastal bays, estuaries, shoals, and reefs are often overwhelmed by pollution. Dead zones and poisonous algal blooms are increasingly widespread. Toxic chemicals, heavy metals, oil, sediment, and plastic refuse affect some of the most attractive and productive ocean regions. The potential losses caused by this pollution amount to billions of dollars each year. In terms of quality of life, the costs are incalculable.

Discarded plastic flotsam and jetsam are becoming a ubiquitous mark of human impact on the oceans. Since plastic is lightweight and nonbiodegradable, it is carried thousands of miles on ocean currents and lasts for years. Even the most remote beaches of distant islands are likely to have bits of polystyrene foam containers or polyethylene packing material that were discarded half a world away. It has been estimated that 6 million metric tons of plastic bottles, packaging material, and other litter are tossed from ships every year into the ocean, where they ensnare and choke seabirds, mammals, and even fish (fig. 10.26).

Oil pollution affects beaches and open seas around the world. Oceanographers estimate that between 3 million and 6 million metric tons of oil are discharged into the world's oceans each year from oil tankers, fuel leaks, intentional discharges of fuel oil, and coastal industries. About half of this amount is due to maritime transport. Of this portion, most is not from dramatic, headline-making accidents such as the 1989 *Exxon Valdez* spill in Alaska but, rather, from routine, open-sea bilge pumping and tank cleaning. These activities are illegal but very common.

The transport of huge quantities of oil creates opportunities for major oil spills through a combination of human and natural hazards. Military conflict in the Middle East destabilizes shipping routes. More important, drilling and transport in stormy seas cause spills. Plans to drill for oil along the seismically active California and Alaska coasts have been controversial because of the damage that spills could cause to these biologically rich coastal ecosystems.

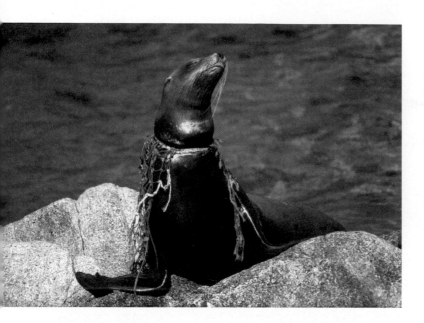

FIGURE 10.26 A deadly necklace. Marine biologists estimate that cast-off nets, plastic beverage yokes, and other packing residue kill hundreds of thousands of birds, mammals, and fish each year.

Fortunately, awareness of ocean pollution is growing. Oil spill cleanup technologies and response teams are improving, although most oil is eventually decomposed by natural bacteria. Efforts are growing to control waste plastic. Many states now require that six-pack yokes be made of biodegradable or photodegradable plastic, limiting their longevity as potential killers. International concern about ocean ship waste is increasing, and some shipping companies have been prosecuted and fined for dumping fuel oil. Beach pollution—mainly plastic debris, but also sewage waste, oil, and chemical contaminants—is becoming more common, but it is also more frequently reported in the mainstream media. Volunteer efforts are helping to reduce beach pollution locally: in one day, volunteers in Texas gathered more than 300 tons of plastic refuse from Gulf Coast beaches.

POLLUTION CONTROL

The cheapest and most effective way to reduce pollution is to avoid producing it or releasing it in the first place. Eliminating lead from gasoline has resulted in a widespread and significant decrease in the amount of lead in U.S. surface waters. Studies have shown that as much as 90 percent less road deicing salt can be used in many areas without significantly affecting the safety of winter roads. Careful handling of oil and petroleum products can greatly reduce the amount of water pollution caused by these materials. Although we still have problems with persistent chlorinated hydrocarbons spread widely in the environment, the banning of DDT and PCBs in the 1970s has resulted in significant reductions in levels in wildlife.

Industry can reduce pollution by recycling or reclaiming materials that otherwise might be discarded in the waste stream.

These approaches usually have economic as well as environmental benefits. Companies can extract valuable metals and chemicals and sell them, instead of releasing them as toxic contaminants into the water system. Both markets and reclamation technologies are improving as awareness of these opportunities grows. In addition, modifying land use is an important component of reducing pollution.

Nonpoint sources are often harder to control than point sources

Farmers have long contributed a huge share of water pollution, especially in the developed world. Increasingly, though, farmers are finding ways to save money and water quality at the same time. Soil conservation practices on farmlands (chapter 7) maintain soil fertility, as well as protect water quality. Precise application of fertilizer, irrigation water, and pesticides saves money and reduces water contamination. Preserving wetlands that act as natural processing facilities for removing sediment and contaminants helps protect surface and groundwaters.

In urban areas, reducing waste that enters storm sewers is essential. It is getting easier for city residents to recycle waste oil and to properly dispose of paint and other household chemicals that they once dumped into storm sewers or the garbage. Urbanites can also minimize use of fertilizers and pesticides. Regular street sweeping greatly reduces nutrient loads (from decomposing leaves and debris) in rivers and lakes. Runoff can also be diverted away from streams and lakes. Many cities are separating storm sewers and municipal sewage lines to avoid overflow during storms.

A good example of the problems of watershed management is seen in Chesapeake Bay, America's largest estuary. Once fabled for its abundant oysters, crabs, shad, striped bass, and other valuable fisheries, the bay had deteriorated seriously by the early 1970s. Citizens' groups, local communities, state legislatures, and the federal government together established an innovative pollution-control program that made the bay the first estuary in America targeted for protection and restoration.

Among the principal objectives of this plan is reducing nutrient loading through land-use regulations in the bay's six watershed states to control agricultural and urban runoff. Pollution-prevention measures, such as banning phosphate detergents, also are important, as are upgrading wastewater treatment plants and improving compliance with discharge and filling permits. Efforts are underway to replant thousands of hectares of sea grasses and to restore wetlands that filter out pollutants. Since the 1980s, annual phosphorous discharges into Chesapeake Bay have dropped 40 percent. Nitrogen levels, however, have remained constant or have even risen in some tributaries. Although progress has been made, the goals of reducing both nitrogen and phosphate levels by 40 percent and restoring viable fish and shellfish populations are still decades away. Still, as EPA Administrator Carol Browner says, it demonstrates the "power of cooperation" in environmental protection. (See related story "Watershed Protection in the Catskills" at <u>www.mhhe.com/cunningham4e</u>.)

Human waste degrades naturally in low concentrations

As we have already seen, human and animal wastes usually create the most serious health-related water pollution problems. More than 500 types of disease-causing (pathogenic) bacteria, viruses, and parasites can travel from human or animal excrement through water.

Natural processes

In the poorer countries of the world, most rural people simply go out into the fields and forests to relieve themselves, as they have always done. Where population densities are low, natural processes eliminate wastes quickly, making this an effective method of sanitation. The high population densities of cities, however, make this practice unworkable. Even major cities of many less-developed countries are often littered with human waste that has been left for rains to wash away or for pigs, dogs, flies, beetles, or other scavengers to consume. This is a major cause of disease, as well as being extremely unpleasant. Studies have shown that a significant portion of the airborne dust in Mexico City is actually dried, pulverized human feces.

Where intensive agriculture is practiced—especially in wet rice paddy farming in Asia—it has long been customary to collect "night soil" (human and animal waste) to be spread on the fields as fertilizer. This waste is a valuable source of plant nutrients, but it is also a source of disease-causing pathogens in the food supply.

Until about 50 years ago, most rural American families and quite a few residents of towns and small cities depended on a pit toilet, or "outhouse," for waste disposal. Untreated wastes tended to seep into the ground, however, and pathogens sometimes contaminated drinking water. The development of septic tanks and properly constructed drain fields represented a considerable improvement in public health. Septic systems allow solids to settle in a tank, where bacteria decompose them; liquids percolate through soil, where soil bacteria presumably purify them. Where population densities are not too high, this can be an effective method of waste disposal. With urban sprawl, however, groundwater pollution often becomes a problem.

Municipal treatment has three levels of quality

Over the past 100 years, sanitary engineers have developed ingenious and effective municipal wastewater treatment systems to protect human health, ecosystem stability, and water quality. This topic is an important part of pollution control, and is a principal responsibility of every municipal government.

Primary treatment physically separates large solids from the waste stream with screens and settling tanks (fig. 10.27a). Settling tanks allow grit and some dissolved (suspended) organic solids to fall out as sludge. Water drained from the top of settling tanks still carries up to 75 percent of the organic matter, including many pathogens. These pathogens and organics are removed by **secondary treatment,** in which aerobic bacteria break down dissolved organic compounds. In secondary treatment, effluent is aerated, often with sprayers or in an aeration tank, in which air is pumped through the microorganism-rich slurry (fig. 10.27b). Fluids can also be stored in a sewage lagoon, where sunlight, algae, and air process waste more cheaply but more slowly. Effluent from secondary treatment processes is usually disinfected with chlorine, UV light, or ozone to kill harmful bacteria before it is released to a nearby waterway.

Tertiary treatment removes dissolved metals and nutrients, especially nitrates and phosphates, from the secondary effluent. Although wastewater is usually free of pathogens and organic material after secondary treatment, it still contains high levels of these inorganic nutrients. If discharged into surface waters, these nutrients stimulate algal blooms and eutrophication. Allowing effluent to flow through a wetland or lagoon can remove nitrates and phosphates. Alternatively, chemicals often are used to bind and precipitate nutrients (fig. 10.27c).

Sewage sludge can be a valuable fertilizer, but it can be controversial because it can contain metals and toxic chemicals. Many cities spread sludge on farms and forest lands, and European cities are beginning to convert it to methane (natural gas). Many cities, however, incinerate or landfill sludge, both expensive options.

In many American cities, sanitary sewers are connected to storm sewers, which carry runoff from streets and parking lots. Storm sewers are routed to the treatment plant rather than discharged into surface waters because runoff from streets, yards, and industrial sites generally contains a variety of refuse, fertilizers, pesticides, oils, rubber, tars, lead (from gasoline), and other undesirable chemicals. Unfortunately heavy storms often overload the system, especially where the system is old and already overburdened. As a result, large volumes of raw sewage and toxic surface runoff are dumped directly into receiving waters. To prevent this overflow, cities are spending hundreds of millions of dollars to separate storm and sanitary sewers.

Constructed wetlands are low-cost treatment

A number of alternative treatment systems have been developed. One of the most attractive is **constructed wetlands,** artificial marshes designed to filter and decompose waste. Arcata, California, for instance, needed an expensive sewer plant upgrade. Instead, the city transformed a 65-ha garbage dump into a series of ponds and marshes that serve as a simple, low-cost, waste treatment facility. Arcata saved millions of dollars and improved the environment simultaneously. The marsh is a haven for wildlife and has become a prized recreation area for the city. Eventually, the purified water flows into Humboldt Bay, where marine life flourishes.

Similar wetland waste treatment systems are now operating in many developing countries. Effluent from these operations can be used to irrigate crops or raise fish for human consumption if care is taken first to destroy pathogens. Usually 20 to 30 days of exposure to sun, air, and aquatic plants is enough to make the water safe. These systems make an important contribution to human food supplies. A 2,500 ha waste-fed aquaculture facility in Kolkata (Calcutta), for example, supplies about 7,000 metric tons of fish annually to local markets.

FIGURE 10.27 Activated sludge wastewater treatment. (a) Primary treatment removes only solids and suspended sediment. (b) Secondary treatment, through aeration of activated sludge, followed by sludge removal and chlorination of effluent, kills pathogens and removes most organic material. (c) During tertiary treatment, passage through a trickling bed evaporator and/or a tertiary filter further removes inorganic nutrients, oxidizes any remaining organics, and reduces effluent volume.

The World Bank estimates that more than 3 billion people will be without sanitation services by the year 2030 under a business-as-usual scenario. With investments in innovative programs, however, sanitation could be provided to about half those people, and a great deal of misery and suffering could be avoided.

Waste treatment can even be done indoors. Ocean Arks International in Falmouth, Massachusetts, has been developing holistic systems for water purification that are combinations of different plants and animals, including algae, rooted aquatic plants, clams, snails, and fish, each chosen to provide a particular service in a contained environment. Technically, the water that has flowed through such a system is drinkable, although few people feel comfortable doing so. More often, the final effluent is used to flush toilets or for irrigation. Called ecological engineering, this novel approach can save resources and money, and it can serve as a valuable educational tool (fig. 10.28).

Conventional treatment misses new pollutants

Countless additional organic compounds enter our water unmonitored. In 2002, the USGS released the first-ever study of pharmaceuticals and hormones in streams. Scientists sampled 130 streams, looking for 95 contaminants, including antibiotics, natural and synthetic hormones, detergents, plasticizers, insecticides, and fire retardants (fig. 10.29). All these substances were found, usually in low concentrations. One stream had 38 of the compounds tested. Drinking water standards exist for only 14 of the 95 substances. A similar study found the same substances in groundwater, which is much harder to clean than surface waters. What are the effects of these widely used chemicals, on our environment or on people consuming the water? Nobody knows. This study is a first step toward filling huge gaps in our knowledge about their distribution, though.

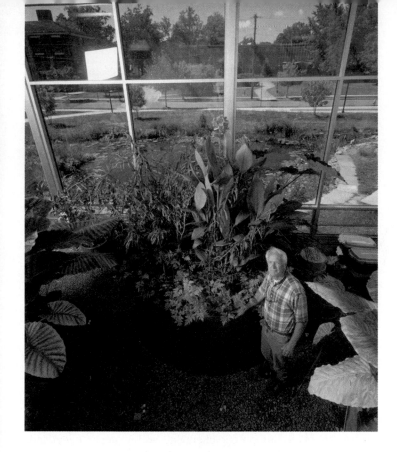

FIGURE 10.28 In-house wastewater treatment in Oberlin College's Environmental Studies building. Constructed wetlands outside, and tanks inside, allow plants to filter water and remove nutrients.

Remediation can involve containment, extraction, or plants

Just as there are many sources of water contamination, there are many ways to clean it up. New developments in environmental engineering are providing promising solutions to many water pollution problems. Containment methods keep dirty water from spreading. Chemicals can be added to toxic wastewater to precipitate, immobilize, or solidify contaminants. Many pollutants can be destroyed or detoxified by chemical reactions that oxidize, reduce, neutralize, hydrolyze, precipitate, or otherwise change their chemical composition. Where chemical techniques are ineffective, physical methods may work. Solvents and other volatile organic compounds, for instance, can be stripped from solution by aeration and then burned in an incinerator. (See related story "International Accord to Clean Up the Rhine River" at www.mhhe.com/cunningham4e.)

Often, living organisms can clean contaminated water effectively and inexpensively. We call this **bioremediation.** Restored wetlands, for instance, along stream banks or lake margins can effectively filter out sediment and remove pollutants. Some plants are very efficient at taking up heavy metals and organic contaminants. Bioremediation offers exciting and inexpensive alternatives to conventional cleanup (see Exploring Science, p. 250).

Lowly duckweed (*Lemna* sp.), the green scum you often see covering the surface of eutrophic ponds, grows fast and can remove large amounts of organic nutrients from water. Large duckweed lagoons are being used as inexpensive, low-tech, sewage treatment plants in developing countries. The duckweed can also be harvested and used as feed, fuel, or fertilizer. Up to 35 percent of its dry mass is protein—about twice as much as alfalfa, a popular animal feed.

WATER LEGISLATION

Water pollution control has been among the most broadly popular and effective of all environmental legislation in the United States. It has not been without controversy, however. Table 10.5 describes some of the most important water legislation in the United States.

The Clean Water Act was ambitious, popular, and largely successful

Passage of the U.S. Clean Water Act of 1972 was a bold, bipartisan step that made clean water a national

FIGURE 10.29 Detection frequency of organic, wastewater contaminants in a recent USGS survey. Maximum concentrations in water samples are shown above the bars in micrograms per liter. Dominant substances included DEET insect repellent, caffeine, and triclosan, which comes from antibacterial soaps.

TABLE 10.5

Some Important Water Quality Legislation

1. *Federal Water Pollution Control Act (1972).* Establishes uniform nationwide controls for each category of major polluting industries.

2. *Marine Protection Research and Sanctuaries Act (1972).* Regulates ocean dumping and established sanctuaries for protection of endangered marine species.

3. *Ports and Waterways Safety Act (1972).* Regulates oil transport and the operation of oil-handling facilities.

4. *Safe Drinking Water Act (1974).* Requires minimum safety standards for every community water supply. Among the contaminants regulated are bacteria, nitrates, arsenic, barium, cadmium, chromium, fluoride, lead, mercury, silver, and pesticides; radioactivity and turbidity also are regulated. This act also contains provisions to protect groundwater aquifers.

5. *Resource Conservation and Recovery Act (RCRA) (1976).* Regulates the storage, shipping, processing, and disposal of hazardous wastes and sets limits on the sewering of toxic chemicals.

6. *Toxic Substances Control Act (TOSCA) (1976).* Categorizes toxic and hazardous substances, establishes a research program, and regulates the use and disposal of poisonous chemicals.

7. *Comprehensive Environmental Response, Compensation, and Liability Act (CERCLA) (1980)* and *Superfund Amendments and Reauthorization Act (SARA) (1984).* Provide for sealing, excavation, or remediation of toxic and hazardous waste dumps.

8. *Clean Water Act (1985) (amending the 1972 Water Pollution Control Act).* Sets as a national goal the attainment of "fishable and swimmable" quality for all surface waters in the United States.

9. *London Dumping Convention (1972).* Calls for an end to all ocean dumping of industrial wastes, tank-washing effluents, and plastic trash. The United States is a signatory to this international convention.

priority. Along with the Endangered Species Act and the Clean Air Act, this is one of the most significant and effective pieces of environmental legislation ever passed by the U.S. Congress. It also is an immense and complex law, with more than 500 sections regulating everything from urban runoff, industrial discharges, and municipal sewage treatment to land-use practices and wetland drainage.

The ambitious goal of the Clean Water Act was to return all U.S. surface waters to "fishable and swimmable" conditions. For point sources, the act requires discharge permits and use of the best practicable control technology (BPT). For toxic substances, the act sets national goals of best available, economically achievable technology (BAT) and zero discharge goals for 126 priority toxic pollutants. As discussed earlier, these regulations have had a positive effect on water quality. While not yet swimmable or fishable everywhere, surface-water quality in the United States has significantly improved on average over the past quarter century. Perhaps the most important result of the act has been investment of $54 billion in federal funds and more than $128 billion in state and local funds for municipal sewage treatment facilities.

Opponents of federal regulation have tried repeatedly to weaken or eliminate the Clean Water Act. They regard restriction of their "right" to dump toxic chemicals and wastes into wetlands and waterways to be an undue loss of freedom. They resent being forced to clean up municipal water supplies and call for cost/benefit analysis that places greater weight on economic interests in all environmental planning.

Supporters of the Clean Water Act would like to see a shift away from an "end-of-the-pipe" focus on effluent removal and more attention to changing industrial processes, so that toxic substances aren't produced in the first place. Many people also would like to see stricter enforcement of existing regulations, mandatory minimum penalties for violations, more effective community right-to-know provisions, and increased powers for citizen lawsuits against polluters.

REVIEW QUESTIONS

1. Describe the path a molecule of water might follow through the hydrologic cycle from the ocean to land and back again.

2. Describe the distribution of rainfall on the earth, in terms of latitude and/or countries and regions that receive most abundant precipitation and least precipitation (see fig. 10.1).

3. What is an acre-foot of water?

4. About what percent of the world's water is liquid, fresh, surface water that supports most terrestrial life (see fig. 10.2)?

5. What is an *aquifer*? How does water get into an aquifer? Explain the idea of an *artesian well* and a *cone of depression*.

6. What is the difference between water *withdrawal* and *consumption*? Which sector of water use (see fig. 10.8) consumes most globally? Overall, has water use increased in the past century? Has efficiency increased or decreased in the three main use sectors?

7. Describe some of the environmental and social costs associated with dam building.

8. Describe at least one example of the environmental costs of water diversion from rivers to farms or cities.

9. Define *water pollution.*

10. Explain the difference between point and nonpoint pollution. Which is harder to control? Why?

11. Why are nutrients considered pollution? Explain the ideas of *eutrophication* and an *oxygen sag* (see fig. 10.18).

12. List three main kinds of inorganic pollutants.

13. When was the Clean Water Act passed, and what were some of its main ideas?

14. Describe primary, secondary, and tertiary water treatment.

15. What are some sources of groundwater contamination? Why is groundwater pollution such a difficult problem?

16. What is *bioremediation*?

17. Why are nutrients important factors in water pollution and eutrophication? What two elements are the most important nutrients in water pollution?

18. Does rainfall contribute to groundwater? Does groundwater contribute to ponds, lakes, and streams? How?

19. What is the difference between water withdrawal and consumption? What sector consumes the most water? Why? Which sector has improved its efficiency least in the past century?

20. Why is the Aral Sea drying? Where is the Aral Sea? (See maps in Appendixes for help.)

CRITICAL THINKING AND DISCUSSION

Apply the principles you have learned in this chapter to discuss these questions with other students.

1. What changes might occur in the hydrologic cycle if our climate were to warm or cool significantly?

2. Why does it take so long for deep ocean waters to circulate through the hydrologic cycle? What happens to substances that contaminate deep ocean water or deep aquifers in the ground?

3. How precise do you think the estimate is that 1.5 billion people lack access to safe drinking water? If, in 2025, two-thirds of the world's population live in countries with severe water shortages, what might be the number of people who lack access to clean water? Why is this number ambiguous?

4. Do you think that water pollution is worse now than it was in the past? What considerations go into a judgment such as this? How do your personal experiences influence your opinion?

5. What additional information would you need to make a judgment about whether conditions are getting better or worse? How would you weigh different sources, types, and effects of water pollution?

6. Under what conditions might sediment in water or cultural eutrophication be beneficial? How should we balance positive and negative effects?

7. Suppose that part of the silt in a river is natural and part is human-caused. How might you evaluate what proportion of the silt is caused by humans?

SUGGESTED READINGS

Ayres, Gene. 2003. Rocket fuel in our food. *Worldwatch* 16(6):12–20.

Baron, J. S., et al. 2002. Meeting ecological and societal needs for freshwater. *Ecological Applications* 12(5):1247–60.

Beman, J. M., et al. 2005. "Agricultural runoff fuels large phytoplankton blooms in vulnerable areas of the ocean." *Nature* 434:211–214.

Boyd, G. R., et al. 2004. "Pharmaceuticals and personal care products (PPCPs) and endocrine disrupting chemicals (EDCs) in stormwater canals and Bayou St. John in New Orleans, Louisiana, USA." *Science of the Total Environment* 222(1–3):137–148.

Bromberg, Gideon. 2004. "Water and Peace" *World Watch* 17(4): 24–30.

Chowdhury, A. M. R. 2004. "Arsenic Crisis in Bangladesh." *Scientific American* 291(2):86–92.

Gleick, Peter H. 2003. Global freshwater resources: soft-path solutions for the 21st century. *Science* 302:1524–28.

Glennon, Robert. 2003. *Water Follies*. Island Press.

Grenoble, Penelope. 2003. Yangtze farewell. *Orion* 22(6):26–33.

Katsoyiannis, I. A., and A. I. Zouboulis. 2004. Application of biological processes for the removal of arsenic from groundwaters. *Water Research* 38(1):17–26.

Katsoyiannis, I. A. and A. I. Zouboulis. 2004. "Application of biological processes for the removal of arsenic from groundwaters." *Water Research* 38(1):17–26.

Leslie, Jacques. 2005. *Deep Water: The Epic Struggle Over Dams, Displaced People, and the Environment*. Farrar, Straus and Giroux.

Nilsson, C., et al. 2005. "Fragmentation and Flow Regulation of the World's Large River Systems." *Science* 308(5720): 405–408.

Olmstead, Sheila M. 2003. Water supply and poor communities: What's price got to do with it? *Environment* 45(10):22–35.

Peglar, M. T., et al. 2004. "Identification of amoebae implicated in the life cycle of Pfiesteria and Pfiesteria-like dinoflagellates." *Journal of Eukaryotic Microbiology* 51(5):542–552.

Pestana, M. H. D., and M. L. L. Formoso. 2003. Mercury contamination in Lavras do Sul, south Brazil: A legacy from past and recent gold mining. *The Science of the Total Environment* 307(1–3):125–40.

Postel, Sandra. 2001. Growing food with less water. *Scientific American* 284(2):46–51.

Reisner, Marc. 1993. *Cadillac Desert: The American West and Its Disappearing Water*. Penguin Books.

Sarkar, S., et al. 2005. "Well-head arsenic removal units in remote villages of Indian subcontinent: Field results and performance evaluation." *Water Research* 39(10):2196–2206.

Shiva, Vandana. 2002. *Water Wars: Privatization, Pollution and Profit.* South End Press.

Webb, S. R., et al. 2005. "Etiology of ulcerative lesions of Atlantic menhaden (*Brevoortia tyrannus*) from James River, Virginia." *Parasitology Research* Aug. 16 PMID 16133297.

Yang, Shinwoo, and Kenneth Carlson. 2003. Evolution of antibiotic occurrence in a river through pristine, urban and agricultural landscapes. *Water Research* 37(19):4645–56.

 DATA analysis   **Examining Pollution Sources**

Understanding the origins of pollution is the first step toward considering policies for reducing it. The chapter you have just read includes several graphs displaying pollution sources. The following questions ask you to think more about these sources:

1. In figure 10.22 (p. 253), which sources do you think are mainly point sources? Which are mainly nonpoint sources? How might you consider urban runoff to be either point or nonpoint? (If you need ideas to start, see figure 10.25, p. 255)

2. Explain the units on the Y-axis in figure 10.22. Does the graph show you the amount of actual pollution, by weight or volume, from different sources? Why might it be useful to show the extent of impacts?

3. Based on your knowledge of water contaminants, describe the kinds of pollutants you would expect from each of the sources shown:

   Agriculture _____

   Sewage treatment _____

   Dams, diversions, channelization _____

   Urban runoff _____

   Mining _____

   Removal of streamside vegetation _____

   Forestry _____

4. How would you design a sampling strategy to assess water pollution on your school campus?

5. Figure 10.29 (p. 259) shows some of the new developments in water pollution assessment. Conventional treatment systems were not designed to remove thousands of newly invented chemical compounds, or increasingly widespread compounds, including those shown in the figure. Explain the units used on the Y-axis. Also explain why the numbers (maximum concentrations found) are shown.

6. How many of the pollutants shown do *you* use? Don't forget to include caffeine (which is classed as a nonprescription drug) and antibacterial soaps (disinfectants). Steroids include cholesterol, which occurs naturally in foods.

7. Try to think of additional substances that you might contribute to wastewater.

8. This graph results from a reconnaissance study done by the U.S. Geological Survey. The researchers wanted to assess whether a list of 95 contaminants could be detected at all in public waterways. If you wanted to design a study like this, what sorts of sites would you select for sampling? How might your results and your conclusions differ if you did a random sample?

For Additional Help in Studying This Chapter, please visit our website at www.mhhe.com/cunningham4e. You will find practice quizzes, a chapter summary, key terms, answers to review questions, additional case studies, regional examples, and an extensive reading list, all of which will help you understand the material in this chapter.

An aerial view of the Jonah Field in the Upper Green River Basin shows some of the thousands of well pads connected by a spider web of service roads and pipelines spread across the landscape. This gas field lies across the annual migration route of 50,000 antelope, the greatest mammal migration in the lower 48 states.

C H A P T E R

Environmental Geology and Earth Resources

11

When we heal the earth, we heal ourselves.
–David Orr–

OBJECTIVES

Studying this chapter will help you answer the following questions:

- What is coal-bed methane? Why is it both important and controversial?
- How do continental and oceanic crust differ?
- Why do volcanoes occur?
- What are some of the environmental and social costs of mining and oil- and gas-drilling?

- What are some solutions to those costs?
- How can we reduce our consumption of geologic resources?
- Why is mass wasting a problem?

Case Study Coal-Bed Methane: A Clean Fuel or a Dirty Business?

Of the many resources we extract from the earth, fossil fuels are among the most valuable. Coal-bed methane, from expansive coal deposits mainly in Western states, is one of the newest and most controversial of these energy sources. Methane (CH_4) is the main component of natural gas. It occurs naturally within coal deposits, formed within ancient swamp muck and peat bogs, as well as by the same geologic heat and pressure that compressed that ancient muck and peat into rock-hard coal.

Traditionally, methane has been a hazardous mine gas that can suffocate miners or explode, or it has been released to the atmosphere, where it contributes to global warming. Most coal-bed methane is held in place by pressure from overlying aquifers (rock formations saturated with water). To release the gas, water is pumped from the aquifer to the surface. As water pressure is released, methane escapes from the coal seam. It rises to the surface with the water and is captured. All that groundwater carries salts and minerals dissolved from the surrounding rock, mainly sodium, barium, bicarbonates, sulfate, boron, and iron. Although none of these is toxic at low levels, they harm both animals and plants at high concentrations.

The volume of water pumped to release coal-bed methane is phenomenal. A typical coal-bed methane well produces 75,000 liters of water per day. Wells are usually distributed thickly across the landscape (see opening photo), multiplying the volume of water produced in a watershed. Much of this water is allowed to seep into groundwater and shallow aquifers or is released into natural waterways. The increased water volume erodes stream banks, and the elevated salt levels poison fish, wildlife, and aquatic ecosystems. In arid Western states, natural streams and groundwater are also needed for irrigating farms and pastures. In some areas, groundwater is contaminated. In others, water tables have fallen as coal-bed aquifers are drawn down, and farmers and ranchers lose their wells and springs. Wetlands also dry up as groundwater falls, and wildlife is left stranded. "It may be a clean fuel," says one rancher, "but it's a dirty business."

The sheer extent and density of operations also make coal-bed methane controversial. In Wyoming's Powder River Basin, energy companies have already installed nearly 14,000 wells and have proposed 39,000 more. Eventually, this area could contain as many as 140,000 wells, together with a sprawling network of roads, pipelines, compressor stations, and wastewater pits. The Green River Basin and the San Juan Basin, with three to five times as much potential gas and oil as the Powder River Basin, may see even greater environmental damage.

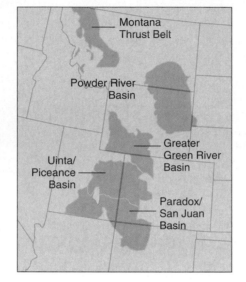

Coal-bed methane deposits occur in five intermountain basins in the western United States.

In 2002 the U.S. Environmental Protection Agency (EPA) gave its worst possible rating to the Environmental Impact Statement (EIS) for the proposed Powder River wells because of concerns over wastewater disposal. Nevertheless, the Bush administration approved the plan and ordered federal land managers across the Rockies to look for ways to remove or reduce environmental restrictions on gas drilling. This order came in spite of a federal study finding that 63 percent of the natural gas in the region was completely open to drilling and only 12 percent was completely protected.

An unlikely coalition of ranchers, hunters, anglers, conservationists, water users, and renewable energy activists have banded together to fight coal-bed gas extraction, calling on Congress to protect private property rights, to enforce the Clean Water Act, and to conserve sensitive public lands. Lifelong Republicans have banded together with tree-hugging environmentalists to protect their way of life.

On the other hand, oil and gas companies argue that they are contributing to American energy independence. As long as we choose to drive large vehicles, commute long distances, and heat and cool large houses, we will need fossil fuels. Methane is the cleanest-burning fossil fuel, producing far less particulates, sulfur oxides, and other pollutants than coal. Brine from the gas wells may be disagreeable, but it isn't nearly as toxic and long-lasting as nuclear waste. The dense networks of roads and drilling pads are unsightly and hazardous to wildlife, but they're better than strip-mining.

Expect to see many new disputes over fossil fuel extraction in the decades ahead. Geologists estimate that at least 346 trillion ft^3 of "technically recoverable" natural gas and 62 billion barrels of petroleum occurs in the five intermountain basins stretching from Montana to New Mexico. Extracting these would involve disturbing vast stretches of country, but it could also provide us with a 15-year supply of gas at present rates of use, and at least four times as much oil as the most optimistic estimates for the Arctic National Wildlife Refuge. The total value of methane and petroleum liquids from the Rocky Mountains could be as much as $200 billion over the next decade. Beyond this, 13 percent of the lower 48 states has some coal, and extractable methane occurs with many of these deposits. Vast deposits of oil-saturated sand and shale also underlie the forests and plains of Alberta and neighboring provinces. The debates over extracting these reserves will not end soon.

Fossil fuels are just one of the essential geologic resources on which we depend. In this chapter we'll examine geologic resources and consider why they're distributed as they are. We will also consider some of the environmental and social costs of extracting them, as well as ways those costs can be reduced.

EARTH PROCESSES SHAPE OUR RESOURCES

Every one of us shares the benefits of geologic resources. Right now you are probably wearing several geologic products: plastics, including glasses and synthetic fabric, are made from oil; iron, copper, and aluminum mines produced your snaps, zippers, and the screws in your glasses; silver, gold, and diamond mines may have produced your jewelry. All of us also share responsibility for the environmental and social devastation that often results from mining and drilling. Fortunately, there are many promising solutions to reduce these costs, including recycling and alternative materials. The question is whether voters will demand that we use these technologies—and whether consumers will share the costs of responsible production.

Why are these resources distributed as they are? To understand how and where earth resources are created, we must examine the earth's structure and the processes that shape it.

Earth is a dynamic planet

Although we think of the ground under our feet as solid and stable, the earth is a dynamic and constantly changing structure. Titanic forces inside the earth cause continents to split, move apart, and then crash into each other in slow but inexorable collisions.

The earth is a layered sphere. The **core,** or interior, is composed of a dense, intensely hot mass of metal—mostly iron—thousands of kilometers in diameter (fig. 11.1). Solid in the center but more fluid in the outer core, this immense mass generates the magnetic field that envelops the earth.

Surrounding the molten outer core is a hot, pliable layer of rock called the **mantle.** The mantle is much less dense than the core because it contains a high concentration of lighter elements, such as oxygen, silicon, and magnesium.

The outermost layer of the earth is the cool, lightweight, brittle rock **crust.** The crust below oceans is relatively thin (8–15 km), dense, and young (less than 200 million years old) because of constant recycling. Crust under continents is relatively thick (25–75 km), light, and as old as 3.8 billion years, with new material being added continually. It also is predominantly granitic, while oceanic crust is mainly dense basaltic rock. Table 11.1 compares the composition of the whole earth (dominated by the dense core) and the crust.

Tectonic processes reshape continents and cause earthquakes

The huge convection currents in the mantle are thought to break the overlying crust into a mosaic of huge blocks called **tectonic plates** (fig. 11.2). These plates slide slowly across the earth's surface like wind-driven ice sheets on water, in some places breaking up into smaller pieces, in other places crashing ponderously into each other to create new, larger landmasses. Ocean basins form where continents crack and pull apart. The Atlantic Ocean, for example, is growing slowly as Europe and Africa move away

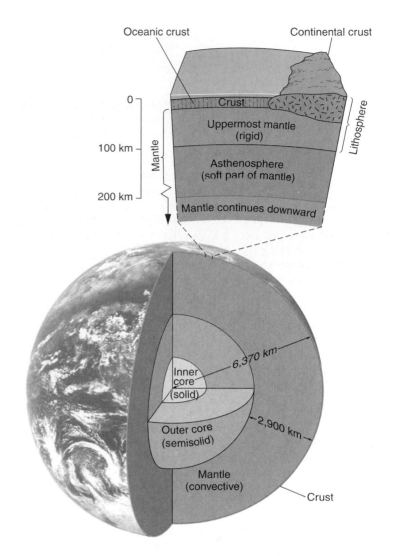

FIGURE 11.1 Earth's cross section. Slow convection in the mantle causes the thin, brittle crust to move.

TABLE 11.1

Eight Most Common Chemical Elements (Percent) in Whole Earth and Crust

Whole Earth		Crust	
Iron	33.3	Oxygen	45.2
Oxygen	29.8	Silicon	27.2
Silicon	15.6	Aluminum	8.2
Magnesium	13.9	Iron	5.8
Nickel	2.0	Calcium	5.1
Calcium	1.8	Magnesium	2.8
Aluminum	1.5	Sodium	2.3
Sodium	0.2	Potassium	1.7

FIGURE 11.2 Map of tectonic plates. Plate boundaries are dynamic zones, characterized by earthquakes, volcanism, and the formation of great rifts and mountain ranges. Arrows indicate direction of subduction where one plate is diving beneath another. These zones are sites of deep trenches in the ocean floor and high levels of seismic and volcanic activity.

Sources: Data from U.S. Department of the Interior and U.S. Geological Survey.

from the Americas. **Magma** (molten rock) forced up through the cracks forms new oceanic crust that piles up underwater in **mid-ocean ridges.** Creating the largest mountain range in the world, these ridges wind around the earth for 74,000 km (46,000 mi) (see fig. 11.2). Although concealed from our view, this jagged range boasts higher peaks, deeper canyons, and sheerer cliffs than any continental mountains. Slowly spreading from these fracture zones, ocean plates push against continental plates.

Earthquakes are caused by grinding and jerking as plates slide past each other. Mountain ranges like those on the west coast of North America and in Japan are pushed up at the margins of colliding continental plates. The Himalayas are still rising as the Indian subcontinent grinds slowly into Asia. Southern California is slowly sailing north toward Alaska. In about 30 million years, Los Angeles will pass San Francisco, if both still exist by then.

When an oceanic plate collides with a continental landmass, the continental plate usually rides up over the seafloor, while the oceanic plate is **subducted,** or pushed down into the mantle, where it melts and rises back to the surface as magma (fig. 11.3). Deep ocean trenches mark these subduction zones, and volcanoes form where the magma erupts through vents and fissures in the overlying crust. Trenches and volcanic mountains ring the Pacific Ocean rim from Indonesia to Japan to Alaska and down the west coast of the Americas, forming a so-called ring of fire where

oceanic plates are being subducted under the continental plates. This ring is the source of more earthquakes and volcanic activity than any other region on the earth.

Over millions of years, continents can drift long distances. Antarctica and Australia once were connected to Africa, for instance, somewhere near the equator and supported luxuriant forests. Geologists suggest that several times in the earth's history most or all of the continents have gathered to form supercontinents, which have ruptured and re-formed over hundreds of millions of years (fig. 11.4). The redistribution of continents has profound effects on the earth's climate and may help explain the periodic mass extinctions of organisms marking the divisions between many major geologic periods (fig. 11.5).

MINERALS AND ROCKS

A **mineral** is a naturally occurring, inorganic solid with a specific chemical composition and a specific internal crystal structure. A mineral is solid; therefore, ice is a mineral (with a distinct composition and crystal structure), but liquid water is not. Similarly molten lava is not crystalline, although it generally hardens to create distinct minerals. Metals (such as iron, copper, aluminum, or gold) come from mineral ores, but once purified, metals are

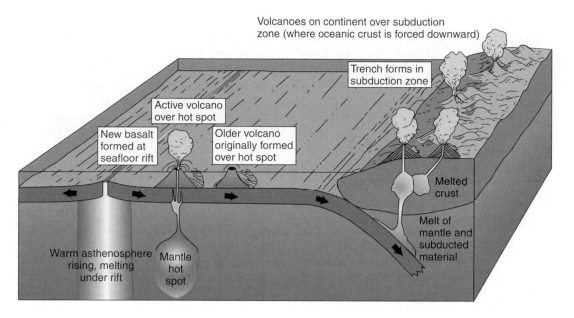

FIGURE 11.3 Plate tectonic movement. Where thin, oceanic plates diverge, upwelling magma forms midocean ridges. A chain of volcanoes, such as the Hawaiian Islands, may form as plates pass over a hot spot. Where plates converge, melting can cause volcanoes, such as the Cascades.

no longer crystalline and thus are not minerals. Depending on the conditions in which they were formed, mineral crystals can be microscopically small, such as asbestos fibers, or huge, such as the tree-size selenite crystals recently discovered in a Chihuahua, Mexico mine.

FIGURE 11.4 Pangaea, an ancient supercontinent of 200 million years ago, combined all the world's continents in a single landmass. Continents have combined and separated repeatedly.

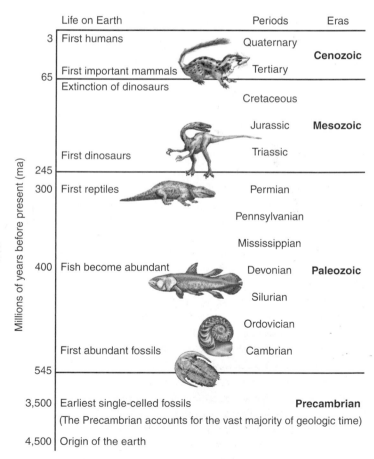

FIGURE 11.5 Periods and eras in geologic time, and major lifeforms that mark some periods.

A **rock** is a solid, cohesive aggregate of one or more minerals. Within the rock, individual mineral crystals (or grains) are mixed together and held firmly in a solid mass. The grains may be large or small, depending on how the rock was formed, but each grain retains its own unique mineral qualities. Each rock type has a characteristic mixture of minerals, grain sizes, and ways in which the grains are mixed and held together. Granite, for example, is a mixture of quartz, feldspar, and mica crystals. Rocks with a granite-like mineral content but much finer crystals are called rhyolite; chemically similar rocks with large crystals are called pegmatite.

The rock cycle creates and recycles rocks

Although rocks appear hard and permanent, they are part of a relentless cycle of formation and destruction. They are crushed, folded, melted, and recrystallized by dynamic processes related to those that shape the large-scale features of the earth's crust. We call this cycle of creation, destruction, and metamorphosis the **rock cycle** (fig. 11.6). Understanding something of how this cycle works helps explain the origin and characteristics of different types of rocks.

There are three major rock classifications: igneous, metamorphic, and sedimentary. **Igneous rocks** (from *igni,* the Latin word for fire) are solidified from hot, molten magma or lava. Most rock in the earth's crust is igneous. Magma extruded to the surface from volcanic vents cools quickly to make finely crystalline rocks, such as basalt, rhyolite, or andesite. Magma that cools slowly in subsurface chambers or is intruded between overlying layers makes coarsely crystalline rocks, such as gabbro (rich in iron and silica) or granite (rich in aluminum and silica), depending on the chemical composition of the magma.

Metamorphic rocks form from the melting, contorting, and recrystallizing of other rocks. Deep in the ground, tectonic forces squeeze, fold, heat, and recrystallize solid rock. Under these conditions, chemical reactions can alter both the composition and the structure of the component minerals. Metamorphic rocks are classified by their chemical composition and by the degree of recrystallization: some minerals form only under extreme pressure and heat (diamonds or jade, for example); others form under more moderate conditions (graphite or talc). Some common metamorphic rocks are marble (from limestone), quartzite (from sandstone), and slate (from mudstone and shale). Metamorphic rocks often have beautiful colors and patterns left by the twisting and folding that created them.

Sedimentary rocks are formed when loose grains of other rocks are consolidated by time and pressure. Sandstone, for example, is solidified from layers of sand, and mudstone consists of extremely hardened mud and clay. Tuff is formed from volcanic ash, and conglomerates are aggregates of sand and gravel. Some sedimentary rocks develop from crystals that precipitate out of extremely salty water. Rock salt, made of the mineral halite, is ground up to produce ordinary table salt (sodium chloride). Salt deposits often form when a body of saltwater dries up, leaving salt crystals behind. Limestone is a rock composed of cemented remains of marine organisms. You can often see the shapes of shells and corals in a piece of limestone. Sedimentary formations often have distinctive layers that show different conditions when they were laid down. Erosion can reveal these layers and inform us of their history (fig. 11.7).

Weathering and sedimentation

Most crystalline rocks are extremely hard and durable, but exposure to air, water, changing temperatures, and reactive chemical agents slowly breaks them down in a process called **weathering** (fig. 11.8). Mechanical weathering is the physical breakup of rocks into smaller particles without a change in chemical composition of the constituent minerals. You have probably seen the results of mechanical weathering in rounded rocks in rivers or on shorelines, smoothed by constant tumbling in waves or currents. On a larger scale, mountain valleys are carved by rivers and glaciers.

Chemical weathering is the selective removal or alteration of specific minerals in rocks. This alteration leads to weakening and disintegration of rock. Among the more important chemical weathering processes are oxidation (combination of oxygen with an element to form an oxide or a hydroxide mineral) and hydrolysis (hydrogen atoms from water molecules combine with other chemicals to form acids). The products of these reactions are more susceptible to both mechanical weathering and dissolving in water. For instance, when carbonic acid (formed when rainwater absorbs CO_2) percolates through porous limestone layers in the ground, it dissolves the rock and creates caves.

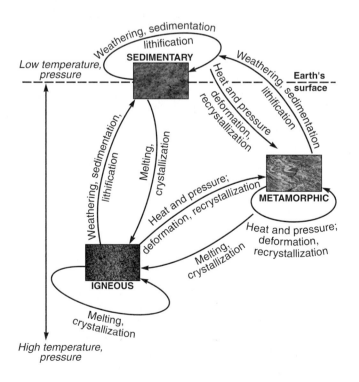

FIGURE 11.6 The rock cycle includes a variety of geologic processes that can transform any rock.

Principles of Environmental Science http://www.mhhe.com/cunningham4e

FIGURE 11.7 Eons of erosion in Arizona's Grand Canyon reveals many sedimentary layers, and some deep metamorphic rocks.

FIGURE 11.8 Weathering slowly reduces an igneous rock to loose sediment. Here, exposure to moisture expands minerals in the rock, and frost may also force the rock apart.

Particles of rock loosened by wind, water, ice, and other weathering forces are carried downhill, downwind, or downstream until they come to rest again in a new location. The deposition of these materials is called **sedimentation.** Water, wind, and glaciers deposit particles of sand, clay, and silt far from their source. Much of the American Midwest, for instance, is covered with hundreds of meters of sedimentary material left by glaciers (till, or rock debris deposited by glacial ice), wind (loess, or fine dust deposits), river deposits of sand and gravel, and ocean deposits of sand, silt, clay, and limestone.

ECONOMIC GEOLOGY AND MINERALOGY

Economic mineralogy is the study of minerals that are valuable for manufacturing and trade. Most economic minerals are metal ores, minerals with unusually high concentrations of metals. Lead, for example, often comes from the mineral galena (PbS),

and copper comes from sulfide ores, such as bornite (Cu_5FeS_4). Nonmetallic geologic resources include graphite, feldspar, quartz crystals, diamonds, and other crystals that are valued for their usefulness or beauty. Metals have been so important in human affairs that major epochs of human history are commonly known by their dominant materials and the technology involved in using those materials (Stone Age, Bronze Age, Iron Age, etc.). The mining, processing, and distribution of these materials have broad implications for both our culture and our environment. Most economically valuable crustal resources exist everywhere in small amounts; the important thing is to find them concentrated in economically recoverable levels.

Public policy in the United States has encouraged mining on public lands as a way of boosting the economy and utilizing natural resources. Today these policies are controversial, and there are efforts to recover public revenue from these publicly owned resources. (See related story "Should We Revise the 1872 Mining Law?" at www.mhhe.com/cunningham4e.)

Metals are essential to our economy

Metals are malleable substances that are useful and valuable because they are strong, relatively light, and can be reshaped for many purposes. The availability of metals and the methods to extract and use them have determined technological developments, as well as economic and political power for individuals and nations.

The metals consumed in greatest quantity by world industry include iron (740 million metric tons annually), aluminum (40 million metric tons), manganese (22.4 million metric tons), copper and chromium (8 million metric tons each), and nickel (0.7 million metric tons). Most of these metals are consumed in the United States, Western Europe, Japan, and China. They are produced primarily in South America, South Africa, and Russia (fig. 11.9). It is easy to see how these facts contribute to a worldwide mineral trade network that has become crucially important to the economic and social stability of all nations involved. Table 11.2 shows the primary uses of these metals.

Nonmetal mineral resources include gravel, clay, glass, and salts

Nonmetal minerals constitute a broad class that covers resources from gemstones to sand, gravel, salts, limestone, and soils. Sand and gravel production for road and building construction comprise by far the greatest volume and dollar value of all nonmetal mineral resources and a far greater volume than all metal ores. Sand and gravel are used mainly in brick and concrete construction, in paving, as loose road filler, and for sandblasting. High-purity silica sand is our source of glass. These materials usually are retrieved from surface pit mines and quarries, where they were deposited by glaciers, winds, or ancient oceans.

Limestone, like sand and gravel, is mined and quarried for concrete and crushed for road rock. It also is cut for building stone, pulverized for use as an agricultural soil additive that

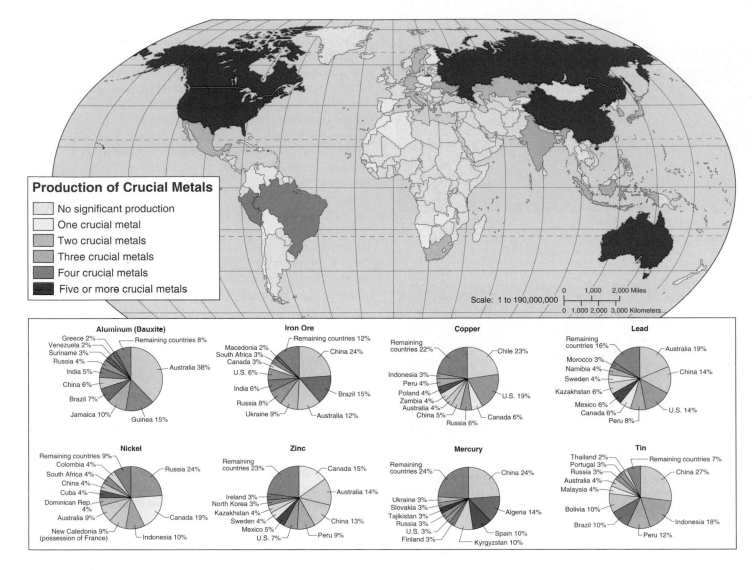

FIGURE 11.9 World production of metals most essential for an industrial economy. Principal consumers are the United States, Western Europe, Japan, and China.

neutralizes acidic soil, and roasted in lime kilns and cement plants to make plaster (hydrated lime) and cement.

Evaporites (materials deposited by evaporation of chemical solutions) are mined for halite, gypsum, and potash. These are often found at or above 97 percent purity. Halite, or rock salt, is used for water softening and ice melting on winter roads in some northern areas. Refined, it is a source of table salt. Gypsum (calcium sulfate) now makes our plaster wallboard, but it has been used to cover walls ever since the Egyptians plastered their frescoed tombs along the Nile River some 5,000 years ago. Potash is an evaporite composed of a variety of potassium chlorides and potassium sulfates. These highly soluble potassium salts have long been used as a soil fertilizer.

Sulfur deposits are mined mainly for sulfuric acid production. In the United States, sulfuric acid use amounts to more than 200 lbs per person per year, mostly because of its use in industry, car batteries, and some medicinal products.

Durable, highly valuable, and easily portable, gemstones and precious metals have long been a way to store and transport wealth. Unfortunately, these valuable materials also have bankrolled despots, criminal gangs, and terrorism in many countries. In recent years, brutal civil wars in Africa have been financed—and often motivated by—gold, diamonds, tantalum ore, and other high-priced commodities. In 2004 investigators discovered that members of the terrorist group Al Qaida used diamonds purchased in Liberia to finance the September 11, 2001 attacks in the United States. Much of this illegal trade ends up in the $100 billion per year global jewelry trade, two-thirds of which sells in the United States. Many people who treasure a diamond ring or a gold wedding band as a symbol of love and devotion are unaware that it may have been obtained through slave labor, torture, and environmentally destructive mining and processing methods. Civil rights organizations are campaigning to require better documentation of the origins of gems and

TABLE 11.2

Primary Uses of Some Major Metals

Metal	Use
Aluminum	Packaging foods and beverages (38%), transportation, electronics
Chromium	High-strength steel alloys
Copper	Building construction, electric and electronic industries
Iron	Heavy machinery, steel production
Lead	Leaded gasoline, car batteries, paints, ammunition
Manganese	High-strength, heat-resistant steel alloys
Nickel	Chemical industry, steel alloys
Platinum group	Automobile catalytic converters, electronics, medical uses
Gold	Medical, aerospace, electronic uses; accumulation as monetary standard
Silver	Photography, electronics, jewelry

precious metals to prevent their use as financing for crimes against humanity (see related story "Conflict Diamonds" at www.mhhe.com/cunningham4e).

In 2004 a group of Nobel Peace Laureates called on the World Bank to overhaul its policies on lending for resource extractive industries. "War, poverty, climate change, and ongoing violations of human rights—all of these scourges are all too often linked to the oil and mining industries," wrote Archbishop Desmond Tutu, winner of the 1984 Nobel Peace Prize for helping eliminate apartheid in South Africa. In response, the World Bank appointed an Extractive Industries Review headed by former Indonesian Environment Minister Emil Salim. Its final report, agreed with many concerns raised by environmental and community organizations.

The earth provides almost all our fuel

Modern society functions largely on energy produced from geologic deposits of oil, coal, and natural gas. Nuclear energy, which runs on uranium, makes up a large portion of our electricity resources. Energy production from these sources is discussed in chapter 12. Oil, coal, and gas are organic, created over millions of years as extreme heat and pressure transformed the remains of ancient organisms. They are not minerals, because they have no crystalline structure, but they can be considered part of economic mineralogy because they are such important geologic resources. In addition to providing energy, oil is the source material for plastics, and natural gas is used to make agricultural fertilizers.

The search for these organic deposits is one of the most important parts of economic mineralogy (see Exploring Science,

p. 273). Debates over exploitation and ownership of these resources play important roles in national and international politics. The Persian Gulf War of 1990 was fought over control of vast underground oil deposits, as has been Russia's war in Chechnya. Debates over exploiting potential oil reserves in the Alaskan National Wildlife Refuge (ANWR) have raged in the United States for decades. (See related stories "Exploiting Oil in ANWR" and "Oil and the War in Chechnya?" at www.mhhe.com/cunningham4e.) Recently, Canada's oil shale and tar sands have become another source of oil (chapter 12).

ENVIRONMENTAL EFFECTS OF RESOURCE EXTRACTION

Each of us depends daily on geologic resources mined or pumped from sites around the world. We use scores of metals and minerals, many of which we've never even heard of, in our lights, computers, watches, fertilizers, and cars. Extracting and purifying these resources can have severe environmental and social consequences. The most obvious effect of mining and well drilling is often the disturbance or removal of the land surface. Farther-reaching effects, though, include air and water pollution. The EPA lists more than 100 toxic air pollutants, from acetone to xylene, released from U.S. mines and wells every year. Nearly 80,000 metric tons of particulate matter (dust) and 11,000 tons of sulfur dioxide are released from nonmetal mining alone. Chemical- and sediment-runoff pollution is a major problem in many local watersheds.

Gold and other metals are often found in sulfide ores that produce sulfuric acid when exposed to air and water. In addition, metal elements often occur in very low concentrations—10 to 20 parts per billion may be economically extractable for gold, platinum, and other metals. Consequently, vast quantities of ore must be crushed and washed to extract metals. Cyanide, mercury, and other toxic substances are used to chemically separate metals from the minerals that contain them, and these substances can easily contaminate lakes and streams. Further, a great deal of water is used in washing crushed ore with cyanide and other solutions. In arid Nevada, the USGS estimates that mining consumes about 230,000 m^3 (60 million gal) per day. After use in ore processing, much of this water contains sulfuric acid, arsenic, heavy metals, and other contaminants. Mine runoff leaking into lakes and streams can damage or destroy aquatic ecosystems (fig. 11.10).

Mining has enormous environmental effects

There are many techniques for extracting geologic materials. The most common methods are open-pit mining, strip-mining, and underground mining. An ancient method of accumulating gold, diamonds, and coal is placer mining, in which pure nuggets are washed from stream sediments. Since the California gold rush of 1849, placer miners have used water cannons to blast away hillsides. This method, which chokes stream ecosystems

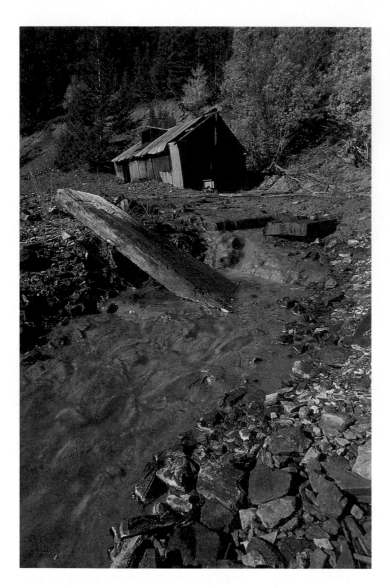

FIGURE 11.10 Thousands of abandoned mines on public lands poison streams and groundwater with acid, metal-laced drainage. This old mine in Montana drains into the Blackfoot River, the trout stream featured in Norman Maclean's book *A River Runs through It.*

burning since 1962; control efforts have cost at least $40 million, but the fire continues to expand. China, which depends on coal for much of its heating and electricity, has hundreds of smoldering mine fires; one has been burning for 400 years. According to a recent study from the International Institute for Aerospace Survey in the Netherlands, these fires consume up to 200 million tons of coal every year and emit as much carbon dioxide as all the cars in the United States. Toxic fumes containing mercury, arsenic, selenium, radon, and other hazardous emissions are also released from these fires.

Open-pit mines are used to extract massive beds of metal ores and other minerals. The size of modern open pits can be hard to comprehend. The Bingham Canyon mine, near Salt Lake City, Utah, is 800 m (2,640 ft) deep and nearly 4 km (2.5 mi) wide at the top. More than 5 billion tons of copper ore and waste material have been removed from the hole since 1906. A chief environmental challenge of open-pit mining is that groundwater accumulates in the pit. In metal mines, a toxic soup results. No one yet knows how to detoxify these lakes, which endanger wildlife and nearby watersheds. (See related story "Death in a Mine Pit" at www.mhhe.com/cunningham4e.)

Half the coal used in the United States comes from strip mines. Since coal is often found in expansive, horizontal beds, the entire land surface can be stripped away to cheaply and quickly expose the coal. The overburden, or surface material, is placed back into the mine, but usually in long ridges called spoil banks. Spoil banks are very susceptible to erosion and chemical weathering. Since the spoil banks have no topsoil (the complex organic mixture that supports vegetation—see chapter 7), revegetation occurs very slowly.

The 1977 federal Surface Mining Control and Reclamation Act (SMCRA) requires better restoration of strip-mined lands, especially where mines replaced prime farmland. Since then, the record of strip-mine reclamation has improved substantially. Complete mine restoration is expensive, often more than $10,000 per hectare. Restoration is also difficult because the developing soil is usually acidic and compacted by the heavy machinery used to reshape the land surface.

with sediment, is still used in Alaska, Canada, and many other regions. Another ancient, and much more dangerous, method is underground mining. Ancient Roman, European, and Chinese miners tunneled deep into tin, lead, copper, coal, and other mineral seams. Mine tunnels occasionally collapse, and natural gas in coal mines can explode. Water seeping into mine shafts also dissolves toxic minerals. Contaminated water seeps into groundwater; it is also pumped to the surface, where it enters streams and lakes.

In underground coal mines, another major environmental risk is fires. Hundreds of coal mines smolder in the United States, China, Russia, India, South Africa, and Europe. The inaccessibility and size of these fires make many impossible to extinguish or control. One mine fire in Centralia, Pennsylvania, has been

How do geologists discover oil and gas deposits? Oil and gas are among the most lucrative of earth resources, but they are hidden deep in the ground, sometimes thousands of meters deep. The main method for finding these hidden reservoirs is seismic exploration and 3-D imaging.

Seismology is the study of how waves of energy (seismic waves) move through the earth. Seismic exploration involves producing, and then detecting, seismic pulses. The way these seismic waves move through the earth tells a great deal about the density of the rocks, where layers of differing density meet, where the earth is fractured, and where hidden pockets of rock are saturated with oil or gas.

To carry out seismic exploration, geologists set out an array of "geophones" to record vibrations in the earth, and then produce a series of intense vibrations along a survey line. Initially, dynamite blasts were used to create seismic waves in the earth. Now "thumper trucks" are usually used (fig. 1). Thumper trucks don't actually thump; instead, they lower a plate to the ground, press on the plate with the truck's 30-ton mass, and then produce low-frequency vibrations that travel through the earth. This process is repeated every 100 m or so, along survey lines that may run many kilometers across the landscape. As the vibrations bounce off of rock layers in the ground, seismic waves travel back to the geophones. The geophones record how long the waves took to travel—thus how dense the rocks were—and send the data to a central computer.

By compiling the data from many geophones, and from hundreds or thousands of vibration events, geologists can reconstruct the shapes of features in the ground. Originally, these shapes were mapped two-dimensionally, along a single track of the truck or dynamite blasts. As computers have improved, they have become able to process the terabytes of data needed to map a grid of survey lines. Using a grid of survey lines, geologists can create three-dimensional models of structures in the earth (fig. 2—3-D image). Increasingly, geologists re-survey sites after pumping begins, to detect how and where oil and gas are depleted. This approach is called "4-dimensional imaging" because repeat visits add a time dimension to the process.

These methods have greatly reduced the number of test wells needed to explore for gas and oil, and they have allowed oil and gas companies to locate small, deep, or unconventional deposits they might never have found with other methods.

Local conservation groups, especially in Wyoming and Utah, however, still object to the damage thumper trucks leave on the landscape. In the dry, open country of Western states, soils and vegetation are easily disturbed and recover very slowly. The deep ruts left by the 30-ton trucks, as they crawl cross-country up hills and across gullies, usually in trains of four or five trucks together, will be visible for many decades, or possibly centuries, to come. Similarly, in the delicate tundra of the Arctic, networks of survey tracks disappear slowly. In the mountainous rainforests of Ecuador, meanwhile, felling trees for survey lines has become a dominant cause of deforestation.

Increasingly, oil and gas exploration are moving offshore, as land-based resources become harder to extract. Most new discoveries are hundreds of meters below the sea, as well as thousands of meters below the seafloor. For undersea exploration, ships carry powerful airguns. Blasts from the airguns produce sound waves that travel into the seafloor, and reflected wavelenghts are picked up by sensors towed behind the ship. Although ships leave no tracks, marine biologists worry about the effects of noise on whales and other marine mammals.

Seismic imaging is one of the huge recent advances in the science of finding and extracting fossil fuels. It has reduced the damage of test wells but also made exploration feasible in wild and roadless areas previously thought useless to oil companies. How do you think we should manage exploration in public lands, as new technology makes it easier? How would your answers differ if you were a rancher in Wyoming, a Senator from Wyoming, or someone from a distant state?

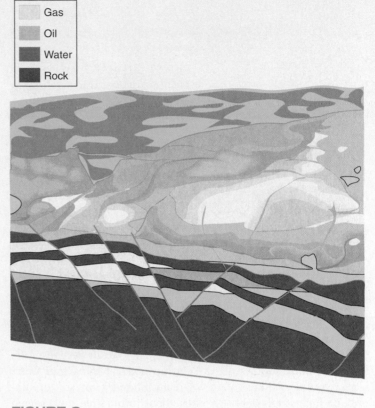

Gas
Oil
Water
Rock

FIGURE 2

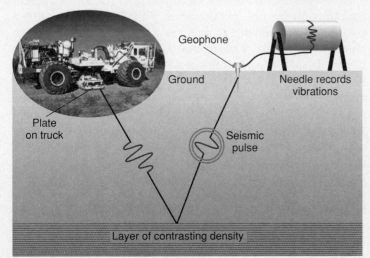

Geophone

Ground

Needle records vibrations

Plate on truck

Seismic pulse

Layer of contrasting density

FIGURE 1

FIGURE 11.11 Mountaintop removal mining is a relatively new, and deeply controversial, method of extracting Appalachian coal.

Bitter controversy has grown recently over mountaintop removal, a coal mining method practiced mainly in Appalachia. Long, sinuous ridge-tops are removed by giant, 20-story-tall shovels to expose horizontal beds of coal (fig. 11.11). Up to 215 m (700 ft) of ridge-top is pulverized and dumped into adjacent river valleys. The debris can be laden with selenium, arsenic, coal, and other toxic substances. At least 900 km (560 mi) of streams have been buried in West Virginia alone. Environmental lawyers have sued to stop the destruction of streams, arguing that it violates the Clean Water Act (chapter 10). In response, the Bush administration issued a "clarification" of the Clean Water Act, rewriting the law to allow stream filling by any sort of mining or industrial debris, rather than forbidding any stream destruction. Environmentalists charge that the new wording subverts the Clean Water Act, giving a green light to all sorts of stream destruction. West Virginians are deeply divided about mountaintop removal because they live in affected stream valleys but also depend on a coal economy.

The Mineral Policy Center in Washington, D.C., estimates that 19,000 km (12,000 mi) of rivers and streams in the United States are contaminated by mine drainage. The EPA estimates that cleaning up impaired streams, along with 550,000 abandoned mines in the United States, may cost $70 billion. Worldwide, mine closing and rehabilitation costs are estimated in the trillions of dollars. Because of the volatile prices of metals and coal, many mining companies have gone bankrupt before restoring mine sites, leaving the public responsible for cleanup. In 2002 more than 500 leading mine executives and their critics convened at the Global Mining Initiative, a meeting in Toronto aimed at improving the sustainability of mining. Executives acknowledged that in the future they will increasingly be held liable for environmental damages, and they said they were seeking ways to improve the industry's social and environmental record. Jay Hair, secretary general of the International Council on Mining and Metals, stated that "environmental protection and social responsibility are important" and that mining companies were interested in participating in sustainable development. Mine executives also recognized that, increasingly, big cleanup bills will cut into company values and stock prices. Finding creative ways to keep mines cleaner from the start will make good economic sense, even if it's not easy to do.

Processing contaminates air, water, and soil

Metals are extracted from ores by heating or by using chemical solvents. Both processes release large quantities of toxic materials that can be even more environmentally hazardous than mining. **Smelting**—roasting ore to release metals—is a major source of air pollution. One of the most notorious examples of ecological devastation from smelting is a wasteland near Ducktown, Tennessee. In the mid-1800s, mining companies began excavating the rich copper deposits in the area. To extract copper from the ore, they built huge, open-air wood fires, using timber from the surrounding forest. Dense clouds of sulfur dioxide released from sulfide ores poisoned the vegetation and acidified the soil over a 50 mi^2 (13,000 ha) area. Rains washed the soil off the denuded land, creating a barren moonscape.

Sulfur emissions from Ducktown smelters were reduced in 1907 after Georgia sued Tennessee over air pollution. In the 1930s the Tennessee Valley Authority (TVA) began treating the soil and replanting trees to cut down on erosion. Recently, upwards of $250,000 per year has been spent on this effort. While the trees and other plants are still spindly and feeble, more than two-thirds of the area is considered "adequately" covered with vegetation. Similarly, smelting of copper-nickel ore in Sudbury, Ontario, a century ago caused widespread ecological destruction that is slowly being repaired following pollution-control measures (see fig. 9.25).

Chemical extraction is used to dissolve or mobilize pulverized ore, but it uses and pollutes a great deal of water. A widely used method is **heap-leach extraction,** which involves piling crushed ore in huge heaps and spraying it with a dilute alkaline-cyanide solution (fig. 11.12). The solution percolates through the pile and dissolves gold. The gold-containing solution is then pumped to a processing plant that removes the gold by electrolysis. A thick clay pad and plastic liner beneath the ore heap is supposed to keep the poisonous cyanide solution from contaminating surface or groundwater, but leaks are common.

Once all the gold is recovered, mine operators may simply walk away from the operation, leaving vast amounts of toxic effluent in open ponds behind earthen dams. A case in point is the Summitville Mine near Alamosa, Colorado. After extracting $98 million in gold, the absentee owners declared bankruptcy in 1992, abandoning millions of tons of mine waste and huge, leaking ponds of cyanide. The Environmental Protection Agency may spend more than $100 million trying to clean up the mess and keep the cyanide pool from spilling into the Alamosa River.

FIGURE 11.12 In a heap-leach operation, huge piles of low-grade ore are heaped on an impervious pad and sprayed continuously with a cyanide solution. As the leaching solution trickles through the crushed ore, it extracts gold and other precious metals. The "pregnant" effluent solution is then pumped to a processing plant, where metals are extracted and purified. This technique is highly profitable but carries large environmental risks.
Source: Data from George Laycock, *Audubon Magazine*, vol. 91(7), July 1989.

CONSERVING GEOLOGIC RESOURCES

Conservation offers great potential for extending our supplies of economic minerals and reducing the effects of mining and processing. The advantages of conservation are significant: less waste to dispose of, less land lost to mining, and less consumption of money, energy, and water resources.

Recycling saves energy as well as materials

Some waste products already are being exploited, especially for scarce or valuable metals. Aluminum, for instance, must be extracted from bauxite by electrolysis, an expensive, energy-intensive process. Recycling waste aluminum, such as beverage cans, on the other hand, consumes one-twentieth of the energy of extracting new aluminum. Today, nearly two-thirds of all aluminum beverage cans in the United States are recycled, up from only 15 percent 20 years ago. The high value of aluminum scrap ($650 a ton versus $60 for steel, $200 for plastic, $50 for glass, and $30 for paperboard) gives consumers plenty of incentive to deliver their cans for collection. Recycling is so rapid and effective that half of all the aluminum cans now on a grocer's shelf will be made into another can within two months. Table 11.3 shows the energy cost of extracting other materials.

Platinum, the catalyst in automobile catalytic exhaust converters, is valuable enough to be regularly retrieved and recycled from used cars (fig. 11.13). Other commonly recycled metals are gold,

TABLE 11.3

Energy Requirements in Producing Various Materials from Ore and Raw Source Materials

Product	Energy Requirement (MJ/kg)[1]	
	New	**From Scrap**
Glass	25	25
Steel	50	26
Plastics	162	n.a.[2]
Aluminum	250	8
Titanium	400	n.a.[2]
Copper	60	7
Paper	24	15

Source: Data from E. T. Hayes, *Implications of Materials Processing*, 1997.
[1]Megajoules per kilogram.
[2]Not available.

silver, copper, lead, iron, and steel. The last four are readily available in a pure and massive form, including copper pipes, lead batteries, and steel and iron auto parts. Gold and silver are valuable enough to warrant recovery, even through more difficult means.

Despite its potential, however, recycling rates remain modest (fig. 11.14). Can you think of some explanations for these rates?

While total U.S. steel production has fallen in recent decades—largely because of inexpensive supplies from new and efficient Japanese steel mills—a new type of mill subsisting entirely on a readily available supply of scrap/waste steel and iron is a growing industry. **Minimills,** which remelt and reshape scrap iron and steel, are smaller and cheaper to operate than traditional integrated mills that perform every process from preparing raw ore to finishing iron and steel products. Minimills produce steel at between $225 and $480 per metric ton, while steel from integrated mills costs $1,425 to $2,250 per metric ton, on average. The energy cost is also lower in minimills: 5.3 million BTU/ton of steel, compared with 16.08 million BTU/ton in integrated mill furnaces. Minimills now produce about half of all U.S. steel. Recycling is slowly increasing as raw materials become more scarce and wastes become more plentiful.

New materials can replace mined resources

Mineral and metal consumption can be reduced by new materials or new technologies developed to replace traditional uses. This is a long-standing tradition; for example, bronze replaced stone technology, and iron replaced bronze. More recently, the introduction of plastic pipe has decreased our consumption of copper, lead, and steel pipes. In the same way, the development of fiber-optic technology and satellite communication reduces the need for copper telephone wires.

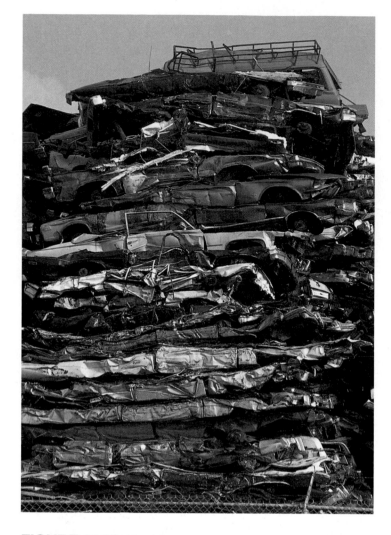

FIGURE 11.13 The richest metal source we have—our mountains of scrapped cars—offers a rich, inexpensive, and ecologically beneficial resource that can be "mined" for a number of metals.

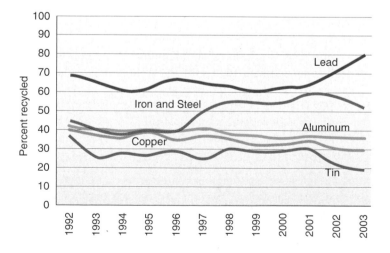

FIGURE 11.14 Rates of metal recycling remain modest for most metals even though recycling saves energy and the environment.
Data Source: U.S. Geological Survey, 2006.

Iron and steel have been the backbone of heavy industry, but we are now moving toward other materials. One of our primary uses for iron and steel has been machinery and vehicle parts. In automobile production, steel is being replaced by polymers (long-chain organic molecules similar to plastics), aluminum, ceramics, and new high-technology alloys. All of these reduce vehicle weight and cost, while increasing fuel efficiency. Some of the newer alloys that combine steel with titanium, vanadium, or other metals wear much better than traditional steel. Ceramic engine parts provide heat insulation around pistons, bearings, and cylinders, keeping the rest of the engine cool and operating efficiently. Plastics and glass fiber–reinforced polymers are used in body parts and some engine components.

Electronics and communications (telephone) technology, once major consumers of copper and aluminum, now use ultra-high-purity glass cables to transmit pulses of light, instead of metal wires carrying electron pulses. Once again, this technology has been developed for its greater efficiency and lower cost, but it also affects consumption of our most basic metals.

GEOLOGIC HAZARDS

Earthquakes, volcanoes, floods, and landslides are normal earth processes, events that have made our earth what it is today. However, when they affect human populations, their consequences can be among the worst and most feared disasters that befall us.

Earthquakes are frequent and deadly hazards

The 2004 earthquake and tsunami in Banda Aceh, Indonesia, stunned the world, killed over 230,000 people, and caused damage as far away as Africa. Less than a year later, an earthquake in Pakistan killed 80,000 people. **Earthquakes** are sudden movements in the earth's crust that occur along faults (planes of weakness), where one rock mass slides past another one. When movement along faults occurs gradually and relatively smoothly, it is called creep or seismic slip and may be undetectable to the casual observer. When friction prevents rocks from slipping easily, stress builds up until it is finally released with a sudden jerk. The point on a fault at which the first movement occurs during an earthquake is called the epicenter.

Earthquakes have always seemed mysterious, sudden, and violent, coming without warning and leaving ruined cities and dislocated landscapes in their wake. Cities such as Kobe, Japan, or Mexico City, parts of which are built on soft landfill or poorly consolidated soil, usually suffer the greatest damage from earthquakes (fig. 11.15). Water-saturated soil can liquify when shaken. Buildings sometimes sink out of sight or fall down like a row of dominoes under these conditions (see related stories on earthquakes at www.mhhe.com/cunningham4e).

Earthquakes frequently occur along the edges of tectonic plates, especially where one plate is being subducted, or pushed down, beneath another. Earthquakes also occur in the centers of continents, however. In fact, one of the largest earthquakes ever recorded

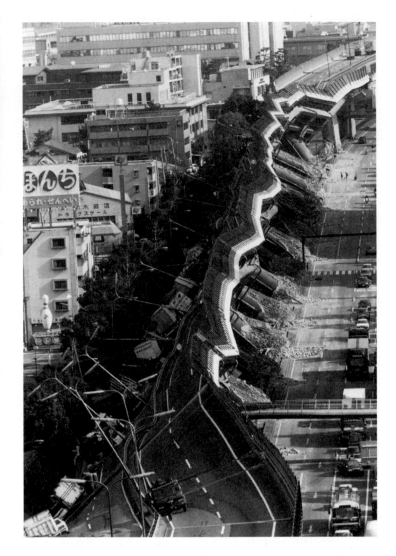

FIGURE 11.15 Earthquakes are most devastating where building methods cannot withstand shaking. The 2005 earthquake in Pakistan may have killed 80,000 people and left thousands more homeless.

in North America was one of an estimated magnitude 8.8 that struck the area around New Madrid, Missouri, in 1812. Fortunately, few people lived there at the time, and the damage was minimal.

Modern contractors in earthquake zones attempt to prevent damage and casualties by constructing buildings that can withstand tremors. The primary methods used are heavily reinforced structures, strategically placed weak spots in the building that can absorb vibration from the rest of the building, and pads or floats beneath the building on which it can shift harmlessly with ground motion.

One of the most notorious effects of earthquakes is the **tsunami,** Japanese for "harbor wave." These giant sea swells, such as the one originating in Banda Aceh, Indonesia in 2005, can move at 1,000 kph (600 mph), or faster, away from the center of an earthquake. When these swells approach the shore, they can create breakers as high as 65 m (nearly 200 ft). Tsunamis also can be caused by underwater volcanic explosions or massive seafloor slumping.

Volcanoes eject deadly gases and ash

Volcanoes and undersea magma vents are the sources of most of the earth's crust. Over hundreds of millions of years, gaseous emissions from these sources formed the earth's earliest oceans and atmosphere. Many of the world's fertile soils are weathered volcanic materials. Volcanoes have also been an ever present threat to human populations (fig. 11.16). One of the most famous historic volcanic eruptions was that of Mount Vesuvius in southern Italy, which buried the cities of Herculaneum and Pompeii in A.D. 79. The mountain had been showing signs of activity before it erupted, but many citizens chose to stay and take a chance on survival. On August 24, the mountain buried the two towns in ash. Thousands were killed by the dense, hot, toxic gases that accompanied the ash flowing down from the volcano's mouth. It continues to erupt from time to time.

Nuees ardentes (French for "glowing clouds") are deadly, denser-than-air mixtures of hot gases and ash like those that inundated Pompeii and Herculaneum. Temperatures in these clouds may exceed 1,000°C, and they move at more than 100 kph (60 mph). *Nuees ardentes* destroyed the town of St. Pierre on the Caribbean island of Martinique on May 8, 1902. Mount Pelee released a cloud of *nuees ardentes* that rolled down through the town, killing between 25,000 and 40,000 people within a few minutes. All of the town's residents died except for a single prisoner being held in the town dungeon.

Disastrous mudslides are also associated with volcanoes. The 1985 eruption of Nevado del Ruíz, 130 km (85 mi) northwest of Bogotá, Colombia, caused mudslides that buried most of the town of Armero and devastated the town of Chinchina. An estimated 25,000 people were killed. Heavy mudslides also accompanied the eruption of Mount St. Helens in Washington in 1980. Sediments mixed with melted snow destroyed roads, bridges, and property, but because of sufficient advance warning, there were

FIGURE 11.16 Lava and ash flow spill down the slopes of Mayon volcano in the Philippines in this September 23, 1984, image. Because more than 73,000 people evacuated danger zones, this eruption caused no casualties.

FIGURE 11.17 Volcanic activity at Mount Rainier has produced at least 12 large mudflows in the past 6,000 years (*brown areas*). Future flows could threaten large populations now in the area.
Source: Data from T. W. Sisson, USGS Open File Report 95-642.

few casualties. Geologists worry that similar mudflows from an eruption at Mount Rainier would threaten much larger populations (fig. 11.17).

Volcanic eruptions often release large volumes of ash and dust into the air. Mount St. Helens expelled 3 km³ of dust and ash, causing ash fall across much of North America. This was only a minor eruption. An eruption in a bigger class of volcanoes was that of Tambora in Indonesia in 1815, which expelled 175 km³ of dust and ash, more than 58 times that of Mount St. Helens. These dust clouds circled the globe and reduced sunlight and air temperatures enough so that 1815 was known as the year without a summer.

It is not just a volcano's dust that blocks sunlight. Sulfur emissions from volcanic eruptions combine with rain and atmospheric moisture to produce sulfuric acid (H_2SO_4). Droplets of H_2SO_4 interfere with solar radiation and can significantly cool the world climate. In 1991 Mount Pinatubo in the Philippines emitted 20 million tons of sulfur dioxide aerosols, which remained in the stratosphere for two years. This thin haze cooled the entire earth by 1°C for about two years. It also caused a 10 to 15 percent reduction in stratospheric ozone, allowing increased ultraviolet light to reach the earth's surface.

Floods are part of a river's land-shaping processes

Like earthquakes and volcanoes, **floods** are normal events that cause damage when people get in the way. As rivers carve and shape the landscape, they build broad **floodplains,** level expanses that are periodically inundated. Large rivers, such as the Mississippi, can have huge floodplains. Many cities have been built on these flat, fertile plains, which are so convenient to the river. Floodplains that flood very irregularly may appear safe for many years, but eventually, most floodplains do flood. The severity of floods can be described by the height of water above the normal stream banks or by how frequently a similar event normally occurs—on average—for a given area. Note that these are statistical averages over long time periods. A "10-year flood" would be expected to occur once in every ten years; a "100-year flood" would be expected to occur once every century. But two 100-year floods can occur in successive years or even in the same year.

The biggest economic loss from floods is usually not the buildings and property they carry away but, rather, the contamination they cause. Virtually everything floodwaters touch in a house—carpets, furniture, drapes, electronics, even drywall and insulation—must be removed and discarded. Although floods usually don't take as many lives as some other natural disasters, they are among the most costly because they occur relatively frequently and in populated river corridors.

Many human activities increase both the severity and frequency of floods. Paving roads and parking lots reduces water infiltration into the soil and speeds the rate of runoff into streams and lakes. Clearing forests for agriculture and filling cities with buildings similarly increase both the volume and rate of water discharge after a storm.

Normally, floodplains store water, but with farming, clearing, and building, floodplains have lost much of their ability to absorb floodwaters.

Even more than development, though, flood-control structures have separated floodplains from rivers. Levees and flood walls are built to contain water within riverbanks, and river channels are dredged and deepened to allow water to recede faster. Every flood-control structure simply transfers the problem downstream, however. The water has to go somewhere. If it doesn't soak into the ground upstream, it will simply exacerbate floods somewhere downstream—leading to more levee development, and then to more flooding farther downstream, and so on.

Flood control

More than $25 billion of river-control systems have been built on the Mississippi and its tributaries. These systems have protected many communities over the past century. In the major floods of 1993, however, this elaborate system helped turn a large flood into a major disaster. Deprived of the ability to spill out over floodplains, the river is pushed downstream to create faster currents and deeper floods until eventually a levee gives way somewhere. Hydrologists calculate that the floods of 1993 were about 3 m (10 ft) higher than they would have been, given the same rainfall in 1900 before the flood-control structures were in place (fig. 11.18).

Under current rules, the government is obligated to finance most levees and flood-control structures. Many people think that it would be much better to spend this money to restore wetlands,

FIGURE 11.18 The Mississippi River inundated downtown Davenport, Iowa, during the summer floods of 1993. Although sympathetic with the heartbreak and economic losses caused by this flooding, many people argue that floodplains such as this never should have been settled in the first place.

replace groundcover on water courses, build check dams on small streams, move buildings off the floodplain, and undertake other nonstructural ways of reducing flood danger. According to this view, floodplains should be used for wildlife habitat, parks, recreation areas, and other uses not susceptible to flood damage.

The National Flood Insurance Program administered by the Federal Emergency Management Agency (FEMA) was intended to aid people who cannot buy insurance at reasonable rates, but its effects have been to encourage building on the floodplains by making people feel that, whatever happens, the government will take care of them. Many people would like to relocate homes and businesses out of harm's way after the recent floods or to improve them so they will be less susceptible to flooding, but owners of damaged property can collect only if they rebuild in the same place and in the same way as before. This perpetuates problems rather than solves them.

Mass wasting includes slides and slumps

Gravity constantly pulls downward on every material everywhere on earth. Hillsides, beaches, even relatively flat farm fields can lose material to erosion. Often water helps mobilize loose material, and catastrophic slumping, beach erosion, and gully development can occur in a storm. A general term for downhill slides of earth is "mass wasting."

Landslides are sudden collapses of hillsides. In the United States alone, landslides and related mass wasting cause over $1 billion in property damage every year. When unconsolidated sediments on a hillside are saturated by a storm or exposed by logging, road building, or house construction, slopes are especially susceptible to sudden landslides.

Often people are unaware of the risks they face by locating on or under unstable hillsides. Sometimes they simply ignore clear

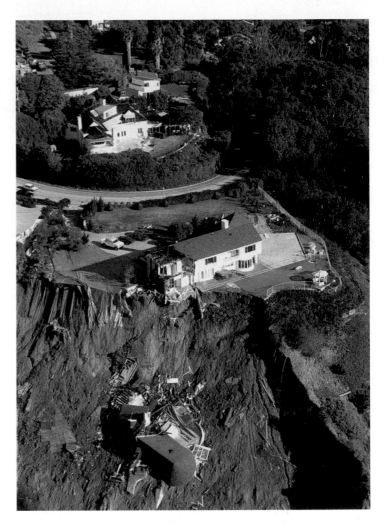

FIGURE 11.19 Parts of an expensive house slide down the hillside in Pacific Palisades, California. Building at the edge of steep slopes made of unconsolidated sediment in an earthquake-prone region is a risky venture.

and obvious danger. In southern California, where land prices are high, people often build expensive houses on steep hills and in narrow canyons. Most of the time, this dry environment appears quite stable, but the chaparral vegetation burns frequently and fiercely. When fires expose the soil in late summer, heavy winter rains cause mudslides and debris flows that destroy whole neighborhoods (fig. 11.19). In developing countries, mudslides have buried entire villages in minutes. **Soil creep,** on the other hand, moves material inexorably downhill at an imperceptibly slow pace.

Erosion destroys fields and undermines buildings

Gullying is the development of deep trenches on relatively flat ground. Especially on farm fields, which have a great deal of loose soil unprotected by plant roots, rainwater running across the surface can dig deep gullies. Sometimes land becomes useless for farming because gullying is so severe and because erosion has removed the fertile topsoil. Agricultural soil erosion has

FIGURE 11.20 Winter storms have eroded the beach and undermined the foundations of homes on this barrier island. Breaking through protective dunes to build such houses damages sensitive plant communities and exposes the whole island to storm sand erosion. Coastal zone management attempts to limit development on fragile sites.

been described as an invisible crisis. Erosion has reduced the fertility of millions of acres of prime farmland in the United States alone.

Beach erosion occurs on all sandy shorelines because the motion of the waves is constantly redistributing sand and other sediments. One of the world's longest and most spectacular sand beaches runs down the Atlantic coast of North America from New England to Florida and around the Gulf of Mexico. Much of this beach lies on some 350 long, thin **barrier islands** that stand between the mainland and the open sea. Behind these barrier islands lie shallow bays or brackish lagoons fringed by marshes or swamps.

Early inhabitants recognized that exposed, sandy shores were hazardous places to live, and they settled on the bay side of barrier islands or as far upstream on coastal rivers as was practical. Modern residents, however, place a high value on living where they have an ocean view and ready access to the beach. And they assume that modern technology makes them immune to natural forces. The most valuable and prestigious property is closest to the shore. Over the past 50 years, more than 1 million acres of estuaries and coastal marshes have been filled to make way for housing or recreational developments.

Construction directly on beaches and barrier islands can cause irreparable damage to the whole ecosystem. Normally fragile vegetative cover holds the shifting sand in place. Damaging this vegetation with construction, building roads, and breaching dunes with roads can destabilize barrier islands. Storms then wash away beaches or even whole islands. A single severe storm in 1962 caused $300 million in property damage along the east coast of the United States and left hundreds of beach homes tottering into the sea (fig. 11.20). FEMA estimates that 25 percent of all coastal homes in the United States will have the ground washed out from under them by 2060.

Cities and individual property owners often spend millions of dollars to protect beaches from erosion. Sand is dredged from the ocean floor or hauled in by the truckload, only to wash away again in the next storm. Building artificial barriers, such as groins or jetties, can trap migrating sand and build beaches in one area, but they often starve downstream beaches and make erosion there even worse (fig. 11.21).

As is the case for inland floodplains, government policies often encourage people to build where they probably shouldn't. Subsidies for road building and bridges, support for water and sewer projects, tax exemptions for second homes, flood insurance, and disaster relief are all good for the real estate and construction businesses but invite people to build in risky places. Flood insurance typically costs $300 per year for $80,000 of coverage. In 1998 FEMA paid out $40 billion in claims, 80 percent of which were flood-related. Settlement usually requires that structures be rebuilt exactly where and as they were before. There is no restriction on how many claims can be made, and policies are rarely canceled, no matter what the risk. Some beach houses have been rebuilt—at public expense—three times in a decade. The General Accounting Office found that 2 percent of federal flood policies were responsible for 30 percent of the claims.

The Coastal Barrier Resources Act of 1982 prohibited federal support, including flood insurance, for development on sensitive islands and beaches. In 1992, however, the U.S. Supreme Court ruled that ordinances forbidding floodplain development amount to an unconstitutional "taking," or confiscation, of private property.

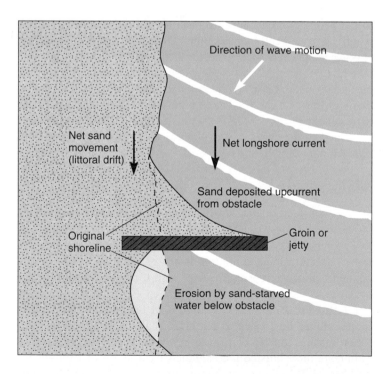

FIGURE 11.21 Groins are structures built perpendicular to a beach to slow erosion and trap sand. They build up beaches toward prevailing wind and wave direction but starve beaches and increase erosion downstream. Dashed line indicates original shoreline.

Principles of Environmental Science

REVIEW QUESTIONS

1. How does tectonic plate movement create ocean basins, mid-ocean ridges, and volcanoes?

2. What is the "ring of fire"?

3. Describe the processes and components of the rock cycle (fig. 11.6).

4. What is the difference between metals and nonmetal mineral resources?

5. What is a *mineral* and a *rock*? Why are pure metals not minerals?

6. Which countries are the single greatest producers of our major metals? (See fig. 11.9.)

7. Describe some of the mining, processing, and drilling methods that can degrade water or air quality.

8. Compare the different mining methods of underground, open-pit, strip, and placer mining, as well as mountaintop removal.

9. What resources, aside from minerals themselves, can be saved by recycling?

10. Give an example of resource substitution.

11. Describe the most deadly risks of volcanoes.

12. What is *mass wasting*? Give three examples and explain why they are a problem.

13. Why is building on barrier islands risky?

14. What is a *floodplain*? Why is building on floodplains controversial?

15. Describe the processes of chemical weathering and mechanical weathering. How do these processes contribute to the recycling of rocks?

16. The Mesozoic period begins and ends with the appearance and disappearance of dinosaurs. What fossils mark the other geologic eras? (See fig. 11.5.)

17. Describe how seismic exploration works in finding oil and gas deposits. What are its costs and benefits?

CRITICAL THINKING AND DISCUSSION

Apply the principles you have learned in this chapter to discuss these questions with other students.

1. Understanding and solving the environmental problems of mining are basically geologic problems, but geologists need information from a variety of environmental and scientific fields. What are some of the other sciences (or disciplines) that could contribute to solving mine contamination problems?

2. Geologists are responsible for identifying and mapping mineral resources. But mineral resources are buried below the soil and covered with vegetation. How do you suppose geologists in the field find clues about the distribution of rock types?

3. If you had an igneous rock with very fine crystals and one with very large crystals, which would you expect to have formed deep in the ground, and why?

4. Heat and pressure tend to help concentrate metal ores. Explain why such ores might often occur in mountains such as the Andes in South America.

5. The idea of tectonic plates shifting across the earth's surface is central to explanations of geologic processes. Why is this idea still called the "theory" of plate tectonic movement?

6. Geologic evidence from fossils and sediments provided important evidence for past climate change. What sorts of evidence in the rocks and landscape around you suggest that the place where you live once looked much different than it does today?

SUGGESTED READINGS

Bawden, G. W., et al. 2001. Tectonic contraction across Los Angeles after removal of groundwater pumping effects. *Nature* 412:812–15.

Bolles, Edmund Blair. 1999. *The Ice Finders: How a Poet, a Professor, and a Politician Discovered the Ice Age.* Counterpoint.

Brimhall, G. 1991. "The Genesis of Ores," *Scientific American* 264(5):84–90.

Gould, Stephen Jay. 1990. *Wonderful Life: The Burgess Shale and the Nature of History.* W. W. Norton.

Kendall, J. M., et al. 2005. "Magma-assisted rifting in Ethiopia." *Nature* 433:146–148.

Kennedy, Danny. 1997. U.S. mine gouges for gold. *Earth Island Journal* 12(2):24.

Knoo, Andrew H. 2003. *When Life Nearly Died: The Greatest Mass Extinction of All Time.* Thames & Hudson.

McAdoo, B. G., and P. Watts, 2004 "Tsunami hazard from submarine landslides on the Oregon continental slope." *Marine Geology* 203, p. 235–245.

McPhee, John. 2000. *Annals of a Former World.* Farrar Strauss Giroux.

Mueller, Karl, et al. 2004. "Analyzing the 1811–1812 New Madrid earthquakes with recent instrumentally recorded aftershocks." *Nature.* 429(6989):284–288.

Rudwick, Martin J. S. 1995. *Scenes from Deep Time: Early Pictorial Representations of the Prehistoric World*. University of Chicago Press.

Rudwick, Martin J. S. 2005. *Bursting the Limits of Time: The Reconstruction of Geohistory in the Age of Revolution*: University of Chicago Press.

Thomas, D. S. G., et al. 2005. "Remobilization of southern African desert dune systems by twenty-first century global warming." *Naturel* 435:1218–1221.

Wilkinson, Bruce H. 2005. "Humans as geologic agents: A deep-time perspective." *Geology* 33(3):161–164.

Winchester, Simon. 2002. *The Map That Changed the World: William Smith and the Birth of Modern Geology*. Perennial Press.

DATA analysis   Exploring Recent Earthquakes

Go to the USGS Earthquake Center (http://earthquake.usgs.gov/eqcenter/). This is a rich source of global earthquake information.

1. Look at the USA map of earthquakes. Where are the greatest concentrations of earthquakes? Can you explain why those locations have lots of earthquakes?

2. Find the World Lists of earthquakes, and click on Magnitude 5+. Where was the largest recent earthquake? What was its magnitude? Its depth? Are listed quakes clustered in a few regions? Where are those regions?

3. Still looking at the world list of large earthquakes, make a graph showing the frequency of different magnitudes for listed earthquakes: On a piece of paper, draw a line with earthquake magnitude classes, then mark an X for each earthquake over the appropriate size class, like this:

```
X
X       X             X
5.0    5.5    6.0    6.5    7.0    7.5    8.0+
```

This kind of graph is called a frequency distribution graph, or a histogram. Would you describe earthquake sizes as mostly small, evenly distributed, mostly large, or mostly in the middle of the range?

Understanding Volcanoes

USGS Hawaii Volcano Observatory is one of the most important places for studying and understanding volcanoes. Visit the observatory's website at http://hvo.wr.usgs.gov/.

1. Which of the volcanoes discussed at this website is the largest in the world? Which is the most active?

2. Why are earthquakes discussed on this volcano website?

3. What are the major hazards to people from these volcanoes?

For a good collection of volcano images, go to the USGS Cascades Volcano Observatory at http://vulcan.wr.usgs.gov/Photo/framework.html.

Evaluating Erosion on Farmland

The Natural Resources Conservation Service (NRCS), part of the U.S. Department of Agriculture, is the agency that monitors agricultural resources and conditions. Among its data-gathering efforts, the NRCS produces a Natural Resources Inventory (NRI), with maps of farmland conditions across the mainland United States. Visit the NRCS/NRI website at www.nrcs.usda.gov/technical/land/erosion.html.

Find the list of erosion maps. Look at several of the maps. First identify the meaning of the colors. Then identify the major concentration of high and low erosion rates

1. Where is wind erosion worst? Where is water erosion worst?

2. Where are the most acres of highly erodible cropland? What kind of physical features might occur there?

For Additional Help in Studying This Chapter, please visit our website at www.mhhe.com/cunningham4e. You will find practice quizzes, a chapter summary, key terms, answers to review questions, additional case studies, regional examples, and an extensive reading list, all of which will help you understand the material in this chapter.

Windmills supply all the electricity used on Denmark's Ærø Island, while solar and biomass energy provide space heating and vehicle fuel. Altogether, 100 percent of the island's energy comes from renewable sources.

12 Energy

America is addicted to oil.

–George W. Bush–

OBJECTIVES

Studying this chapter will help you answer the following questions:

- Where do we now get most of our energy?
- How does our energy use compare to that of people in other countries?
- How long are our supplies of oil, coal, and gas going to last?
- Should we build more nuclear power plants?

- Could solar, wind, hydropower, and other renewable energy sources replace fossil fuels?
- What are photovoltaic and fuel cells, and how do they work?
- What are biofuels? Do they provide a net energy output?
- What would be an ideal energy future?

Case Study   Renewable Energy Islands

A 19,000 m^2 array of solar water heaters provides space heating for the town of Marstal on Ærø Isand.

Denmark has substantial oil and gas supplies under the North Sea, but the Danes have chosen to wean themselves away from dependence on fossil fuels. Currently the world leader in renewable energy, Denmark now gets 20 percent of its power from solar, wind, and biomass. Some parts of this small, progressive country have moved even further toward sustainability. One of the most inspiring examples of these efforts are the small islands of Samsø and Ærø, which now get 100 percent of their energy from renewable sources.

Samsø and Ærø lie between the larger island of Zealand (home to Copenhagen) and the Jutland Peninsula. The islands are mostly agricultural. Together, they have an area of about 200 km^2 (77 mi^2) and a population of about 12,000 people. In 1997, Samsø and Ærø were chosen in a national competition to be renewable energy demonstration projects. The first step in energy independence is conservation. As you'll learn in this chapter, Denmark uses roughly half as much energy per person as the United States, although by most measures the Danes have a higher standard of living than most Americans. Danish energy conservation is achieved with high-efficiency appliances, superior building insulation, high-mileage vehicles, and other energy-saving measures. Most homes are clustered in small villages, both to save agricultural land and to facilitate district heating. Living closely together also makes having a private automobile less necessary.

Some 30 large wind generators provide 100 percent of Samsø and Ærø's electricity. Two-thirds of these windmills are located offshore, and are publicly owned. The 11 onshore wind turbines are mostly privately owned, but a share of the profits is used to finance other community energy projects. Space heating accounts for about one-third of the energy consumption on the islands. District heating systems provide most of this energy. Several large solar collector arrays supply about half the hot water for space heating and household use. Biomass-based (straw, wood chips, manure) systems supply the remainder of the island's heating needs. Some of this biomass comes from energy crops (fast-growing elephant grass and hybrid poplars, for example, are grown on marginal farmland), while the rest comes from agricultural waste. Biodiesel (primarily from rapeseed oil) fuels farm tractors and ferries, while most passenger vehicles are electric.

Geothermal pumps supplement the solar water heaters, and in one village a recently closed landfill produces methane that is used to run a small electric generator. Nuclear power is considered an unacceptable option in Denmark and doesn't feature in current energy plans. Samsø and Ærø have won numerous prizes and awards for their pioneering conversion to renewable energy. Over the past 20 years, as a result of other projects like those on Samsø and Ærø, both Denmark's fossil fuel consumption and their greenhouse gas emissions have remained constant. All of us could learn from their example.

Many other countries, both in Europe and elsewhere in the world, are turning to renewable energy to reduce their dependence on environmentally damaging and politically unstable fossil fuels. In this chapter, we'll look at world energy resources as well as how we currently obtain the energy we use, and what our options are for finding environmentally and socially sustainable ways to meet our energy needs.

ENERGY SOURCES AND USES

Being able to utilize external energy to do useful work is one of the most unique characteristics of humans. Access to a source of high-quality energy is essential for continuation of the current way of life. But what do these terms mean? **Work** is the application of force through a distance. **Energy** is the capacity to do work. **Power** is the rate of flow of energy, or the rate at which work is done. The energy we use to move our muscles, to think, and to carry out metabolic functions comes from stored chemical energy (or potential energy) in our food. Food energy is generally measured in calories (cal). One cal is the amount of energy to heat 1 g of water 1°C. A kilocalorie (or food Calorie) is 1,000 cal. In physics, the basic metric unit of force is a newton, which is the force necessary to accelerate 1 kilogram 1 meter per second. A **joule (J)** is the amount of work done when a force of 1 newton is exerted over 1 meter, or 1 amp per second flows through 1 ohm. One J equals 0.238 cal. Some other common energy units are presented in table 12.1.

What are the sources of our energy?

Fossil fuels (petroleum, natural gas, and coal) now provide about 87 percent of all commercial energy in the world (fig. 12.1). Renewable sources—solar, wind, geothermal, and hydroelectricity—make up about 7 percent of our commercial power (but hydro accounts

TABLE 12.1

Some Energy Units

1 joule (J) = the force exerted by a current of 1 amp per second flowing through a resistance of 1 ohm

1 watt (W) = 1 joule (J) per second

1 kilowatt-hour (kWh) = 1 thousand (10^3) watts exerted for 1 hour

1 megawatt (MW) = 1 million (10^6) watts

1 gigawatt (GW) = 1 billion (10^9) watts

1 petajoule (PJ) = 1 quadrillion (10^{15}) joules

1 British thermal unit (Btu) = energy to heat 1 lb of water 1°F

1 PJ = 947 billion Btu, or 278 million kWh

1 standard barrel (bbl) of oil = 42 gal (160 liters) or 5.8 million Btu

1 metric ton of standard coal = 27.8 million Btu or 4.8 bbl oil

FIGURE 12.1 Worldwide commercial energy consumption. This does not include energy collected for personal use or traded in informal markets.
Source: Data from British Petroleum, 2006.

Perhaps the most important facts about fossil fuel consumption are that we in the 20 richest countries consume nearly 80 percent of the natural gas, 65 percent of the oil, and 50 percent of the coal produced each year. Although we make up less than one-fifth of the world's population, we use more than one-half of the commercial energy supply. The United States, for instance, constitutes only about 4.5 percent of the world's population but consumes about one-quarter of all fossil fuels. To be fair, however, some of that energy goes to grow crops or manufacture goods that are later shipped to developing countries.

How much energy do you use every year? Most of us don't think about it much, but maintaining the lifestyle we enjoy requires an enormous energy input. On average, each person in the United States and Canada uses more than 300 gigajoules (GJ) (equivalent to about 60 barrels of oil) per year. By contrast, in some of the poorest countries of the world, such as Ethiopia, Nepal, and Bhutan, each person generally consumes less than 1 GJ per year. This means that each of us consumes, on average, almost as much energy in a single day as a person in one of these countries consumes in a year.

Clearly, energy consumption is linked to the comfort and convenience of our lives (fig. 12.2). Those of us in the richer countries enjoy many amenities not available to most people in the world. The link is not absolute, however. Several European countries, including Sweden, Denmark, and Finland, have higher standards of living than does the United States by almost any measure but use about half as much energy. These countries have

for almost all of that). Although not included in these data, private systems (individual houses, businesses, industries) may capture nearly as much renewable energy (excluding hydropower) as that reported in commercial transactions. This summary probably also underrepresents the importance of biomass because these fuels often are gathered by the people using them, or they are sold in the informal economy and not reported internationally. In many poorer countries, such as Haiti, Bhutan, and Malawi, biomass supplies more than 90 percent of the energy used for heating and cooking. This is an important energy source for poor people but can be a serious cause of forest destruction (chapter 6).

Nuclear power is roughly equal to hydroelectricity worldwide (about 6 percent of all commercial energy), but it makes up about 20 percent of all electric power in more-developed countries, such as Canada and the United States. We have enough nuclear fuel to produce power for a long time, and it has the benefit of not contributing to global warming (chapter 9), but as we discuss later in this chapter, safety concerns make this option unacceptable to most people.

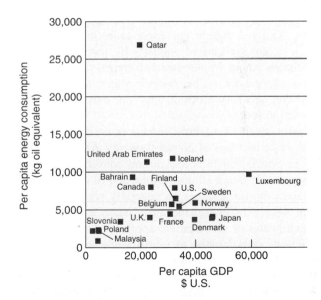

FIGURE 12.2 Per capita energy consumption and GDP. In general, income and standard of living increase with energy availability, but some energy-rich countries, such as Qatar, use vast amounts of energy without improving human development very much. Some countries, such as Norway, Denmark, and Finland, have a much higher human development index than the United States while consuming only about half as much energy. Because of the scale of this graph, low-income, low-energy countries aren't visible.
Source: Data from UNDP, 2004.

had effective energy conservation programs for many years. Japan also has a far lower energy consumption rate than might be expected for its industrial base and income level. Because Japan has few energy resources of its own, it has developed very efficient energy conservation measures.

How do we use energy?

The largest share (32.6 percent) of the energy used in the United States is consumed by industry (fig. 12.3). Mining, milling, smelting, and forging of primary metals consume about one-quarter of the industrial energy share. The chemical industry is the second largest industrial user of fossil fuels, but only half of its use is for energy generation. The remainder is raw material for plastics, fertilizers, solvents, lubricants, and hundreds of thousands of organic chemicals in commercial use. The manufacture of cement, glass, bricks, tile, paper, and processed foods also consumes large amounts of energy. Residential and commercial customers use roughly 37 percent of the primary energy consumed in the United States, mostly for space heating, air conditioning, lighting, and water heating. Small motors and electronic equipment take an increasing share of residential and commercial energy. In many office buildings, computers, copy machines, lights, and human bodies liberate enough waste heat that a supplementary source is not required, even in the winter.

Transportation consumes about 27 percent of all energy used in the United States each year. About 98 percent of that energy comes from petroleum products refined into liquid fuels, and the remaining 2 percent is provided by natural gas and electricity. Almost three-quarters of all transport energy is used by motor vehicles. Nearly 3 trillion passenger miles and 600 billion ton miles of freight are carried annually by motor vehicles in the United States. About 75 percent of all freight traffic in the United States is carried by trains, barges, ships, and pipelines, but because they are very efficient, they use only 12 percent of all transportation fuel.

Finally, analysis of how energy is used has to take into account waste and loss of potential energy. About half of all the energy in primary fuels is lost during conversion to more useful forms, while being shipped to the site of end use, or during use. Electricity, for instance, is generally promoted as a clean, efficient source of energy because, when it is used to run a resistance heater or an electrical appliance, almost 100 percent of its energy is converted to useful work and no pollution is given off. What happens before then, however? We often forget that huge amounts of pollution are released during mining and burning of the coal that fires power plants. Furthermore, nearly two-thirds of the energy in the coal that generated that electricity was lost in thermal conversion in the power plant. About 10 percent more is lost during transmission and stepping down to household voltages. Similarly, about 75 percent of the original energy in crude oil is lost during distillation into liquid fuels, transportation of that fuel to market, storage, marketing, and combustion in vehicles.

(a) U.S. energy consumption

(b) U.S. oil suppliers

(c) Energy use

FIGURE 12.3 Fossil fuels supply 85 percent of the energy used in the United States, and oil (three-quarter's of it imported) makes up half that amount. Canada is now the largest oil supplier for the United States, and industry is the largest energy-use sector. Transportation, however, uses the vast bulk of all oil. **Source:** Data from US Department of Energy, 2006.

Natural gas is our most efficient fuel. Only 10 percent of its energy content is lost in shipping and processing, since it moves by pipelines and usually needs very little refining. Ordinary gas-burning furnaces are about 75 percent efficient, and high-economy furnaces can be as much as 95 percent efficient. Because natural gas has more hydrogen per carbon atom than oil or coal, it produces about half as much carbon dioxide—and therefore contributes half as much to global warming—per unit of energy.

FOSSIL FUELS

Fossil fuels are organic chemicals created by living organisms millions of years ago and buried in sediments, where high pressures and temperatures concentrated and transformed them into energy-rich compounds. Coal is solid, petroleum (crude oil) is liquid, oil shales and tar sands are semisolid tars trapped in rock, and natural gas (methane and other gaseous hydrocarbons) is, as its name suggests, a gas. Most of these fuels were laid down during the Carboniferous period (286 million to 360 million years ago), when the earth's climate was warmer and wetter than it is now, but some are thought to date as far back as Precambrian times, perhaps a billion years ago. Because fossil fuels take so long to form, they are essentially nonrenewable resources.

Coal resources are vast

World coal deposits are vast, ten times greater than conventional oil and gas resources combined. Coal seams can be 100 m thick and can extend across tens of thousands of square kilometers that were vast, swampy forests in prehistoric times. The total resource is estimated to be 10 trillion metric tons. If all this coal could be extracted, and if coal consumption continued at present levels, this would amount to several thousand years' supply. At present rates of consumption, these **proven-in-place reserves**—those explored and mapped but not necessarily economic at today's prices—will last about 200 years. Proven reserves are generally a small fraction of a total resource (fig. 12.4).

Where are these coal deposits located? They are not evenly distributed throughout the world (fig. 12.5). North America has about one-fourth of all proven reserves. China and Russia have nearly as much. Eastern and Western Europe have fairly large

FIGURE 12.5 Proven-in-place coal reserves by region, 2006.
Source: British Petroleum, 2006.

coal supplies in spite of their relatively small size. Africa and Latin America have very little coal despite their large size. Antarctica is thought to have large coal deposits, but they would be difficult, expensive, and ecologically damaging to mine.

It would seem that the abundance of coal deposits is a favorable situation. But do we really want to use all of the coal? Coal mining is a dirty, dangerous activity. Underground mines are notorious for cave-ins, explosions, and lung diseases, such as black-lung suffered by miners. Surface mines (called strip mines, where large machines strip off overlying sediment to expose coal seams) are cheaper and generally safer for workers than tunneling, but leave huge holes where coal has been removed and vast piles of discarded rock and soil (fig. 12.6). An especially damaging technique employed in Appalachia is called mountaintop removal. Typically, the whole top of a mountain ridge is scraped off to access buried coal. The waste rock pushed down into the nearest valley buries forests, streams, houses, and farms (chapter 11 for further discussion of this issue). Mine reclamation is now mandated in the United States, but efforts often are only partially successful.

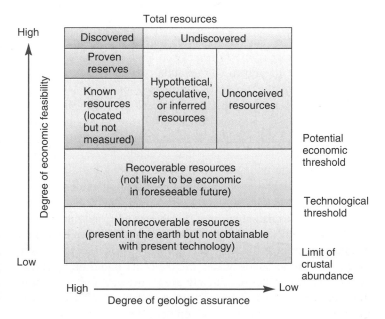

FIGURE 12.4 Categories of natural resources according to economic and technological feasibility, as well as geologic assurance.

FIGURE 12.6 Some giant mining machines stand as tall as a 20-story building and can scoop up thousands of cubic meters of rock per hour. Note the pick-up trucks in the machine's shadow.

Coal comes in a variety of forms with varying chemical composition, harness, and energy content. Lignite, the softest of this family, is not much more concentrated than peat and has about 4,000 Btu/lb. It can have very low sulfur content (0.06 to 0.4 percent by weight) but is high in ash. Subbituminous and bituminous coal have energy contents between 8,000 and 15,000 Btu/lb and generally have 1 to 5 percent sulfur content. Anthracite, which is the hardest coal, can be 96 percent carbon and have more than 15,000 Btu/lb. Coal can also have dangerous concentrations of toxic minerals, such as mercury, arsenic, chromium, and lead.

Coal burning releases large amounts of air pollution. Every year the roughly one billion tons of coal burned in the United States (83 percent for electric power generation) releases 18 million metric tons of sodium dioxide (SO_2), 5 million metric tons of nitrogen oxides (NO_x), 4 million metric tons of airborne particulates, 600,000 metric tons of hydrocarbons and carbon monoxide, 40 tons of mercury, and close to a trillion metric tons of carbon dioxide (CO_2). This is about three-quarters of the SO_2, one-third of the NO_x, and about half of the industrial CO_2 released in the United States each year. As described in chapter 9, sulfur and nitrogen oxides combine with water in the air to form sulfuric and nitric acid, making coal burning the largest single source of acid rain in many areas. These air pollutants have many deleterious effects, including human health costs, injury to domestic and wild plants and animals, and damage to buildings and property. Total losses from air pollution are estimated to be between $5 billion and $10 billion per year in the United States alone. By some accounts, at least 5,000 excess human deaths per year can be attributed to coal production and burning.

For many years, there have been proposals for "clean coal" technology. Partial combustion of a coal slurry releases gases that can be stripped of pollutants before being burned in a gas turbine. This cycle promises cleaner operation and better efficiency than ordinary furnaces. A great deal of money has been spent on this technology, but it has never reached commercial application. Recent record prices for oil, however, have reawakened interest in innovative uses for coal (see related story on "Gasification" at www.mhhe.com/cunningham4e).

China and India, both of which have very large coal resources, now burn about half of all coal mined annually in the world. Both of these countries plan to increase coal production greatly in the next few decades to fuel their rapidly growing economies. If they do so, we could have run-away global climate change due to the enormous amount of CO_2 they will emit. We desperately need to persuade them to switch to renewable energy sources.

Oil resources aren't evenly distributed

The total amount of liquid oil in the world is estimated to have been originally about 4 trillion bbl (600 billion metric tons), half of which is thought to be ultimately recoverable. Some 465 billion bbl of oil already have been consumed. In 2006 the proven reserves were roughly 1.15 trillion bbl, enough to last only 40 years at the current consumption rate of 28.5 billion bbl per year. It is estimated that another 800 billion bbl either remain to be discovered or are

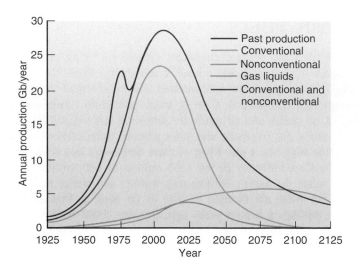

FIGURE 12.7 Worldwide production of crude oil with predicted Hubbert production. Gb = billion barrels.
Source: Jean Laherrère, www.hubbertpeak.org.

not recoverable at current prices with present technology. Many geologists expect that within a decade or so world oil production will peak and then begin to decline. As this happens, oil prices will rise, which will make other energy sources more economically attractive. Dr. M. King Hubbert, a Shell geophysicist, first predicted in the 1940s a bell-shaped curve for U.S. oil production, peaking in the 1970s, which it did. So far, world oil production seems to be following a similar path (fig. 12.7). These estimates don't take into account the very large potential from unconventional semiliquid hydrocarbon resources, such as shale oil and tar sands, which might double the total reserve if they can be mined with acceptable social, economic, and environmental impacts.

Rapidly growing economies in China, India, South Korea, and other developing countries coupled with rising standards of living are already causing fierce competition for limited oil supplies, driving prices to more than $75 per barrel. High gasoline prices may stimulate a new interest in conservation, as they did in the 1970s.

By far the largest supply of proven-in-place oil is in Saudi Arabia, which claimed 262.7 billion bbl in 2005, about one-fourth of the total proven world reserve (fig. 12.8). Altogether, the countries of the Middle East control nearly two-thirds of all known oil reserves—a fact that explains a great deal about American foreign policy in recent years. Among the places considered most likely to contain undiscovered oil, in addition to the Middle East, are the northeast Greenland Shelf, the western Siberian and Caspian areas, and the Niger and Congo delta areas of Africa. Sediments under and around the Caspian Sea have been estimated to hold 200 billion bbl of oil. If this is true, this region is second only to Saudi Arabia in size of its liquid oil reserves.

The United States has already used more than half of its original recoverable petroleum resource. Of the 120 billion bbl thought to remain, about 30.7 billion bbl are proven-in-place. If we stopped importing oil and depended exclusively on indigenous supplies, our proven reserves would last only four years at current

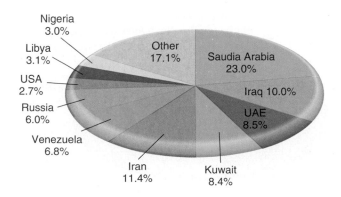

FIGURE 12.8 Proven oil reserves. Ten countries account for nearly 84 percent of all known recoverable oil.
Source: Data from British Petroleum, 2006.

rates of consumption. The regions of North America with the greatest potential for substantial new oil discoveries are portions of the continental shelf along the California coast, the Arctic Ocean, and the Grand Banks, all of which are prime wildlife areas. Proposals to do exploratory drilling in these areas have been strongly opposed by many people concerned about the potential for long-term environmental damage from drilling activities and oil spills like that of the *Exxon Valdez* in Prince William Sound. Other people argue that having an inexpensive, plentiful energy supply, even if only for a few decades, is worth the social and environmental costs (see related story "Oil Drilling in the Arctic" at www.mhhe.com/cunningham4e).

Like coal, petroleum often contains high sulfur levels. Because it is highly corrosive, the sulfur in oil is usually stripped out before the fuel is shipped to market and thus doesn't end up in urban air nearly as often as that from coal. Oil is used primarily for transportation (the gasoline in your car or truck) and heating. Because about one-quarter of all energy used in the United States fuels transportation, and oil provides more than 90 percent of that transportation energy, oil combustion is responsible for between one-quarter and two-thirds of the volatile organic compounds, nitrogen oxides, and carbon monoxide emitted each year.

Oil drilling is far less destructive to the earth's surface than coal mining, but the huge amounts of oil shipped around the world and used as lubricants and fuels result in very serious soil and water pollution. Oceanographers estimate that between 3 million and 6 million metric tons of oil are discharged into the world's rivers and oceans every year. About half of this comes from oil tanker accidents and dumping of ballast water. The rest is mostly deliberate dumping of used motor oil. Many people believe that a storm sewer is a good place to discard waste oil and don't realize that the sewer probably leads directly into a lake, river, or seashore.

Oil shales and tar sands contain huge amounts of petroleum

Estimates of our recoverable oil supplies usually don't account for the very large potential from unconventional resources. The World Energy Council estimates that oil shales, tar sands, and other unconventional deposits contain ten times as much oil as liquid petroleum reserves. **Tar sands** are composed of sand and shale particles coated with bitumen, a viscous mixture of long-chain hydrocarbons. Shallow tar sands are excavated and mixed with hot water and steam to extract the bitumen, then fractionated to make useful products. For deeper deposits, superheated steam is injected to melt the bitumen, which can then be pumped to the surface, like liquid crude. Once the oil has been retrieved, it still must be cleaned and refined to be useful.

Canada and Venezuela have the world's largest and most accessible tar sand resources. Canadian deposits in northern Alberta are estimated to be equivalent to 1.7 trillion bbl of oil, and Venezuela has nearly as much. Together, these deposits are three times as large as all conventional liquid oil reserves. By 2010 Alberta is expected to increase its flow to 2 million bbl per day, or twice the maximum projected output of the Arctic National Wildlife Refuge (ANWR). Furthermore, because Athabascan tar sand beds are 40 times larger and much closer to the surface than ANWR oil, the Canadian resource will last longer and may be cheaper to extract. Canada is already the largest supplier of oil to the United States, having surpassed Saudi Arabia in 2000.

There are severe environmental costs, however, in producing this oil. A typical plant producing 125,000 bbl of oil per day creates about 15 million m^3 of toxic sludge, releases 5,000 tons of greenhouse gases, and consumes or contaminates billions of liters of water each year. Surface mining in Canada could destroy millions of hectares of boreal forest. Native Cree, Chipewyan, and Metis people worry about effects on traditional ways of life if forests are destroyed and wildlife and water are contaminated. Many Canadians dislike becoming an energy colony for the United States, and environmentalists argue that investing billions of dollars to extract this resource simply makes us more dependent on fossil fuels.

Oil shales are fine-grained sedimentary rock rich in solid organic material called kerogen. Like tar sands, the kerogen can be heated, liquefied, and pumped out like liquid crude oil (fig. 12.9). Oil shale beds up to 600 m (1,800 ft) thick underlie much of Colorado, Utah, and Wyoming. If these deposits could be extracted at a reasonable price and with acceptable environmental impacts, they might yield the equivalent of several trillion barrels of oil. Mining and extraction of oil shale use vast amounts of water (a scarce resource in the arid western United States) and create enormous quantities of waste. The rock matrix expands when heated, resulting in two or three times the volume of waste as was dug out of the ground. Billions of dollars were spent in the 1980s on pilot projects to produce synthetic oil. When oil prices dropped, these schemes were abandoned.

With rapidly rising crude oil prices in recent years, interest in oil shale has rekindled. In 2005, the Bureau of Land Management began selling leases once again for pilot projects in oil shale recovery. One technique that might release less air and water pollution is *in situ* refining, in which oil is melted underground and

FIGURE 12.9 Oil shale, a sedimentary rock that contains a solid organic material called kerogen, underlies large areas in Colorado and Utah. It could contain several trillion barrels of oil, more than the entire world supply of liquid oil, but extracting that resource could cause severe environmental and social problems.

then pumped out to be sent to a refinery. Whether this vast resource can ever be extracted with acceptable environmental, economic, and social costs remains to be seen.

Natural gas

Natural gas is the world's third largest commercial fuel, making up 24 percent of global energy consumption. Because natural gas produces only half as much CO_2 as an equivalent amount of coal, substitution could help reduce global warming (chapter 9).

Russia has 26.7 percent of known natural gas reserves (mostly in Siberia and the Central Asian republics) and account for about 35 percent of all global production. Both Eastern and Western Europe buy substantial quantities of gas from these wells. Figure 12.10 shows the distribution of proven natural gas reserves in the world.

The total ultimately recoverable natural gas resources in the world are estimated to be 10,000 trillion ft^3, corresponding to about 80 percent as much energy as the recoverable reserves of crude oil. The proven world reserves of natural gas are 6,200 trillion ft^3 (176 trillion metric tons). Because gas consumption rates are only about half of those for oil, current gas reserves represent roughly a 60-year supply at present usage rates. Proven reserves in the United States are about 185 trillion ft^3, or 3 percent of the world total. This is about a ten-year supply at current rates of consumption. Known reserves are more than twice as large.

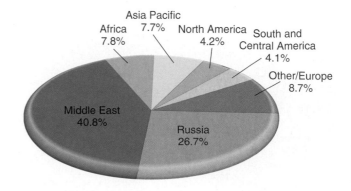

FIGURE 12.10 Proven natural gas reserves by region, 2002.
Source: Data from British Petroleum, 2006.

Large amounts of methane are released from coal deposits. The Rocky Mountain front in Colorado, Wyoming, and Montana could have 10 percent of the total world methane supply. Proposals for up to hundreds of thousands of coal-bed methane wells in Wyoming and Utah, for example, have raised concerns about water pollution and land damage (see opening Case Study, chapter 11). Other states could face similar problems.

World consumption of natural gas is growing by about 2.2 percent per year, considerably faster than either coal or oil. Much of this increase is in the developing world, where concerns about urban air pollution encourage the switch to cleaner fuel. Gas can be shipped easily and economically through buried pipelines. The United States has been fortunate to have abundant gas resources accessible by an extensive pipeline system. It is difficult and dangerous, however, to ship and store gas between continents. To make the process economic, gas is compressed and liquefied. At $-160°C$ ($-260°F$) the liquid takes up about one-six-hundredth the volume of gas. Special refrigerated ships transport liquefied natural gas (LNG) (fig. 12.11). LNG weighs only about half as much as water, so the ships are very buoyant. Finding sites for terminals to load and unload these ships is difficult. Many cities are unwilling to accept the risk of an explosion of the volatile cargo. A fully loaded LNG ship contains about as much energy as a medium-size atomic bomb. Furthermore, huge amounts of seawater are used to warm and re-gasify the LNG. This can have deleterious effects on coastal ecology. However, in 2005, the U.S. Senate voted to eliminate state or local authority in siting LNG terminals.

LNG shipments now supply only about 3 percent of the U.S. gas supply, but this is expected to climb to 20 percent of U.S. consumption by 2025 as Canadian resources are exhausted or diverted to domestic use. Dr. Robert Socolow and his Princeton colleagues estimate that to eliminate 1 billion tons of carbon emissions in 2054 by switching from coal to natural gas would require 50 large LNG tankers docking and unloading every day somewhere in the world (see opening Case Study chapter 9). This would take about three times as many LNG ships as there are now crude oil tankers in the world.

FIGURE 12.11 As domestic supplies of natural gas dwindle, the United States is turning increasingly to shipments of liquefied gas in specialized ships, such as this one at an Australian terminal. An explosion of one of these ships would release about as much energy as a medium-size atomic bomb.

NUCLEAR POWER

In 1953 President Dwight Eisenhower presented his "Atoms for Peace" speech to the United Nations. He announced that the United States would build nuclear-powered electrical generators to provide clean, abundant energy. He predicted that nuclear energy would fill the deficit caused by predicted shortages of oil and natural gas. It would provide power "too cheap to meter" for continued industrial expansion of both the developed and the developing world. It would be a supreme example of "beating swords into plowshares." Technology and engineering would tame the evil genie of atomic energy and use its enormous power to do useful work.

However, rapidly increasing construction costs, declining demand for electric power, and safety fears have made nuclear energy much less attractive than promoters expected. Of the 140 reactors on order in 1975, 100 were subsequently canceled. Nuclear power now produces about 8 percent of the U.S. energy supply (1 percent more than the world average). All of it is used to generate electricity.

The nuclear power industry has been campaigning for greater acceptance, arguing that reactors don't release greenhouse gases that cause global warming. About half the existing nuclear reactors in the United States have had their licenses renewed for an additionl 20 years beyond their original 40-year design, and several new plants are now on the drawing board. A number of prominent environmentalists are now promoting nuclear power as a solution to global climate change. Professor Robert Socolow calculates that we would need to double the total global nuclear capacity (to about 700 large power plants) to eliminate 1 billion

FIGURE 12.12 Two nuclear reactors (domes) at the San Onofre Nuclear Generating Station sit between the beach and Interstate 5, the major route between Los Angeles and San Diego.

tons of annual CO_2 emissions. However, worries about accidents and susceptiblity to terrorist attacks make many people fearful of this solution (fig. 12.12).

How do nuclear reactors work?

The most commonly used fuel in nuclear power plants is U^{235}, a naturally occurring radioactive isotope of uranium. Ordinarily, U^{235} makes up only about 0.7 percent of uranium ore—too little to sustain a chain reaction in most reactors. It must be purified and concentrated by mechanical or chemical procedures. When the U^{235} concentration reaches about 3 percent, the uranium is formed into cylindrical pellets slightly thicker than a pencil and about 1.5 cm long. Although small, these pellets pack an amazing amount of energy. Each 8.5 g pellet is equivalent to a ton of coal or 4 bbl of crude oil.

The pellets are stacked in hollow metal rods approximately 4 m long. About 100 of these rods are bundled together to make a **fuel assembly.** Thousands of fuel assemblies containing about 100 tons of uranium are bundled in a heavy steel vessel called the reactor core. Radioactive uranium atoms are unstable—that is, when struck by a high-energy subatomic particle called a neutron, they undergo **nuclear fission** (splitting), releasing energy and more neutrons. When uranium is packed tightly in the reactor core, the neutrons released by one atom will trigger the fission of another uranium atom and the release of still more neutrons (fig. 12.13). Thus, a self-sustaining **chain reaction** is set in motion, and vast amounts of energy are released.

The chain reaction is moderated (slowed) in a power plant by a neutron-absorbing cooling solution that circulates between the fuel rods. In addition, **control rods** of neutron-absorbing

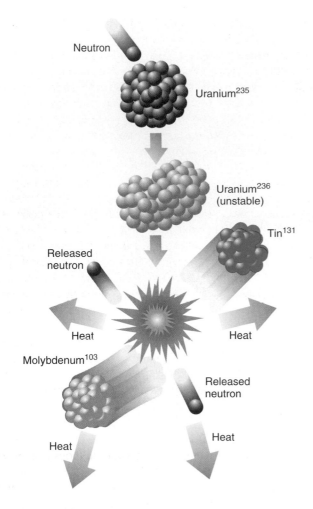

FIGURE 12.13 The process of nuclear fission is carried out in the core of a nuclear reactor. In the sequence shown here, the unstable isotope uranium-235 absorbs a neutron and splits to form tin-131 and molybdenum-103. Two or three neutrons are released per fission event and continue the chain reaction. The total mass of the reaction product is slightly less than the starting material. The residual mass is converted to energy (mostly heat).

material, such as cadmium or boron, are inserted into spaces between fuel assemblies to shut down the fission reaction or are withdrawn to allow it to proceed. Water or some other coolant is circulated between the fuel rods to remove excess heat. The greatest danger in one of these complex machines is a cooling system failure. If the pumps fail or pipes break during operation, the nuclear fuel quickly overheats, and a "meltdown" can result that releases deadly radioactive material. Although nuclear power plants cannot explode like a nuclear bomb, the radioactive releases from a worst-case disaster, such as the meltdown of the Chernobyl reactor in the Soviet Ukraine in 1986, are just as devastating as a bomb. (See related story on Chernobyl at www.mhhe.com/cunningham4e.)

Nuclear reactor design

Seventy percent of the world's nuclear plants are pressurized water reactors (PWR). Water circulates through the core, absorbing heat

FIGURE 12.14 Pressurized water nuclear reactor. Water is superheated and pressurized as it flows through the reactor core. Heat is transferred to nonpressurized water in the steam generator. The steam drives the turbogenerator to produce electricity.

as it cools the fuel rods (fig. 12.14). This primary cooling water is heated to 317°C (600°F) and reaches a pressure of 2,235 psi. It then is pumped to a steam generator, where it heats a secondary water-cooling loop. Steam from the secondary loop drives a high-speed turbine generator that produces electricity. Both the reactor vessel and the steam generator are contained in a thick-walled, concrete-and-steel containment building that prevents radiation from escaping and is designed to withstand high pressures and temperatures in case of accidents. Engineers operate the plant from a complex, sophisticated control room containing many gauges and meters that indicate how the plant is running.

Overlapping layers of safety mechanisms are designed to prevent accidents, but these fail-safe controls make reactors very expensive and very complex. A typical nuclear power plant has 40,000 valves, compared with only 4,000 in a fossil fuel–fired plant of similar size. In some cases, the controls are so complex that they confuse operators and cause accidents, rather than prevent them. Under normal operating conditions, however, a PWR releases very little radioactivity and is probably less dangerous for nearby residents than a coal-fired power plant.

In Britain, France, and the former Soviet Union, a common reactor design uses graphite, both as a moderator and as the structural material for the reactor core. In the British MAGNOX design (named after the magnesium alloy used for its fuel rods), gaseous carbon dioxide is blown through the core to cool the fuel assemblies and carry heat to the steam generators. In the Soviet design, called RBMK (the Russian initials for a graphite-moderated, water-cooled reactor), low-pressure cooling water circulates through the core in thousands of small metal tubes.

These designs were originally thought to be very safe because graphite has a high capacity for both capturing neutrons and dissipating heat. Designers claimed that these reactors could not possibly run out of control; unfortunately, they were proven wrong. The small cooling tubes are quickly blocked by steam if the cooling system fails, and the graphite core burns when exposed to air. Burning graphite in the Chernobyl nuclear plant made the fire much more difficult to control than it might have been in another reactor design.

There are other reactor designs, such as the high-temperature, gas-cooled reactor (HTGCR), or pebble bed reactor, that are inherently safer than PWR or RBMK reactors. What might our history have been if we had chosen one of these safer designs originally?

What will we do with radioactive waste?

One of the most difficult problems associated with nuclear power is the disposal of wastes produced during mining, fuel production, and reactor operation. How these wastes are managed may ultimately be the overriding obstacle to nuclear power.

Enormous piles of mine wastes and abandoned mill tailings in uranium-producing countries represent another serious waste disposal problem. Production of 1,000 tons of uranium fuel typically generates 100,000 tons of tailings and 3.5 million liters of liquid waste. There now are approximately 200 million tons of radioactive waste in piles around mines and processing plants in the United States. This material is carried by the wind or washes into streams, contaminating areas far from its original source. Canada has even more radioactive mine waste on the surface than does the United States.

In addition to the leftovers from fuel production, there are about 100,000 tons of low-level waste (contaminated tools, clothing, building materials, etc.) and about 15,000 tons of high-level (very radioactive) wastes in the United States. The high-level wastes consist mainly of spent fuel rods from commercial nuclear power plants and assorted wastes from nuclear weapons production. For the past 20 years, spent fuel assemblies from commercial reactors have been stored in deep, water-filled pools at the power plants. These pools were originally intended only as temporary storage until the wastes were shipped to reprocessing centers or permanent disposal sites.

With internal waste storage pools now full but neither reprocessing nor permanent storage available, a number of utility companies are beginning to store nuclear waste in large, metal dry casks placed outside power plants (fig. 12.15). These projects are meeting with fierce opposition from local residents, who fear the casks will leak. Most nuclear power plants are built near rivers, lakes, or seacoasts. Extremely toxic radioactive materials could spread quickly over large areas if leaks occur. A hydrogen gas explosion and fire in 1997 in a dry storage cask at Wisconsin's Point Beach nuclear plant intensified opponents' suspicions about this form of waste storage.

In 1987 the U.S. Department of Energy announced plans to build the first high-level waste repository on a barren desert ridge near Yucca Mountain, Nevada. After 15 years of research and $4 billion in exploratory drilling, it still isn't certain that the site is safe (see related story on Yucca Mountain at www.mhhe.com/cunningham4e). Intensely radioactive wastes are to be buried deep in the ground, where it is hoped that they will remain unexposed to groundwater and earthquakes for the thousands of years required for the radioactive materials to decay to a safe level. Although the area is very dry now, we can't be sure that it will always remain that way. Total costs now are expected to be at least $35 billion. Although the facility was supposed to open in 1998, the earliest possible date is now 2010.

Some nuclear experts believe that **monitored, retrievable storage** would be a much better way to handle wastes. This method involves holding wastes in underground mines or secure surface facilities where they can be watched. If canisters begin to leak, they could be removed for repacking. Safeguarding the wastes would be expensive, and the sites might be susceptible to wars and terrorist attacks. We might need a perpetual priesthood of nuclear guardians to ensure that the wastes are never released into the environment.

FIGURE 12.15 Spent fuel is being stored temporarily in large, aboveground "dry casks" at many nuclear power plants.

ENERGY CONSERVATION

One of the best ways to avoid energy shortages and to relieve environmental and health effects of our current energy technologies is simply to use less (see What Can You Do? p. 294). Conservation offers many benefits both to society and to the environment.

Using energy more efficiently

Much of the energy we consume is wasted. This statement is not a simple admonishment to turn off lights and turn down furnace

Some Things You Can Do to Save Energy

1. Drive less: make fewer trips, use telecommunications and mail instead of going places in person.
2. Use public transportation, walk, or ride a bicycle.
3. Use stairs instead of elevators.
4. Join a car pool or drive a smaller, more efficient car; reduce speeds.
5. Insulate your house or add more insulation to the existing amount.
6. Turn thermostats down in the winter and up in the summer.
7. Weatherstrip and caulk around windows and doors.
8. Add storm windows or plastic sheets over windows.
9. Create a windbreak on the north side of your house; plant deciduous trees or vines on the south side.
10. During the winter, close windows and drapes at night; during summer days, close windows and drapes if using air conditioning.
11. Turn off lights, television sets, and computers when not in use.
12. Stop faucet leaks, especially hot water.
13. Take shorter, cooler showers; install water-saving faucets and showerheads.
14. Recycle glass, metals, and paper; compost organic wastes.
15. Eat locally grown food in season.
16. Buy locally made, long-lasting materials.

FIGURE 12.16 Per capita energy consumption in the United States rose rapidly in the 1960s. Price shocks in the 1970s encouraged conservation. Although GDP continued to grow in the 1980s and 90s, higher efficiency kept per capita consumption relatively constant.
Source: Data from U.S. Department of Energy.

thermostats in winter; it is a technological challenge. Our ways of using energy are so inefficient that most potential energy in fuel is lost as waste heat, becoming a form of environmental pollution. More efficient and less energy-intensive industry, transportation, and domestic practices could save large amounts of energy. Oil price shocks in the 1970s encouraged conservation. Although population and GDP have continued to grow since then, the **energy intensity,** or amount of energy needed to provide goods and services has declined (fig. 12.16). In response to federal regulations and high gasoline prices, automobile gas-mileage averages in the United States more than doubled from 13 mpg in 1975 to 28.8 mpg in 1988. Unfortunately, the oil glut and falling fuel prices of the 1990s discouraged further conservation. By 2004 the average slipped to only 20.4 mpg.

Much more could be done. High-efficiency automobiles are already available. Low-emission, hybrid gas-electric vehicles get up to 30.3 km/liter (72 mpg) on the highway (see related story on hybrid engines at www.mhhe.com/cunningham4e). Amory B. Lovins of the Rocky Mountain Institute in Colorado estimates that raising the average fuel efficiency of the U.S. car and light-truck fleet by 1 mpg would cut oil consumption about 295,000 bbl per day. In one year, this would equal the total amount the U.S. Department of the Interior hopes to extract from the Arctic National Wildlife Refuge in Alaska.

Many improvements in domestic energy efficiency have occurred in the past decade. Today's average new home uses one-half the fuel required in a house built in 1974, but much more can be done. Household energy losses can be reduced by one-half to three-fourths by using better insulation, installing double- or triple-glazed windows, purchasing thermally efficient curtains or window coverings, and sealing cracks and loose joints. Reducing air infiltration is usually the cheapest, quickest, and most effective way of saving energy because it is the largest source of losses in a typical house. It doesn't take much skill or investment to caulk around doors, windows, foundation joints, electrical outlets, and other sources of air leakage. Mechanical ventilation is needed to prevent moisture buildup in tightly sealed homes.

For even greater savings, new houses can be built with extra-thick, superinsulated walls; air-to-air heat exchangers to warm incoming air; and even double-walled sections that create a "house within a house." The R-2000 program in Canada details how energy conservation can be built into homes. Special double-glazed windows that have internal reflective coatings and that are filled with an inert gas (argon or xenon) have an insulation factor of R11, the same as a standard 4-inch-thick insulated wall, or ten times as efficient as a single-pane window. Superinsulated houses now being built in Sweden require 90 percent less energy for heating and cooling than the average American home.

Orienting homes so that living spaces have passive solar gain in the winter and are shaded by trees or roof overhang in the summer also helps conserve energy. Earth-sheltered homes built into the south-facing side of a slope or protected on three sides by an earth berm are exceptionally efficient energy savers because they maintain relatively constant subsurface temperatures (fig. 12.17). Sod roofs provide good insulation, prevent rain runoff, and last longer than asphalt shingles. Because they are heavier, however, they need stronger supports.

Straw-bale construction offers both high insulating qualities and a renewable, inexpensive building material that can be

FIGURE 12.17 Earth-sheltered homes take advantage of the stable temperatures and insulating qualities of the earth. This house has south-facing windows for maximum solar gain and high celerestory windows that give light to the back of the house as well as summer ventilation.

assembled by amateurs (fig. 12.18). This isn't a new technique. Settlers on the Great Plains built straw-bale houses a century ago because they didn't have wood. Some of those houses are still standing. The bales are strong and will support the roof without any additional timber framing. They must be thoroughly water-proofed, however, with stucco, adobe, or plaster both inside and out so the straw doesn't decay. It's also important to seal them so mice and other vermin can't take up residence. The thick walls are terrific sound insulators as well as highly energy-efficient. The cost can be far less than a conventionally built home.

FIGURE 12.18 Carolyn Roberts and her sons build a straw bale house near Tucson, Arizona.
Courtesy of Carolyn Roberts/A House of Straw.

FIGURE 12.19 Typical standby energy consumption by household electrical appliances.
Source: Data from U.S. Department of Energy.

One of the most direct and immediate ways that individuals can save energy is to turn off appliances. Few of us realize how much electricity is used by appliances in a standby mode. You may think you've turned off your TV, DVD player, cable box, or printer, but they're really continuing to draw power in an "instant-on" mode. For the average home, standby appliances can represent up to 25 percent of the monthly electric bill. Home office equipment, including computers, printers, cable modems, and copiers, usually are the biggest energy consumers (fig. 12.19). Putting your computer to sleep saves about 90 percent of the energy it uses when fully on, but turning it completely off is even better.

Industrial energy savings are another important part of our national energy budget. More efficient electric motors and pumps, new sensors and control devices, advanced heat-recovery systems, and material recycling have reduced industrial energy requirements significantly. In the early 1980s, U.S. businesses saved $160 billion per year through conservation. When oil prices collapsed, however, many businesses returned to wasteful ways.

Cities can make surprising contributions to energy conservation. New York City has become a leader in this effort, replacing 11,000 traffic signals with more-efficient LEDs (light-emitting diodes), and 180,000 old refrigerators with energy-saving models.

Cogeneration

One of the fastest growing sources of new energy is **cogeneration,** the simultaneous production of both electricity and steam or hot water in the same plant. By producing two kinds of useful energy in the same facility, the net energy yield from the primary fuel is increased from 30–35 percent to 80–90 percent. In 1900 half the electricity generated in the United States came from plants that also provided industrial steam or district heating. As power plants became larger, dirtier, and less acceptable as neighbors, they were forced to move away from their customers. Waste heat from the turbine generators became an unwanted pollutant to be disposed of in the environment. Furthermore, long

FIGURE 12.20 A technician adjusts a gas microturbine that produces on-site heat and electricity for businesses, industry, or multiple housing units.

transmission lines, which are unsightly and lose up to 20 percent of the electricity they carry, became necessary.

By the 1970s, cogeneration had fallen to less than 5 percent of our power supplies, but interest in this technology is being renewed. The capacity for cogeneration more than doubled in the 1980s to about 30,000 megawatts. District heating systems are being rejuvenated, and plants that burn municipal wastes are being studied. New combined-cycle coal-gasification plants, or "mini-nukes," offer high efficiency and clean operation that may be compatible with urban locations. Small neighborhood- or apartment building-sized power-generating units are being built that burn methane (from biomass digestion), natural gas, diesel fuel, or coal (fig. 12.20). The Fiat Motor Company makes a small generator for about $10,000 that produces enough electricity and heat for four or five energy-efficient houses. These units are especially valuable for facilities, such as hospitals or computer centers, that need uninterrupted power.

SOLAR ENERGY

The sun is a giant nuclear furnace in space, constantly bathing our planet with a free energy supply. Solar heat drives winds and the hydrologic cycle. All biomass, as well as fossil fuels and our food (both of which are derived from biomass), results from con-

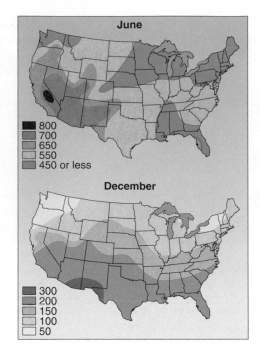

FIGURE 12.21 Average daily solar radiation in the United States in June and December. One langley, the unit for solar radiation, equals 1 cal/cm^2 of earth surface (3.69 Btu/ft^2).
Source: Data from National Weather Bureau, U.S. Department of Commerce.

version of light energy (photons) into chemical bond energy by photosynthetic bacteria, algae, and plants.

The average amount of solar energy arriving at the top of the atmosphere is 1,330 watts m^2. About half of this energy is absorbed or reflected by the atmosphere (more at high latitudes than at the equator), but the amount reaching the earth's surface is some 10,000 times all the commercial energy used each year. However, this tremendous infusion of energy comes in a form that, until recently, has been too diffuse and low in intensity to be used except for environmental heating and photosynthesis. Figure 12.21 shows solar energy levels over the United States for typical summer and winter days.

Solar collectors can be passive or active

Our simplest and oldest use of solar energy is **passive heat absorption,** using natural materials or absorptive structures with no moving parts to simply gather and hold heat. For thousands of years, people have built thick-walled stone and adobe dwellings that slowly collect heat during the day and gradually release it at night. After cooling at night, these massive building materials maintain a comfortable daytime temperature within the house, even as they absorb external warmth.

A modern adaptation of this principle is a glass-walled "sunspace," or greenhouse, on the south side of a building. Massive energy-storing materials, such as brick walls, stone floors, or barrels of heat-absorbing water, incorporated into buildings collect heat that is released slowly at night. An interior, heat-absorbing wall called a Trombe wall is an effective passive heat collector. Some Trombe walls are built of glass blocks enclosing a water-filled space or water-filled circulation tubes so heat from solar rays can be absorbed and stored, while light passes through to inside rooms.

Active solar systems generally pump a heat-absorbing fluid medium (air, water, or an antifreeze solution) through a relatively small collector, rather than passively collecting heat in a stationary medium, such as masonry. Active collectors can be located adjacent to or on top of buildings, rather than being built into the structure (see p. 284).

A flat, black surface sealed with a double layer of glass makes a good solar collector. A fan circulates air over the hot surface and into the house through ductwork of the type used in standard forced-air heating. Alternatively, water can be pumped through the collector to pick up heat for space heating or to provide hot water. Water heating consumes 15 percent of the U.S. domestic energy budget, so savings in this area alone can be significant. A simple, flat panel with about 5 m^2 of surface can reach 95°C (200°F) and can provide enough hot water for an average family of four almost anywhere in the United States. In California, 650,000 homes now heat water with solar collectors. In Greece, Italy, Israel, and other countries where fuels are very expensive, up to 70 percent of domestic hot water comes from solar collectors. And, as the opening Case Study for this chapter shows, the Danish islands of Samsø and Ærø meet 100 percent of their space heating needs with solar and biomass.

Sunshine doesn't reach us all the time, of course. How can solar energy be stored for times when it is needed? A number of options are available. In a climate where sunless days are rare and seasonal variations are small, a small, insulated water tank is a good solar energy storage system. In areas where clouds block the sun for days at a time or where energy must be stored for winter use, a large, insulated bin containing a heat-storing mass, such as stone, water, or clay, provides solar energy storage. During the summer months, a fan blows the heated air from the collector into the storage medium. In the winter, a similar fan at the opposite end of the bin blows the warm air into the house. During the summer, the storage mass is cooler than the outside air, and it helps cool the house by absorbing heat. During the winter, it is warmer and acts as a heat source by radiating stored heat. In many areas, six or seven months' worth of thermal energy can be stored in 10,000 gal of water or 40 tons of gravel, about the amount of water in a small swimming pool or the gravel in two average-size dump trucks.

Solar energy can produce high temperatures

Parabolic mirrors are curved reflecting surfaces that collect light and focus it into a concentrated point. There are two ways to use mirrors to collect solar energy to generate high temperatures. One technique uses long, curved mirrors focused on a central tube

FIGURE 12.22 Parabolic mirrors focus sunlight on steam-generating tubes at this power plant in the California desert.

containing a heat-absorbing fluid (fig. 12.22). Fluid flowing through the tubes reaches much higher temperatures than possible in a basic flat-panel collector. Another high-temperature system uses thousands of smaller mirrors arranged in concentric rings around a tall, central tower. The mirrors, driven by electric motors, track the sun and focus its light on a heat absorber at the top of the "power tower," where molten salt is heated to temperatures as high as 500°C (1,000°F), which then drives a steam-turbine electric generator.

Under optimum conditions, a 50-ha (130-acre) mirror array should be able to generate 100 MW of clean, renewable power. The only power tower in the United States is Southern California Edison's Solar II plant in the Mojave Desert east of Los Angeles. Its 2,000 mirrors focused on a 100-m (300-ft)-tall tower generate 10 MW, or enough electricity for 5,000 homes at an operating cost far below that of nuclear power or oil. We haven't had enough experience with these facilities to know how reliable the mirrors, motors, heat absorbers, and other equipment will be over the long run.

Photovoltaic cells generate electricity directly

The **photovoltaic cell** offers an exciting potential for capturing solar energy in a way that will provide clean, versatile, renewable energy. This simple device has no moving parts and negligible maintenance costs, produces no pollution, and has a lifetime equal to that of a conventional fossil fuel or nuclear power plant.

Photovoltaic cells capture solar energy and convert it directly to electrical current by separating electrons from their parent atoms and accelerating them across a one-way electrostatic barrier formed

FIGURE 12.23 The operation of a photovoltaic cell. Boron impurities incorporated into the upper silicon crystal layers cause electrons (e-) to be released when solar radiation hits the cell. The released electrons move into the lower layer of the cell, thus creating a shortage of electrons, or a positive charge, in the upper layer and an oversupply of electrons, or negative charge, in the lower layer. The difference in charge creates an electric current in a wire connecting the two layers.

FIGURE 12.24 Solar roof tiles (*shiny area*) can generate enough electricity for a house full of efficient appliances. On sunny days, this array can produce a surplus to sell back to the utility company, making it even more cost-efficient.

by the junction between two different types of semiconductor material (fig. 12.23). The photovoltaic effect, which is the basis of these devices, was first observed in 1839 by French physicist Alexandre-Edmond Becquerel, who also discovered radioactivity. His discovery didn't lead to any useful applications until 1954, when researchers at Bell Laboratories in New Jersey learned how to carefully introduce impurities into single crystals of silicon.

These handcrafted, single-crystal cells were much too expensive for any practical use until the advent of the U.S. space program. In 1958, when *Vanguard I* went into orbit, its radio was powered by six palm-sized photovoltaic cells that cost $2,000 per peak watt of output, more than 2,000 times as much as conventional energy at the time. Since then, prices have fallen dramatically. In 1970 they cost $100 per watt; in 2006 they were about $5 per watt. This makes solar energy cost-competitive with other sources in some areas. A photovoltaic array of about 30–40m^2 will generate enough electricity for an efficient house (fig. 12.24).

By 2020 photovoltaic cells could be less than $1 per watt of generating capacity. With 15 percent efficiency and a 30-year life, they should be able to produce electricity for around 6¢ per kilowatt hour. At that time, coal-fired steam power will probably cost about half again as much, and nuclear power will likely cost twice as much as photovoltaic energy.

During the past 25 years, the efficiency of energy captured by photovoltaic cells has increased from less than 1 percent of incident light to more than 10 percent under field conditions and over 75 percent in the laboratory. Promising experiments are underway using exotic metal alloys, such as gallium arsenide, and semiconducting polymers of polyvinyl alcohol, which are more efficient in energy conversion than silicon crystals. Every year college students from all over North America race solar-powered cars across the country to test solar technology and publicize its potential. Perhaps you and your classmates would like to build a solar car and enter the race.

One of the most promising developments in photovoltaic cell technology in recent years is the invention of **amorphous silicon collectors.** First described in 1968 by Stanford Ovshinky, a self-taught inventor from Detroit, these noncrystalline silicon semiconductors can be made into lightweight, paper-thin sheets that require much less material than conventional photovoltaic cells. They also are vastly cheaper to manufacture and can be made in a variety of shapes and sizes, permitting ingenious applications. Roof tiles with photovoltaic collectors layered on their surface already are available. Even flexible films can be coated with amorphous silicon collectors. Silicon collectors already are providing power to places where conventional power is unavailable, such as lighthouses, mountaintop microwave repeater stations, villages on remote islands, and ranches in the Australian outback.

You probably already use amorphous silicon photovoltaic cells. They are being built into solar-powered calculators, watches, toys, photosensitive switches, and a variety of other consumer products. Japanese electronic companies presently lead in this field, having foreseen the opportunity for developing a market for photovoltaic cells. This market is already more than $100 million per year. Japanese companies now have home-roof arrays capable of providing all the electricity needed for a typical home at prices in some areas competitive with power purchased from a utility. And Shanghai, China, recently announced a plan to install photovoltaic collectors on 100,000 roofs. This is expected to generate 430 million kWh annually and replace 20,000 tons of coal per year.

Think about how solar power could affect your future energy independence. Imagine the benefits of being able to build a house anywhere and having a cheap, reliable, clean, quiet source of energy with no moving parts to wear out, no fuel to purchase, and little equipment to maintain. You could have all the energy you need without commercial utility wires or monthly energy bills. Coupled with modern telecommunications and information technology, an independent energy source would make it possible to live in the countryside and yet have many of the employment and entertainment opportunities and modern conveniences available in a metropolitan area.

Electrical energy is difficult and expensive to store or ship

Electrical energy is difficult and expensive to store or to transport over long distances. If we come to depend on solar, wind, or other renewable energy sources, we may need many more high-voltage power lines to transport electricity to distant markets (fig. 12.25). Many property owners resist having power lines cross their land because they are unsightly and may create hazardous electromagnetic fields. An alternative would be to use electricity at remote sites to electrolyze water, then ship hydrogen through buried pipelines. Hydrogen is the ideal fuel for fuel cells, discussed later in this chapter.

Storage is a problem for photovoltaic generation as well as other sources of electric power. Traditional lead-acid batteries are heavy and have low energy densities; that is, they can store only moderate amounts of energy per unit mass or volume. Acid from batteries is hazardous, and lead from smelters or battery manufacturing is a serious health hazard for workers who handle these materials. A typical lead-acid battery array sufficient to store several days of electricity for an average home would cost about $5,000 and weigh 3 or 4 tons. All the components for an electric car are readily available, but battery technology limits how far they can go between charges.

FIGURE 12.25 As we come to depend more on wind, solar, and hydropower energy, we will probably need more transmission lines to get that energy to market.

Another strategy is to store energy in a form that can be turned back into electricity when needed. **Pumped-hydro storage** involves pumping water to an elevated reservoir at times when excess electricity is available. The water is released to flow back down through turbine generators when extra energy is needed. Using a similar principle, pressurized air can be pumped into such reservoirs as natural caves, depleted oil and gas fields, abandoned mines, or special tanks. An Alabama power company uses off-peak electricity to pump air at night into a deep salt mine. By day, the air flows back to the surface through turbines, driving a generator that produces electricity. Cool night air is heated to 1,600°F by compression plus geothermal energy, increasing pressure and energy yield.

Flywheels also are the subject of current experimentation for energy storage. Massive, high-speed flywheels, spinning in a nearly friction-free environment, store large amounts of mechanical energy in a small area. This energy is easily convertible to electrical energy. It is difficult, however, to find materials strong enough to hold together when spinning at high speed. Flywheels have a disconcerting tendency to fail explosively and unexpectedly, sending shrapnel in all directions.

Utilities are promoting renewable energy

Utility restructuring currently being planned in the United States could include policies to encourage conservation and alternative energy sources. Among the proposed policies are (1) "distributional surcharges" in which a small per kilowatt-hour charge is levied on all utility customers to help finance renewable energy research and development, (2) "renewables portfolio" standards to require power suppliers to obtain a minimum percentage of their energy from sustainable sources, and (3) **green pricing** that allows utilities to profit from conservation programs and charge premium prices for energy from renewable sources.

Some states already are pursuing these policies. For example, Iowa has a Revolving Loan Fund supported by a surcharge on investor-owned gas and electric utilities. This fund provides low-interest loans for renewable energy and conservation. Several states have initiated green pricing programs as a way to encourage a transition to sustainable energy. One of the first was in Colorado, where 1,000 customers agreed to pay $2.50 per month above their regular electric rates to help finance a 10 MW wind farm being built on the Colorado–Wyoming border. Buying a 100 kW "block" of wind power provides the same environmental benefits as planting a half acre of trees or not driving an automobile 4,000 km (2,500 mi) per year. Not all green pricing plans are as straightforward as this, however. Some utilities collect the premium rates for facilities that already exist or for renewable energy bought from other utilities at much lower prices.

FUEL CELLS

Rather than store and transport energy, another alternative is to generate it locally, on demand. **Fuel cells** are devices that use ongoing electrochemical reactions to produce an electrical current. They are very similar to batteries except that, rather than recharging them with an electrical current, you add more fuel for the chemical reaction.

All fuel cells consist of a positive electrode (the cathode) and a negative electrode (the anode) separated by an electrolyte, a material that allows the passage of charged atoms, called ions, but is impermeable to electrons (fig. 12.26). In the most common

FIGURE 12.26 Fuel cell operation. Electrons are removed from hydrogen atoms at the anode to produce hydrogen ions (protons) that migrate through a semipermeable electrolyte medium to the cathode, where they reunite with electrons from an external circuit and oxygen atoms to make water. Electrons flowing through the circuit connecting the electrodes create useful electrical current.

systems, hydrogen or a hydrogen-containing fuel is passed over the anode, while oxygen is passed over the cathode. At the anode, a reactive catalyst, such as platinum, strips an electron from each hydrogen atom, creating a positively charged hydrogen ion (a proton). The hydrogen ion can migrate through the electrolyte to the cathode, but the electron is excluded. Electrons flow through an external circuit, and the electrical current generated by their passage can be used to do useful work. At the cathode, the electrons and protons are reunited and combined with oxygen to make water.

Fuel cells provide direct-current electricity as long as they are supplied with hydrogen and oxygen. For most uses, oxygen is provided by ambient air. Hydrogen can be supplied as a pure gas, but storing hydrogen gas is difficult and dangerous because of its explosive nature. Liquid hydrogen takes far less space than the gas but must be kept below –250°C (–400°F), not a trivial task for most mobile applications. The alternative is a device called a **reformer** or converter that strips hydrogen from fuels such as natural gas, methanol, ammonia, gasoline, ethanol, or even vegetable oil. Many of these fuels can be derived from sustainable biomass crops. Even methane effluents from landfills and wastewater treatment plants can be used as a fuel source. Where a fuel cell can be hooked permanently to a gas line, hydrogen can be provided by solar, wind, or geothermal facilities that use electricity to hydrolyze water.

A fuel cell that runs on pure oxygen and hydrogen produces no waste products except drinkable water and radiant heat. When a reformer is coupled to the fuel cell, some pollutants are released (most commonly carbon dioxide), but the levels are typically far less than conventional fossil fuel combustion in a power plant or an automobile engine. Although the theoretical efficiency of electrical generation of a fuel cell can be as high as 70 percent, the actual yield is closer to 40 or 45 percent. This is not much better than a very good fossil fuel power plant or a gas turbine electrical generator. On the other hand, the quiet, clean operation and variable size of fuel cells make them useful in buildings where waste heat can be captured for water heating or space heating. A 45-story office building at 4 Times Square, for example, has two 200 kW fuel cells on its fourth floor that provide both electricity and heat. The same building has photovoltaic panels on its façade, natural lighting, fresh-air intakes to reduce air conditioning, and a number of other energy conservation features.

Fuel cells could be ideal zero-emission power sources for vehicles. Fuel cell buses are being tested in Canada, and Honda introduced an experimental fuel cell automobile in California in 2002.

The current from a fuel cell is proportional to the size (area) of the electrodes, while the voltage is limited to about 1.23 volts per cell. A number of cells can be stacked together until the desired power level is reached. A fuel cell stack that provides almost all of the electricity needed by a typical home (along with hot water and space heating) would be about the size of a refrigerator. A 200 kW unit fills a medium-size room and provides enough energy for 20 houses or a small factory (fig. 12.27). Tiny fuel cells running on methanol might soon be used in cell phones,

Principles of Environmental Science

http://www.mhhe.com/cunningham4e

FIGURE 12.27 The Long Island Power Authority has installed 75 stationary fuel cells to provide reliable backup power.

FIGURE 12.28 Harvesting marsh reeds (*Phragmites* sp.) as a source of biomass fuel. On the Danish Islands of Samsø and Ærø, about half of all space heating is provided by biomass.

pagers, toys, computers, videocameras, and other appliances now run by batteries. Rather than buy new batteries or spend hours recharging spent ones, you might just add an eyedropper of methanol every few weeks to keep your gadgets operating.

ENERGY FROM BIOMASS

Wood fires have been a primary source of heating and cooking for thousands of years. As recently as 1850, wood supplied 90 percent of the fuel used in the United States. Wood now provides less than 1 percent of the energy in the United States, but in many of the poorer countries of the world, wood and other biomass fuels still provide up to 95 percent of all energy used. The 1,500 million m^3 of fuelwood collected in the world each year is about half of all wood harvested.

In northern industrialized countries, wood burning has increased since 1975 in an effort to avoid rising oil, coal, and gas prices. Most northern areas have adequate wood supplies to meet demands at current levels, but problems associated with wood burning may limit further expansion of this use. Inefficient and incomplete burning of wood in fireplaces and stoves produces smoke laden with fine ash and soot and hazardous amounts of carbon monoxide and hydrocarbons. In valleys where inversion layers trap air pollutants, the effluent from wood fires can be a major source of air quality degradation and health risk. Polycyclic aromatic compounds produced by burning are especially worrisome because they are carcinogenic (cancer-causing). Some communities have passed ordinances limiting the use of woodstoves.

Biofuels could replace oil-based fuels

As the opening Case Study in this chapter shows, biomass can make a substantial contribution to renewable energy supplies. The islands of Samsø and Ærø get about half their space heating from biomass, both from straw and other crop wastes and from energy crops, such as reeds and elephant grass growing on land unsuitable for crops (fig. 12.28). Burning these crops in an industrial boiler for district heating makes it easier to install and maintain pollution-control equipment than in individual stoves. Most plant material has low sulfur, so it doesn't contribute to acid rain. And because it burns at a lower temperature than coal, it doesn't create as much nitrogen oxides. Of course, these crops are carbon neutral—that is, they absorb as much CO_2 in growing as they emit when burned. Wood chips and straw make great fuel for a boiler, but what about fuel for vehicles? You don't want to drive around with a hay bale in the backseat.

Biofuels (ethanol or methanol, which are made from plant materials, or biodiesel fuel, which is made from vegetable oils or animal fat) could meet much of our transportation needs. Alcohols can be burned directly in engines adapted to use these fuels, or they can be mixed with gasoline to be used in any normal engine. Ethanol makes gasoline burn cleaner, and most states now require that 5 to 10 percent of an oxygenated additive, such as ethanol, be added to gasoline. Methanol can be made from wood chips, straw, or any cellulosic material.

Ethanol production could be a solution to utilize grain surpluses and bring a higher price for farmers than the food market offers. As one Midwestern politician pointed out, "The WTO is putting pressure on Europe and America to reduce farm subsidies, but there's no rule that forbids subsidies for energy crops." There has been much debate about whether ethanol represents a net energy gain (see What Do You Think? p. 302). Modern state-of-the-art ethanol refineries claim a 50 percent gain in

What Do You Think?

Can Biofuels Be Sustainable?

Biofuels (ethanol refined from plant material, and diesel fuel made from vegetable oils or animal fats) are thought by many people to be the answer to both our farm crisis and our fuel needs. Brazil is the current world leader in crop-based ethanol production. In 2006, Brazil refined some 18 billion l (4.7 bill gal) of ethanol, mostly from sugar cane waste called bagasse). By 2012, Brazil expects to double this output and export billions of liters of biofuels to other countries. Almost all new vehicles in Brazil now have flexfuel engines. Most motorists mix their own combination (about 80 percent ethanol is most common) right at the pump. They are motivated to do so, because ethanol is about 40 percent cheaper than gasoline. Rapidly increasing soy and corn production in Brazil (chapter 7) could also provide vegetable oils for diesel fuel.

Alcohol, mostly from sugarcane waste, is cheaper than either gasoline or diesel fuel in Brazil. All passenger vehicles in Brazil are required to use at least 20 percent ethanol.

The United States is currently second in world ethanol production. In 2006, the U.S. produced 15.2 billion l of ethanol (4 billion gal), mostly from shelled corn. Like Brazil, the U.S. hopes to double this yield by 2012. Almost all American states now require small amounts of oxygenated fuel, such as ethanol, to improve air quality. Some states mandate higher ethanol content as a way of reducing oil dependence and supporting farmers. The highest of these is in Minnesota, which requires 20 percent ethanol in gasoline, the highest concentration thought to be feasible without modifying ordinary engines.

But do these crops represent a net energy gain? Does it take more fossil fuel energy to grow, harvest, and process crop-based biofuels than you get back in the finished product? Experts hotly debate this question. For 15 years, David Pimental from Cornell University has repeatedly published calculations showing a net energy loss in biofuels. In 2005, he was joined by Tad Patzek from the University of California–Berkeley, in claims that it takes 29 percent more energy to refine ethanol from corn than it yields. Soy-based biodiesel is equally inefficient, these authors maintain, and cellulose-based biofuels are even worse. Switchgrass and woodchips take at least 50 percent more energy than they produce as ethanol, according to their calculations.

On the other side of this debate are scientists, such as Bruce Dale of Michigan State University and John Sheehan from the U.S. National Renewable Energy Laboratory, who argue that biofuels produced by modern techniques represent a positive energy return. They say that Pimental and Patzek used outdated data and unreasonable assumptions in making their estimates. Dramatic improvements in farming productivity, coupled with much greater efficiency in ethanol fermentation now yield about 50 percent more energy in ethanol from corn, and about 60 percent more energy in ethanol from cellulose crops than in the inputs, they calculate. A major difference between the two sides in this debate is process energy for fermentation and purification. Pimental and Patzek assume this energy would come from fossil fuels, but Dale and Sheehan argue that fermentation waste can be burned to make manufacturing self-sufficient, as is already being done in Brazil.

Another difference in the calculations between these two sides is how to account for crop inputs. Pimental and Patzek include not only the fuel to cultivate and harvest crops, but also the energy to manufacture the equipment used to grow and process the crops. Dale and Sheehan view this as an unfair handicap on biofuels. How far back do you carry the input? Do you also include the energy to feed and house the farmers who grow the crops, and the workers who make the machinery? Using Pimental and Patzek's formulas, Dale calculates that ethanol and soy diesel are much more energy efficient than gasoline (-39 percent) and coal-based electricity (-235 percent). By 2025, according to Dale and Sheehan, biofuels crops could provide farmers with profits of more than $5 billion per year, could save consumers about $20 billion in fuel costs, and could reduce U.S. greenhouse emissions by 1.7 billion tons per year—equal to more than 80 percent of transportation-related emissions and 22 percent of total emissions in 2002.

Of course hydrogen fuel would probably be a better choice than any of these fuels if it could be generated by solar or wind energy, but cost-effective fuel cells and a hydrogen economy appear to be decades, at best, away. In the meantime, how much do you think we should invest in crop-based biofuels?

---

ethanol made from corn over the energy content of the inputs, and a somewhat better net yield from cellulose crops. Removing high-value nutraceuticals, such as L-ribose and levoglucoseon before fermentation makes ethanol production much more profitable. Ethanol currently can be made for about $1 per gal (3.8 l) in the United States, and about half that amount in Brazil. Local residents sometimes complain, however, about odors from ethanol plants.

Just about anything organic, from turkey entrails to cow dung or sugar cane waste can be used to make biodiesel, and with oil over $75 a barrel in recent years, just about everything now is. Factories have been set up next to meat processing plants to convert waste material into biodiesel. Many people are converting diesel engines to burn discarded restaurant deep-frying fat. They say that these "greasel" engines leave a pleasant whiff of french fries or egg rolls as they roll along.

A much greater potential for diesel fuel is from vegetable oil. The European Union currently consumes about 4 million metric tons (1 billion gal) of biodiesel fuel, mostly made from rapeseed oil (called canola oil in North America). Plans are to triple this production by 2012 to 10 percent of all fuel consumed in Europe. Planners expect this move to create 250,000 jobs, reduce oil imports, improve air quality, save consumers billions of Euros, and reduce CO_2 emissions by 200 million

Principles of Environmental Science

tons per year. Already, however, this increased use of vegetable oils has reduced the amount available for human consumption. Tropical countries, such as Indonesia and Malaysia, hope that biodiesel demand will stimulate palm oil exports, which have fallen out of favor for human diets. Conservationists, however, worry that increasing oil palm plantations will lead to more tropical forest destruction.

Wouldn't a major conversion to biofuels lead to other food shortages? Not necessarily. Energy crops, such as switch grass, cattails, and hybrid poplar could be grown on land not suitable for food production. Furthermore, most of the corn and soybeans now grown in the United States is now used for animal food. A pilot plant in Argentina now produces 1,000 l of biodiesel and 8 tons of animal feed from 10 tons of soy. Using a system like this could meet much of our fuel needs without reducing food production very much.

Water is another worry about the sustainability of biofuels. Currently, it takes 3 to 6 liters of water for each liter of ethanol produced. In many plains states where grain is grown, there isn't enough water for both agriculture and fuel production. Plans for new processing facilities are being scaled back because of water shortages. So, although biofuels could make a useful contribution to our fuel supplies, especially in the interim between the age of fossil fuels and whatever comes next, there are concerns about the social and environmental impacts of a sudden switch to these alternate energy sources.

The United States already has about 5 million **flexfuel vehicles** that can burn variable mixtures of ethanol and gasoline, but few of them currently use anything beyond the standard 5 or 10 percent ethanol blend. Owners either don't know that they could burn up to 85 percent ethanol or they don't have access to biofuel. Vehicle manufacturers continue to make these engines because a special clause in the U.S. Energy Bill allows them to sell more gas guzzlers if they also produce flexfuel vehicles. The ironic effect is that the existence of these alternative fuel vehicles actually results in more oil consumption.

Fuelwood is scarce in many developing countries

Two billion people—about one-third of the total world population—depend on firewood and charcoal as their primary energy source. Of these people, three-quarters (1.5 billion) do not have an adequate, affordable supply. Many people in the less-developed countries face a daily struggle to find enough fuel to warm their homes and cook their food. The problem is intensifying because rapidly growing populations in many developing countries create increasing demands for firewood and charcoal from a diminishing supply (fig. 12.29).

As firewood becomes increasingly scarce, women and children, who do most of the domestic labor in many cultures, spend more and more hours searching for fuel. In some places, it now takes eight hours or more just to walk to the nearest fuelwood supply and even longer to walk back with a load of sticks and branches that will last only a few days.

FIGURE 12.29 A charcoal market in Ghana. Firewood and charcoal provide the main fuel for billions of people. Forest destruction results in wildlife extinction, erosion, and water loss.

Currently, about half of all wood harvested each year worldwide is used as fuel. The poorest countries, such as Ethiopia, Bhutan, Burundi, and Bangladesh, depend on biomass for 90 percent of their energy. It's estimated that the 1,700 million tons of fuelwood now harvested each year globally is at least 500 million tons less than is needed. By 2025 the worldwide demand for fuelwood is expected to be about twice current harvest rates, while supplies will not have expanded much beyond current levels. Some places will be much worse than this average. In some African countries, such as Mauritania, Rwanda, and the Sudan, firewood demand already is ten times the sustainable yield. Reforestation projects, agroforestry, community woodlots, and inexpensive, efficient, locally produced woodstoves could help alleviate expected fuelwood shortages in many places.

Dung and methane make good fuels

Where wood and other fuels are in short supply, people often dry and burn animal manure. This may seem like a logical use of waste biomass, but it can intensify food shortages in poorer countries. Not putting this manure back on the land as fertilizer reduces crop production and food supplies.

When cow dung is burned in open fires, more than 90 percent of the potential heat and most of the nutrients are lost. Compare that with the efficiency of using dung to produce methane gas, an excellent fuel. In the 1950s, simple, cheap methane digesters were designed for villages and homes, but they were not widely used. In China, 6 million households use biogas for cooking and lighting. Two large municipal facilities in Nanyang will soon provide fuel for more than 20,000 families. Perhaps other countries will follow China's lead. As one Indian politician remarked, "We may be running out of oil, but we'll always have dung."

Methane gas, the main component of natural gas, is produced by anaerobic decomposition (digestion by anaerobic [oxygen-free] bacteria) of any moist organic material. Under these conditions, organic materials are decomposed by anaerobic rather than aerobic (oxygen-using) bacteria, producing flammable gases instead of

FIGURE 12.30 Continuous unit for converting organic material to methane by anaerobic fermentation. One kilogram of dry organic matter will produce 1–1.5 m^3 of methane, or 2,500–3,600 million calories per metric ton.

FIGURE 12.31 Hydropower dams produce clean renewable energy but can be socially and ecologically damaging.

carbon dioxide. The same process can be reproduced artificially by placing organic wastes in a container and providing warmth and water (fig. 12.30). Bacteria are ubiquitous enough to start the culture spontaneously.

Burning methane produced from manure provides more heat than burning the dung itself, and the sludge left over from bacterial digestion is a rich fertilizer, containing bacterial biomass as well as most of the nutrients originally in the dung. Whether the manure is of livestock or human origin, airtight digestion also eliminates some health hazards associated with direct use of dung, such as exposure to fecal pathogens and parasites.

Municipal sewage treatment plants routinely use anaerobic digestion as a part of their treatment process, and many facilities collect the methane they produce and use it to generate heat or electricity for their operations. Although this technology is well developed, its utilization could be much more widespread.

ENERGY FROM THE EARTH'S FORCES

The winds, waves, tides, ocean thermal gradients, and geothermal areas are renewable energy sources. Although available only in selected locations, these sources could make valuable contributions to our total energy supply.

Falling water has been used as an energy source since ancient times

The invention of water turbines in the nineteenth century greatly increased the efficiency of hydropower dams (fig. 12.31). By 1925 falling water generated 40 percent of the world's electric power.

Since then, hydroelectric production capacity has grown 15-fold, but fossil fuel use has risen so rapidly that water power is now only one-quarter of total electrical generation. Still, many countries produce most of their electricity from falling water. Norway, for instance, depends on hydropower for 99 percent of its electricity; Brazil, New Zealand, and Switzerland all produce at least three-quarters of their electricity with water power. Canada is the world's leading producer of hydroelectricity, running 400 power stations with a combined capacity exceeding 60,000 MW. First Nations people protest that their rivers are being diverted and lands flooded to generate electricity, most of which is sold to the United States.

The total world potential for hydropower is estimated to be about 3 million MW. If all of this capacity were put to use, the available water supply could provide between 8 and 10 terrawatt hours (10^{12} watt-hours) of electrical energy. Currently, we use only about 10 percent of the potential hydropower supply. The energy derived from this source in 1994 was equivalent to about 500 million tons of oil, or 8 percent of the total world commercial energy consumption.

Much of the hydropower development in recent years has been in enormous dams. There is a certain efficiency of scale in giant dams, and they bring pride and prestige to the countries that build them, but they can have unwanted social and environmental effects that spark protests in many countries. China's Three Gorges Dam on the Yangtze River, for instance, spans 2.0 km (1.2 mi) and is 185 m (600 ft) tall. The reservoir it creates is 644 km (400 mi) long and has displaced more than 1 million people (see related story "Three Gorges Dam" at www.mhhe.com/cunningham4e).

In tropical climates, large reservoirs often suffer enormous water losses. Lake Nasser, formed by the Aswan High Dam in Egypt, loses 15 billion m³ each year to evaporation and seepage. Unlined canals lose another 1.5 billion m³. Together, these losses represent one-half of the Nile River flow, or enough water to irrigate 2 million ha of land. The silt trapped by the Aswan High Dam formerly fertilized farmland during seasonal flooding and provided nutrients that supported a rich fishery in the delta region. Farmers now must buy expensive chemical fertilizers, and the fish catch has dropped almost to zero. Schistosomiasis, spread by snails that flourish in the reservoir, is an increasingly serious problem.

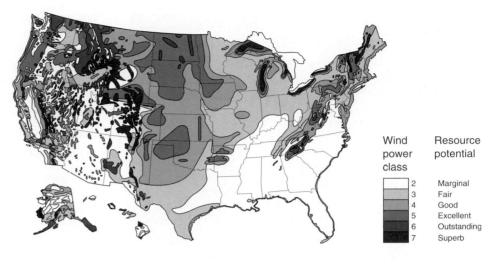

Wind power class	Resource potential
2	Marginal
3	Fair
4	Good
5	Excellent
6	Outstanding
7	Superb

FIGURE 12.32 United States wind resource map. Mountain ranges and areas of the High Plains have the highest wind potential, but much of the country has a fair to good wind supply. **Source:** Data from U.S. Department of Energy.

Wind energy is our fastest growing renewable resource

The air surrounding the earth has been called a 20-billion-km³ storage battery for solar energy. The World Meteorological Organization has estimated that 20 million MW of wind power could be commercially tapped worldwide, not including contributions from windmill clusters at sea. This is about 50 times the total present world nuclear generating capacity. Wind power has advantages and disadvantages, as do other nontraditional technologies. Like solar power and hydropower, wind power taps a natural physical force. Like solar power (its ultimate source), wind power is an abundant, nonpolluting resource, and it causes minimal environmental disruption. Also like solar power, however, it requires expensive storage during peak production times to offset nonwindy periods.

As the world's conventional fuel prices rise, interest in wind energy is growing rapidly. In the 1980s the United States was a world leader in wind technology, and California hosted 90 percent of all existing wind power generators. Some 17,000 windmills marched across windy mountain ridges at Altamont, Tehachapi, and San Gorgino Passes. Poor management, technical flaws, and overdependence on subsidies, however, led to bankruptcy of major corporations, including Kenetech, once the largest turbine producer in the United States. Now Danish, German, and Japanese wind machines are capturing the rapidly growing world market. Wind technology is now Denmark's second-largest export, employing 20,000 people and bringing in about $1 billion (U.S.) per year.

The 47,000 MW of installed wind power currently in operation worldwide demonstrates the economy of wind turbines. Theoretically up to 60 percent efficient, windmills typically produce about 35 percent efficiency under field conditions. Where conditions are favorable, wind power is now cheaper than any other new energy source, with electric prices as low as 3¢ per kilowatt-hour in places with steady winds averaging at least 24 kph (15 mph). Large areas of western North America meet this requirement (fig. 12.32). Some energy experts predict that North America's Great Plains could become the Saudi Arabia of wind power.

The World Energy Council predicts that wind could account for 200,000 MW of electricity by 2020, depending on how seriously politicians take global warming and how many uneconomical nuclear reactors go offline. One thousand MW meets the energy needs of about 50,000 typical U.S. households or is equivalent to about 6 million bbl of oil. Shell Oil suggests that half of all the world's energy could be wind and solar generated by the middle of this century.

In places like Denmark, which has winds above 16 mph an average of 245 days per year, windmills have long been recognized as a valuable energy source for pumping water and grinding grain. Germany, with 17,000 MW of installed capacity, now gets one-third of its electricity from wind power, and is the world leader in this technology. Spain is second with 10,000 MW, and the United States is third with 9,200 MW. The World Energy Council predicts that wind generating capacity could grow another tenfold by 2020.

Wind farms are large concentrations of wind generators producing commercial electricity (fig. 12.33). California continues to lead the United States in wind power, with 2,096 MW installed capacity in 2006, or about one-fifth of the U.S. total. Texas with 1,294 MW is second in installed capacity. Minnesota, Iowa, Oregon, Washington, Wyoming, Colorado, and New Mexico all have between 250 and 650 MW of installed wind power.

Do wind farms have any negative impacts? They generally occupy places with wind and weather too severe for residential or other development. Most wind farms are too far from residential areas to be heard or seen. But they do interrupt the view in remote, isolated places and destroy the sense of isolation and natural beauty. Bird and bat kills have been a problem for some wind farms. Careful placement outside of migration corridors and the addition of warning devices can reduce mortality greatly.

Wouldn't wind power take up a huge land area if we were to depend on it for a major part of our energy supply? As table 12.2 shows, the actual space taken up by towers, roads, and other structures on a wind farm is only about one-third as much as would be

FIGURE 12.33 Renewable energy sources, such as wind, solar energy, geothermal power, and biomass crops, could eliminate our dependence on fossil fuels and prevent extreme global climate change, if we act quickly.

TABLE 12.2

Jobs and Land Required for Alternative Energy Sources

Technology	Land Use (m² per gigawatt-hour for 30 years)	Jobs (per terawatt-hour per year)
Coal	3,642	116
Photovoltaic	3,237	175
Solar thermal	3,561	248
Wind	1,335	542

Source: Data from Lester R. Brown et al., *Saving the Planet,* 1991. W. W. Norton & Co., Inc.

consumed by a coal-fired power plant or solar thermal energy system to generate the same amount of energy over a 30-year period. Furthermore, the land under windmills is more easily used for grazing or farming than is a strip-mined coal field or land under solar panels. Farmers are finding wind energy to be a lucrative crop. A single tower sitting on 0.1 ha (0.25 acre) can pay $100,000 per year.

When a home owner or community invests independently in wind generation, the same question arises as with solar energy: what should be done about energy storage when electricity production exceeds use? Besides the storage methods mentioned earlier, many private electricity producers believe the best use for excess electricity is to sell it to the public utility grid. In states that allow net energy pricing, you sell excess power from wind or solar systems back to a local utility. Ideally, the utility pays you something close to its average wholesale price. The 1978 Public Utilities Regulatory Policies Act required utilities to buy power generated by small hydro, wind, cogeneration (simultaneous production of useful heat and electricity), and other privately owned technologies at a fair price. Not all utilities yet comply, but some—notably in California, Oregon, Maine, and Vermont—are purchasing significant amounts of private energy.

Geothermal heat, tides, and waves could supply substantial amounts of energy in some places

The earth's internal temperature can provide a useful source of energy in some places. High-pressure, high-temperature steam fields exist below the earth's surface. Around the edges of continental plates or where the earth's crust overlays magma (molten rock) pools close to the surface, this **geothermal energy** is expressed in the form of hot springs, geysers, and fumaroles. Yellowstone National Park is the largest geothermal region in the United States. Iceland, Japan, and New Zealand also have high concentrations of geothermal springs and vents. Depending on the shape, heat content, and access to groundwater, these sources produce wet steam, dry steam, or hot water.

While few places have geothermal steam, the earth's warmth can help reduce energy costs nearly everywhere. Pumping water through deeply buried pipes can extract enough heat so that a heat pump will operate more efficiently. Similarly, the relatively uniform temperature of the ground can be used to augment air conditioning in the summer (fig. 12.34).

FIGURE 12.34 Geothermal energy can cut heating and cooling costs by half in many areas. In summer (*shown here*), warm water is pumped through buried tubing (*earth loops*), where it is cooled by constant underground temperatures. In winter, the system reverses and the relatively warm soil helps heat the house. Where space is limited, earth loops can be vertical. If more space is available, the tubing can be laid in shallow horizontal trenches, as shown here.

Principles of Environmental Science

Ocean tides and waves contain enormous amounts of energy that can be harnessed to do useful work. A tidal station works like a hydropower dam, with its turbines spinning as the tide flows through them. A high-tide/low-tide differential of several meters is required to spin the turbines. Unfortunately, variable tidal periods often cause problems in integrating this energy source into the electric utility grid. Nevertheless, demand has kept some plants running for many decades.

Ocean wave energy can easily be seen and felt on any seashore. The energy that waves expend as millions of tons of water are picked up and hurled against the land, over and over, day after day, can far exceed the combined energy budget for both insolation (solar energy) and wind power in localized areas. Captured and turned into useful forms, that energy could make a substantial contribution to meeting local energy needs.

Dutch researchers estimate that 20,000 km of ocean coastline are suitable for harnessing wave power. Among the best places in the world for doing this are the west coasts of Scotland, Canada, the United States (including Hawaii), South Africa, and Australia. Wave energy specialists rate these areas at 40 to 70 kW per meter of shoreline. Altogether, it's calculated, if the technologies being studied today become widely used, wave power could amount to as much as 16 percent of the world's current electrical output.

Some of the designs being explored include oscillating water columns that push or pull air through a turbine, and a variety of floating buoys, barges, and cylinders that bob up and down as waves pass, using a generator to convert mechanical motion into electricity. It's difficult to design a mechanism that can survive the worst storms.

An interesting new development in this field is the Pelamis wave-power generator developed by the Scottish start-up company Ocean Power Delivery (fig. 12.35). The first application of this technology is now being built 5 km off the coast of Portugal. It will use three units capable of producing some 2.25 MW of electricity, or enough to supply 1,500 Portuguese households. If preliminary trials go well, plans are to add 40 more units in a year or two. Each of the units consists of four cylindrical steel sections linked by hinged joints. Anchored to the seafloor at its nose, the snakelike machine points into the waves and undulates up and down and side to side as swells move along its 125 m length. This motion pumps fluid to hydraulic motors that drive electrical generators to produce power, which is carried to shore by underwater cables.

Pelamis's inventor, Richard Yemm, says that survivability is the most important feature of a wave-power device. Being offshore, the Pelamis isn't exposed to the pounding breakers that destroy shore-based wave-power devices. If waves get too steep, the Pelamis simply dives under them, much as a surfer dives under a breaker. These wave converters lie flat in the water and are positioned far offshore, so they are unlikely to stir up as much opposition as do the tall towers of wind generators.

WHAT'S OUR ENERGY FUTURE?

None of the renewable energy sources discussed in this chapter are likely to completely replace fossil fuels and nuclear power in the near future. They could, however, make a substantial collective contribution toward providing us with the conveniences we crave in a sustainable, environmentally friendly manner. They could also make us energy independent and balance our international payment deficit.

The World Energy Council projects that renewables could provide about 40 percent of world cumulative energy consumption under an idealized "ecological" scenario assuming that political leaders take global warming seriously and pass taxes to encourage conservation and protect the environment (fig. 12.36). This scenario also envisions measures to shift wealth from the north to the south and to enhance economic equity. By the end of the twenty-first century, renewable sources could provide all our energy needs if we take the necessary steps to make this happen.

Environmentalists point to the dangers of air pollution, global climate change, and other environmental problems associated with burning fossil fuels. Businesses stress the importance

FIGURE 12.35 The Pelamis wave converter (named after a sea snake) is a 125 m long and 3.5 m diameter tube-hinged, so it undulates as ocean swells pass along it. This motion drives pistons that turn electrical generators. Energy experts calculate that capturing just 1 to 2 percent of global wave power could supply at least 16 percent of the world's electrical demand.

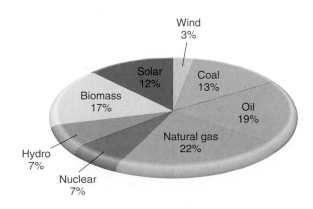

FIGURE 12.36 Idealized "ecological" scenario for cumulative world energy consumption, 2000 to 2100.
Source: Data from World Energy Council, 2002.

of a reliable energy supply for economic growth. While both call for a new policy, they disagree on what it should contain. Conservatives tend to favor increasing production and easing regulations on power plant operation and transmission-line siting, rather than limiting demand. Progressives, on the other hand, prefer conservation measures, such as forcing automakers to increase average fuel efficiency of cars and light trucks and providing heating bill assistance for low-income households.

The Republican-sponsored Energy Bill, based largely on the recommendations of a secret task force headed by Vice President Dick Cheney, and made up entirely of oil, coal, gas and nuclear company representatives, calls for $30 billion in subsidies for traditional energy, but only $5 billion for conservation or renewables.

Speaking for his task force, Mr. Cheney said that oil, coal, and natural gas would remain the United States' primary energy resources for "years down the road." He strongly supported the use of nuclear power and claimed that the United States needs to build one large electric-generating plant each week for the next 20 years. He dismissed as 1970s-era thinking the notion that "we could simply conserve or ration our way out" of an energy crisis. "Conservation," he said, "may be a sign of personal virtue, but it is not a sufficient basis for a sound, comprehensive energy policy."

In light of what you've read in this chapter about current and potential sources, what sort of energy policy would you suggest if you were invited to advise government officials? How much conservation is realistic or fair to ask people to adopt? What's a reasonable percentage of our energy that might come from renewable sources in 20 or 50 years? If you wanted to move toward the "ecological scenario" described by the World Energy Council, how would you go about implementing that plan? What government actions do you think would be most effective in accomplishing these goals?

REVIEW QUESTIONS

1. Where are Samsø and Ærø islands, and how do they supply their energy needs?

2. Define *energy* and *power*.

3. What are the major sources of global commercial energy?

4. How does energy consumption in the United States compare to that in other countries?

5. Who is the leading supplier of oil to the United States?

6. What are *proven-in-place reserves*?

7. How much coal do we have, and how long will it last?

8. Why don't we want to use all the coal in the ground?

9. How does the amount of liquid oil compare to coal resources?

10. How long is this oil likely to last?

11. Where is most liquid oil located?

12. What are *tar sands* and *oil shales*? How much oil could they produce?

13. Why might we not want to exploit tar sands and oil shales?

14. Why is natural gas considered to be a superior fuel to either coal or oil?

15. How are nuclear wastes now being stored?

16. What's the difference between active and passive solar energy?

17. How do photovoltaic cells work?

18. What's a *fuel cell,* and how does it work?

19. What are *biofuels,* and how could they contribute to sustainability?

20. How much wind capacity is there worldwide?

CRITICAL THINKING AND DISCUSSION

Apply the principles you have learned in this chapter to discuss these questions with other students.

1. If you were the energy czar of your state or country, where would you invest your budget? Why?

2. We have discussed a number of different energy sources and energy technologies in this chapter. Each has advantages and disadvantages. If you were an energy policy analyst, how would you compare such different problems as the risk of a nuclear accident versus air pollution effects from burning coal?

3. If your local utility company were going to build a new power plant in your community, what kind would you prefer? Why?

4. The nuclear industry is placing ads in popular magazines and newspapers, claiming that nuclear power is environmentally friendly, since it doesn't contribute to the greenhouse effect. How do you respond to that claim?

5. How would you evaluate the debate about net energy loss or gain in biofuels? What questions would you ask the experts on each side of this question? What worldviews or hidden agendas do you think might be implicit in this argument?

6. Although we have used vast amounts of energy resources in the process of industrialization and development, some would say that it was a necessary investment to get to a point at which we can use energy more efficiently and sustainably. Do you agree? Might we have followed a different path?

SUGGESTED READINGS

Aleklett, Kjell. 2006. Oil: A bumpy road ahead. *World Watch* 19 (1):10–12.

Dias de Oliveira, M. E., et al. 2005. Ethanol as fuel: Energy, carbon dioxide balances, and ecological footprint. *BioScience* 55(July):593–602.

Goodstein, David. 2004. *Out of Gas: The End of the Age of Oil.* W. W. Norton.

Kim, S., and B. E. Dale. 2004. Cumulative energy and global warming impact from the production of biomass for biobased products. *Journal of Industrial Ecology* 7 (3–4):147–62.

Lovins, Amory B., et al. 2004. *Winning the Oil Endgame: Innovation for Profits, Jobs, and Security.* Rocky Mountain Institute.

Pimental, David, and Tad W. Patzek. 2005. Ethanol production using corn, switchgrass, and wood; Biodesel production using soybean and sunflower. *Natural Resources Research* 14 (1):65–76.

Reece, Erik. 2005. Death of a mountain: Radical strip mining and the leveling of Appalachia. *Harper's* 310 (1859):41–60.

Scigliano, Eric. 2005. Wave energy: Can a mechanical snake that surfs the ocean squeeze enough watts from water? *Discover* 26 (12):42–45.

Scott H. P., et al. 2004. Generation of methane in the earth's mantle: *In situ* high-pressure–temperature measurements of carbonate reduction. *Proc. Natl. Acud. Sci.* 101 (39): 14023–26.

Simmons, M. R. 2005. *Twilight in the Desert: The Coming Saudi Oil Shock and the World Economy.* Wiley.

 DATA analysis   **Energy Calculations**

Most college students either already own or are likely to buy an automobile and a computer sometime soon. How do these items compare in energy usage? Suppose that you were debating between a high-mileage car, such as the Honda Insight, or a sport utility vehicle, such as a Ford Excursion. How do the energy requirements of these two purchases measure up? To put it another way, how long could you run a computer on the energy you would save by buying an Insight rather than an Excursion?

Here are some numbers you need to know. The Insight gets about 75 mpg, while the Excursion gets about 12 mpg. A typical American drives about 15,000 mi per year. A gallon of regular, unleaded gasoline contains about 115,000 Btu on average. Most computers use about 100 watts of electricity. One kilowatt-hour (kWh) = 3,413 Btu.

1. How much energy does the computer use if it is left on continuously? (You really should turn it off at night or when it isn't in use, but we'll simplify the calculations.)

    100 watt/h × 24 h/day × 365 days/yr = _____ kWh/yr

2. How much gasoline would you save in an Insight, compared with an Excursion?

    a. Excursion:

       15,000 mi/yr ÷ 12 mpg   = _____ gal/yr

    b. Insight:

       15,000 mi/yr ÷ 75 mpg   = _____ gal/yr

    c. Gasoline savings (*a* − *b*)   = _____ gal/yr

    d. Energy savings:

       (gal × 115,000 Btu)   = _____ Btu/yr

    e. Converting Btu to kWh:

       (Btu × 0.00029 Btu/kWh) = _____ kWh/yr saved

3. How long would the energy saved run your computer?

    kWh/yr saved by Insight ÷ kWh/yr consumed by computer = _____

For Additional Help in Studying This Chapter, please visit our website at www.mhhe.com/cunningham4e. You will find practice quizzes, a chapter summary, key terms, answers to review questions, additional case studies, regional examples, and an extensive reading list, all of which will help you understand the material in this chapter.

A crane unloads a garbage barge at Fresh Kills on Staten Island, the world's largest landfill before it closed in 2001.

13
Solid and Hazardous Waste

We are living in a false economy where the price of goods and services does not include the cost of waste and pollution.

–Lynn Landes, founder and director of Zero Waste America–

OBJECTIVES

Studying this chapter will help you answer the following questions:

- What are the major components of the waste stream? How do we dispose of most of our waste?

- How does a sanitary landfill operate? Why are we searching for alternatives to landfills?

- Why is ocean dumping a problem?

- What are the benefits of recycling?

- What are the "three Rs" of waste reduction, and which is most important?

- What are some steps *you* can take to reduce waste production?

- How can biomass waste be converted to natural gas?

- What are toxic and hazardous waste? How do we dispose of them?

- What is bioremediation? Is it a promising solution in waste management?

- What is the Superfund, and has it shown progress?

Case Study The New Alchemy: Creating Gold from Garbage

Most people think of recycling in terms of newspapers, plastic bottles, and other household goods. Your daily household recycling is the bedrock of recycling programs, but another growing and exciting area of recycling is done at commercial and industrial scales. The 230 million tons of garbage the United States produces each year includes some knotty problems: old furniture and carpeting, appliances and computers, painted wood, food waste. It's no wonder that we've simply dumped it all in landfills as long as we could. But landfill space is diminishing rapidly.

Incinerators are a common alternative, but they are expensive to build and operate, and they can produce dangerous air contaminants, including dioxins from burned plastics, and heavy metals.

One of our largest sources of waste is construction and demolition debris—the rubble left over when a building is torn down, remodeled, or built. Construction and demolition account for over 140 million tons of waste per year, about 1.5 kg per person per day—on top of the 230 million tons per year of municipal solid waste. All this mixed debris is normally trucked to landfills, but alternatives have emerged in recent years.

A slowly growing number of cities and companies are sending their waste, including construction debris, to commercial recyclers. One example is Taylor Recycling, based in Montgomery, New York. Starting as a tree removal business, Taylor has expanded to construction and demolition waste and now operates in four states. The company recycles and sells 97 percent of the mixed debris it receives, well above the industry average of 30 to 50 percent. Trees are ground and converted to mulch for landscaping. Dirt from stumps is screened and sold as clean garden soil. Mixed materials are sorted into recyclable glass, metals, and plastics. Construction debris is sorted and ground: broken drywall is ground to fresh gypsum, which is sold to drywall producers; wood is composted or burned; bricks are crushed for fill and construction material. Organic waste that can't be separated, such as food-soaked paper, is sent to a gassifier. The gassifier is like an enclosed, oxygen-free pressure cooker, which converts biomass to natural gas. The gas runs electric generators for the plant, and any extra gas can be sold. Waste heat warms the recycling facility. The 3 percent of incoming waste that doesn't get recycled is mainly mixed plastics, which are currently landfilled.

From their base outside of New York City, recycling is clearly a good idea. New York has used up most of its landfill space and now ships garbage to Virginia, Ohio, Pennsylvania, and South Carolina. Fuel and trucking costs alone drive up disposal costs, and with landfill capacity shrinking, tipping fees are climbing. According to Jim Taylor, the 1,000 garbage trucks leaving New York City each day travel an average of 300 miles round-trip, at less than 4 miles per gallon of fuel.

The story of garbage processing is changing globally. Garbage has long been one of the United States' largest exports, but increasingly the country is exporting sorted recycled materials, as well. Chinese manufacturers are finding valuable material sources in American waste. In Western Europe, where environmental regulation and landfill space are both tight, recycling, composting, and conversion of biomass to gas are booming businesses. The Swiss company Kompogas, one of many companies processing garbage, ferments organic waste in giant tanks, producing methane, compost, and fertilizer.

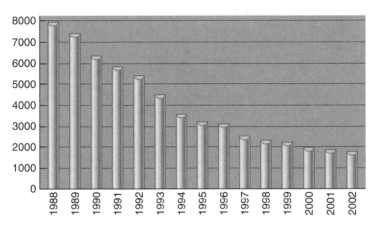

The number of landfills in the United States has declined steadily as space has become scarce and safety rules have increased.
Data Source: U.S. EPA.

New technologies are providing innovative strategies for recycling waste. One method that is getting attention is Changing World Technologies' thermal depolymerization process ("thermal" means it involves heating; "depolymerization" essentially means breaking down organic molecules) that converts almost any kind of organic waste into clean, usable diesel oil. Animal by-products, sewage sludge, shredded tires, and other waste can be converted with 85 percent energy efficiency. Pilot plants have been built for specialized sources, such as a turkey processing plant in Colorado, but the technology could accommodate mixed sources as well.

Recycling is a rapidly growing industry because it makes money coming and going. Recyclers are paid to haul away waste, which they turn into marketable products. The business is also exciting because these companies, like Taylor Recycling, see the huge social and economic benefits of environmental solutions. Often when we discuss environmental problems, businesses are part of the problem. But these examples show that business owners can be just as excited as anybody about environmental quality. With a good business model, being green can be very rewarding.

Garbage disposal may be one of the most exciting stories in environmental science, because it shows so much promise and innovation. So far, the examples discussed here make up a minority of our waste management strategies, but they are expanding. In this chapter we'll look at our other waste management methods, the composition of our waste, and some of the differences between solid waste and hazardous waste.

WASTE

Waste is everyone's business. We all produce unwanted by-products and residues in nearly everything we do. According to the Environmental Protection Agency (EPA), the United States produces 11 billion tons of solid waste each year. Nearly half of that amount consists of agricultural waste, such as crop residues and animal manure, which are generally recycled into the soil on the farms where they are produced. They represent a valuable resource as groundcover to reduce erosion and as fertilizer to nourish new crops, but they also constitute the single largest source of nonpoint air and water pollution in the country. About one-third of all solid wastes are mine tailings, overburden from strip mines, smelter slag, and other residues produced by mining and primary metal processing. Much of this material is stored in or near its source of production and isn't mixed with other kinds of wastes. Improper disposal practices, however, can result in serious and widespread pollution.

Industrial waste—other than mining and mineral production—amounts to some 400 million metric tons per year in the United States. Most of this material is recycled, converted to other forms, destroyed, or disposed of in private landfills or deep injection wells. About 60 million metric tons of industrial waste fall in a special category of hazardous and toxic waste, which we will discuss later in this chapter.

Municipal waste—a combination of household and commercial refuse—amounts to about 230 million metric tons per year in the United States. That's just over 2 kg (4.6 lbs) per person per day—twice as much per capita as Europe or Japan, and five to ten times as much as most developing countries (fig. 13.1).

The waste stream is everything we throw away

Does it surprise you to learn that you generate that much garbage? Think for a moment about how much we discard every year. There are organic materials, such as yard and garden wastes, food wastes, and sewage sludge from treatment plants; junked cars; worn-out furniture; and consumer products of all types. Newspapers, magazines, advertisements, and office refuse make paper one of our major wastes (fig. 13.2). In spite of recent progress in recycling, many of the 200 billion metal, glass, and plastic food and beverage containers used every year in the United States end up in the trash. Although plastic makes up only about 10 percent of our waste by weight, it comprises about 20 percent by volume. Wood, concrete, bricks, and glass come from construction and demolition sites, dust and rubble from landscaping and road building. All of this varied and voluminous waste has to arrive at a final resting place somewhere.

The **waste stream** is a term that describes the steady flow of varied wastes that we all produce, from domestic garbage and yard wastes to industrial, commercial, and construction refuse. Many of the materials in our waste stream would be valuable resources if they were not mixed with other garbage. Unfortunately, our collecting and dumping processes mix and crush

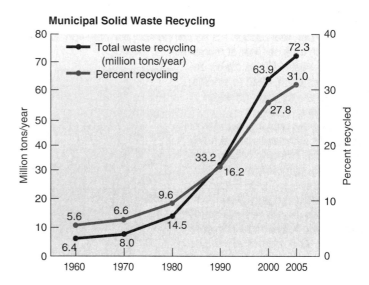

FIGURE 13.1 Bad news and good news in solid waste production. Per capita waste has risen steadily to more than 2 kg per person per day. Recycling rates are also rising, however. Recycling data include composting. **Source:** Data from Environmental Protection Agency, 2006.

everything together, making separation an expensive and sometimes impossible task. In a dump or incinerator, much of the value of recyclable materials is lost.

Another problem with refuse mixing is that hazardous materials in the waste stream get dispersed through thousands of tons of miscellaneous garbage. This mixing makes the disposal or burning of what might have been rather innocuous stuff a difficult, expensive, and risky business. Spray-paint cans, pesticides, batteries (zinc, lead, or mercury), cleaning solvents, smoke detectors containing radioactive material, and plastics that produce dioxins and PCBs (polychlorinated biphenyls) when burned are mixed with paper, table scraps, and other nontoxic materials. The best thing to do with household toxic and hazardous materials is to separate them for safe disposal or recycling, as we will see later in this chapter.

Principles of Environmental Science http://www.mhhe.com/cunningham4e

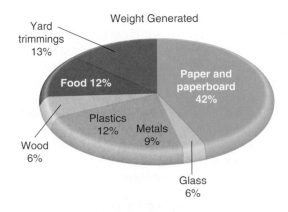

Weight Generated

Yard trimmings 13%

Food 12%

Plastics 12%

Wood 6%

Metals 9%

Glass 6%

Paper and paperboard 42%

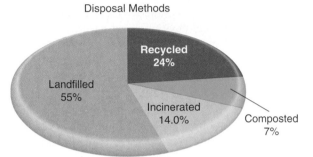

Disposal Methods

Recycled 24%

Landfilled 55%

Incinerated 14.0%

Composted 7%

FIGURE 13.2 Composition of municipal solid waste in the United States by weight, before recycling, and disposal methods.
Source: Data from U.S. Environmental Protection Agency Office of Solid Waste Management, 2006.

WASTE DISPOSAL METHODS

Where do our wastes go now? In this section, we will examine some historic methods of waste disposal, as well as some future options. Notice that our presentation begins with the least desirable—but most commonly used—measures and proceeds to discuss some preferable options. Keep in mind as you read this that modern waste management reverses this order and stresses the "three Rs" of reduction, reuse, and recycling before destruction or, finally, secure storage of wastes.

Open dumps release hazardous materials into the air and water

Often, the way people dispose of waste is to simply drop it someplace. Open, unregulated dumps are still the predominant method of waste disposal in most developing countries. The giant developing world megacities have enormous garbage problems. Mexico City, one of the largest cities in the world, generates some 10,000 tons of trash each day. Until recently, most of this torrent of waste was left in giant piles, exposed to the wind and rain, as well as rats, flies, and other vermin. Manila, in the Philippines, has at least ten huge open dumps. The most notorious is called "Smoky Mountain" because of its constant smoldering fires (fig. 13.3). Thousands of people live and work on this 30 m high heap of refuse. They spend their days sorting through the garbage for edible or recyclable materials. Health conditions are

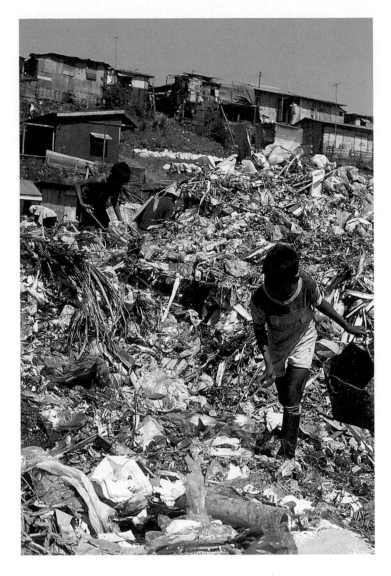

FIGURE 13.3 Scavengers sort through the trash at "Smoky Mountain," one of the huge metropolitan dumps in Manila, Philippines. Some 20,000 people live and work on these enormous garbage dumps. The health effects are tragic.

abysmal, but these people have nowhere else to go. The government would like to close these dumps, but how will the residents be housed and fed? Where else will the city put its garbage?

Most developed countries forbid open dumping, at least in metropolitan areas, but illegal dumping is still a problem. You have undoubtedly seen trash accumulating along roadsides and in vacant, weedy lots in the poorer sections of cities. Is this just a question of aesthetics? Consider the problem of waste oil and solvents. An estimated 200 million liters of waste motor oil are poured into the sewers or allowed to soak into the ground every year in the United States. This is about five times as much as was spilled by the *Exxon Valdez* in Alaska in 1989! No one knows the volume of solvents and other chemicals disposed of by similar methods.

Increasingly, these toxic chemicals are showing up in the groundwater supplies on which nearly half the people in America

FIGURE 13.4 Dumping of trash at sea is a global problem. Even on the most remote islands, beaches are covered with plastic flotsam and jetsam.

depend for drinking (chapter 10). An alarmingly small amount of oil or other solvents can pollute large quantities of drinking or irrigation water. One liter of gasoline, for instance, can make a million liters of water undrinkable. The problem of illegal dumping is likely to become worse as acceptable sites for waste disposal become more scarce and costs for legal dumping escalate. We clearly need better enforcement of antilittering laws, as well as a change in our attitudes and behavior.

Ocean dumping is nearly uncontrollable

The oceans are vast, but not so large that we can continue to treat them as carelessly as has been our habit. Every year some 25,000 metric tons (55 million lbs) of packaging, including half a million bottles, cans, and plastic containers, are dumped at sea. Beaches, even in remote regions, are littered with the nondegradable flotsam and jetsam of industrial society (fig. 13.4). About 150,000 tons (330 million lbs) of fishing gear—including more than 1,000 km (660 mi) of nets—are lost or discarded at sea each year. Environmental groups estimate that 50,000 northern fur seals are entangled in this refuse and drown or starve to death every year in the North Pacific alone.

Until recently, many cities in the United States dumped municipal refuse, industrial waste, sewage, and sewage sludge into the ocean. Federal legislation now prohibits this dumping. New York City, the last to stop offshore sewage sludge disposal, finally ended this practice in 1992. Still, 60 million to 80 million m^3 of dredge spoil—much of it highly contaminated—are disposed of at sea. Some people claim that the deep abyssal ocean plain is the most remote, stable, and innocuous place to dump our wastes. Others argue that we know too little about the values of these remote places or the rare species that live there to smother them with sludge and debris.

Landfills receive most of our waste

Over the past 50 years, most American and European cities have recognized the health and environmental hazards of open dumps. Increasingly, cities have turned to **sanitary landfills,** where solid waste disposal is regulated and controlled. To decrease smells and litter and to discourage insect and rodent populations, landfill operators are required to compact the refuse and cover it every day with a layer of dirt (fig. 13.5). This method helps control pollution, but the dirt fill also takes up as much as 20 percent of landfill space. Since 1994, all operating landfills in the United States have been required to control such hazardous substances as oil, chemical compounds, toxic metals, and contaminated rainwater that seep through piles of waste. An impermeable clay and/or plastic lining underlies and encloses the storage area. Drainage systems are installed in and around the liner to catch drainage and to help monitor chemicals that may be leaking. Modern municipal solid waste landfills now have many of the safeguards of hazardous waste repositories described later in this chapter.

More careful attention is now paid to the siting of new landfills. Sites located on highly permeable or faulted rock formations are passed over in favor of sites with less leaky geologic foundations. Landfills are being built away from rivers, lakes, floodplains, and aquifer recharge zones, rather than near them, as was often

Compacted waste filling trench

Original terrain

Daily 6-inch earth cover

FIGURE 13.5 In a sanitary landfill, trash and garbage are crushed and covered each day to prevent accumulation of vermin and spread of disease. A waterproof lining is now required to prevent leaching of chemicals into underground aquifers.

Life Cycle Analysis

One step toward understanding your place in the waste stream is to look at the life cycle of the materials you buy. Here is a rough approximation of the process. With another student, choose one item that you use regularly. On paper, list your best guess for the following: (1) a list of the major materials in it; (2) the original sources (geographic locations and source materials) of those materials; (3) the energy needed to extract/convert the materials; (4) the distances the materials traveled; (5) the number of businesses involved in getting the item to you; (6) where the item will go when you dispose of it; (7) what kinds of reused/recycled products could be made from the materials in it.

done in the past. More care is being given to a landfill's long-term effects, so that costly cleanups and rehabilitation can be avoided.

Historically, landfills have been a convenient and relatively inexpensive waste-disposal option in most places, but this situation is changing rapidly. Rising land prices and shipping costs, as well as increasingly demanding landfill construction and maintenance requirements, are making this a more expensive disposal method. The cost of disposing of a ton of solid waste in Philadelphia went from $20 in 1980 to more than $100 in 1990. Union County, New York, experienced an even steeper price rise. In 1987, it paid $70 to get rid of a ton of waste; a year later, the same ton cost $420, or about $10 for a typical garbage bag. The United States now spends about $10 billion per year to dispose of trash. A decade from now, it may cost us $100 billion per year to dispose of our trash and garbage.

Currently, 55 percent of all municipal solid waste in the United States is landfilled, 30 percent is recycled, and 15 percent is incinerated. Suitable places for waste disposal are becoming scarce in many areas. Other uses compete for open space. Citizens have become more concerned and vocal about health hazards, as well as aesthetics. It is difficult to find a neighborhood or community willing to accept a new landfill. Since 1984, when stricter financial and environmental protection requirements for landfills took effect, about three-quarters of all existing landfills in the United States have closed. In many cases, this means that old, small, uneconomical landfills closed, while larger, more modern ones took over. Nevertheless, many major cities are running out of local landfill space. They export their trash, at enormous expense, to neighboring communities and even other states. More than half the solid waste from New Jersey goes out of state, some of it up to 800 km (500 mi) away.

Exporting waste exposes poor nations to hazards

Although most industrialized nations in the world have agreed to stop shipping hazardous and toxic waste to less-developed

countries, the practice still continues. In 1999, for example, 3,000 tons of incinerator waste from a plastics factory in Taiwan were unloaded from a ship in the middle of the night and dumped in a field near the small coastal Cambodian village of Bet Trang. The village residents thought they had been blessed with a windfall. They emptied out chunks of crumbling residue so they could use the white, plastic shipping bags as bedding and roofing material. They rinsed out bags to use for rice storage, and they ripped them open with their teeth to get string to use as clotheslines and lashing for their oxcarts. Children played happily on the big pile of dusty, white material.

In the following weeks, unfortunately, the villagers discovered that, rather than a treasure, they had a calamity. The first sign of trouble was when one of the dock workers who unloaded the waste died and five others were hospitalized with symptoms of nerve damage and respiratory distress. Villagers also began to complain of a variety of illnesses. The village was evacuated, and about 1,000 residents of the nearby city of Sihanoukville fled in panic. Subsequent analysis found high levels of mercury and other toxic metals in the residue. The Formosa Plastics Corporation, which shipped the waste, admitted paying a $3 million bribe to Cambodian officials to permit its dumping. They said they couldn't dispose of the waste in Taiwan because of a threat of public protest. Following an international uproar, the plastics company agreed to go back and pick up the waste. But the villagers who handled the toxic wastes face an uncertain future. Is it safe to reinhabit their homes? Is it wise to have children? Will they suffer long-term health effects from exposure to this material? (See related story "The Voyage of the Khian Sea" at www.mhhe.com/cunningham/4e.) The United Nations' "Basel Convention," signed by 148 countries in 1989, prohibits transboundary shipping of hazardous wastes. Still, "toxic traders" find loopholes and illegal avenues to continue this business.

As we will discuss later in this chapter, "garbage imperialism" also operates within richer countries as well. Poor neighborhoods and minority populations are much more likely than richer ones to be the recipients of dumps, waste incinerators, and other locally unwanted land uses (LULUs). In recent years, attention has turned, in the United States, to Indian reservations, which are exempt from some state and federal regulations concerning waste disposal. Virtually every tribe in America has been approached with schemes to store wastes on their reservation (see Exploring Science, p. 316).

Another method of disposing of toxic wastes is to "recycle" them as asphalt or concrete filler for building highways. This is considered a beneficial use, but what happens to the toxins as the roadway is slowly worn away by traffic? Similar waste products are "land farmed," or sold as fertilizer and soil amendments. There are no safety standards for fertilizer composition because it's not intended for human consumption, but much of it is used on crops that humans will eat or that will be fed to livestock that are part of our food chain. For example, Florida has about a billion cubic meters of phosphogypsum, a waste product of phosphate mining, that producers want to market as a soil amendment. While it's true

When a new landfill, petrochemical factory, incinerator, or other unwanted industrial facility is proposed for a minority neighborhood, charges of environmental racism often are raised by those who oppose this siting. Everyday experiences tell us that minority neighborhoods are much more likely to have high pollution levels and facilities that you wouldn't want to live near than are middle- or upper-class white neighborhoods. But does this prove that land-use decisions are racist or just that minorities are less politically powerful than middle- or upper-class residents? Could it be that land prices are simply cheaper and public resistance to

Native Americans march in protest of toxic waste dumping on tribal lands.

locating a polluting facility in a place that's already polluted is less than putting it in a cleaner environment? Or does this distinction matter? Perhaps showing that a disproportionate number of minorities live in dirtier places is evidence enough of racism. How would you decide?

One of the first systematic studies showing this inequitable distribution of environmental hazards based on race in the United States was conducted by Robert D. Bullard in 1978. Asked for help by a predominantly black community in Houston that was slated for a waste incinerator, Bullard discovered that all five of the city's existing landfills and six of eight incinerators were located in African-American neighborhoods. In a book entitled *Dumping on Dixie,* Bullard showed that this pattern of risk exposure in minority communities is common throughout the United States. Among his findings are:

- Three of the five largest commercial hazardous waste landfills, accounting for about 40 percent of all hazardous waste disposal in the United States,

are located in predominantly African-American or Hispanic communities.

- Sixty percent of African Americans and Latinos and nearly half of all Asians, Pacific Islanders, and Native Americans live in communities with uncontrolled toxic waste sites.

- The average percentage of the population made up by minorities in communities without a hazardous waste facility is 12 percent. By contrast, communities with one hazardous waste facility have, on average, twice as high (24 percent) a minority population, while those with two or more such facilities average three times as high a minority population (38 percent) as those without one.

But does this prove that race, not class or income, is the strongest determinant of who is exposed to environmental hazards? What additional information might you look for to make this distinction? One of the lines of evidence Dr. Bullard raises is the fact that the discrepancy between the pollution exposure

of middle-class blacks and that of middle-class whites is even greater than the difference between poorer whites and blacks. While upper-class whites can "vote with their feet" and move out of polluted and dangerous neighborhoods, Bullard argues, minorities are restricted by color barriers and prejudice to less desirable locations.

Some additional evidence uncovered by this research is variation in the way toxic waste sites are cleaned up and how polluters are punished in different neighborhoods. White communities see faster responses and get better results once toxic wastes are discovered than do minority communities. For instance, the EPA takes 20 percent longer to place a hazardous waste site in a minority community on the Superfund National Priority List than it does for one in a white community. Penalties assessed against polluters of white communities average six times higher than those against polluters of minority communities. Cleanup is more thorough in white communities as well. Most toxic wastes in white communities are treated—that is, removed or destroyed. By contrast, waste sites in minority neighborhoods are generally only "contained" by putting a cap over them, leaving contaminants in place to potentially resurface or leak into groundwater at a later date.

How would you evaluate these findings? Do they convince you that racism is at work, or do you think that other explanations might be equally likely? Which of these arguments do you find most persuasive, or what other evidence would you need to make a reasoned judgment about whether or not environmental racism is a factor in determining who gets exposed to pollution and who enjoys a cleaner, more pleasant environment?

that phosphate is an essential plant nutrient, this particular product is also radioactive. In another case in Oregon, metal-rich dust and ash from steel mills is classified as hazardous waste when it leaves the mill. After being mixed with other minerals, however, it becomes fertilizer that will be spread on farm fields. Manufacturers are required to report the "active" ingredient—such as nitrogen, phosphorus, and phosphate—content of their product, but much can go unreported as "inert" matter.

Incineration produces energy but causes pollution

Faced with growing piles of garbage and a lack of available landfills at any price, many cities have built waste incinerators to burn municipal waste. Another term commonly used for this technology is **energy recovery,** or waste-to-energy, because the heat derived from incinerated refuse is a useful resource. Burning

FIGURE 13.6 A diagram of a municipal "mass burn" garbage incinerator. Steam produced in the boiler can be used to generate electricity or to heat nearby buildings.

garbage can produce steam used directly for heating buildings or generating electricity. Internationally, well over 1,000 waste-to-energy plants in Brazil, Japan, and Western Europe generate much-needed energy while reducing the amount that needs to be landfilled. In the United States more than 110 waste incinerators burn 45,000 metric tons of garbage daily. Some of these are simple incinerators; others produce steam and/or electricity.

Types of incinerators

Municipal incinerators are specially designed burning plants capable of burning thousands of tons of waste per day. In some plants, refuse is sorted as it comes in to remove unburnable or recyclable materials before combustion. This is called **refuse-derived fuel** because the enriched burnable fraction has a higher energy content than the raw trash. Another approach, called **mass burn,** is to dump everything smaller than sofas and refrigerators into a giant furnace and burn as much as possible (fig. 13.6). This technique avoids the expensive and unpleasant job of sorting through the garbage for nonburnable materials, but it often causes greater problems with air pollution and corrosion of burner grates and chimneys.

Incinerators have serious negative consequences. Residual ash and unburnable residues representing 10 to 20 percent of the original volume are usually taken to a landfill for disposal. Because the volume of burned garbage is reduced by 80 to 90 percent, disposal is a smaller task. However, the residual ash usually contains a variety of toxic components that make it an environmental hazard if not disposed of properly. Ironically, one worry about incinerators is whether enough garbage will be available to feed them. Often, incinerators compete with recycling programs for paper, plastics, and other materials. Some communities in which recycling has been really successful have had to buy garbage from neighbors to meet contractual obligations to waste-to-energy facilities. In other places, fears that this might happen have discouraged recycling efforts.

Incinerator cost and safety

The cost-effectiveness of garbage incinerators is the subject of heated debates. Initial construction costs are high—usually between $100 million and $300 million for a typical municipal

facility. Tipping fees at an incinerator (the fee charged to haulers for each ton of garbage dumped) are often much higher than those at a landfill. As landfill space near metropolitan areas becomes more scarce and more expensive, however, landfill rates are certain to rise. It may pay in the long run to incinerate refuse so that the lifetime of existing landfills will be extended.

Environmental safety of incinerators is another point of concern. The EPA has found alarmingly high levels of dioxins, furans, lead, and cadmium in incinerator ash. These toxic materials are more concentrated in the fly ash (lighter, airborne particles capable of penetrating deep into the lungs) than in heavy bottom ash. Dioxin levels can be as high as 780 ppb (parts per billion). One part per billion of TCDD, the most toxic dioxin, is considered a health concern. All of the incinerators studied exceeded cadmium standards, and 80 percent exceeded lead standards. Proponents of incineration argue that, if they are run properly and equipped with appropriate pollution-control devices, incinerators are safe for the general public. Opponents counter that neither public officials nor pollution-control equipment can be trusted to keep the air clean. They argue that recycling and source reduction efforts are better ways to deal with waste problems.

The EPA, which generally supports incineration, acknowledges the health threat of incinerator emissions but holds that the danger is very slight. The EPA estimates that dioxin emissions from a typical municipal incinerator may cause one death per million people in 70 years of operation. Critics of incineration claim that a more accurate estimate is 250 deaths per million in 70 years.

One way to reduce these dangerous emissions is to remove batteries containing heavy metals and plastics containing chlorine before wastes are burned. Bremen, Germany, is one of several European cities now trying to control dioxin emissions by keeping all plastics out of incinerator waste. Bremen is requiring households to separate plastics from other garbage. This is expected to eliminate nearly all dioxins and other combustion by-products and prevent the expense of installing costly pollution-control equipment that otherwise would be necessary to keep the burners operating. Minneapolis has initiated a recycling program for the small "button" batteries used in hearing aids, watches, and calculators in an attempt to lower mercury emissions from its incinerator.

SHRINKING THE WASTE STREAM

Having less waste to discard is obviously better than struggling with disposal methods, all of which have disadvantages and drawbacks. In this section we will explore some of our options for recycling, reuse, and reduction of the wastes we produce.

Recycling captures resources from garbage

The term *recycling* has two meanings in common usage. Sometimes we say we are recycling when we really are reusing something, such as refillable beverage containers. In terms of solid waste management, however, **recycling** is the reprocessing of discarded materials into new, useful products (fig. 13.7). Some recycling processes reuse materials for the same purposes; for instance, old aluminum cans and glass bottles are usually melted and recast into new cans and bottles. Other recycling processes turn old materials into entirely new products. Old tires, for instance, are shredded and turned into rubberized road surfacing. Newspapers become cellulose insulation, kitchen wastes become a valuable soil amendment, and steel cans become new automobiles and construction materials.

There have been some dramatic successes in recycling in recent years. New Jersey, for instance, now claims a 60 percent recycling rate, something thought unattainable a decade ago. The high value of aluminum scrap (as much as $1,200 per ton in recent years) has spurred a large percentage of aluminum recycling nearly everywhere. About two-thirds of all aluminum cans are now recycled, up from only 15 percent 20 years ago. This recycling is so rapid that half of all the aluminum cans now on grocery shelves will be made into another can within two months.

Recycling challenges

Despite encouraging gains in recycling rates, major challenges exist. Despite the value of aluminum, Americans still throw away nearly 350,000 metric tons of aluminum beverage containers each year. That is enough to make 3,800 Boeing 747 airplanes. Plastics recyclers have developed innovative methods and products, but the low price of virgin plastic, made from oil, is usually less than the cost of transporting and storing used plastics. Consequently, less than 5 percent of the United States' 24 million tons of plastic waste is recycled each year.

FIGURE 13.8 Creating a stable, economically viable market for recycled products is essential for recycling success. Consumers can help by buying recycled products.

Wild fluctuations in commodity prices make it still harder to develop a market for recycled materials. Newsprint, for example, cost $160 a ton in 1995; by 1999 it dropped to just $42 per ton and then climbed to $95 per ton in 2005 (fig. 13.8).

Contamination is a major obstacle in plastics recycling. Most of the 24 billion plastic soft drink bottles sold every year in the United States are made of PET (polyethylene terphthalate), which can be remanufactured into carpet, fleece clothing, plastic strapping, and nonfood packaging. However, even a trace of vinyl—a single PVC (polyvinyl chloride) bottle in a truckload, for example—can make PET useless. Although most bottles are now marked with a recycling number, it's hard for consumers to remember which is which. Because single-use beverage containers are so costly to recycle, they have been outlawed in Denmark and Finland. A looming worry is the prospect of single-use plastic beer bottles. Already being test-marketed, these bottles are made of PET but are amber colored to block sunlight and have a special chemical coating to keep out oxygen, which would ruin the beer. The special color, interior coating, and vinyl cap lining will make these bottles incompatible with regular PET, and it will probably cost more to remove them from the waste stream than the reclaimed plastic is worth. Plastic recycling already is down 50 percent from a decade ago because so many soft drink bottles are sold and consumed on the go, and never make it into recycling bins. Throwaway beer bottles could badly set back recycling programs.

The present U.S. public policies often tend to favor extraction of new raw materials. Energy, water, and raw materials are often sold to industries below their real cost to create jobs and stimulate the economy. For instance, in 1999 a pound of recycled clear PET, the material in most soft-drink bottles, sold for about 40¢. By contrast, a pound of off-grade, virgin PET cost 25¢. Setting the prices of natural resources at their real cost would tend to encourage efficiency and recycling. State, local, and national statutes requiring government agencies to purchase a minimum

FIGURE 13.7 Trucks with multiple compartments pick up residential recyclables at curbside, greatly reducing the amount of waste that needs to be buried or burned.

Principles of Environmental Science

http://www.mhhe.com/cunningham4e

amount of recycled material have helped create a market for used materials. Each of us can play a role in creating markets as well. If we buy things made from recycled materials—or ask for them if they aren't available—we will help recycling programs succeed.

Recycling saves money, energy, and space

Recycling is usually a better alternative to either dumping or burning wastes. It saves money, energy, raw materials, and land space, while also reducing pollution. Recycling also encourages individual awareness and responsibility for the refuse produced (fig. 13.9). Curbside pickup of recyclables costs around $35 per ton, as opposed to the $80 paid to dispose of them at an average metropolitan landfill. Many recycling programs cover their own expenses with materials sales and may even bring revenue to the community.

Recycling drastically reduces pressure on landfills and incinerators. Philadelphia is investing in neighborhood collection centers that will recycle 600 tons a day, enough to eliminate the need for a previously planned, high-priced incinerator. New York City, down to one available landfill but still producing 27,000 tons of garbage a day, set a target of 50 percent waste reduction to be accomplished by recycling office paper and household and commercial waste. In 2002 Mayor Michael Bloomberg discontinued most recycling, arguing that the program was too expensive. The city quickly found that disposing of waste was more expensive than recycling, and most programs were reinstated.

Japan probably has the most successful recycling program in the world. Half of all household and commercial wastes in Japan are recycled, while the rest are about equally incinerated or landfilled. The country has begun a push to increase recycling, because incineration costs almost as much. Some communities have raised recycling rates to 80 percent, and others aim to reduce

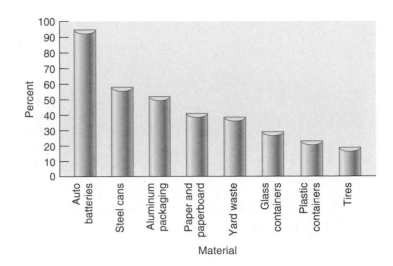

FIGURE 13.10 Recycling rates of selected materials in the United States. **Source:** Data from Environmental Protection Agency, 2003.

waste altogether by 2020. This level of recycling is most successful when waste is well sorted. In Yokohama, a city of 3.5 million, there are now 10 categories of recyclables, including used clothing and sorted plastics. Some communities have 30 or 40 categories for sorting recyclables.

Recycling lowers demands for raw resources (fig. 13.10). The United States cuts down 2 million trees every day to produce newsprint and paper products, a heavy drain on its forests. Recycling the print run of a single Sunday issue of the *New York Times* would spare 75,000 trees. Every piece of plastic made in the United States reduces the reserve supply of petroleum and makes the country more dependent on foreign oil. Recycling 1 ton of aluminum saves 4 tons of bauxite (aluminum ore) and 700 kg of petroleum coke and pitch, as well as keeping 35 kg of aluminum fluoride out of the air.

Recycling also reduces energy consumption and air pollution. Plastic bottle recycling could save 50 to 60 percent of the energy needed to make new ones. Making new steel from old scrap offers up to 75 percent energy savings. Producing aluminum from scrap instead of bauxite ore cuts energy use by 95 percent, yet the United States still throws away more than a million tons of aluminum every year. If aluminum recovery were doubled worldwide, more than a million tons of air pollutants would be eliminated every year.

Reducing litter is an important benefit of recycling. Ever since disposable paper, glass, metal, foam, and plastic packaging began to accompany nearly everything we buy, these discarded wrappings have collected on our roadsides and in our lakes, rivers, and oceans. Litter is a costly as well as unsightly problem. Americans pay an estimated 32¢ for each piece of litter picked up by crews along state highways, which adds up to $500 million every year. "Bottle bills" requiring deposits on bottles and cans have reduced littering in many states.

Composting recycles organic waste

Pressed for landfill space, many cities have banned yard waste from municipal garbage. Rather than bury this valuable organic

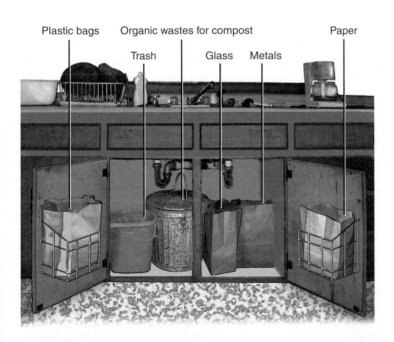

FIGURE 13.9 Source separation in the kitchen—the first step in a strong recycling program.

FIGURE 13.11 Composting is a good way to convert yard waste, vegetable scraps, and other organic materials into useful garden mulch. Mix everything together, keep it moist and well aerated, and in a few weeks, you will have a rich, odor-free mulch. This three-bin system allows you to have batches in different stages.

material, they are turning it into a useful product through **composting**: biological degradation or breakdown of organic matter under aerobic (oxygen-rich) conditions. The organic compost resulting from this process makes a nutrient-rich soil amendment that aids water retention, slows soil erosion, and improves crop yields. A home compost pile is an easy and inexpensive way to dispose of organic waste in an interesting and environmentally friendly way. All you need to do is to pile up lawn clippings, vegetable waste, fallen leaves, wood chips, or other organic matter in an out-of-the way place, keep it moist, and turn it over every week or so (fig. 13.11). Within a few months, naturally occurring microorganisms will decompose the organic material into a rich, pleasant-smelling compost that you can use as a soil amendment.

Some cities and counties have developed successful composting programs. Residents pay a small amount to drop off yard waste, and they can buy back rich, inexpensive soil amendment.

Energy from waste

Every year we throw away the energy equivalent of 80 million barrels of oil in organic waste in the United States. In developing countries up to 85 percent of the waste stream is food, textiles, vegetable matter, and other biodegradable materials. Worldwide, at least one-fifth of municipal waste is organic kitchen and garden refuse. In a landfill, much of this matter is decomposed by microorganisms generating billions of cubic meters of methane ("natural gas"), which contributes to global warming if allowed to escape into the atmosphere (chapter 9). Many cities are drilling methane wells in their landfills to capture this valuable resource.

Methane capture is relatively inefficient, though. An even better strategy is converting organic waste to methane in a contained, anaerobic digester. While composting breaks down organic waste in oxygen-rich environments, an oxygen-free environment produces methane. Many new technologies are being developed or are in use: most use steam and pressure to "cook" waste and release clean methane.

Anaerobic digestion also can be done on a small scale. Millions of household methane generators provide fuel for cooking and lighting for homes in China and India (chapter 12). In the United States some farmers produce all the fuel they need to run their farms—both for heating and for running trucks and tractors—by generating methane from animal manure.

Demanufacturing

Demanufacturing is the disassembly and recycling of obsolete consumer products, such as television sets, computers, telephones, refrigerators, washing machines, and air conditioners. There are at least two billion televisions and personal computers in use globally. Televisions often are discarded after only about five years, and computers, play-stations, cellular telephones, and other electronics become obsolete even faster. Stoves, refrigerators, and other "white goods" have a much longer lifetime—typically about 12 years—but the EPA estimates that Americans dispose of 54 million of these household appliances every year. Many of these consumer products contain both valuable materials and toxins that must be kept out of the environment. Older refrigerators and air conditioners, for example, have chloroflurocarbons (CFCs) that destroy stratospheric ozone and cannot be released into the air. Because new production has been banned, recycled CFCs are worth about $100 per pound. For both reasons, it pays to recycle them.

Between 300 million and 600 million computers are in use in the United States (up to two per person), and most will be discarded in the next few years. Only about 10 percent of computers are recycled. These computers contain at least 2.5 billion kg of lead, as well as (mercury, gallium, germanium, nickel, palladium, beryllium, selenium, arsenic) and valuable scrap, such as gold, silver, copper, and steel. It's estimated that 90 percent of the cadmium, lead, and mercury contamination in our solid waste stream comes from consumer electronics.

About 80 percent of U.S. e-waste is shipped overseas, mostly to China and developing countries in Asia and Africa. There, villagers, including young children, break them apart to retrieve copper, silver, gold, lead, and other metals. Effects on workers handling toxic materials may be severe, especially for growing children. Soil, groundwater, and surface water contamination at these sites has been found to be as much as 200 times the World Health Organization's standards. An estimated 100,000 workers demanufacture electronics in southeastern China alone (fig. 13.12).

In response to such problems, the European Union now requires that recycling fees be built into the purchase price of many electronics. This adds about $20 to the purchase price, less than 1 percent for many computers. Increasingly, local governments are requiring electronics recycling. Massachusetts was the first state to require that both households and businesses recycle used computer parts. Recycling companies in the United States are beginning to develop comprehensive electronics recycling systems, and these offer growing alternatives to shipping our hazardous waste overseas.

Principles of Environmental Science

FIGURE 13.12 A Chinese woman smashes a cathode ray tube from a computer monitor in order to remove valuable metals. This kind of unprotected demanufacturing is highly hazardous to both workers and the environment.

FIGURE 13.13 Reusing discarded products is a creative and efficient way to reduce wastes. This recycling center in Berkeley, California, is a valuable source of used building supplies and a money saver for the whole community.

In 2006, Maine instituted an innovative, controversial, but market-based law to manage this waste: make the producers pay. Most of Maine's communities are small and can't afford to collect, ship, or process the tons of miscellaneous appliances generated each year. Now cities and towns can collect appliances and ship them to "consolidators," which sort items, identify manufacturers, and send a bill to the manufacturers for the cost of disposal. The aim of this plan is to give manufacturers an incentive to make items with fewer toxic components or with parts that are easier to separate and recycle. Opponents argue that companies will just pass on the costs to consumers, but currently taxpayers pay the cost of disposal, in reduced landfill space and pollution. California, the only other state thus far to regulate e-waste, charges a recycling fee at the time of purchase. At least 15 other states have proposed similar e-waste laws.

Reuse is even more efficient than recycling

Even better than recycling or composting is cleaning and reusing materials in their present form, thus saving the cost and energy of remaking them into something else. We do this already with some specialized items. Auto parts are regularly sold from junkyards, especially for older car models. In some areas stained-glass windows, brass fittings, fine woodwork, and bricks salvaged from old houses bring high prices. Some communities sort and reuse a variety of materials received in their dumps (fig. 13.13).

In many cities, glass and plastic bottles are routinely returned to beverage producers for washing and refilling. The reusable, refillable bottle is the most efficient beverage container we have. It is better for the environment than remelting and more profitable for local communities. A reusable glass container makes an average of 15 round-trips between factory and customer before it becomes so scratched and chipped that it has to be recycled. Reusable containers also favor local bottling companies and help preserve regional differences.

Since the advent of cheap, lightweight, disposable food and beverage containers, many small, local breweries, canneries, and bottling companies have been forced out of business by huge national conglomerates. These big companies can afford to ship food and beverages great distances, as long as it is a one-way trip. If they had to collect their containers and reuse them, canning and bottling factories serving large regions would be uneconomical. Consequently, the national companies favor recycling rather than refilling because they prefer fewer, larger plants and don't want to be responsible for collecting and reusing containers. In some circumstances, life-cycle assessment shows that washing and decontaminating containers takes as much energy and produces as much air and water pollution as manufacturing new ones.

In many less-affluent nations, reuse of all sorts of manufactured goods is an established tradition. Where most manufactured products are expensive and labor is cheap, it pays to salvage, clean, and repair products. Cairo, Manila, Mexico City, and many other cities have large populations of poor people who make a living by scavenging. Entire ethnic populations may survive on scavenging, sorting, and reprocessing scraps from city dumps. Eliminating waste shipments can reduce pollution but may threaten livelihoods. How can we resolve this dilemma?

Reducing waste is often the cheapest option

Most of our attention in waste management focuses on recycling. But slowing the consumption of throw-away products is by far the most effective way to save energy, materials, and money. The "three Rs" waste hierarchy—reduce, reuse, recycle—lists the most important strategy first. Industries are increasingly finding that reducing saves money. Soft-drink makers use less aluminum per can than they did 20 years ago, and plastic bottles use less plastic. 3M has saved over $500 million in the past 30 years by reducing its use of raw materials, reusing waste products, and increasing efficiency. Individual action is essential, too (What Can You Do? on this page).

Excess packaging of food and consumer products is one of our greatest sources of unnecessary waste. Paper, plastic, glass, and metal packaging material makes up 50 percent of our domestic trash by volume (see related story "South Africa's National Flower?" at www.mhe.com/cunningham4e). Much of that packaging is primarily for marketing and has little to do with product protection (fig. 13.14). Manufacturers and retailers might be persuaded to reduce these wasteful practices if consumers ask for products

FIGURE 13.14 How much more do we need? Where will we put what we already have?
Reprinted with special permission of King Features Syndicate.

without excess packaging. Canada's National Packaging Protocol (NPP) recommends that packaging minimize depletion of virgin resources and production of toxins in manufacturing. The preferred hierarchy is (1) no packaging, (2) minimal packaging, (3) reusable packaging, and (4) recyclable packaging. This plan sets an ambitious target of 50 percent reduction in excess packaging.

Where disposable packaging is necessary, we still can reduce the volume of waste in our landfills by using materials that are compostable or degradable. **Photodegradable plastics** break down when exposed to ultraviolet radiation. **Biodegradable plastics** incorporate such materials as cornstarch that microorganisms can decompose. Several states have introduced legislation requiring biodegradable or photodegradable six-pack beverage yokes, fast-food packaging, and disposable diapers. These degradable plastics often don't decompose completely; they only break down to small particles that remain in the environment. In doing so, they can release toxic chemicals. And in modern, lined landfills, they don't decompose at all. Furthermore, they make recycling less feasible and may lead people to believe that littering is okay.

HAZARDOUS AND TOXIC WASTES

The most dangerous aspect of the waste stream is that it often contains highly toxic and hazardous materials that are injurious to both human health and environmental quality (fig. 13.15). We now produce and use a vast array of flammable, explosive, caustic, acidic, and highly toxic chemical substances for industrial, agricultural, and domestic purposes. According to the EPA, U.S. industries generate about 265 million metric tons of officially classified hazardous

FIGURE 13.15 Modern society produces large amounts of toxic and hazardous waste.

APPLICATIONS

A Personal Hazardous Waste Inventory

Make a survey of your house or apartment to see how many toxic materials you can identify. If you live in a dorm, you may need to survey your parents' house. If you are in an art class, consider paints and solvents. Read the cautionary labels. Which of the products are considered hazardous? When you use them, do you usually follow all the safety procedures recommended? Where Is the nearest hazardous waste collection site for disposal of unwanted household products? If you don't know, call your city administrator or mayor's office to find out.

wastes each year, slightly more than 1 metric ton for each person in the country. In addition, considerably more toxic and hazardous waste material is generated by industries or processes not regulated by the EPA. Shockingly, at least 40 million metric tons (22 billion lbs) of toxic and hazardous wastes are released into the air, water, and land in the United States each year. The biggest sources of these toxins are the chemical and petroleum industries (fig. 13.16).

Hazardous waste includes many dangerous substances

Legally, a **hazardous waste** is any discarded material, liquid or solid, that contains substances known to be (1) fatal to humans or laboratory animals in low doses; (2) toxic, carcinogenic, mutagenic, or teratogenic to humans or other life-forms; (3) ignitable with a flash point less than 60°C; (4) corrosive; or (5) explosive or highly reactive (undergoes violent chemical reactions either by itself or when mixed with other materials). Notice that this definition includes both toxic and hazardous materials, as defined in chapter 8. Certain compounds are exempt from regulation as hazardous waste if they are accumulated in less than 1 kg (2.2 lbs) of commercial chemicals or 100 kg of contaminated soil, water, or debris.

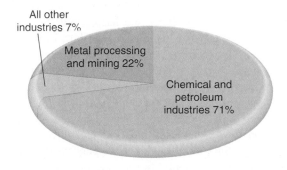

All other industries 7%

Metal processing and mining 22%

Chemical and petroleum industries 71%

FIGURE 13.16 Producers of hazardous wastes in the United States. **Source:** Data from the U.S. Environmental Protection Agency.

Even larger amounts (up to 1,000 kg) are exempt when stored at an approved waste treatment facility for the purpose of being beneficially used, recycled, reclaimed, detoxified, or destroyed.

Most hazardous waste is recycled, converted to nonhazardous forms, stored, or otherwise disposed of on-site by the generators—chemical companies, petroleum refiners, and other large industrial facilities—so that it doesn't become a public problem. Still, the hazardous waste that does enter the waste stream or the environment represents a serious environmental problem. And orphan wastes left behind by abandoned industries remain a serious threat to both environmental quality and human health. For years little attention was paid to this material. Wastes stored on private property, buried, or allowed to soak into the ground were considered of little concern to the public. An estimated 5 billion metric tons of highly poisonous chemicals were improperly disposed of in the United States between 1950 and 1975 before regulatory controls became more stringent.

Federal legislation regulates hazardous waste

Two important federal laws regulate hazardous waste management and disposal in the United States. The Resource Conservation and Recovery Act (RCRA, pronounced "rickra") of 1976 is a comprehensive program that requires rigorous testing and management of toxic and hazardous substances. A complex set of rules requires generators, shippers, users, and disposers of these materials to keep meticulous account of everything they handle and what happens to it from generation (cradle) to ultimate disposal (grave) (fig. 13.17).

The Comprehensive Environmental Response, Compensation, and Liability Act (CERCLA or Superfund Act), passed in 1980 and modified in 1984 by the Superfund Amendments and Reauthorization Act (SARA), is aimed at rapid containment, cleanup, or remediation of abandoned toxic waste sites. This statute authorizes the EPA to undertake emergency actions when a threat exists that toxic material will leak into the environment. The EPA is empowered to bring suit for the recovery of its costs from potentially responsible parties, such as site owners, operators, waste generators, or transporters.

SARA also established (under title III) a community right to know and state emergency response plans that give citizens access

FIGURE 13.17 Toxic and hazardous wastes must be tracked from "cradle to grave" by detailed shipping manifests.

Labels in figure: Hazardous waste generator; EPA or state agency office; Transporter; Storage; Tracking of hazardous waste through manifest system; Transporter; Treatment facility; Secure landfill; Transporter

to information about what is present in their communities. One of the most useful tools in this respect is the **Toxic Release Inventory,** which requires 20,000 manufacturing facilities to report annually on releases of more than 300 toxic materials. You can find specific information in the inventory about what is in your neighborhood.

The government does not have to prove that anyone violated a law or what role he or she played in a Superfund site. Rather, liability under CERCLA is "strict, joint, and several," meaning that anyone associated with a site can be held responsible for the entire cost of cleaning it up, no matter how much of the mess they made. In some cases property owners have been assessed millions of dollars for removal of wastes left there years earlier by previous owners. This strict liability has been a headache for the real estate and insurance businesses.

CERCLA was amended in 1995 to make some of its provisions less onerous. In cases where treatment is unavailable or too costly and it is likely that a less costly remedy will become available within a reasonable time, interim containment is now allowed. The EPA also now has the discretion to set site-specific cleanup levels, rather than adhere to rigid national standards.

Superfund sites are those listed for federal cleanup

The EPA estimates that there are at least 36,000 seriously contaminated sites in the United States. The General Accounting

Office (GAO) places the number much higher, perhaps more than 400,000 when all are identified. By 2004, some 1,671 sites had been placed on the National Priority List (NPL) for cleanup with financing from the federal Superfund program. The **Superfund** is a revolving pool designed to (1) provide an immediate response to emergency situations that pose imminent hazards and (2) to clean up or remediate abandoned or inactive sites. Without this fund, sites would languish for years or decades while the courts decided who was responsible for paying for the cleanup. Originally a $1.6 billion pool, the fund peaked at $3.6 billion. From its inception, the fund was financed by taxes on producers of toxic and hazardous wastes. Industries opposed this "polluter pays" tax, because current manufacturers are often not the ones responsible for the original contamination. In 1995 Congress agreed to let the tax expire. Since then the Superfund has dwindled, and the public has picked up an increasing share of the bill. In the 1980s the public covered less than 20 percent of the Superfund. In 2004, however, general revenues (public funds) paid the entire cost off a greatly reduced program, and the industry share was zero.

Total costs for hazardous waste cleanup in the United States are estimated to be between $370 billion and $1.7 trillion, depending on how clean sites must be and what methods are used. For years, Superfund money was spent mostly on lawyers and consultants, and cleanup efforts were often bogged down in disputes over liability and best cleanup methods. During the 1990s, however, progress improved substantially, with a combi-

Principles of Environmental Science

http://www.mhhe.com/cunningham4e

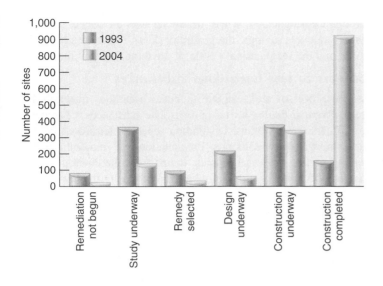

FIGURE 13.18 Progress on Superfund National Priority List (NPL) sites. After years of little progress, the number of completed sites jumped from 155 in 1993 to 926 in 2004. Over 90 percent of the 1,529 currently listed NPL sites are under construction or completed.
Source: Data from Environmental Protection Agency, 2004.

FIGURE 13.19 Old electrical transformers filled with PCBs await disposal. If they leak, it could be a million-dollar problem.

nation of rule adjustments and administrative commitment to cleanup. From 1993 to 2004, the number of completed NPL cleanups jumped from 155 to 926 of the 1,529 listed sites (fig. 13.18). In recent years, though, progress has slowed because of underfunding and a lower priority among administrators.

What qualifies a site for the NPL? These sites are considered to be especially hazardous to human health and environmental quality because they are known to be leaking or have a potential for leaking supertoxic, carcinogenic, teratogenic, or mutagenic materials (chapter 8). The ten substances of greatest concern or most commonly detected at Superfund sites are lead, trichloroethylene, toluene, benzene, PCBs, chloroform, phenol, arsenic, cadmium, and chromium. These and other hazardous or toxic materials are known to have contaminated groundwater at 75 percent of the sites now on the NPL. In addition, 56 percent of these sites have contaminated surface waters, and airborne materials are found at 20 percent of the sites. Seventy million Americans, including 10 million children, live within 6 km of a Superfund site.

Where are these thousands of hazardous waste sites, and how did they get contaminated? Old industrial facilities, such as smelters, mills, petroleum refineries, and chemical manufacturing plants, are highly likely to have been sources of toxic wastes. Regions of the country with high concentrations of aging factories, such as the "rust belt" around the Great Lakes or the Gulf Coast petrochemical centers, have large numbers of Superfund sites. Mining districts also are prime sources of toxic and hazardous waste. Within cities, factories and places such as railroad yards, bus repair barns, and filling stations, where solvents, gasoline, oil, and other petrochemicals were spilled or dumped on the ground, often are highly contaminated.

Some of the most infamous toxic waste sites were old dumps where many different materials were mixed together indiscriminately. For instance, Love Canal in Niagara Falls, New York, was an open dump that both the city and nearby chemical factories used as a disposal site. More than 20,000 tons of toxic chemical waste were buried under what later became a housing development. Another infamous example occurred in Hardeman County, Tennessee, where about a quarter of a million barrels of chemical waste were buried in shallow pits that subsequently leaked toxins into the groundwater.

Brownfields present both liability and opportunity

Among the biggest problems in cleaning up hazardous waste sites are questions of liability and the degree of purity required (fig. 13.19). In many cities, these problems have created large areas of contaminated properties, known as **brownfields,** that have been abandoned or are not being used to their potential because of real or suspected pollution. Up to one-third of all commercial and industrial sites in the urban core of many big cities fall in this category. In heavy industrial corridors the percentage typically is higher.

For years no one was interested in redeveloping brownfields because of liability risks. Who would buy a property, knowing that they might be forced to spend years in litigation and negotiations and be forced to pay millions of dollars for pollution they didn't create? Even if a site has been cleaned to current standards, there is a worry that additional pollution might be found in the future or that more stringent standards might be applied.

In many cases, property owners complain that unreasonably high levels of purity are demanded in remediation programs. Consider the case of Columbia, Mississippi. For many years a 35-ha (81-acre) site in Columbia was used for turpentine and pine tar manufacturing. Soil tests showed concentrations of phenols and other toxic organic compounds exceeding federal safety standards. The site was added to the Superfund NPL, and remediation was ordered. Some experts recommended that the best solution was to simply cover the surface with clean soil and enclose the property with a fence to keep people out. The total costs would have been about $1 million. Instead, the EPA ordered Reichhold Chemical, the last known property owner, to excavate more than 12,500 tons of soil and haul it to a commercial hazardous waste dump in Louisiana at a cost of some $4 million. The intention is to make the site safe enough to be used for any purpose, including housing—even though no one has proposed building anything there. According to the EPA, the dirt must be clean enough for children to play in—even eat—without risk.

Similarly, in places where contaminants have seeped into groundwater, the EPA generally demands that cleanup be carried to drinking-water standards. Many critics believe that these pristine standards are unreasonable. Former Congressman Jim Florio, a principal author of the original Superfund Act, says, "It doesn't make any sense to clean up a rail yard in downtown Newark so it can be used as a drinking water reservoir." Depending on where the site is, what else is around it, and what its intended uses are, much less stringent standards may be perfectly acceptable.

Brownfield redevelopment is increasingly seen as an opportunity for rebuilding cities, creating jobs, increasing the tax base, and preventing needless destruction of open space at urban margins. In 2002 the EPA established a new brownfields revitalization fund designed to encourage restoration of more sites, as well as more kinds of sites. In some communities former brownfields are being turned into "eco-industrial parks" that feature environmentally friendly businesses and bring in much-needed jobs to inner-city neighborhoods (chapter 14).

Hazardous waste must be processed or stored permanently

What shall we do with toxic and hazardous wastes? In our homes, we can reduce waste generation and choose less toxic materials. Buy only what you need for the job at hand. Use up the last little bit, or share leftovers with a friend or neighbor. Many common materials that you probably already have make excellent alternatives to commercial products.

Produce less waste

As with other wastes, the safest and least expensive way to avoid hazardous waste problems is to avoid creating the wastes in the first place. Manufacturing processes can be modified to reduce or eliminate waste production. In Minnesota, the 3M Company reformulated products and redesigned manufacturing processes to eliminate more than 140,000 metric tons of solid and hazardous wastes, 4 billion liters (1 billion gal) of wastewater, and 80,000 metric tons of air pollution each year. It frequently found that these new processes not only spared the environment but also saved money by using less energy and fewer raw materials.

Recycling and reusing materials also eliminates hazardous wastes and pollution. Many waste products of one process or industry are valuable commodities in another. Already, about 10 percent of the wastes that would otherwise enter the waste stream in the United States are sent to surplus material exchanges, where they are sold as raw materials for use by other industries. This figure could probably be raised substantially with better waste management. In Europe at least one-third of all industrial wastes are exchanged

through clearinghouses, where beneficial uses are found. This represents a double savings: the generator doesn't have to pay for disposal, and the recipient pays little, if anything, for raw materials.

Convert to less hazardous substances

Several processes are available to make hazardous materials less toxic. *Physical treatments* tie up or isolate substances. Charcoal or resin filters absorb toxins. Distillation separates hazardous components from aqueous solutions. Precipitation and immobilization in ceramics, glass, or cement isolate toxins from the environment, so that they become essentially nonhazardous. One of the few ways to dispose of metals and radioactive substances is to fuse them in silica at high temperatures to make a stable, impermeable glass that is suitable for long-term storage. Plants, bacteria, and fungi can also concentrate or detoxify contaminants (see Exploring Scence p. 327).

Incineration is applicable to mixtures of wastes. A permanent solution to many problems, it is quick and relatively easy, but not necessarily cheap—nor always clean—unless done correctly. Wastes must be heated to over 1,000°C (2,000°F) for a sufficient period of time to complete destruction. The ash resulting from thorough incineration is reduced in volume up to 90 percent and often is safer to store in a landfill or another disposal site than the original wastes. Nevertheless, incineration remains highly controversial (fig. 13.20).

Chemical processing can transform materials to make them nontoxic. Included in this category are neutralization, removal of metals or halogens (chlorine, bromine, etc.), and oxidation. The Sunohio Corporation of Canton, Ohio, for instance, has developed a process called PCBx, in which chlorine in such molecules as PCBs is replaced with other ions that render the compounds less toxic. A portable unit can be moved to the location of the hazardous wastes, eliminating the need for shipping them.

FIGURE 13.20 Actor Martin Sheen joins local activists in a protest in East Liverpool, Ohio, site of the largest hazardous waste incinerator in the United States. About 1,000 people marched to the plant to pray, sing, and express their opposition. Involving celebrities draws attention to your cause. A peaceful, well-planned rally builds support and acceptance in the broader community.

Exploring SCIENCE

Bioremediation

Cleaning up the thousands of hazardous waste sites at factories, farms, and gas stations is an expensive project. In the United States alone, waste cleanup is projected to cost at least $700 billion. Usually hazardous waste remediation (cleanup) involves digging up soil and incinerating it, potentially releasing toxins into the air, or trucking it to a secure landfill. Contaminated groundwater is frequently pumped out of the ground; hopefully, contaminants are retrieved at the same time.

A promising alternative to these methods involves **bioremediation,** or biological waste treatment. Microscopic bacteria and fungi can absorb, accumulate, and detoxify a remarkable variety of toxic compounds. They can also accumulate heavy metals, and some have been developed that can metabolize (break down) PCBs. Aquatic plants such as water hyacinths and cattails can also be used to purify contaminated effluent.

Recently, an increasing variety of plants have been used in phytoremediation (cleanup using plants). Some types of mustard can extract lead, arsenic, zinc, and other metals from contaminated soil. Radioactive strontium and cesium have been extracted from soil near the Chernobyl nuclear power plant using common sunflowers. Poplar trees can absorb and break down toxic organic chemicals. Natural bacteria in groundwater. When provided with plenty of oxygen, can neutralize contaminants in aquifers. Experiments have shown that pumping air *into* groundwater can be a more effective cleanup method than pumping water *out*.

How do plants, bacteria, and fungi do all this? Many of the biophysical details are poorly understood, but in general, plant roots are designed to efficiently extract nutrients, water, and trace minerals from soil and groundwater. The mechanisms involved may aid extraction of metallic and organic contaminants. Some plants also use toxic elements as a defense against herbivores: locoweed, for example, selectively absorbs elements such as selenium, concentrating toxic levels in its leaves. Absorption can be extremely effective. Bracken fern growing in Florida has been found to contain arsenic at concentrations more than 200 times higher than in the soil in which it was growing.

Genetically modified plants are also being developed to process toxins. Poplars have been developed to process toxins, using a gene borrowed from bacteria that transforms a toxic compound of mercury into a safer form. In another experiment, a gene for producing mammalian liver enzymes, which specialize in breaking down toxic organic compounds, was inserted into tobacco plants. The plants succeeded in producing the liver enzymes and breaking down toxins absorbed through their roots.

These remediation methods are not without risks. Insects could consume leaves containing concentrated substances, allowing contaminants to enter the food chain. Some absorbed contaminants are volatilized, or emitted in gaseous form, through pores in plant leaves. Once contaminants are absorbed into plants, the plants themselves are usually toxic and must be landfilled. But the cost of phytoremediation can be less than half the cost of landfilling or treating toxic soil, and the volume of plant material requiring secure storage is a fraction of the volume of the contaminated dirt.

Cleaning up hazardous and toxic waste sites will be a big business for the foreseeable future, in North America and around the world. Innovations such as bioremediation offer promising prospects for business development, as well as for environmental health and saving taxpayer money.

Fast-growing poplar trees have been developed that can absorb and break down toxic compounds.

Store permanently

Inevitably, there will be some materials that we can't destroy, make into something else, or otherwise cause to vanish. We will have to store them out of harm's way. There are differing opinions about how best to do this.

Retrievable Storage Dumping wastes in the ocean or burying them in the ground generally means that we have lost control of them. If we learn later that our disposal technique was a mistake, it is difficult, if not impossible, to go back and recover the wastes. For many supertoxic materials, the best way to store them may be in **permanent retrievable storage.** This means placing waste storage containers in a secure building, salt mine, or bedrock cavern, where they can be inspected periodically and retrieved, if necessary, for repacking or for transfer if a better means of disposal is developed. This technique is more expensive than burial in a landfill because the storage area must be guarded and monitored continu-

ously to prevent leakage, vandalism, or other dispersal of toxic materials. Remedial measures are much cheaper with this technique, however, and it may be the best system in the long run.

Secure Landfills One of the most popular solutions for hazardous waste disposal has been landfilling. Although, as we saw earlier in this chapter, many such landfills have been environmental disasters, newer techniques make it possible to create safe, modern **secure landfills** that are acceptable for disposing of many hazardous wastes. The first line of defense in a secure landfill is a thick bottom cushion of compacted clay that surrounds the pit like a bathtub (fig. 13.21). Moist clay is flexible and resists cracking if the ground shifts. It is impermeable to groundwater and will safely contain wastes. A layer of gravel is spread over the clay liner, and perforated drainpipes are laid in a grid to collect any seepage that escapes from the stored material. A thick polyethylene liner, protected from

CHAPTER 13   Solid and Hazardous Waste

327

Leachate removal standpipes
Methane removal
Topsoil cover
Clay cap
Plastic liner
Recompacted clay
Existing clay
Testing well
Monitoring underdrains
Bedrock

FIGURE 13.21 A secure landfill for toxic waste. A thick plastic liner and two or more layers of impervious compacted clay enclose the landfill. A gravel bed between the clay layers collects any leachate, which can then be pumped out and treated. Well samples are tested for escaping contaminants and methane is collected for combustion.

punctures by soft padding materials, covers the gravel bed. A layer of soil or absorbent sand cushions the inner liner, and the wastes are packed in drums, which then are placed into the pit,

separated into small units by thick berms of soil or packing material.

When the landfill has reached its maximum capacity, a cover much like the bottom sandwich of clay, plastic, and soil—in that order—caps the site. Vegetation stabilizes the surface and improves its appearance. Sump pumps collect any liquids that filter through the landfill, either from rainwater or leaking drums. This leachate is treated and purified before being released. Monitoring wells check groundwater around the site to ensure that no toxins have escaped.

Most landfills are buried below ground level to be less conspicuous; however, in areas where the groundwater table is close to the surface, it is safer to build above-ground storage. The same protective construction techniques are used as in a buried pit. An advantage to such a facility is that leakage is easier to monitor because the bottom is at ground level.

Transportation of hazardous wastes to disposal sites is of concern because of the risk of accidents. Emergency-preparedness officials conclude that the greatest risk in most urban areas is not nuclear war or natural disaster but crashes involving trucks or trains carrying hazardous chemicals through densely packed urban corridors. Another worry is who will bear financial responsibility for abandoned waste sites. Hazardous wastes remain toxic long after the businesses that created them are gone. As is the case with nuclear wastes (chapter 12), we may need new institutions for perpetual care of these wastes.

REVIEW QUESTIONS

1. List some items that can be recycled from construction and demolition waste.

2. What are *solid wastes* and *hazardous wastes*? What is the difference between them?

3. Describe the difference between an open dump, a sanitary landfill, and a modern, secure, hazardous waste disposal site.

4. Why are landfill sites becoming limited around most major urban centers in the United States? What steps are being taken to solve this problem?

5. Describe some concerns about waste incineration.

6. List some benefits and drawbacks of recycling wastes. What are the major types of materials recycled from municipal waste, and how are they used?

7. What is *e-waste*? How is most of it disposed of, and what are some strategies for improving recycling rates?

8. What is *composting,* and how does it fit into solid waste disposal?

9. Describe some ways that we can reduce the waste stream to avoid or reduce disposal problems.

10. What materials are most recycled in the United States?

11. What societal problems are associated with waste disposal? Why do people object to waste handling in their neighborhoods?

12. What is the *National Priority List (NPL)*? Does it include all badly contaminated sites in the country? Are most sites on the list treated or untreated? Who pays for NPL cleanup?

13. What are *brownfields,* and why do cities want to redevelop them?

14. What are *bioremediation* and *phytoremediation*? What are some advantages to these methods?

CRITICAL THINKING AND DISCUSSION

Apply the principles you have learned in this chapter to discuss these questions with other students.

1. If you were planning an archeological dig of a modern garbage dump, would you record weight or volume of the objects uncovered? Why would this matter?

2. A toxic waste disposal site has been proposed for the Pine Ridge Indian Reservation in South Dakota. Many tribal members oppose this plan, but some favor it because of the jobs and income it will bring to an area with 70 percent unemployment. If local people choose immediate survival over long-term health, should we object or intervene?

3. Should industry officials be held responsible for dumping chemicals that were legal when they did it but are now known to be extremely dangerous? At what point can we argue that they should have known about the hazards involved?

4. Suppose that your brother or sister has decided to buy a house next to a toxic waste dump because it costs $20,000 less than a comparable house elsewhere. What do you say to him or her?

5. Is there a fundamental difference between incinerating municipal, medical, or toxic industrial waste? Would you oppose an incinerator in your neighborhood for one type of waste but not others? Why or why not?

6. The Netherlands incinerates much of its toxic waste at sea by a shipborne incinerator. Would you support this as a way to dispose of our wastes as well? What are the critical considerations for or against this approach?

7. Some scientists argue that permanent retrievable storage of toxic and hazardous wastes is preferable to burial. How can we be sure that material that will be dangerous for thousands of years will remain secure? If you were designing such a repository, how would you address this question?

SUGGESTED READINGS

Bergman, B. J. 1999. The hidden life of computers. *Sierra* 84(4):32–33.

Blumberg, Louis, and Robert Gottlieb. 1989. *War on Waste: Can America Win Its Battle with Garbage?* Island Press.

Kazuhiro, Ueta, and Harumi Koizumi. 2001. Reducing household waste: Japan learns from Germany. *Environment* 43(9):20–32.

McDonough, William, and Michael Braungart. 2002. *Cradle to Cradle: Remaking the Way We Make Things.* North Point Press.

Rathje, William, and Cullen Murphy. 1992. *Rubbish! The Archaeology of Garbage.* HarperCollins.

Taylor, David. 1999. Talking trash: The economic and environmental issues of landfills. *Environmental Health Perspectives* 107(8):A404–A409.

Warren-Rhodes, Kimberley, and Albert Koenig. 2001. Escalating trends in the urban metabolism of Hong Kong: 1971–1997. *AMBIO: A Journal of the Human Environment* 30(7):429–38.

 DATA analysis How Much Waste Do You Produce, and How Much Do You Know How to Manage?

As people become aware of waste disposal problems in their communities, more people are recycling more materials. Some things are easy to recycle, such as newsprint, office paper, or aluminum drink cans. Other things are harder to classify. Most of us give up pretty quickly and throw things in the trash if we have to think too hard about how to recycle them.

1. Take a poll to find out how many people in your class know how to recycle the items in the table at right. Once you have taken your poll, convert the numbers to percentages: divide the number who know how to recycle each item by the number of students in your class, and then multiply by 100.

2. Now find someone on your campus who works on waste management. This might be someone in your university/college administration, or it might be someone who actually empties trash containers. (You might get more interesting and straightforward answers from the latter.) Ask the following questions: (1) Can this person fill in the items your class didn't know about? (2) Is there a college/university policy about recycling? What are some of the points on that policy? (3) How much does the college spend each year on waste

disposal? How many tuition payments does that total? (4) What are the biggest parts of the waste stream? (5) Does the school have a plan for reducing that largest component?

Item	Percentage Who Know How to Recycle
Newspapers	
Paperboard (cereal boxes)	
Cardboard boxes	
Cardboard boxes with tape	
Plastic drink bottles	
Other plastic bottles	
Styrofoam food containers	
Food waste	
Plastic shopping bags	
Plastic packaging materials	
Furniture	
Last year's course books	
Left-over paint	

For Additional Help in Studying This Chapter, please visit our website at www.mhhe.com/cunningham4e. You will find practice quizzes, a chapter summary, key terms, answers to review questions, additional case studies, regional examples, and an extensive reading list, all of which will help you understand the material in this chapter.

Tanguá Park, Curitiba's newest, provides green space for recreation, protects the city watershed, and preserves a large area of native forest. The tubular section of this building is a reference to other structures designed by Jamie Lerner, including bus stations, the city airport, the opera house, and the botanical garden.

14

Economics and Urbanization

What kind of world do you want to live in? Demand that your teachers teach you what you need to know to build it.

–Peter Kropotkin–

OBJECTIVES

Studying this chapter will help you answer the following questions:

- How have the size and location of the world's largest cities changed over the past century?
- Define slum and shantytown, and describe the conditions you might find in them.
- What is urban sprawl? How have automobiles contributed to sprawl?

- What are some principles of smart growth and new urbanism?
- Describe sustainable development and why it's important.
- What value do we get from free ecological services?
- What's the difference between GNP and GPI?
- What do we mean by internalizing external costs?

Case Study   Curitiba: A Model Sustainable City

A few decades ago, Curitiba, Brazil, like many cities in the developing world, faced rapid population growth, air pollution, congested streets, and inadequate waste disposal systems. In 1969, Jamie Lerner, a landscape architect and former student activist, ran for mayor on a platform of urban renewal, social equity, and environmental protection. For the next 30 years, first as mayor, and then governor of Parana State, Lerner instituted social reforms and urban planning that have made the city an outstanding model of sustainable urban development. Today, with a population of about 1.8 million, Curitiba has an international reputation for progressive social programs, environmental protection, cleanliness, and livability. An international conference of mayors and city planners called Curitiba the most innovative city in the world.

The master plan instituted by Lerner called for rational municipal zoning and land-use regulations—rare provisions for a city in the developing world. An extensive network of parks and open space protected from future development extends throughout the city and provides access to nature as well as recreation opportunities. It also helps protect the watershed and prevent floods that once afflicted the city. Tanguá Park, for example, Curitiba's newest, is built on land once destined to be a garbage dump (see photo opposite page). The dump was moved elsewhere, however, and the park now encompasses about 250 ha of open space, including a rare stand of Brazilian pine (*Araucaria agustafolia*), the symbol of Parana State. Together, the many parks, bicycle trails, city gardens, and public riverbanks make the city a beautiful and pleasant place to live. Curitiba has more open space per capita than most American cities.

The heart of Curitiba's environmental and social plan is education for everyone. The first federal university in Brazil was built here in 1912. Free municipal schools give the city the highest literacy rate in the country. Environmental education reaches to everyone as well. School children study ecology along with Portuguese and math. With the help of children, who encourage their parents, the city has instituted a complex program that separates organic waste, trash, plastic, glass, and metal for recycling. Everything reusable is salvaged from the waste stream and sold as raw material to local industries. The city calculates that 1,200 trees per day are saved by paper recycled in this program. "Imagine if the whole of Brazil did this," Lerner exclaims. "We could save 26 million trees per year!" More than 70 percent of the city residents now participate in recycling.

Environmental protection and resource conservation also contribute to social welfare programs. Low-income Curitibanos are employed to pick up and sort recyclables, as well as to work in civic gardens and a city-run farm. Along with the many street sweepers in their distinctive orange suits, these programs provide jobs and pride while also keeping the city clean and beautiful. More than 1,100 municipal facilities provide services to the entire population. Free day care is available to all working mothers with children under 6 years old. Eldercare programs provide psychological services as well as food and housing to seniors. New job-training and small-business incubators run by the city help Curitibanos learn technical skills and launch new enterprises. The city has also built a technology park to attract new-economy businesses.

One of the crises that originally motivated Mayor Lerner was preservation of the historic central business district of his city. Traffic choked narrow streets and air pollution drove wealthy residents and businesses out of the central city, leaving beautiful old buildings to decay and eventually be torn down. In one of his boldest moves, Lerner banned most vehicles from the central business district and turned the streets into pedestrian malls. Rather than depend on private automobiles to move people around, Curitiba built a remarkably successful and economical bus-based rapid transit system. More than 340 feeder routes throughout the metropolitan area link to high-speed articulated buses that travel on dedicated busways. Everyone in the city lives within walking distance of frequent, economical, rapid public transportation. Today more than three-quarters of all trips made each day in the city utilize this innovative system, a percentage beyond the fondest imagination of most American city planners.

Much of Curitiba's city center has been converted to pedestrian shopping streets, while historic buildings have been converted to new uses. Black and white tile mosaics are featured throughout the city.

Benches, fountains, landscaping, distinctive lighting, and other amenities along with European-style shops and sidewalk cafes now make Curitiba's central district a pleasant place to stroll, shop, or simply gather with friends. Historic buildings have been refurbished, and many have been recycled for new purposes. Throughout the city center, walking streets are paved with characteristic black and white ceramic mosaics that draw on the city's colonial past.

Curitiba illustrates a number of ways that we can live sustainably with our environment and with each other. In this chapter, we'll look at other aspects of city planning and urban environments as well as some principles of ecological economics that help us understand the nature of resources and the choices we face both as individuals and communities.

FIGURE 14.1 In less than 20 years, Shanghai, China, has built Pudong, a new city of 1.5 million residents and 500 skyscrapers on former marshy farmland across the Huang Pu River from the historic city center. This kind of rapid urban growth is occurring in many developing countries.

URBANIZATION

For most of human history, the vast majority of people have lived in rural areas where they engaged in hunting and gathering, farming, fishing, or other natural-resource based occupations. Since the beginning of the Industrial Revolution about 300 years ago, however, cities have grown rapidly in both size and power (fig. 14.1). Now, for the first time ever, more people live in urban areas than in the country. In 1950, only 38 percent of the world population lived in cities (table 14.1). By 2030, that percentage is expected to nearly double. This means that over the next three decades about 3 billion people will crowd into cities. Some areas—Europe, North America, and Latin America—are already highly urbanized. Only Africa and Asia are below 45 percent urbanized.

Demographers predict that 90 percent of the human population growth in this century will occur in developing countries, and that almost all of that growth will occur in cities (fig. 14.2). Already huge **urban agglomerations** (mergers of multiple municipalities) are forming throughout the world. Some of these **megacities** (urban areas with populations over 10 million people) are already truly enormous, some claiming up to 30 million people. Can cities this size—especially in poorer countries—supply all the public services necessary to sustain a civilized life?

If we are to learn to live sustainably—that is, to depend on renewable resources while also protecting environmental quality, biodiversity, and the ecological services on which all life depends—

TABLE 14.1

Urban Share of Total Population (Percentage)

	1950	2000	2030*
Africa	18.4	40.6	57.0
Asia	19.3	43.8	59.3
Europe	56.0	75.0	81.5
Latin America	40.0	70.3	79.7
North America	63.9	77.4	84.5
Oceania	32.0	49.5	60.7
World	38.3	59.4	70.5

*Projected.

Source: Data from United Nations Population Division, 2003.

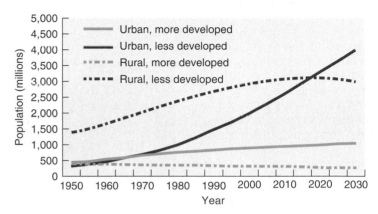

FIGURE 14.2 Growth of urban and rural populations in more-developed regions and in less-developed regions.
Source: United Nations Population Division. *World Urbanization Prospects,* 2004.

that challenge will have to be met primarily in the cities of the world, where most people will live. Many of us dream of moving to a secluded hideaway in the country, where we could grow our own food, chop wood, and carry water. But it probably isn't possible for 6.5 billion people (let alone the 8 to 9 billion expected by the end of this century) to live rural, subsistence lifestyles. Learning to live together in cities is probably the only way we'll survive.

Cities can be engines of economic progress and social reform. Some of the greatest promise for innovation comes from cities like Curitiba, where innovative leaders can focus knowledge and resources on common problems. Cities can be efficient places to live, where mass transportation can move people around and goods and services are more readily available than in the country. Concentrating people in urban areas leaves open space available for farming and biodiversity. But cities can also be dumping grounds for poverty, pollution, and unwanted members of society. Providing food, housing, transportation, jobs, clean water, and sanitation to the 2 or 3 billion new urban residents expected to crowd into cities—especially those in the developing world—in this century may be one of the preeminent challenges of this century.

As Curitiba shows, there is much we can do to make our cities more livable. It's especially encouraging that a Brazilian city, where

Principles of Environmental Science

per capita income is only one-fifth that of the United States, can make so much progress. But what of countries where personal wealth is only one-tenth that of Brazil? What hope is there for them?

Large cities are expanding rapidly

You can already see the dramatic shift in size and location of big cities. In 1900 only 13 cities in the world had populations over 1 million (table 14.2). All of those cities, except Tokyo and Peking were in Europe or North America. London was the only city in the world with more than 5 million residents. By 2007, there were at least 300 cities—100 of them in China alone—with more than 1 million residents. Of the 13 largest of these metropolitan areas, none are in Europe. Only New York City and Los Angeles are in a developed country. By 2025, it's expected that at least 93 cities will have populations over 5 million, and three-fourths of those cities will be in developing countries (fig. 14.3). In just the next 25 years, Mumbai, India; Delhi, India; Karachi, Pakistan; Manila, Philippines; and Jakarta, Indonesia all are expected to grow by at least 50 percent.

China represents the largest demographic shift in human history. Since the end of Chinese collectivized farming and factory work in 1986, around 250 million people have moved from rural areas to cities. And in the next 25 years an equal number is expected to join this vast exodus. In addition to expanding existing cities, China plans to build 400 new urban centers with populations of

TABLE 14.2

The World's Largest Urban Areas (Populations in Millions)

1900		2015**	
London, England	6.6	Tokyo, Japan	31.0
New York, USA	4.2	New York, USA	29.9
Paris, France	3.3	Mexico City	21.0
Berlin, Germany	2.4	Seoul, Korea	19.8
Chicago, USA	1.7	São Paolo, Brazil	18.5
Vienna, Austria	1.6	Osaka, Japan	17.6
Tokyo, Japan	1.5	Jakarta, Indonesia	17.4
St. Petersburg, Russia	1.4	Delhi, India	16.7
Philadelphia, USA	1.4	Los Angeles, USA	16.6
Manchester, England	1.3	Beijing, China	16.0
Birmingham, England	1.2	Cairo, Egypt	15.5
Moscow, Russia	1.1	Manila, Philippines	13.5
Peking, China*	1.1	Buenos Aires, Brazil	12.9

*Now spelled Beijing.

**Projected.

Source: Data from T. Chandler, *Three Thousand Years of Urban Growth*, 1974, Academic Press; and *World Gazetter*, 2003.

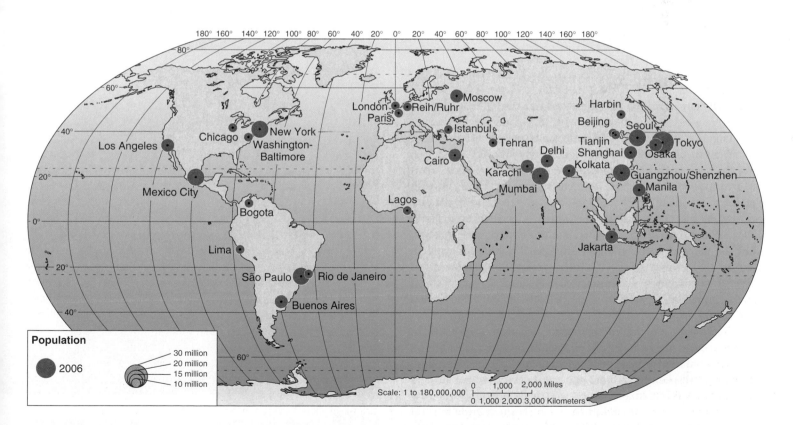

FIGURE 14.3 Some of the world's largest urban agglomerations in 2006. Sizes of many of these metropolitan areas are only educated guesses. As much as half the population of some cities in developing countries can be undocumented migrant workers or shanty town residents who are difficult to count. Most population growth in this century is expected to be in the megacities of the developing world.

at least 500,000 over the next 20 years. Already at least half of the concrete and one-third of the steel used in construction around the world each year is consumed in China.

Consider Shanghai, for example. In 1985, the city had a population of about 10 million. It's now about 19 million—including at least 4 million migrant laborers. In the past decade, Shanghai has built 4,000 skyscrapers (buildings with more than 25 floors). The city already has twice as many tall buildings as Manhattan, and proposals have been made for 1,000 more. The problem is that most of this growth has taken place in a swampy area called Pudong, across the Huang Pu River from the historic city center (see fig. 14.1). Pudong is now sinking about 1.5 cm per year due to groundwater drainage and the weight of so many buildings.

Other Chinese cities have plans for similar massive building projects to revitalize blighted urban areas. Harbin, a city of about 9 million people and the capital of Heilongjiang Province, for example, recently announced plans to relocate across the Songhua River on 740 km^2 (285 mi^2, or roughly the size of New York City) of former farmland. Residents hope these new towns will be both more livable for their residents and more ecologically sustainable than the old cities they're replacing. In 2005, the Chinese government signed a long-term contract with a British engineering firm to build at least five "eco-cities," each the size of a large Western capital. Plans calls for these cities to be self-sufficient in energy, water, and most food products, with the aim of zero emissions of greenhouse gases from transportation.

Immigration is driven by push and pull factors

People migrate to cities for many reasons. In China over the past 20 years—or in America during the twentieth century—mechanization eliminated jobs and drove people off the land. Where cropland is owned by a minority of wealthy landlords, as is the case in many developing countries, subsistence farmers are often ejected when new cash crops or cattle grazing become economically viable. Many people also move to the city because of the opportunities and independence offered there. Cities offer jobs, better housing, entertainment, and freedom from the constraints of village traditions. Possibilities exist in the city for upward social mobility, prestige, and power not ordinarily available in the country. Cities support specialization in arts, crafts, and professions for which markets don't exist elsewhere.

Government policies often favor urban over rural areas in ways that both push and pull people into cities. Developing countries commonly spend most of their budgets on improving urban areas (especially around the capital city, where leaders live). This gives the major cities a virtual monopoly on new jobs, housing, education, and finance, all of which bring in rural people searching for a better life. Lima, for example, has only 20 percent of Peru's population, but has 50 percent of the national wealth, 60 percent of the manufacturing, 65 percent of the retail trade, 73 percent of the industrial wages, and 90 percent of all banking in the country. Similar statistics pertain to many national capitals.

Congestion, pollution, and water shortages plague many cities

First-time visitors to a supercity—particularly in a developing country—often are overwhelmed by the immense crush of pedestrians and vehicles of all sorts jostling for space in the streets. The noise, congestion, and confusion of traffic make it seem suicidal to venture onto the street. Jakarta, Indonesia, for instance, is one of the most densely populated cities in the world (fig. 14.4). Traffic is chaotic almost all the time. People often spend three or four hours each way commuting to work from outlying areas.

Pollution from burgeoning traffic and from unregulated factories degrades air quality in many urban areas. China's spectacular economic growth has resulted in an flood of private automobiles mainly in cities. Beijing, for instance, has doubled the number of cars on its streets in just the past five years to 2.5 million. China is the world's second-largest producer of greenhouse gases, and the World Bank warns that it is home to 16 of the world's 20 cities with the worst air pollution. Chinese health authorities say that a third of the country's urban residents are exposed to harmful air pollution levels. They blame this pollution for more than 400,000 premature deaths each year.

FIGURE 14.4 Motorized rickshaws, motor scooters, bicycles, street vendors, and pedestrians all vie for space on the crowded streets of Jakarta. The heat, noise, smells, and sights are overpowering. In spite of the difficulties of living here, people work hard and have hope for the future.

Other megacities have similar problems. It's estimated that 60 percent of Kolkata's (Calcutta's) residents suffer from respiratory diseases. Mexico City, which sits in a high mountain bowl with abundant sunshine, little rain, and high traffic levels, has one of the world's highest levels of photochemical smog.

Few cities in developing countries can afford to build modern waste treatment systems for their rapidly growing populations. The World Bank estimates that only one-third of urban residents in developing countries have satisfactory sanitation services. In Latin America, only 2 percent of urban sewage receives any treatment. In Egypt, Cairo's sewer system was built about 50 years ago to serve a population of 2 million people. It is now being overwhelmed by more than five times that many residents. Less than 1 percent of India's 500,000 towns and villages have even partial sewage systems or water treatment facilities.

It's often difficult to find clean drinking water for urban areas. According to Qiu Baoxing, Chinese minister of construction, 70 percent of his country's surface water is so polluted by industrial toxins, human waste, and agricultural chemicals that it is unsuited for human consumption. One hundred of China's 660 cities face severe water shortages, he reported. Worldwide, according to the United Nations, at least 1.1 billion people don't have safe drinking water, and twice that many don't have adequate sanitation. It may not be so bad to have primitive sanitation if you live in a sparsely populated rural area, but imagine what it's like to live in a densely crowded megacity, such as Jakarta, with 10 million people, fewer than half of whom are served by a municipal sanitary system.

Many cities lack sufficient housing

The United Nations estimates that at least 1 billion people live in crowded, unsanitary slums of the central cities or in the vast shantytowns and squatter settlements that ring the outskirts of most major cities in the developing world. Around 100 million people have no home at all. In Mumbai (formerly Bombay), India, for example, it's estimated that half a million people sleep on the streets, sidewalks, and traffic circles because they can find no other place to live (fig. 14.5).

Slums are generally legal but inadequate multifamily tenements or rooming houses, often converted from some other use. Families live crowded in small rooms with inadequate ventilation and sanitation. Often these structures are rickety and unsafe. In 1999, for example, a 7.4 magnitude earthquake hit eastern Turkey, killing more than 14,000 people when shoddy, poorly built apartments collapsed.

Shantytowns, with shacks built of corrugated metal, discarded packing crates, brush, plastic sheets, and other scavenged materials grow on the outskirts of many cities in the developing world. They can house millions of people but generally lack clean water, sanitation, or safe electrical power. Shantytowns are usually illegal, but they quickly fill in the empty space in towns where squatters can build shelters close to jobs. With little or no public services, shantytowns often fill with trash and debris.

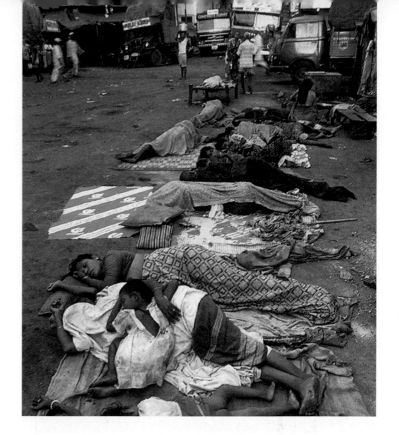

FIGURE 14.5 In Mumbai, India, as many as half a million people sleep on the streets because they have no other place to live. Ten times as many live in crowded, dangerous slums and shantytowns throughout the city.

Many governments try to clean out illegal settlements by torching or bulldozing the huts and sending riot police to drive out residents, but people either move back in or relocate to another shantytown. In 2005, the government of Zimbabwe destroyed the homes of some 700,000 people in shantytowns around the capital of Harare. Families were evicted in the middle of the night during the coldest weather of the year, often with only minutes to gather their belongings. President Robert Mugabe justified this blitzkreig as necessary to control crime, but critics claimed it was mainly to remove political opponents.

Two-thirds of the population of Kolkata are thought to live in unplanned squatter settlements, and nearly half the 25 million residents of Mexico City occupy the unauthorized *colonias* around the city. Often, shantytowns occupy the most polluted, dangerous parts of cities where no one else wants to live. In Bhopal, India, and Mexico City, for example, squatter settlements were built next to deadly industrial sites. In Brazil, shantytowns called *favelas* perch on steep hillsides unwanted for other building (fig. 14.6). As desperate and inhumane as conditions are in these slums and shantytowns, many people do more than merely survive there. They work hard, raise families, educate their children, and often improve their living standard little by little as they make some money.

Many countries are recognizing that the only way they can house all their citizens is to cooperate with shantytown dwellers. Recognizing land rights, providing financing for home improvements, and supporting community efforts to provide water, sewers, and power can greatly improve living conditions for many poor people.

FIGURE 14.6 Shantytowns called *favelas* perch on hillsides above Rio de Janeiro.

URBAN PLANNING

How can we live together in cities in ways that are environmentally sound, socially just, and economically sustainable? Starting with Greek cities thousands of years ago, planners have debated the best ways for us to organize ourselves.

Transportation is crucial in city development

Most of the world's major cities grew up around a port, a river crossing, a railroad hub, or some other focal point for transportation. Often, those original reasons for city placement no longer apply, and a location that made sense for a small, backwoods community no longer works for a major metropolitan center.

Getting people around within a large urban area has become one of the most difficult problems that many city officials face. A century ago, most American cities were organized around transportation corridors. First horse-drawn carriages, then electric streetcars provided a way for people to get to work, school, and shops. Everyone, rich or poor, wanted to live as close to the city center as possible, and within easy walking distance of the trolley or streetcar line. When Henry Ford introduced the first affordable, mass-produced automobile, it allowed people to build houses on larger lots in areas served only by streets. Freeway construction, which began in America in the 1950s, allowed people to move even further out into the country. Cities that were once compact began to spread over the landscape, consuming space and wasting

resources. This pattern of development is known as **sprawl.** While there is no universally accepted definition of the term, sprawl generally includes the characteristics outlined in table 14.3.

In most American metropolitan areas, the bulk of new housing is in large, tract developments that leapfrog out beyond the city edge in a search for inexpensive rural land with few restrictions on land use or building practices (fig. 14.7). The U.S. Department of Housing and Urban Development estimates that urban sprawl consumes some 200,000 ha (roughly 500,000 acres) of farmland every

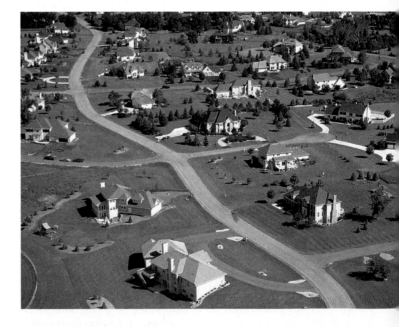

FIGURE 14.7 Huge houses on sprawling lots consume land, alienate us from our neighbors, and make us ever more dependent on automobiles.
© 2003 Regents of the University of Minnesota. All rights reserved. Used with permission of the Design Center for American Urban Landscape.

year. Although the price of tract homes often is less than comparable urban property, there are external costs in the form of new roads, sewers, water mains, power lines, schools, shopping centers, and other infrastructure required by this low-density development. Ironically, people who move to the country to escape from urban problems such as congestion, crime, and pollution often find that they have simply brought those problems with them.

Because many Americans live far from where they work, shop, or recreate, they consider it essential to own a private automobile. The average U.S. driver spends 443 hours per year behind a steering wheel, or the equivalent of one full 8-hour day per week in an automobile. The freeway system was designed to allow drivers to travel at high speeds from source to destination without ever having to stop. As more and more vehicles clog highways, however, the reality is far different. In Los Angeles, for example, which has the worst congestion in the United States, the average speed in 1982 was 58 mph (93 kph), and the average driver spent less than 4 hours per year in traffic jams. In 2004, the average speed was only 35.6 mph (57.3 kph), and the typical driver spent 97 hours in bumper-to-bumper traffic.

Altogether, it's estimated that traffic congestion costs the United States $78 billion per year in wasted time and fuel. Some people argue that the existence of traffic jams in cities shows that more highways are needed. Often, however, building more traffic lanes simply encourages more people to drive farther and put more cars on the road. Meanwhile, about one-third of Americans are too young, too old, or too poor to drive. For these people, car-oriented development causes isolation and makes daily tasks like grocery shopping difficult. Parents spend long hours transporting young children. Teenagers and aging grandparents are forced to drive, often presenting a hazard on public roads.

As the opening Case Study for this chapter shows, Curitiba provides an excellent example of successful mass transit and environmentally sound, socially just, and economically sustainable urban planning. Curitiba's bus rapid transit system focuses around five transportation corridors radiating from the city center (fig. 14.8). Within these corridors, high-speed, articulated buses, each of which can carry 270 passengers, travel on dedicated roadways closed to all other vehicles. These bus-trains make limited stops at transfer terminals that link to 340 feeder routes extending throughout the city. Everyone in the city is within walking distance of a bus stop that has frequent, convenient, affordable service.

Each bus station along the dedicated busway is made up of tubular structures elevated to be at the same height as the bus floor, providing access for disabled persons and making entry and exit quicker and easier for everyone (fig. 14.9). Passengers pay as they go through a turnstile to enter the terminal. When a bus pulls up, it opens multiple doors through which passengers can enter or exit. Dozens of passengers might disembark and an equal number get on the bus during the 60-second stop, making a trip over many kilometers remarkably fast. The system charges a single fee for a trip in one direction, regardless of the number of transfers involved. This makes the system equitable for those who can't afford to live in expensive neighborhoods close to the city center.

Express route
Interdistrict
Rapid buses
Feeder lines
Commuter routes

FIGURE 14.8 Curitiba's transit system has more than 340 interlinked bus routes. Neighborhood feeder lines connect to high-volume, express buses running on dedicated busways. Interdistrict routes carry passengers between suburbs, while specialized commuter routes carry workers directly from distant neighborhoods to the city center. Altogether this system moves more than 1.9 million people per day, or more than 75 percent of all personal transportation in the city.

FIGURE 14.9 High-speed, bi-articulated buses travel on dedicated transit ways in Curitiba and make limited stops at elevated tubular bus terminals that allow many passengers to disembark and reload during 60-second stops.

Curitiba's buses make more than 21,000 trips per day, traveling more than 440,000 km (275,000 mi) and carrying 1.9 million passengers, or about three-quarters of all personal trips within the city. One of the best things about this system is its economy. Working with existing roadways for the most part, the city was able to construct this system for one-tenth the cost of a light rail system or freeway system, and one-hundredth the cost of a subway. The success of bus rapid transit has allowed Curitiba to remain relatively compact and to avoid the sprawl engendered by an American-style freeway system.

Traditionally, most ecologists have studied pristine, natural ecosystems, trying to understand how nature works without human disturbance. In recent years, however, we have come to recognize that cities are ecological systems, too. If ecology is the relationship among organisms and their environment, what could be more ecological than studying the organism (humans) with the greatest impact on other species and our environment? As the human population becomes increasingly urbanized, perhaps the most important place to study human ecology is in cities. Rather than merely seeing humans as disturbing factors, some scientists are beginning to gather data on how urban ecosystems work.

Urban areas share many characteristics with natural ecosystems. Energy flows into and out of the city, and is degraded and dispersed as it is used to do work. Materials are used to create structures, some living and some nonliving. Species compete for shelter, food, habitat, and other resources. The most successful species flourish and proliferate; less fit ones dwindle and eventually vanish. Every day, a typical American city of a million people consumes roughly 1 million metric tons (roughly 2 billion lbs) of raw materials. This urban material flow includes 500,000 m^3 (1.3 million gal) of water, the energy equivalent of about 20,000 metric tons of fossil fuels, 13,000 metric tons of other minerals (including food packaging and construction materials), 12,000 metric tons of farm products, and 10,000 metric tons of wood and paper products.

To remove municipal waste from this city of a million residents, about 1,000 fully loaded garbage trucks travel every day to rural areas carrying the discarded remnants of this urban metabolism. In addition, the city's municipal wastewater treatment plants discharge somewhat more than 500,000 m^3 of water (that lost to evaporation is offset by precipitation and solids added to the wastewater stream). You may remember from chapter 2 that most natural ecosystems recycle materials. Recycling and reuse also occur in urban ecology, although rarely with the speed and efficiency of natural systems.

Urban biodiversity and wildlife also can be important. While many biologists have tended to assume that cities are devoid of life other than humans and our domesticated companion animals and plants, we are now beginning to appreciate that urban open space can be essential refuges for wild creatures as well. As cities grow, it's not uncommon for consolidated urban areas to be more than 100 miles (160 km) across. This would present an insurmountable barrier to many migrating animals if the city consists of nothing but concrete, glass, and metal. Urban parks, forests, corridors along rivers, and even ordinary yards, if managed properly, can provide a refuge for migrating wildlife and can sustain local populations that contribute significantly to overall biodiversity. Planting native vegetation and designing landscapes to provide maximum food and shelter are important steps in protecting and promoting urban biodiversity.

Long-term ecological research (LTER) sites are ecologically significant sites identified and funded by the National Science Foundation to allow long-term, in-depth research on ecological problems and questions. Although most of the LTER sites are in remote locations, where nature has minimal human interference, two LTER sites were set up in Phoenix and Baltimore. Funded by the National Science Foundation for an initial period of six years at nearly a million dollars per year, but with an understanding that the research will go on much beyond that, these projects use modern technology to study every aspect of urban ecology. How does the city shape its own weather? What plants and animals live in the city, and where, and how? What ecological processes cycle materials and energy? How do human activities influence those processes?

One of the first things that scientists are discovering is that cities are not homogenous; like other ecosystems, they have shifting patches of contrasting conditions. Using computer models and sophisticated mapping techniques, ecologists can explore how patches change in space and time. Elsewhere, environmental scientists have investigated important problems of pollution and environmental health in cities. Where are toxic and hazardous materials generated, stored, and released in the city? How do they move around and where do they accumulate?

In Detroit, a group of students worked with experts to map data on more than 5,000 children with elevated blood lead levels. Not surprisingly, they found a correlation between low incomes, old housing, incidence of poisoning, and concentration of special-education students. Sometimes public awareness of a problem is one of the best outcomes of this research. Another large group of students in the Detroit area used geographic information systems (digital maps) together with chemical and biological analysis to prepare a detailed study of water quality and ecosystem health in the Rouge River. Perhaps students at your school could organize a similar study of the urban ecology in your area.

We can make our cities more livable

Are there alternatives to unplanned sprawl and wasteful resource use? One option proposed by many urban planners is **smart growth,** which makes effective use of land resources and existing infrastructure by encouraging in-fill development that avoids costly duplication of services and inefficient land use (table 14.4). Smart growth aims to provide a mix of land uses to create a variety of affordable housing choices and opportunities. It also attempts to provide a variety of transportation choices, including pedestrian-friendly neighborhoods. This approach to planning also seeks to maintain a unique sense of place by respecting local cultural and natural features.

By making land-use planning open and democratic, smart growth makes urban expansion fair, predictable, and cost-effective. All stakeholders are encouraged to participate in creating a vision for the city and to collaborate with rather than confront each other. Goals are established for staged and managed growth in urban transition areas with compact development patterns. This approach is not opposed to growth. It recognizes that the goal is not to block growth but to channel it to areas where it can be sustained over the long term. Smart growth strives to enhance

TABLE 14.4

Goals for Smart Growth

1. Create a positive self-image for the community.
2. Make the downtown vital and livable.
3. Alleviate substandard housing.
4. Solve problems with air, water, toxic waste, and noise pollution.
5. Improve communication between groups.
6. Improve community member access to the arts.

Source: Data from Vision 2000, Chattanooga, Tennessee.

access to equitable public and private resources for everyone and to promote the safety, livability, and revitalization of existing urban and rural communities.

Smart growth protects environmental quality. It tries to reduce traffic and to conserve farmlands, wetlands, and open space. As cities grow and transportation and communications enable more community interaction, the need for regional planning becomes greater and more pressing. Community and business leaders must make decisions based on a clear understanding of regional growth needs and how infrastructure can be built most efficiently and for the greatest good.

One of the best examples of successful urban land-use planning in the United States is Portland, Oregon, which has rigorously enforced a boundary on its outward expansion, requiring instead that development be focused on in-filling unused space within the city limits. Because of its many urban amenities, Portland is considered one of the best cities in America. Between 1970 and 1990 the Portland population grew by 50 percent, but its total land area grew only 2 percent. During this time, Portland property taxes decreased 29 percent and vehicle miles traveled increased only 2 percent. By contrast, Atlanta, which had similar population growth, experienced an explosion of urban sprawl that increased its land area three-fold, drove up property taxes 22 percent, and increased traffic miles by 17 percent. A result of this expanding traffic and increasing congestion was that Atlanta's air pollution increased by 5 percent, while Portland, which has one of the best public transit systems in the nation, saw a decrease of 86 percent.

New urbanism incorporates smart growth

Rather than abandon the cultural history and infrastructure investment in existing cities, a group of architects and urban planners is attempting to redesign metropolitan areas to make them more appealing, efficient, and livable. European cities such as Stockholm, Sweden; Helsinki, Finland; Leichester, England; and Neerlands, the Netherlands, have a long history of innovative urban planning. In the United States, Andres Duany, Elizabeth Plater-Zyberk, Peter Calthorpe, and Sym Van Der Ryn have been leaders in this movement. Using what is sometimes called a neo-traditionalist approach,

these designers attempt to recapture some of the best features of small towns and livable cities of the past. They are designing urban neighborhoods that integrate houses, offices, shops, and civic buildings. Ideally, no house should be more than a five-minute walk from a neighborhood center with a convenience store, a coffee shop, a bus stop, and other amenities. A mix of apartments, townhouses, and detached houses in a variety of price ranges ensures that neighborhoods will include a diversity of ages and income levels. Some design principles of this movement include

- Limit city size or organize cities in modules of 30,000 to 50,000 people—large enough to be a complete city but small enough to be a community.

- Maintain greenbelts in and around cities. These provide recreational space and promote efficient land use, as well as help ameliorate air and water pollution.

- Determine in advance where development will take place. This protects property values and prevents chaotic development. Planning can also protect historical sites, agricultural resources, and ecological services of wetlands, clean rivers, and groundwater replenishment.

- Locate everyday shopping and services so people can meet daily needs with greater convenience, less stress, less automobile dependency, and less use of time and energy (fig. 14.10). This might be accomplished by encouraging small-scale commercial development in or close to residential areas.

- Encourage walking or the use of small, low-speed, energy-efficient vehicles (microcars, motorized tricycles, bicycles,

FIGURE 14.10 This walking street in Queenstown, New Zealand, provides opportunities for shopping, dining, and socializing in a pleasant outdoor setting.

FIGURE 14.11 Many cities have large amounts of unused open space that could be used to grow food. Residents often need help decontaminating soil and gaining access to the land.

etc.) for many local trips now performed in full-size automobiles. Creating special traffic lanes, reducing the number or size of parking spaces, and closing shopping streets to big cars might encourage such alternatives.

• Promote more diverse, flexible housing as an alternative to conventional detached, single-family houses. In-fill building between existing houses saves energy, reduces land costs, and might help provide a variety of living arrangements. Allowing single-parent families or groups of unrelated adults to share housing and to use facilities cooperatively also provides alternatives to those not living in a traditional nuclear family.

• Make cities more self-sustainable by growing food locally, recycling wastes and water, using renewable energy sources, reducing noise and pollution, and creating a cleaner, safer environment. Encourage community gardening (fig. 14.11). Reclaimed inner-city space or a greenbelt of agricultural and forestland around the city provides food and open space, and also contributes valuable ecological services, such as purifying air, supplying clean water, and protecting wildlife habitat and recreation land.

• Equip buildings with "green roofs" or rooftop gardens that improve air quality, conserve energy, reduce stormwater runoff, reduce noise, and help reduce urban heat island effects. Intensive gardens can include large trees, shrubs, flowers and may require regular maintenance (fig. 14.12). Extensive gardens require less soil, add less weight to the building, and usually have simple plantings of prairie plants or drought-resistant species, such as sedum, that require minimum care. They can last twice as long as conventional roofs. In Europe more than 1 million m^2 of green roofs are installed every year. Urban roofs arc also a good place for solar collectors or wind turbines.

• Plan cluster housing, or open-space zoning, which preserves at least half of a subdivision as natural areas, farmland, or other forms of open space. Studies have shown that people who move to the country don't necessarily want to live miles from the nearest neighbor; what most desire is long views across an interesting landscape and an opportunity to see wildlife. By carefully clustering houses on smaller lots, a conservation subdivision can provide the same number of buildable lots as a conventional subdivision and still preserve 50 to 70 percent of the land as open space (fig. 14.13). This not only reduces development costs (less distance to

FIGURE 14.12 This award-winning green roof on the Chicago City Hall is functional as well as beautiful. It reduces rain runoff by about 50 percent, and keeps the surface as much as 30°F cooler than a conventional roof on hot summer days.

FIGURE 14.13 This conservation development clusters houses on one-third of its property, and the rest is preserved as native prairie and oak woodland. Being close together, neighbors develop a sense of community, yet everyone has expansive views and access to open space.

What Do You Think?

The Architecture of Hope

How sustainable and self-sufficient can urban areas be? An exciting experiment in minimal impact in London gives us an image of what our future may be. BedZED, short for the Beddington Zero Energy Development is an integrated urban project built on the grounds of an old sewage plant in South London. BedZED's green strategies begin with recycling the ground on which it stands. Designed by architect Bill Dunster and his colleagues, the complex demonstrates dozens of energy-saving and water-saving ideas. BedZED has been occupied, and winning awards, since it was completed in 2003.

Most of BedZED's innovations involve the clever combination of simple, even conventional ideas. Expansive, south-facing, triple-glazed windows provide abundant light, minimize the use of electric lamps, and provide passive solar heat in the winter. Thick, superinsulated walls keep interiors warm in winter and cool in summer. Rotating "wind cowls" on roofs turn to catch fresh breezes, which cool spaces in summer. In winter, heat exchangers use the heat of stale, outgoing air to warm fresh, incoming air. Energy used in space heating is nearly eliminated. Building materials are recycled, reclaimed, or renewable, which reduces the "embodied energy" invested in producing and transporting them.

BedZED does use energy, but the complex generates its own heat and electricity with a small, on-site, superefficient plant that uses local

South-facing windows heat homes, and colorful, rotating "wind cowls" ventilate rooms at BedZED, an ecological housing complex in South London, U.K.

tree trimmings for fuel. Thus BedZED uses no *fossil* fuels, and it is "carbon-neutral" because the carbon dioxide released by burning wood was recently captured from the air by trees. In addition, photovoltaic cells on roofs provide enough free energy to power 40 solar cars. Fuel bills for BedZED residents can be as little as 10 percent of what other Londoners pay for similar-sized homes.

Water-efficient appliances and toilets reduce water use. Rainwater collection systems provide "green water" for watering gardens, flushing toilets, and other nonconsumptive uses. Reed-bed filtration systems purify used water without chemicals. Water meters allow residents to see how much water they use. Just knowing about consumption rates helps encourage conservation. Residents use about half as much water per person as other Londoners.

BedZED residents can save money and time by not using, or even owning a car. Office space is available on-site, so some residents can work where they live, and the commuter rail station is just a ten-minute walk away. The site is also linked to bicycle trails that facilitate bicycle commuting. Car pools and rent-by-the-hour auto memberships allow many residents to avoid owning (and parking) a vehicle altogether.

Building interiors are flooded with natural light, ceilings are high, and most residences have rooftop gardens. Community events and common spaces encourage humane, healthy lifestyles and community ties. Child-care services, shops, entertainment, and sports facilities are built into the project. The approximately 100 housing units are designed for a range of income levels ensuring a racially, ethnically, and age-diverse community. Prices are lower than many similar-sized London homes, and few in this price range or inner-city location have abundant sunlight or gardens.

Similar projects are being built across Europe and even in some developing countries, such as China. Architect Dunster says that BedZED-like developments on cleaned-up brownfields could provide all the 3 million homes that the U.K. expects to need in the next decade with no sacrifice of open space. And as green building techniques, designs, and materials become standard, he argues, they will cost no more than conventional, energy-wasting structures.

What do you think? Would you enjoy living in a dense, urban setting such as BedZED? Would it involve a lower standard of living than you now have? How much would it be worth to avoid spending 8 to 10 hours per week not fighting bumper-to-bumper traffic while commuting to school or work? The average cost of owning and driving a car in the United States is about $9,000 per year. What might you do with that money if owning a vehicle were unnecessary? Try to imagine what urban life might be like if most private automobiles were to vanish. If you live in a typical American city, how much time do you have to enjoy the open space that the suburbs and freeways were supposed to provide? Perhaps, most importantly, how will we provide enough water, energy, and space for the 3 billion people expected to crowd into cities worldwide over the next few decades if we don't adopt some of the sustainable practices and approaches represented by BedZED?

build roads, lay telephone lines, sewers, power cables, etc.) but also helps to foster a greater sense of community among new residents.

- Preserve urban habitat. It can make a significant contribution toward saving biodiversity as well as improving mental health and giving us access to nature.

These planning principles aren't just a matter of esthetics. Dr. Richard Jackson, former director of the National Center for Environmental Health in Atlanta, points out a strong association between urban design and our mental and physical health. As our cities have become ever more spread out and impersonal, we have fewer opportunities for healthful exercise and socializing. Chronic diseases, such as cardiovascular diseases, asthma, diabetes, obesity, and depression, are becoming the predominant health concerns in the United States.

"Despite common knowledge that exercise is healthful," Dr. Jackson says, "fewer than 40% of adults are regularly active, and 25% do no physical activity at all. The way we design our communities makes us increasingly dependent on automobiles for the shortest trip, and recreation has become not physical but observational." Long commutes and a lack of reliable mass transit and walkable neighborhoods mean that we spend more and more time in stressful road congestion. "Road rage" isn't imaginary. Every commuter can describe unpleasant encounters with rude drivers. Urban design that offers the benefits of more walking, more social contact, and surroundings that include water and vegetation can provide healthful physical exercise and psychic respite.

ECONOMICS AND SUSTAINABLE DEVELOPMENT

Like many of our environmental issues, improving urban conditions will ultimately be decided by economics and policy decisions. We'll discuss policy in chapter 15. In the next half of this chapter, we'll review some of the principles of environmental economics.

Can development be sustainable?

By now it is clear that security and living standards for the world's poorest people are inextricably linked to environmental protection. One of the most important questions in environmental science is how we can continue improvements in human welfare within the limits of the earth's natural resources. *Development* means improving people's lives. *Sustainability* means living on the earth's renewable resources without damaging the ecological processes that support us all. **Sustainable development** is an effort to marry these two ideas. A popular definition describes this goal as "meeting the needs of the present without compromising the ability of future generations to meet their own needs."

But is this possible? As you've learned elsewhere in this book, many people argue that our present population and economic levels are exhausting the world's resources. There's no way, they insist, that more people can live at a higher standard without irreversibly degrading our environment. Others claim that there's enough for everyone if we just share equitably and live modestly. Let's look a little deeper into this important debate.

Our definitions of resources shape how we use them

To understand the problems and promise of sustainability, you need to understand the different kinds of resources we use. The way we treat resources depends largely on how we view and define them. **Classical economics,** developed in the 1700s by philosophers such as Adam Smith (1723–1790) and Thomas Malthus (1766–1834), assumes that natural resources are finite—that resources such as iron, gold, water, and land exist in fixed amounts. According to this view, as populations grow, scarcity of these resources reduces quality of life, increases competition, and ultimately causes populations to fall again. In a free market, where fully informed buyers and sellers make free, independent decisions to buy and sell, the price of a pound of butter depends

FIGURE 14.14 Classic supply/demand curves. When price is low, supply is low and demand is high. As prices rise, supply increases but demand falls. Market equilibrium is the price at which supply and demand are equal.

on the supply of available butter (it is cheap when plenty is available) and the demand for butter (buyers pay more when they must compete to get the resource, fig. 14.14).

This price mechanism has been modified by the idea of **marginal costs.** Buyers and sellers evaluate the added return for buying (or making and selling) slightly more of a product, or the marginal cost of slightly greater production or purchase. If the return is greater than the marginal cost, then a sale is made.

The nineteenth-century economist John Stuart Mill assumed that most resources are finite, but he developed the idea of a **steady-state economy.** Rather than boom-and-bust cycles of population and resource use, as envisioned by Malthus, Mill proposed that economies can achieve an equilibrium of resource use and production. Intellectual and moral development continues, he argued, once this stable, secure state is achieved.

Neoclassical economics, developed in the nineteenth century, expanded the idea of resources to include labor, knowledge, and capital. Labor and knowledge are resources because they are necessary to create goods and services; they are not finite because every new person can add more labor and energy to an economy. **Capital** is any form of wealth that contributes to the production of more wealth. Money can be invested to produce more money. Mineral resources can be developed and manufactured into goods that return more money. Economists distinguish several kinds of capital:

1. Natural capital: goods and services provided by nature
2. Human capital: knowledge, experience, human enterprise
3. Manufactured (built) capital: tools, buildings, roads, technology

To this list some social theorists would add social capital, the shared values, trust, cooperation, and organization that can develop in a group of people but cannot exist in one individual alone.

Because the point of capital is the production of more capital (that is, wealth), neoclassical economics emphasizes the idea of growth. Growth results from the flow of resources, goods, and

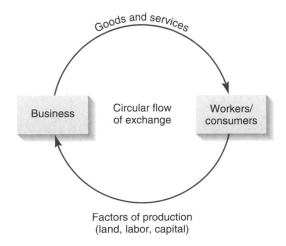

FIGURE 14.15 The neoclassical model of the economy focuses on the flow of goods, services, and factors of production (land, labor, capital) between business and individual workers/consumers. The social and environmental consequences of these relationships are irrelevant in this view.

TABLE 14.5

Important Ecological Services

We depend on our environment to continually provide

1. A regulated global energy balance and climate; chemical composition of the atmosphere and oceans; water catchment and groundwater recharge; production, and recycling of organic and inorganic materials; maintenance of biological diversity.
2. Space and suitable substrates for human habitation, crop cultivation, energy conversion, recreation, and nature protection.
3. Oxygen, fresh water, food, medicine, fuel, fodder, fertilizer, building materials, and industrial inputs.
4. Aesthetic, spiritual, historic, cultural, artistic, scientific, and educational opportunities and information.

Source: Data from R. S. de Groot, *Investing in Natural Capital*, 1994.

services (fig. 14.15). Continued growth is always necessary for continued prosperity, according to this view. Natural resources contribute to production and growth, but they are not critical supplies that limit growth. They are not limiting because resources are considered to be interchangeable and substitutable. As one resource becomes scarce, neoclassical economics predicts that a substitute will be found.

Since production of wealth is central to neoclassical economics, an important measure of growth and wealth is consumption. If a society consumes more oil, more minerals, and more food, it is presumably becoming wealthier. This idea has extended to the idea of *throughput,* the amount of resources a society uses and discards. More throughput is a measure of greater consumption and greater wealth, according to this view. Throughput is commonly measured in terms of **gross national product (GNP),** the sum of all products and services bought and sold in an economy. Because GNP includes activities of offshore companies, economists sometimes prefer **gross domestic product (GDP),** which more accurately reflects the local economy by accounting for only those goods and services bought and sold locally.

Natural resource economics extends the neoclassical viewpoint to treat natural resources as important waste sinks (absorbers), as well as sources of raw materials. Natural capital (resources) is considered more abundant, and therefore cheaper, than built or human-made capital.

Ecological economics incorporates principles of ecology

Ecological economics, applies ecological ideas of system functions and recycling to the definition of resources. This school of thought also recognizes efficiency in nature, and it acknowledges the importance of ecosystem functions for the continuation of human economies and cultures. In nature, one species' waste is another's

food, so that nothing is wasted. We need an economy that recycles materials and uses energy efficiently, much as a biological community does. Ecological economics also treats the natural environment as part of our economy, so that natural capital becomes a key consideration in economic calculations. Ecological functions, such as absorbing and purifying wastewater, processing air pollution, providing clean water, carrying out photosynthesis, and creating soil, are known as **ecological services** (table 14.5). These services are free: we don't pay for them directly (although we often pay indirectly when we suffer from their absence). Therefore, they are often excluded from conventional economic accounting, a situation that ecological economists attempt to rectify (fig. 14.16).

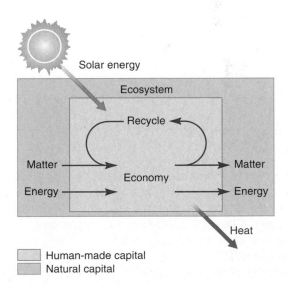

FIGURE 14.16 An ecological economics view considers natural capital and recycling integral to the economy. Human-made capital is created using limited supplies of natural capital.
Source: Data from Herman Daly in A. M. Jansson et al., *Investing in Natural Capital*, ISEE.

Many ecological economists also promote the idea of a steady-state economy. As with John Stuart Mill's original conception of steady states, these economists argue that economic health can be maintained without constantly growing consumption and throughput. Instead, efficiency and recycling of resources can allow steady prosperity where there is little or no population growth. Low birth rates and death rates (like *K*-adapted species, see chapter 3), political and social stability, and reliance on renewable energy would characterize such a steady-state economy. Like Mill, these economists argue that human and social capital—knowledge, happiness, art, life expectancies, and cooperation—can continue to grow even without constant expansion of resource use.

Both ecological economics and neoclassical economics distinguish between renewable and nonrenewable resources. **Non-renewable resources** exist in finite amounts: minerals, fossil fuels, and also groundwater that recharges extremely slowly are all fixed, at least on a human timescale. **Renewable resources** are naturally replenished and recycled at a fairly steady rate. Fresh water, living organisms, air, and food resources are all renewable (fig. 14.17).

These categories are important, but they are not as deterministic as you might think. Nonrenewable resources, such as iron and gold, can be extended through more efficient use: cars now use less steel than they once did, and gold is mixed in alloys to extend its use. Substitution also reduces demand for these resources: car parts once made of iron are now made of plastic and ceramics; copper wire, once stockpiled to provide phone lines, is now being replaced with cheap, lightweight fiber-optic cables made from silica (sand). Recycling also extends supplies of nonrenewable resources. Aluminum, platinum, gold, silver, and many other valuable metals are routinely recycled now, further reducing the demand for extracting new sources. The only

limit to recycling is usually the relative costs of extracting new resources compared with collecting used materials. Recoverable sources of nonrenewable resources are also expanded by technological improvements. New methods make it possible to mine very dilute metal ores, for example. Gold ore of extremely low concentrations is now economically recoverable—that is, you can make money on it—even though the price of gold has fallen because of greater efficiency, more discoveries, and resource substitution. Scarcity of resources, seen by classical economists as the trigger for conflict and suffering, can actually provide the impetus for much of the innovation that leads to substitution, recycling, and efficiency.

On the other hand, renewable resources can become exhausted if they are managed badly. This is especially apparent in biological resources, such as the passenger pigeon, American bison, and Atlantic cod. All these species once existed in extraordinary numbers, but within a few years, each was brought to the brink of extinction (or eliminated entirely) by overharvesting.

Scarcity can lead to innovation

Are we about to run out of essential natural resources? It stands to reason that, if we consume a fixed supply of nonrenewable resources at a constant rate, we'll eventually use up all the economically recoverable reserves. There are many warnings in the environmental literature that our extravagant depletion of nonrenewable resources sooner or later will result in catastrophe, misery, and social decay. Models for exploitation rates of nonrenewable resources—called Hubbert curves after Stanley Hubbert, who developed them—often closely match historic experience for natural resource depletion (fig. 14.18).

Many economists, however, contend that human ingenuity and enterprise often allow us to respond to scarcity in ways that postpone or alleviate dire effects of resource use. The question

FIGURE 14.17 Biological resources are renewable in that they replace themselves by reproduction, but if overused or misused, populations die. If a species is lost, it cannot be re-created. It is permanently lost as a component of its ecosystem and as a resource to humans.

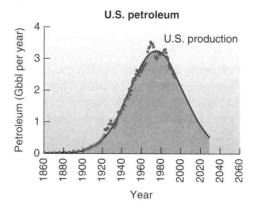

FIGURE 14.18 U.S. petroleum production. Dots indicate actual production. The bell-shaped curve is a theoretical Hubbert curve for a nonrenewable resource. The shaded area under the curve, representing 220 Gbbl (Gbbl = Gigabarrels or billions of standard 42-gal barrels), is an estimate of the total economically recoverable resource.

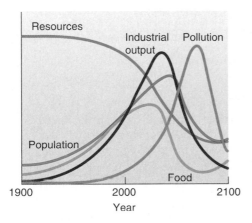

FIGURE 14.19 A run of one of the world models in *Limits to Growth*. This model assumes business-as-usual for as long as possible until Malthusian limits cause industrial society to crash. Notice that pollution continues to increase well after industrial output, food supplies, and population have all plummeted.

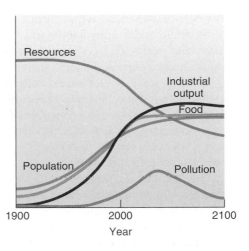

FIGURE 14.20 A run of the world model from *Beyond the Limits*. This model assumes that population and consumption are curbed, new technologies are introduced, and sustainable environmental policies are embraced immediately, rather than after resources are exhausted.

of whether this view is right has to do with the important theme of **limits to growth.**

In the early 1970s an influential study of resource limitations was funded by the Club of Rome, an organization of wealthy business owners and influential politicians. The study was carried out by a team of scientists, from the Massachusetts Institute of Technology, headed by Donnela Meadows. The results of this study were published in the 1972 book *Limits to Growth*. A complex computer model of the world economy was used to examine various scenarios of different resource depletion rates, growing population, pollution, and industrial output. Given the Malthusian assumptions built into this model, catastrophic social and environmental collapse seemed inescapable.

Figure 14.19 shows one example of the world model. Food supplies and industrial output rise as population grows and resources are consumed. Once past the carrying capacity of the environment, however, a crash occurs as population, food production, and industrial output all decline precipitously. Pollution continues to grow as society decays and people die, but eventually it also falls. Notice the similarity between this set of curves and the "boom-and-bust" population cycles described in chapter 3.

Many economists criticized this model because it underestimated technological development and factors that might mitigate the effects of scarcity. In 1992 the Meadows group published updated computer models in *Beyond the Limits* that include technological progress, pollution abatement, population stabilization, and new public policies that work for a sustainable future. If we adopt these changes sooner rather than later, the models show an outcome like that in figure 14.20, in which all factors stabilize sometime in this century at an improved standard of living for everyone. Of course, none of these computer models shows what will happen, only what some possible outcomes might be, depending on the choices we make.

Communal property resources are a classic problem in economics

One of the difficulties of economics and resource management is that there are many resources we all share but nobody owns. Clean air, fish in the ocean, clean water, wildlife, and open space are all natural amenities that we exploit but that nobody clearly controls.

In 1968 biologist Garret Hardin wrote **"The Tragedy of the Commons,"** an article describing how commonly held resources are degraded and destroyed by self-interest. Using the metaphor of the "commons," or community pastures in colonial New England villages, Hardin theorized that it behooves each villager to put more cows on the pasture. Each cow brings more wealth to the individual farmer, but the costs of overgrazing are shared by the entire community. The individual farmer, then, suffers only part of the cost, but gets to keep all the profits from the extra cows s/he puts on the pasture. Consequently, the commons becomes overgrazed, exhausted, and depleted. This dilemma is also known as the "free rider problem." The best solution, Hardin argued, is to give coercive power to the government or to privatize resources so that a single owner controls resource use.

This metaphor has been applied to many resources, especially to human population growth. It benefits every poor villager to produce a few more children, but collectively these children consume all the resources available, making us all poorer in the end. The same argument has been applied to many resource overuse problems, such as depletion of ocean fisheries, pollution, African famines, and urban crime.

Recent critics have pointed out that what Hardin was really describing was not a commons, or collectively owned and managed resource, but an **open access system,** in which there are no rules to manage resource use. In fact, many common resources have been managed successfully for centuries by cooperative agreements among users. Native American management of wild rice beds, Swiss village-owned mountain forests and pastures, Maine lobster fisheries, and communal irrigation systems in Spain, Bali, Laos, and many other countries have all remained viable for centuries under communal management.

Each of these "commons," or **communal resource management systems,** shares a number of features: (1) community members have lived on the land or used the resource for a long time and anticipate that their children and grandchildren will as well, thus giving them a strong interest in sustaining the resource and maintaining bonds with their neighbors; (2) the resource has clearly defined boundaries; (3) the community group size is known and enforced; (4) the resource is relatively scarce and highly variable, so that the community is forced to be interdependent; (5) the management strategies appropriate for local conditions have evolved over time and are collectively enforced; that is, those affected by the rules have a say in them; (6) the resource and its use are actively monitored, discouraging anyone from cheating or taking too much; (7) conflict resolution mechanisms reduce discord; and (8) incentives encourage compliance with rules, while sanctions for noncompliance keep community members in line.

Rather than being the only workable solution to problems in common pool resources, privatization and increasing external controls often prove to be disastrous. Where small villages have owned and operated local jointly held forests and fishing grounds for generations, nationalization and commodification of resources generally have led to rapid destruction of both society and ecosystems. Where communal systems once enforced restraint over harvesting, privatization encouraged narrow self-interest and allowed outsiders to take advantage of the weakest members of the community.

NATURAL RESOURCE ACCOUNTING

Decision making about sustainable resource use often entails **cost-benefit analysis (CBA),** the process of accounting and comparing the costs of a project and its benefits. Ideally, this process assigns values to social and environmental effects of a given undertaking, as well as the value of the resources consumed or produced. However, the results of CBA often depend on how resources are accounted for and measured in the first place. CBA is one of the main conceptual frameworks of resource economics, and it is used by decision makers around the world as a way of justifying the building of dams, roads, and airports, as well as in considering what to do about biodiversity loss, air pollution, and global climate change. CBA is a useful way of rational decision making about these projects. It is also widely disputed because it tends to discount the value of natural resources, ecological services, and human communities, and it is used to justify projects that jeopardize all these resources.

In CBA the monetary value of all benefits of a project are counted up and compared with the monetary costs of the project. Usually, the direct expenses of a project are easy to ascertain: how much will you have to pay for land, materials, and labor?

The monetary worth of lost opportunities—to swim or fish in a river or to see birds in a forest—on the other hand, is much harder to appraise, as are inherent values of the existence of wild species or wild rivers. What is a bug or a bird worth, for instance, or the opportunity for solitude or inspiration? Eventually, the decision maker compares all the costs and benefits to see whether the project is justified or whether an alternative action might bring greater benefit at less cost.

Critics of CBA point out its absence of standards, inadequate attention to alternatives, and the placing of monetary values on intangible and diffuse or distant costs and benefits. Who judges how costs and benefits will be estimated? How can we compare things as different as the economic gain from cheap power with loss of biodiversity or the beauty of a free-flowing river? Critics claim that placing monetary values on everything could lead to a belief that only money and profits count and that any behavior is acceptable as long as you can pay for it. Sometimes speculative or even hypothetical results are given specific numerical values in CBA and then treated as if they were hard facts.

Figure 14.21 shows an example of a cost-benefit analysis for reducing particulate air pollution (soot) in Poland. As you can see, removing the highest 40 percent of particulates is highly cost effective. Approaching 70 percent particulate removal, however, the costs may exceed benefits. Data such as these can be useful to decision makers. On the other hand, this same study showed that controlling sulfur emissions had high costs and negligible benefits. Might this conclusion result from the ways "benefits" were evaluated?

Values such as wildlife, nonhuman ecological systems, and ecological services can be incorporated with natural resource accounting. In theory this accounting contributes to sustainable

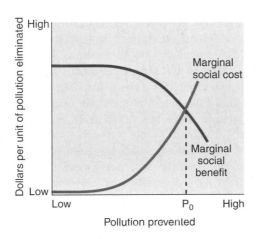

FIGURE 14.21 To achieve maximum economic efficiency, regulations should require pollution prevention up to the optimum point (P_o) at which the costs of eliminating pollution just equal the social benefits of doing so.

resource use because it can put a value on long-term or intangible goods that are necessary but often disregarded in economic decision making. One important part of natural resource accounting is assigning a value to ecological services (table 14.6). The total value of nature's services of $33.3 trillion per year is about twice the entire world GNP. Another is using alternative measures of wealth and development. As mentioned earlier, GNP is a widely used measure of wealth that is based on rates of consumption and throughput. GNP doesn't account for natural

TABLE 14.6

Estimated Annual Value of Ecological Services

Ecosystem Services	Value (trillion $ U.S.)
Soil formation	17.1
Recreation	3.0
Nutrient cycling	2.3
Water regulation and supply	2.3
Climate regulation (temperature and precipitation)	1.8
Habitat	1.4
Flood and storm protection	1.1
Food and raw materials production	0.8
Genetic resources	0.8
Atmospheric gas balance	0.7
Pollination	0.4
All other services	1.6
Total value of ecosystem services	33.3

Source: Adapted from R. Costanza et al., "The Value of the World's Ecosystem Services and Natural Capital," in *Nature,* vol. 387, 1997.

resource depletion or ecosystem damage, however. The World Resources Institute, for example, estimates that soil erosion in Indonesia reduces the value of crop production about 40 percent per year. If natural capital were taken into account, Indonesian GNP would be reduced by at least 20 percent annually. Similarly, Costa Rica experienced impressive increases in timber, beef, and banana production between 1970 and 1990. But decreased natural capital during this period, represented by soil erosion, forest destruction, biodiversity losses, and accelerated water runoff, added up to at least $4 billion, or about 25 percent of annual GNP. A number of countries, including Canada and China, now use a "green GDP" that measures environmental costs as part of economic accounting.

New approaches measure real progress

A number of systems have been proposed as alternatives to GNP that reflect genuine progress and social welfare. In their 1989 book, Herman Daly and John Cobb proposed a **genuine progress index (GPI),** which takes into account real per capita income, quality of life, distributional equity, natural resource depletion, environmental damage, and the value of unpaid labor. They point out that, while per capita GDP in the United States nearly doubled between 1970 and 2000, per capita GPI increased only 4 percent (fig. 14.22). Some social service organizations would add to this index the costs of social breakdown and crime, which would decrease real progress even further over this time span.

The United Nations Development Program (UNDP) uses a benchmark called the **human development index (HDI)** to track social progress. HDI incorporates life expectancy, educational attainment, and standard of living as critical measures of

FIGURE 14.22 Although per capita GDP in the United States nearly doubled between 1970 and 2000 in inflation-adjusted dollars, a genuine progress index that takes into account natural resource depletion, environmental damage, and options for future generations hardly increased at all.

Source: Data from *Redefining Progress,* 2006.

What Can You Do?

Personally Responsible Consumerism

Each of us can do many things to lower our ecological impacts and support "green" businesses through responsible consumerism and ecological economics.

- Practice living simply. Ask yourself if you really need more material goods to make your life happy and fulfilled.
- Rent, borrow, or barter when you can. Can you reduce the amount of stuff you consume by renting, instead of buying, machines and equipment you actually use only rarely?
- Recycle or reuse building materials: doors, windows, cabinets, appliances. Shop at salvage yards, thrift stores, yard sales, or other sources of used clothes, dishes, appliances, etc.
- Consult the *National Green Pages* from Co-Op America for a list of eco-friendly businesses. Write to companies from which you buy goods or services, and ask them what they are doing about environmental protection and human rights.
- Buy green products. Look for efficient, high-quality materials that will last and that are produced in the most environmentally friendly manner possible. Subscribe to clean-energy programs if they are available in your area. Contact your local utility and ask that it provide this option if it doesn't now.
- Buy locally grown or locally made products made under humane conditions by workers who receive a fair wage.
- Think about the total life-cycle costs of the things you buy, especially big purchases, such as cars. Try to account for the environmental impacts, energy use, and disposal costs, as well as initial purchase price.
- Stop junk mail. Demand that your name be removed from mass-mailing lists.
- Invest in socially and environmentally responsible mutual funds or "green" businesses when you have money for investment.

---

development. Gender issues are accounted for in the gender development index (GDI), which is simply HDI adjusted or discounted for inequality or achievement between men and women.

In its annual Human Development Report, the UNDP compares country-by-country progress. As you might expect, the highest development levels are generally found in North America, Europe, and Japan. In 2003 Norway ranked first in the world in both HDI and GDI. The United States ranked seventh and Canada was eighth, although subtracting the HDI rank from GDP index placed Canada first in the world on an equity-adjusted basis. The 25 countries with the lowest HDI in 2003 were all in Africa. Haiti ranked the lowest in the Western Hemisphere.

Although poverty remains widespread in many places, encouraging news also can be found in development statistics. Poverty has fallen more in the past 50 years, the UNDP reports,

than in the previous 500 years. Child death rates in developing countries as a whole have been more than halved. Average life expectancy has increased by 30 percent, while malnutrition rates have declined by almost a third. The proportion of children who lack primary school has fallen from more than half to less than a quarter. And the share of rural families without access to safe water has fallen from nine-tenths to about one-quarter.

Some of the greatest progress has been made in Asia. China and a dozen other countries with populations that add up to more than 1.6 billion have decreased the proportion of their people living below the poverty line by half. Still, in the 1990s the number of people with incomes less than $1 (U.S.) per day increased by almost 100 million to 1.3 billion—and the number appears to be growing in every region except Southeast Asia and the Pacific. Even in industrial countries, more than 100 million people live below the poverty line and 37 million are chronically unemployed.

Internalizing external costs

One of the factors that can make resource-exploiting enterprises look good in cost-benefit analysis is externalizing costs. **Externalizing costs** is the act of disregarding or discounting resources or goods that contribute to producing something but for which the producer does not actually pay. Usually, external costs are diffuse and difficult to quantify. Generally, they belong to society at large, not to the individual user. When a farmer harvests a crop in the fall, for example, the value of seeds, fertilizer, and the sale of the crop is tabulated; the value of soil lost to erosion, water quality lost to nonpoint-source pollution, and depleted fish populations is not accounted for. These are most often costs shared by the whole society, rather than borne by the resource user. They are external to the accounting system, and they are generally ignored in cost-benefit analysis—or when the farmer evaluates whether the year was profitable. Larger enterprises, such as dam building, logging, and road building, generally externalize the cost of ecological services lost along the way.

One way to use the market system to optimize resource use is to make sure that those who reap the benefits of resource use also bear all the external costs. This is referred to as **internalizing costs.** Calculating the value of ecological services or diffuse pollution is not easy, but it is an important step in sustainable resource accounting.

TRADE, DEVELOPMENT, AND JOBS

A sustainable society requires some degree of equitable resource distribution: if most wealth is held by just a few people, the misery and poverty of the majority eventually lead to social instability and instability of resource supplies. Accordingly, wealthy industrial nations have worked harder in recent decades to assist in developing the economies in poorer countries.

International trade can stimulate growth but externalize costs

Expanding trade relations has been promoted as a way to distribute wealth, stimulate economies around the world, and at the same time satisfy the desires of consumers in wealthy countries. According to the economic theory of *comparative advantage,* each place has some sort of goods or services it can make and sell cheaper, or better, than others can. International trade allows us to take advantage of all the best or cheapest products from around the world. If Egypt can produce cotton cheaper than Texas, then we should buy our cotton from Egypt. If Malaysia can manufacture athletic shoes with labor that costs a few cents an hour, then we should buy our shoes from Malaysia. Building factories in Malaysia also stimulates the Malaysian economy, even though workers earn only a small fraction of what a worker in a wealthy country would earn.

A problem with international trade is that it externalizes costs on a grand scale. Tropical hardwood products can be sold extremely cheaply in the United States—as lumber, plywood, shipping pallets, and so on. The environmental costs of producing those hardwood products occur far from the consumer who buys a piece of cheap Brazilian plywood. Making matters worse, the environmental costs of the plywood are usually exported to places where there are few legal controls on pollution and resource extraction. A factory in the United States, for example, is legally bound to minimize its production of air and water pollution. Pollution control can be expensive, and internalizing this cost makes a factory less profitable. A similar factory in Mexico might have far less responsibility for pollution control, making it cheaper, at least in the short term, to produce goods there than in the United States. Ongoing protests in the United States and elsewhere around the world against the World Trade Organization (WTO) and other forces of globalization have been largely about such exporting and externalizing of environmental and social costs of production. (See related story "The Battle in Seattle" at www.mhhe.com/cunningham4e.)

Another criticism of international trade is that the international banking systems that finance it are set up by and for the wealthy countries. The WTO and the General Agreement on Tariffs and Trade (GATT), for example, regulate 90 percent of all international trade. The WTO and GATT are both made up of relationships and agreements between corporations in a few very wealthy countries. Representatives of less powerful countries often charge that these agreements trap poorer regions into the role of suppliers of natural resources—timber, mineral ores, fruit, and cheap labor. These countries are forced to mine their natural capital for only small returns in wealth.

Socially responsible development can help people and protect their environment

The World Bank has more influence on the financing and policies of developing countries than any other institution. Of some $25 billion loaned each year for Third World projects by multinational development banks, about two-thirds comes from the World Bank. This institution was founded in 1945 to provide aid to war-torn Europe and Japan. In the 1950s its emphasis shifted to development aid for Third World countries. This aid was justified on humanitarian grounds, but it also conveniently provided markets and political support for growing American and European multinational corporations.

The World Bank is jointly owned by 150 countries, but one-third of its support comes from the United States. Its president has always been an American. The bulk of its $66.8 billion capital comes from private investors. The bank makes loans for development projects, and the investors hope to make a profit on the investment, usually from interest paid on the loans.

Many World Bank projects have been environmentally destructive and highly controversial. In Botswana, for example, $18 million was provided to increase beef production for export by 20 percent, despite already severe overgrazing on fragile grasslands. The project failed, as had two previous beef production projects in the same area. In Ethiopia, rich floodplains in the Awash River Valley were flooded to provide electric power and irrigation water for cash export crops. More than 150,000 subsistence farmers were displaced, and food production was seriously reduced.

Recently, the World Bank has begun to attempt some environmental and social review of its loans. In part this change comes from demands from the U.S. Congress that projects be environmentally and socially benign. Whether this will improve the World Bank's track record remains to be seen.

The World Bank deals in huge loans for massive projects. These are impressive to investors and to the countries that borrow the money. They are also a huge economic gamble, and on average the economic return on large loans has been quite low.

Recently, smaller, local development programs, often called **microlending,** have begun to develop. These are aimed at small-scale, widespread development, and their results have been very promising. The first of these was Bangladesh's Grammeen (village) Bank network. These banks make small loans, averaging just $67, to help poor people buy a sewing machine, a bicycle, a loom, a cow, or some other commodity that will help them start, or improve, a home business (fig. 14.23). Ninety percent of the customers are women, usually with no collateral or steady income. Still, loan repayment rates are 98 percent—compared with only 30 percent at a conventional bank. This program enhances dignity, respect, and cooperation in a village community, and it teaches individual responsibility and enterprise.

Comparable programs have now sprung up around the world. In the United States, more than a hundred organizations have begun providing microloans and small grants for training. The Women's Self-Employment Project in Chicago, for instance, teaches job skills to single mothers in housing projects. Similarly, "tribal circle" banks on Native American reservations successfully finance microscale economic development projects.

FIGURE 14.23 A small loan can help start a business like this Balinese woman's food stand.

GREEN BUSINESS AND GREEN DESIGN

Businesspeople and consumers are increasingly aware of the unsustainability of producing the goods we use every day. Recently, a number of business innovators have tried to develop green businesses, which produce environmentally and socially sound products. Environmentally conscious, or "green," companies, such as the Body Shop, Patagonia, Aveda, Malden Mills, Johnson and Johnson, and others, have shown that operating according to the principles of sustainable development and environmental protection can be good for public relations, employee morale, and sales (table 14.7).

Green business works because consumers are becoming aware of the ecological consequences of their purchases. Increasing

TABLE 14.7

Goals for an Eco-Efficient Economy

- Introduce no hazardous materials into the air, water, or soil.
- Measure prosperity by how much natural capital we can accrue in productive ways.
- Measure productivity by how many people are gainfully and meaningfully employed.
- Measure progress by how many buildings have no smokestacks or dangerous effluents.
- Make the thousands of complex governmental rules that now regulate toxic or hazardous materials unnecessary.
- Produce nothing that will require constant vigilance from future generations.
- Celebrate the abundance of biological and cultural diversity.
- Live on renewable solar income rather than fossil fuels.

interest in environmental and social sustainability has caused an explosive growth of green products. The *National Green Pages* published by Co-Op America currently lists more than 2,000 green companies. You can find eco-travel agencies, telephone companies that donate profits to environmental groups, entrepreneurs selling organic foods, shade-grown coffee, straw-bale houses, paint thinner made from orange peels, sandals made from recycled auto tires, and a plethora of hemp products, including burgers, ale, clothing, shoes, rugs, and shampoo. Although these eco-entrepreneurs represent a tiny sliver of the $7 trillion per year U.S. economy, they often are pioneers in developing new technologies and offering innovative services. Markets also grow over time: organic food marketing has grown from a few funky local co-ops to a $7 billion market segment. Most supermarket chains now carry some organic food choices. Similarly, natural-care health and beauty products reached $2.8 billion in sales in 1999 out of a $33 billion industry. By supporting these products, you can ensure that they will continue to be available and, perhaps, even help expand their penetration into the market.

Corporations committed to eco-efficiency and clean production include such big names as Monsanto, 3M, DuPont, and Duracell. Applying the famous three *R*s—reduce, reuse, recycle—these firms have saved money and gotten welcome publicity (see related story "Eco-Efficient Carpeting" at www.mhhe.com/cunningham4e). Savings can be substantial. Pollution-prevention programs at 3M, for example, have saved $857 million over the past 25 years. In a major public relations achievement, DuPont has cut its emissions of airborne cancer-causing chemicals almost 75 percent since 1987. Small operations can benefit as well. Stanley Selengut, owner of three eco-tourist resorts in the U.S. Virgin Islands, attributes $5 million worth of business to free press coverage about the resorts' green building features and sustainable operating practices.

Green design is good for business and the environment

Architects are starting to get on board the green bandwagon, too. Acknowledging that heating, cooling, lighting, and operating buildings is one of our biggest uses of energy and resources, architects such as William McDonough are designing "green office" projects. Among McDonough's projects are the Environmental Defense Fund headquarters in New York City; the Environmental Studies Center at Oberlin College in Ohio; the European headquarters for Nike in Hilversum, the Netherlands; and the Gap corporate offices in San Bruno, California (fig. 14.24). Each uses a combination of energy-efficient designs and technologies, including natural lighting and efficient water systems. The Gap office building, for example, is intended to promote employee well-being and productivity, as well as efficiency. It has high ceilings, abundant skylights, windows that open, a full-service fitness center (including pool), and a landscaped atrium for each office bay that brings the outside in. The roof is covered with native grasses. Warm interior tones and

FIGURE 14.24 The award-winning Gap, Inc. corporate offices in San Bruno, California, demonstrate some of the best features of environmental design. A roof covered with native grasses provides insulation and reduces runoff. Natural lighting, an open design, and careful relation to its surroundings make this a pleasant place to work.

Reason

	Percent of total job loss
Seasonal work	
Falling product demand	
Contract completion	
Business ownership change	
Bankruptcy	
Labor-management dispute	
Domestic relocation	
Model changeover	
Import competition	
Weather-related curtailment	
Contract cancellation	
Plant or machine repairs	
Vacation period	
Material shortages	
Overseas relocation	
Automation	
Energy-related disruption	
Environment- or safety- related	
Natural disaster	
Other (including reorganization)	
Not reported	

0 5 10 15 20 25 30
Percent of total job loss

FIGURE 14.25 Although opponents of environmental regulation often claim that protecting the environment costs jobs, studies by economist E. S. Goodstein show that only 0.1 percent of all large-scale layoffs in the United States were the result of environmental laws.
Source: Data from E. S. Goodstein, Economic Policy Institute, Washington, D.C.

natural wood surfaces (all wood used in the building was harvested by certified sustainable methods) give a friendly feeling. Paints, adhesives, and floor coverings are low-toxicity, and the building is one-third more energy-efficient than strict California laws require. The pleasant environment helps improve employee effectiveness and retention. Gap, Inc., estimates that the increased energy and operational efficiency will have a four- to eight-year payback.

Environmental protection creates jobs

For years business leaders and politicians have portrayed environmental protection and jobs as mutually exclusive. They claim that pollution control, protection of natural areas and endangered species, and limits on use of nonrenewable resources will strangle the economy and throw people out of work. Ecological economists dispute this claim, however. Their studies show that only 0.1 percent of all large-scale layoffs in the United States in recent years were due to government regulations (fig. 14.25). Environmental protection, they argue, is not only necessary for a healthy economic system; it actually creates jobs and stimulates business.

Green businesses often create far more jobs and stimulate local economies far more than environmentally destructive ones. Wind energy, for example, provides about five times as many jobs per kWh of electricity than does coal-fired power (chapter 12).

Japan, already a leader in efficiency and environmental technology, has recognized the multibillion-dollar economic potential of "green" business. The Japanese government is investing $4 billion per year on research and development, and now it is selling about $12 billion worth of equipment and services per year worldwide. It is already marketing advanced waste incinerators, pollution-control equipment, alternative energy sources, and water treatment systems. Superefficient "hybrid" gas-electric cars are helping Japanese automakers flourish, while U.S. corporations that once dominated the world slide toward bankruptcy. Unfortunately, the United States has been resisting international pollution-control conventions, rather than recognizing the potential for economic growth and environmental protection in the field of "green" business.

REVIEW QUESTIONS

1. Who was the mayor and chief city planner for Curitiba, Brazil?
2. How many people now live in urban areas?
3. How many cities were over 1 million in 1900? How many are now?
4. Why do people move to urban areas?
5. What is the difference between a shantytown and a slum?

6. Define *sprawl*.
7. How has Curitiba avoided sprawl?
8. In what ways are cities ecosystems?
9. Define *smart growth*.
10. Describe a "green" roof.
11. Describe a few ways in which BedZED is self-sufficient and sustainable.

12. Define *sustainable development.*

13. Briefly summarize the differences in how neoclassical and ecological economics view natural resources.

14. What is the estimated economic value of all the world's ecological services?

15. How is it that nonrenewable resources can be extended indefinitely, while renewable resources are exhaustible?

16. Why isn't the model in figure 14.19 the path that all societies will follow, sooner or later?

17. In your own words, describe what is shown in figure 14.21.

18. What's the difference between open access and communal resource management?

19. Describe the genuine progress index (GPI).

20. What is *microlending*?

CRITICAL THINKING AND DISCUSSION

Apply the principles you have learned in this chapter to discuss these questions with other students.

1. This chapter presents a number of proposals for suburban redesign. Which of them would be appropriate or useful for your community? Try drawing up a plan for the ideal design of your neighborhood.

2. A city could be considered an ecosystem. Using what you learned in chapters 2 and 3, describe the structure and function of a city in ecological terms.

3. If you were doing a cost-benefit study, how would you assign a value to the opportunity for good health or the existence of rare and endangered species in faraway places? Is there a danger or cost in simply saying some things are immeasurable and priceless and therefore off limits to discussion?

4. What would be the effect on the developing countries of the world if we were to change to a steady-state economic system? How could we achieve a just distribution of resource benefits while still protecting environmental quality and future resource use?

5. When an ecologist warns that we are using up irreplaceable natural resources and an economist rejoins that ingenuity and enterprise will find substitutes for most resources, what underlying premises and definitions shape their arguments?

SUGGESTED READINGS

Anderson, Gordon, and Ying, Ge. 2005. The size distribution of Chinese cities. *Regional Science and Urban Economics* 35 (6):756–76.

Brown, Lester. 2001. *Ecoeconomy: Building an Economy for the Earth.* W. W. Norton & Co.

Costanza, Robert, et al. 2000. Managing our environmental portfolio. *BioScience* 50 (2):149–55.

Daley, Herman E., and Joshua Farley. 2004. *Ecological Economics: Principles and Applications.* Island Press.

Duany, Andres, Jeff Speck, and Elizabeth Plater-Zyberk. 2002. *Smart Growth Manual.* McGraw-Hill Co.

Kotkin, Joel. 2005. *The City: A Global History.* Modern Library.

McManus, Reed. 2006. Imagine a city with 30 percent fewer cars. *Sierra* 91 (1):48–49.

Prugh, Thomas, and Erik Assadourian. 2003. What is sustainability, anyway? *World Watch* 16 (5):10–21.

Rabinovitch, Jonas, and Josef Leitman. 1996. Urban planning in Curitiba. *Scientific American* 274 (3):46–53.

Taylor, J. Gary, and Patricia J. Scharlin. 2004. *Smart Alliance: How a Global Corporation and Environmental Activists Transformed a Tarnished Brand.* Yale University Press.

DATA analysis   Using a Logarithmic Scale

We've often used very large numbers in this book. Millions of people suffer from common diseases. Hundreds of millions are moving from the country to the city. Billions of people will probably be added to the world population in the next half century. Cities that didn't exist a few decades ago now have millions of residents. How can we plot such rapid growth and such huge numbers? If you use ordinary graph paper, making a scale that goes to millions or billions will run off the edge of the page unless you make the units very large.

Figure 1, for example, shows the growth of Mumbai, India, over the past 150 years plotted with an **arithmetic scale** (showing constant intervals) for the Y-axis. It looks as if there is very little growth in the first third of this series and then explosive growth during the last few decades, yet we know that the *rate* of growth was

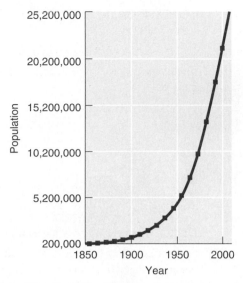

FIGURE 1 The Growth of Mumbai.

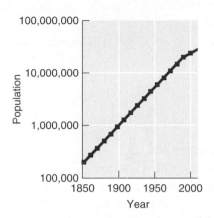

FIGURE 2 The Growth of Mumbai.

actually greater at the beginning than at the end of this time. How could we display this differently? One way to make the graph easier to interpret is to use a **logarithmic scale.** A logarithmic scale, or "log scale," progresses by factor of 10. So the Y-axis would be numbered 0, 1, 10, 100, 1,000. . . . The effect on a graph is to spread out the smaller values and compress the larger values. In figure 2, the same data are plotted using a log scale for the Y-axis, which makes it much easier to see what happened throughout this time period.

Do these two graphing techniques give you a different impression of what's happening in Mumbai? How might researchers use one or the other of these scales to convey a particular message or illustrate details in a specific part of the growth curve?

For Additional Help in Studying This Chapter, please visit our website at <u>www.mhhe.com/cunningham4e</u>. You will find practice quizzes, a chapter summary, key terms, answers to review questions, additional case studies, regional examples, and an extensive reading list, all of which will help you understand the material in this chapter.

Ecotourism and whale watching provide jobs for local people
and help protect the Laguna San Ignacio.

Environmental Policy and Sustainability

You must be the change you wish to see in the world.

–Mahatma Gandhi–

OBJECTIVES

Studying this chapter will help you answer the following questions:

- What is environmental policy, and how is it formed?

- Why are public policies sometimes unfair or irrational?

- Why do some people want to modify or do away with the NEPA?

- What are some of the most important U.S. environmental laws, and what do they do?

- What are adaptive management and ecosystem management? How do they work?

- Why are international environmental laws and conventions sometimes ineffective?

- What is citizen science, and what opportunities does it offer?

- What can individuals do to contribute to environmental protection?

- How can we work together for these same ends?

- What is sustainability, and why is it important?

Case Study   Saving a Gray Whale Nursery

At first glance, the Laguna San Ignacio may merely look like a shallow bay surrounded by a barren, rocky desert. But to many people, this lagoon on the west coast of Mexico's Baja Peninsula is a biological treasure. It's the last relatively pristine place where gray whales congregate each winter to mate, give birth, and nurse their calves. Pacific gray whales (*Eschrichtius robustus*) make a round trip of about 16,000 km (10,000 mi) every year between their summer feeding grounds north of the Arctic Circle and the Baja. The warm, salty water of bays like San Ignacio give calves extra buoyancy that helps them swim and nurse, while also sheltering them from predators and winter storms.

In the nineteenth century, a whaling captain named Scammon discovered the winter calving areas in Baja. The enclosed bays that once protected the whales became killing grounds. In a short time, the Pacific population was reduced from an estimated 25,000 animals to only a few thousand. Whaling bans have allowed the species to rebound to nearly its prehunting population, a great success story in endangered species protection. In 1994, Pacific gray whales were removed from the U.S. endangered species list.

In 1954, the same year that Mexico banned commercial whaling, a sea salt extraction facility was built in Guerro Negro bay (formerly Scammon's Lagoon) and the nearby Ojo de Liebre just north of Laguna San Ignacio. These saltworks, which are now operated by Expotadora de Sal and jointly owned by the Mitsubishi Company and the Mexican government, are the largest in the world, producing 6.5 million metric tons of salt per year. Concern about the effects of this huge industrial development on both the whales and the surrounding desert caused Mexican President Miguel de La Madrid to establish the Vizcaino Biosphere Reserve in 1988, including all three lagoons plus 2.4 million ha (6 million acres) of surrounding desert.

In 1994, however, Expotadora de Sal announced intentions to build an even bigger saltworks at Laguna San Ignacio. Plans called for 300 km² (116 mi²) of salt evaporation ponds carved out of the shoreline and filled by diesel engines that would pump 23,000 l (nearly 6,000 gal) of seawater per second. A 1.6 km (1 mi) long concrete pier built across the lagoon would transport the salt to an offshore loading area that would fill more than 120 salt tankers per year. The threat to whale survival from this immense operation was evident.

One of Mexico's leading environmental groups, el Grupo de los Cien (the Group of 100) started a campaign to stop this huge industrial development. They joined with other nongovernmental groups (NGOs), including the Natural Resources Defense Council (NRDC) and the International Fund for Animal Welfare (IFAW), to raise public awareness and to lobby the Mexican government. The campaign took a number of different approaches. One of these was to organize whale-watching trips featuring movie stars, such as Glenn Close and Pierce Brosnan, to gain attention and educate the public about the issue. Newspaper ads and magazine articles criticized the industrialization of San Ignacio. One of these, entitled "An Unacceptable Risk," presented the scientific value of the lagoon and was signed by 33 of the world's most famous scientists, including several Nobel laureates.

Environmentalists also lobbied Mitsubishi directly, threatening to boycott their cars, TVs, electronics, and other products. A 1998 UN World Heritage Conference in Kyoto, Japan, provided an excellent opportunity to meet face-to-face with company leaders. Activists said to Hajime Koga, manager of Mitsubishi's "Salt Team," "You would never contemplate such a project in a World Heritage site in Japan. Why would you destroy one in another country?" The company was amazed to receive more than 1 million petitions, letters, and emails from all over the world, criticizing their expanded saltworks. Although the environmental NGOs weren't successful in obtaining "In Danger" designation for the biosphere reserve at the conference in Kyoto, they did get this classification at the next meeting of the World Heritage Committee in Marrakech, Morocco, in 2000.

In 2002, Expotadora de Sal announced that it was abandoning plans for Laguna San Ignacio. Mitsubishi said it was the first time in its history that it had changed its policy because of environmental concerns. Simply blocking development isn't enough; long-term solutions need to be economically sustainable as well as scientifically sound and socially just. The 35,000 Mexicans who live within the biosphere reserve need to make a living. In 2005, local residents and environmental NGOs signed an agreement to preserve 50,000 ha (124,000 acres) of land around Laguna San Ignacio. The Ejido Luis Echeverria, a land cooperative, which owns the land, will limit development in exchange for a $25,000 annual payment to be used for low-impact projects, such as ecotourism and whale watching. Eventually, conservationists hope to reach a similar agreement with five other *ejidos* to extend protection to 4,000 km² (1 million acres) of the Vizcaino Biosphere Reserve. This will cost about $10 million.

This case study demonstrates some of the steps in influencing public policy. First, you have to gather information and understand the science that informs your issue. You should recognize how, and by whom, policy is made. You must evaluate which of the many techniques for educating the public and shaping opinion can be effective. And you need to learn how to work with other groups, and to reason with those whose opinions you hope to sway. In this chapter, we'll study the process of policy formation and implementation, as well as how the public can work to affect this process.

Gray whale migration route from Alaska to Baja, California.

ENVIRONMENTAL POLICY AND LAW

A policy is a rule or decision about how to act or deal with problems. On a personal, informal level, you might have a policy always to get your homework in on time. On a national, formal level, we have policies such as the Clean Air Act, a set of rules, agreed upon by a majority of the U.S. Congress, restricting air pollutants and setting fines for those who exceed legal limits. There are international policies as well: the 1987 Montreal Protocol set agreed-upon limits to the production of ozone-depleting chemicals (chapter 9), and the Convention on International Trade in Endangered Species (CITES) is an international agreement to restrict trade in endangered species (chapter 5).

Environmental policy is both the official rules and regulations concerning the environment that are adopted, implemented, and enforced by government agencies, as well as the general public opinion about environmental issues. National policies are established through negotiation and compromise in a democratic society. Sometimes this wrangling can take decades. Theoretically, it allows all voices to be heard, and the resulting policy serves the interest of the majority.

Making policy isn't always fair or just. Rules can seem harsh and unfair to minorities. The unfair influence of money and power in government also can be discouraging to voters. But sound policy for the public good can be formed—especially when the public is active and vocal in defending its interests. Events don't always follow principles, however. The best way to make sure they do is to develop an informed and involved populace.

How is policy created?

How do policy issues and options make their way onto the stage of public debate? Problems are identified and acted upon in a **policy cycle** that acts to continually define and improve the public agenda (fig. 15.1). The first stage in this process is problem identification. Sometimes the government identifies issues for groups that have no voice or don't recognize problems themselves. In other cases, the public identifies a problem, such as loss of biodiversity or health effects of exposure to toxic waste, and demands redress by the government. In either case, proponents describe the issue—either privately or publicly—and characterize the risks and benefits of their preferred course of action.

Seizing the initiative in issue identification often allows a group to define terms, set the agenda, organize stakeholders, choose tactics, aggregate related issues, and legitimate (or delegitimate) issues and actors. It can be a great advantage to set the format or choose the location of a debate. Next, stakeholders develop proposals for preferred policy options, often in the form of legislative proposals or administrative rules. Proponents build support for their position through media campaigns, public education, and personal lobbying of decision makers. By following the legislative or administrative process through its many steps, interest groups ensure that their proposals finally get enacted into law or established as a rule or regulation.

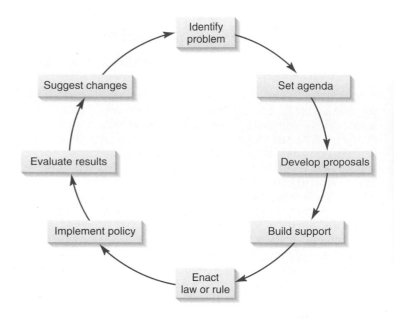

FIGURE 15.1 The policy cycle.

The next step is implementation. Ideally, government agencies faithfully carry out policy directives as they organize bureaucracies, provide services, and enforce rules and regulations, but often it takes continued monitoring to make sure the system works as it should. Evaluating the results of policy decisions is as important as establishing them in the first place. Measuring impacts on target and nontarget populations shows us whether the intended goals, principles, and course of action are being attained. Finally, suggested changes or adjustments are considered that will make the policy fairer or more effective.

There are different routes by which this cycle is carried out. Special economic interest groups, such as industry associations, labor unions, or wealthy and powerful individuals, don't need (or often want) much public attention or support for their policy initiatives. They generally carry out the steps of issue identification, agenda setting, and proposal development privately because they can influence legislative or administrative processes directly through their contacts with decision makers. Public interest groups, on the other hand, often lack direct access to corridors of power and need to rally broad general support to legitimate their proposals.

An important method for getting public interest on the table is to attract media attention. Organizing a dramatic protest or media event can generate a lot of free publicity (fig. 15.2). Announcing a dire threat or sensational claim is a good way to gain attention. The problem is that it takes ever increasing levels of hysteria and hyperbole to get yourself heard in the flood of shocking news with which we are bombarded every day. Ironically, many groups that bemoan the overload of rhetorical overstatement that engulfs us contribute to it to be heard above the din. Nevertheless, an overwhelming majority of Americans believe that environmental protection is an important priority (fig. 15.3).

FIGURE 15.2 Protestors demonstrate against monetary policies and international institutions that threaten livelihoods and environmental quality.

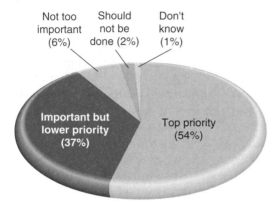

Not too important (6%)

Should not be done (2%)

Don't know (1%)

Important but lower priority (37%)

Top priority (54%)

FIGURE 15.3 More than half the respondents in a Princeton University survey in 2000 said that environmental protection should be a top priority for the president and congress.

Policy formation can be complicated

The policies we establish depend to a great extent on the system within which they operate. For many of us, the ideal political system is one that is open, honest, transparent, and reaches the best possible decisions to maximize benefits to everyone. In a pluralistic, democratic society, we aim to give everyone an equal voice in policy making. Ideally, many separate interests put forward their solutions to public problems that are discussed, debated, and evaluated fairly and equally. Facts and access are open to everyone. Policy choices are made democratically but compassionately; implementation is reasonable, fair, and productive. Unfortunately, this isn't always the way our system works. Our political system usually gives those with the most power and money the greatest voice in public affairs. Still, there have been altruistic movements in the past to ensure human rights and environmental protection that reveal an impulse for generosity and goodwill among the general public.

Another model for public decision making is rational choice and science-based management. In this utilitarian approach, no policy should have greater total costs than benefits. In choosing among policy alternatives, we should always prefer those with the greatest cumulative welfare and the least negative impacts. Professional administrators would weigh various options and make an objective, methodical decision that would bring maximum social gain. There are many reasons, however, that rational choice doesn't always work in the public arena.

- Many conflicting values and needs cannot be compared because they aren't comparable or we don't have perfect information.

- There are few generally agreed-upon broad societal goals, but rather, benefits to specific groups and individuals, many of which are in conflict.

- Policymakers generally aren't motivated to make decisions on the basis of societal goals, but rather to maximize their own rewards: power, status, money, or reelection.

- Large investments in existing programs and policies create "path dependence" and "sunken costs" that prevent policymakers from considering good alternatives foreclosed by previous decisions.

- Uncertainty about the consequences of various policy options compels decision makers to stick as closely as possible to previous policies to reduce the likelihood of adverse, calamitous, unanticipated consequences.

- Policymakers, even if well meaning, don't have sufficient intelligence or adequate data or models to calculate accurate costs and benefits when large numbers of diverse political, social, economic, and cultural values are at stake.

- The segmented nature of policy making in large bureaucracies makes coordinating decision making difficult.

APPLICATIONS

Environment, Science, and Policy in Your Community

Take several minutes with your classmates and make a list of five current, environment-related policy issues in your community or state. (Hint: consider energy, economic developments, population changes, traffic congestion, water and air quality, or public health issues in the news.) Discuss which branch of government—legislative, executive, or judicial—you think should take the lead in addressing each issue. Why did you designate these authorities as you did?

Select one issue from your list, and discuss the following: What are some of the sources of uncertainty involved in the decision-making process? How could scientific research help identify or solve this policy issue? Where in the rule-making process could scientific data help inform decision makers?

Does U.S. environmental policy need an overhaul?

Signed into law by President Nixon in 1970, the **National Environmental Policy Act (NEPA)** is the cornerstone of U.S. environmental policy. Conservationists see this act as a powerful tool for environmental protection, but commercial interests blame it for gridlock and consider it an impediment to business. President Bush has created a task force to study streamlining NEPA and make it less burdensome for industry. Conservationists worry this is really an effort to weaken the law and return to laissez-faire resource management.

NEPA does three important things: (1) it authorizes the Council on Environmental Quality (CEQ), the oversight board for general environmental conditions: (2) it directs federal agencies to take environmental consequences into account in decision making; and (3) it requires an **environmental impact statement (EIS)** be published for every major federal project likely to have an important impact on environmental quality. NEPA doesn't forbid environmentally destructive activities if they comply otherwise with relevant laws, but it demands that agencies admit publically what they plan to do. Once embarrassing information is revealed, however, few agencies will bulldoze ahead, ignoring public opinion. And an EIS can provide valuable information about government actions to public interest groups that wouldn't otherwise have access to these resources.

What kinds of projects require an EIS? The activity must be federal and it must be major, with a significant environmental impact (fig. 15.4). Evaluations are always subjective as to whether specific activities meet these characteristics. Each case is unique and depends on context, geography, the balance of beneficial versus harmful effects, and whether any areas of special cultural, scientific, or historical importance might be affected. A complete EIS for a project is usually time-consuming and costly. The final document is often hundreds of pages long and generally takes six to nine months to prepare. Sometimes just requesting an EIS is enough to sideline a questionable project. In other cases, the EIS process gives adversaries time to rally public opposition and information with which to criticize what's being proposed. If agencies don't agree to prepare an EIS voluntarily, citizens can petition the courts to force them to do so.

Every EIS must contain the following elements: (1) purpose and need for the project, (2) alternatives to the proposed action (including taking no action), (3) a statement of positive and negative environmental impacts of the proposed activities. In addition, an EIS should make clear the relationship between short-term resources and long-term productivity, as well as any irreversible commitment of resources resulting from project implementation.

Among the areas in which lawmakers have tried recently to ignore or limit NEPA include forest policy, energy exploration, and marine wildlife protection. The "Healthy Forest Initiative," for example, called for bypassing EIS reviews for logging or thinning projects, and prohibited citizen appeals of forest management plans. Similarly, when the Bureau of Land Management proposed 77,000 coal-bed methane wells in Wyoming and Montana, promoters claimed that water pollution and aquifer depletion associated with this technology didn't require environmental review (chapter 12). And in the 2005 Energy Bill, Congress inserted a clause that exempts energy companies from NEPA requirements in a number of situations, with the aim of speeding energy development on federal land.

Interestingly, at the same time that the United States seems intent on weakening NEPA, a new Chinese law that requires environmental impact statements for major projects was used recently to oppose a series of dams on the Nu River in southwestern China. The Nu, which was one of only two free-flowing rivers in China, had been targeted for 13 large hydropower dams in spite of the fact that UNESCO designated it a World Heritage Site. The area has been described as the most biologically diverse temperate ecosystem in the world. At first, it appeared that this huge development would go through without any public scrutiny. This new law came into force during dam planning, however, and the spotlight it cast on environmental and social costs appears to have doomed the project. In autocratic governments, environmental protests are often the only form of government criticism allowed, and often are the first step in spreading democracy. This was true in the former Soviet Union, as well as in China.

To be informed environmental citizens, we all need to know something about how policies and laws like NEPA are created and applied. What do you think? Is NEPA an essential safeguard for our environmental quality, or merely an antiquated impediment for business and personal property rights? What information would you need to judge the effectiveness of this important law?

Laws affirm public policy

Laws are rules set by authority, society, or custom. Church laws, social mores, administrative regulations, and a variety of other codes of behavior can be considered laws if they are backed by

FIGURE 15.4 Every major federal project in the United States must be preceded by an environmental impact statement (EIS).

some enforcement power. **Environmental law** includes official rules, decisions, and actions concerning environmental quality, natural resources, and ecological sustainability. In the United States a wide variety of environmental laws are promulgated at both the local and national levels. Because every country has different legislative and legal processes, this chapter will focus primarily on the U.S. system in the interest of simplicity and space. Environmental law can be established or modified in each of the three branches of government: legislative, judicial, and executive—statutory, case, and administrative law, respectively. Understanding how these systems work is an important step in becoming an environmentally literate person.

The legislative branch establishes statutory law

Establishing laws at either the state or the federal level is one of the most important ways of protecting our environment. Many environmental groups spend a good deal of their time and resources trying to influence the legislative process. In this section we'll look at how that system works.

Federal laws (statutes) are enacted by Congress and must be signed by the president. They originate as legislative proposals called bills, which are usually drafted by the congressional staff, often in consultation with representatives of various interest groups. Thousands of bills are introduced every year in Congress. Some are very narrow, providing funds to build a specific section of road or to help a particular person, for instance. Others are extremely broad, perhaps overhauling the Social Security system or changing the entire tax code. Similarly, environmental legislation might deal with a very specific local problem or a national or international issue. Often a number of competing bills on a single issue may be introduced as proponents from different sides attempt to incorporate their views into law. A bill may have a single sponsor if it is the pet project of a particular legislator, or it may have 100 or more co-authors if it is an issue of national importance. Table 15.1 lists some of the most important U.S. environmental legislation.

Citizens can be involved in this process either by writing or calling their elected representatives, or by appearing at public hearings (fig. 15.5). A personal letter or statement is always more

TABLE 15.1

Major U.S. Environmental Laws

Legislation	Provisions
National Environmental Policy Act of 1969	Declared national environmental policy, required environmental impact statements, created Council on Environmental Quality.
Clean Air Act of 1970	Established national primary and secondary air quality standards. Required states to develop implementation plans. Major amendments in 1977 and 1990.
Clean Water Act of 1972	Set national water quality goals and created pollutant discharge permits. Major amendments in 1977 and 1996.
Federal Pesticides Control Act of 1972	Required registration of all pesticides in U.S. commerce. Major modifications in 1996.
Marine Protection Act of 1972	Regulated dumping of waste into oceans and coastal waters.
Coastal Zone Management Act of 1972	Provided funds for state planning and management of coastal areas.
Endangered Species Act of 1973	Protected threatened and endangered species. Directed FWS to prepare recovery plans.
Safe Drinking-Water Act of 1974	Set standards for safety of public drinking-water supplies and to safeguard ground water. Major changes made in 1986 and 1996.
Toxic Substances Control Act of 1976	Authorized EPA to ban or regulate chemicals deemed a risk to health or the environment.
Federal Land Policy and Management Act of 1976	Charged the BLM with long-term management of public lands. Ended homesteading and most sales of public lands.
Resource Conservation and Recovery Act of 1976	Regulated hazardous-waste storage, treatment, transportation, and disposal. Major amendments in 1984.
National Forest Management Act of 1976	Gave statutory permanence to national forests. Directed USFS to manage forests for "multiple use."
Surface Mining Control and Reclamation Act of 1977	Limited strip-mining on farmland and steep slopes. Required restoration of land to original contours.
Alaska National Interest Lands Act of 1980	Protected 40 million ha (100 million acres) of parks, wilderness, and wildlife refuges.
Comprehensive Environmental Response, Compensation and Liability Act of 1980	Created $1.6 billion "Superfund" for emergency response, spill prevention, and site remediation for toxic wastes. Established liability for cleanup costs.
Superfund Amendments and Reauthorization Act of 1994	Increased Superfund to $8.5 billion. Shared responsibility for cleanup among potentially responsible parties. Emphasized remediation and public "right to know."

Source: Data from N. Vig and M. Kraft, *Environmental Policy in the 1990s,* 3rd Congressional Quarterly Press.

FIGURE 15.5 Citizens line up to testify at a legislative hearing. By getting involved in the legislative process, you can be informed and have an impact on government policy.

persuasive than simply signing a petition. Some politicians treat all the copies of a similar form letter or email as a single statement, because they all contain the same information. Still, a petition with a million signatures will probably catch the attention of a legislator—especially if they are all potential voters.

Being involved in local election campaigns can greatly increase your access to legislators. Writing letters or making telephone calls also are highly effective ways to get your message across. All legislators now have email addresses, but letters and phone calls are usually taken more seriously. Getting media attention can sway the opinions of decision makers. Organizing protests, marches, demonstrations, or other kinds of public events can call attention to your issue. Public education campaigns, press conferences, TV ads, and a host of other activities can be helpful. Joining together with other like-minded groups can greatly increase your clout and ability to get things done. It's hard for a single individual or even a small group to have much impact, but if you can organize a mass movement, you may be very effective in influencing public policy.

The judicial branch shapes case law

The judicial branch of government establishes environmental law by ruling on the constitutionality of statutes and interpreting their meaning. We describe the body of legal opinions built up by many court cases as **case law.** Often legislation is written in vague and general terms so as to make it widely enough accepted to gain passage. Congress, especially in the environmental area, often leaves it to the courts to "fill in the gaps." When trying to interpret a law, the courts depend on the legislative record from hearings and debates to determine congressional intent. What was a particular statute meant to do by those who wrote and passed it? In ruling on cases, the court builds up a body of precedent that influences subsequent deliberations. Judges usually respect the rulings of their predecessors unless some new information becomes available.

Criminal law derives from those federal and state statutes that prohibit wrongs against the state or society. Serious crimes, like murder or rape, that are punishable by long jail terms or heavy fines are called felonies. Lesser crimes, such as shoplifting or vandalism, that result in smaller fines or shorter sentences in a county or city jail are labeled misdemeanors. A criminal case is always initiated by a government prosecutor. Guilt or innocence of the defendant is determined by a jury of peers, but the sentence is imposed, often in consultation with the jury, by the judge. The judge is responsible for keeping order in the hearings and for determining points of law (fig. 15.6). The jury acts as a fact-finding body that weighs the truth and reliability of the witnesses and evidence.

Violation of many environmental statutes constitutes a criminal offense. In 1975 the U.S. Supreme Court ruled that corporate officers can be held criminally liable for violations of environmental laws if they were grossly negligent, or the illegal actions can be considered willful and knowing violations. In one of the toughest criminal sentences imposed so far, the president of a Colorado company was sentenced in 1999 to 14 years in prison for knowingly dumping chlorinated solvents that contaminated the water table.

Civil law is a body of laws regulating relations between individuals or between individuals and corporations. Issues such as property rights and personal dignity and freedom are protected by civil law. The defendant in a civil case has a right to be tried by a jury, but in highly technical issues, this right often is waived and the case is heard only by the judge. Being found guilty of a civil offense can result in financial penalties but not jail time.

In contrast to a criminal case, where the burden of proof lies with the prosecution and defendants are considered innocent until proven guilty, civil cases can be decided on a "preponderance of evidence." This makes civil cases considerably easier to win than criminal cases when evidence is ambiguous. A number of mitigating factors also are taken into account in determining guilt and assigning penalties in civil cases. Guilt or innocence is based on

FIGURE 15.6 In a trial court, the judge presides over trials, rules on motions, and interprets the law.

whether the defendant could reasonably have anticipated and avoided the offense. A "good faith effort" to comply or solve the problem can be a factor. The compliance history is important. Is this a first-time offender or a habitual repeater? Finally, is there evidence of economic benefit to the perpetrator? That is, did the violator gain personally from the action? If so, it is more likely that willful intent was involved.

Civil cases can be brought in both state and federal court. In 2000, the Koch Oil Company, one of the largest pipeline and refinery operators in the United States, agreed to pay $35 million in fines and penalties to state and federal authorities for negligence in more than 300 oil spills in Texas, Oklahoma, Kansas, Alabama, Louisiana, and Missouri over the previous decade. Koch also agreed to spend more than $1 billion on cleanup and improved operations.

Sometimes the purpose of a civil suit is to seek an injunction or some other protection from the actions of an individual, a corporation, or a governmental agency. You might ask the courts, for example, to order the government to cease and desist from activities that are in violation of either the spirit or the letter of the law. This sort of civil action is heard only by a judge; no jury is present. Environmental groups have been very successful in asking courts to stop logging and mining operations, to enforce implementation of the Endangered Species Act, to require agencies to enforce air and water pollution laws, and a host of other efforts to protect the environment and conserve natural resources.

The executive branch enforces administrative law

More than 100 federal agencies and thousands of state and local boards and commissions have environmental oversight. They usually have power to set rules, adjudicate disputes, and investigate misconduct.

Agency rule-making and standard-setting can be an important way for environmentally concerned citizens and public interest groups to have an impact on environmental policy. Rule-making is often a complex, highly technical process that is difficult for citizen groups to understand and monitor. The proceedings are usually less dramatic and colorful than criminal trials, and yet can be very important for environmental protection. The Bush administration has made profound changes in U.S. environmental policy through rule-making that avoids open debate or public scrutiny.

The EPA is the primary agency with responsibility for protecting environmental quality in the United States. Created in 1970 at the same time as NEPA, the EPA has more than 18,000 employees and ten regional offices. Often in conflict with Congress, other agencies of the executive branch, and environmental groups, the EPA has to balance many competing interests and conflicting opinions. Greatly influenced by politics, the agency changes dramatically, depending on which party is in power and what attitudes toward the environment prevail at any given time.

The Departments of the Interior and Agriculture are to natural resources what the EPA is to pollution. Interior is home to the National Park Service, which is responsible for more than

FIGURE 15.7 Smokey Bear symbolizes the Forest Service's role in extinguishing forest fires.

376 national parks, monuments, historic sites, and recreational areas. It also houses the Bureau of Land Management (BLM), which administers some 140 million ha (350 million acres) of land, mostly in the western United States. In addition, Interior is home to the U.S. Fish and Wildlife Service, which operates more than 500 national wildlife refuges and administers endangered species protection.

The Department of Agriculture is home to the U.S. Forest Service, which manages about 175 national forests and grasslands, totaling some 78 million ha (193 million acres). With 39,000 employees, the Forest Service is nearly twice as large as the EPA (fig. 15.7). The Department of Labor houses the Occupational Safety and Health Administration (OSHA), which oversees workplace safety. Research that forms the basis for OSHA standards is carried out by the National Institute for Occupational Safety and Health (NIOSH). In addition, several independent agencies that are not tied to any specific department also play a role in environmental protection and public health. The Consumer Products Safety Commission passes and enforces regulations to protect consumers, and the Food and Drug Administration is responsible for the purity and wholesomeness of food and drugs.

How can we manage complex systems?

Managing ecosystems is a complicated and controversial affair. A classic example of the difficulty brought about by clashing human and natural interconnections can be seen in the Florida Everglades. L. Gunderson, C. Holling, and S. Light discuss ways in which policymakers find solutions for difficult problems in their 1995 book *Barriers and Bridges*.

Over the past century, the Everglades ecosystem has been transformed from an immense subtropical wetland into a highly managed, multiple-use system as the result of one of the world's

| Historic flow | Current flow | Planned flow |

FIGURE 15.8 Planned results of the Comprehensive Everglades Restoration Plan. Red dashed outline shows the national park boundary.

largest public works projects. This huge hydraulic system stretches from just south of Orlando through the many springs and lakes that feed the Kissimmee River, through Lake Okeechobee to the Everglades National Park at the southern tip of Florida. In 1947, Marjorie Stoneman Douglas aptly named this the "river of grass" to describe the broad, slowly moving sheet of water that trickles through the vast wetlands of South Florida.

Management policies have lurched from one approach to another as environmental surprises and human intervention have transformed the area. Natural crises include floods, droughts, fires, and storms. Human effects, including agricultural pollution, invasive species, and the increasing water demands of rapidly growing coastal cities, further complicate the picture.

The earliest settlers were intent on "reclaiming" wetlands in order to farm the rich muck soils. They dug canals to drain the land as fast as possible. This prevented the wetland from absorbing and storing excess rainfall. The result was severe flooding in abnormally wet years, such as 1947, when more than 90 percent of the crop was destroyed. The U.S. Army Corps of Engineers responded by creating a massive water control plan, including 2,200 km of canals, 1,000 km of levees, and pumping stations with 3.8 billion liters/day capacity (fig. 15.8).

In 1971, however, the worst drought in 40 years hit the region. This occurred just as the population of southern Florida topped the 2 million mark, and sugarcane production had more than tripled. Competition for water between the Everglades National Park, agricultural interests, and rapidly growing cities demanded new approaches to water management. Stakeholders debated for years about how the system should be managed. In 2000, the state of Florida and the Army Corps agreed to begin a restoration project that would remove many levees and canals, build new reservoirs to store water, dechannelize rivers, and construct 14,000 ha of new filtration wetlands. Despite billions of dollars already spent, however, this project is behind schedule and criticized by all sides as being inadequate (see related story on "The World's Biggest Restoration Project" at www.mhhe.com/cunningham4e).

Some preservationists suggest that we should simply leave ecosystems, such as the Everglades, alone to let nature take its own course. Technological optimists, on the other hand, claim that we need to apply even more engineering to make nature do whatever we want. The fact is that we have intervened for so long and to such a great extent in the Everglades that we can't just walk away now and hope that things will work out for the best. At the same time, we need to recognize that the Everglades are so complex and highly variable that we will probably never understand or control them completely.

Ecologists applaud the attempt to restore natural functions of the ecosystem and to manage it as an entire system rather than to tackle problems as if they were each unique and unconnected to any others. They suggest **adaptive management,** in which action plans are continuously monitored, evaluated, and adjusted to fit changing realities and understandings of the system (fig. 15.9). We also need adaptive models, they propose, for social and institutional change. In some ways, human behavior may be more difficult to understand and predict than that of nature.

Ecosystem management, recognizes that we often need to take action in the face of uncertainty. Decision makers who work at multiple scales should see the interconnections between different levels of organization. If we hope to preserve biodiversity, defend ecological integrity, and maintain ecological processes, we have to recognize ecological boundaries.

Plan — Act — Monitor — Evaluate — Adjust

Adaptive Management

FIGURE 15.9 Adaptive management recognizes that we need to treat management plans for ecosystems as a scientific experiment in which we monitor, evaluate, and adjust our policies to fit changing conditions and knowledge.

TABLE 15.2

Principles of Ecosystem Management

- Plan at multiple scales and time periods
- Use ecological boundaries rather than political ones
- Protect essential ecological processes
- Monitor the system and use adaptive management
- Allow for organizational change
- Make human dimensions part of the plan
- Identify values of nature and stakeholders

Humans can't be separated from nature. We inevitably affect ecological processes and are in turn affected by them. We also need to manage over time periods much longer than the normal planning cycle or term-in-office for most human institutions (table 15.2).

Of course it takes good science to understand and manage complex ecosystems, but what is good science? Advocates from across the political spectrum tend to promote evidence that supports their own agenda, but to label data that contradicts their ideology as "junk science." Does that mean that all research is equally valid? As we discussed in chapter 1, there are objective standards to identify good science. In the first place, it needs to be impartial and open to public scrutiny. Secret, proprietary research may be useful in business, but it shouldn't be used to formulate public policy. Research in which some special interest can determine what questions will be asked, or how data are collected, analyzed, and published, doesn't usually yield sound science. For evidence to be considered valid, it must be reproducible. Materials, methods, and results should be published in peer-reviewed journals in enough detail so that others can independently replicated them. Scientists

have a right—as does everyone—to their own personal opinions and to participate in public debates. Sometimes issues, such as protecting the Laguna San Ignacio whale nursery, described at the beginning of this chapter, are so important that leading scientists feel compelled to speak out. But they have to be careful to distinguish when they are speaking as scientific experts, and when they're speaking as interested citizens.

INTERNATIONAL TREATIES AND CONVENTIONS

As recognition of the interconnections in our global environment has advanced, the willingness of nations to enter into protective **international treaties and conventions** has grown (table 15.3). The earliest of these conventions had no nations as participants; they were negotiated entirely by panels of experts. Not only the number of parties taking part in these negotiations has grown, but the rate at which parties are signing on and the speed at which agreements take force also have increased rapidly (fig. 15.10). The Convention on International Trade in Endangered Species (CITES), for example, was not enforced until 14 years after ratification, but the Convention on Biological Diversity was enforceable after just one year and had 160 signatories only four years after introduction. Over the past 25 years, more than 170 treaties and conventions have been negotiated to protect our global environment. Designed to regulate activities ranging from intercontinental shipping of hazardous waste, to deforestation, overfishing, trade in endangered species, global warming, and wetland protection, these agreements theoretically cover almost every aspect of human impacts on the environment.

Unfortunately, many of these environmental treaties constitute little more than vague good intentions. Even though we often call them laws, there is no body that can legislate or enforce

TABLE 15.3

Some Major International Treaties and Conventions

Date	Title	Covers
1971	Convention on Wetlands of International Importance (Ramsar)	Protects wetlands, especially as waterfowl habitat
1972	Convention to Protect World Cultural and Natural Heritage	Protects cultural sites and natural resources
1973	Convention on International Trade in Endangered Species (CITES)	Restricts trade in endangered plants and animals
1979	Convention on Migratory Species (CMS)	Protects migratory species, especially birds
1982	United Nations Law of the Sea (UNCLOS)	Declares the oceans international commons
1985	Protocol on Substances That Deplete the Ozone Layer (Ozone)	Initiates phaseout of chlorofluorocarbons
1989	Convention on the Transboundary Movements of Hazardous Waste (Basel)	Bans shipment of hazardous waste
1992	Convention on Biodiversity (CBD)	Protects biodiversity as national resources
1992	United Nations Framework Convention on Climate Change (UNFCCC)	Rolls back carbon dioxide production in industrialized countries
1994	Convention to Combat Desertification (CCD)	Provides assistance to fight desertification, especially in Africa

FIGURE 15.10 Additions of participating parties to some major international environmental treaties. The thick portion of each line shows when the agreement went into effect (*date in parentheses*). See table 15.3 for complete treaty names.
Source: Data from United Nations Environment Programme from Global Environment Outlook–2000.

international environmental protection. The United Nations and a variety of regional organizations bring stakeholders together to negotiate solutions to a variety of problems, but the agreed-upon solutions generally rely on moral persuasion and public embarrassment for compliance. Most nations are unwilling to give up sovereignty. There is an international court, but it has no enforcement power. Nevertheless, there are creative ways to strengthen international environmental protection.

One of the problems with most international treaties is the tradition that they must be passed by unanimous consent. A single recalcitrant nation effectively has veto power over the wishes of the vast majority. For instance, more than 100 countries at the UN Conference on Environment and Development (UNCED), held in Rio de Janeiro in 1992, agreed to restrictions on the release of greenhouse gases. At the insistence of U.S. negotiators, however, the climate convention was reworded so that it only urged—but did not require—nations to stabilize their emissions.

As a way of avoiding this problem, some treaties incorporate innovative voting mechanisms. When a consensus cannot be reached, they allow a qualified majority to add stronger measures in the form of amendments that do not need ratification. All members are legally bound to the whole document unless they expressly object. This approach was used in the Montreal Protocol, passed in 1987 to halt the destruction of stratospheric ozone by chlorofluorocarbons (CFCs). The agreement allowed a vote of two-thirds of the 140 participating nations to amend the protocol. Although initially the protocol called for only a 50 percent reduction in CFC production, subsequent research showed that ozone was being depleted faster than previously thought (chapter 9). The protocol was strengthened by amendment to an outright ban on CFC production, in spite of the objection of a few countries.

When strong accords with meaningful sanctions cannot be passed, sometimes the pressure of world opinion generated by revealing the sources of pollution can be effective. Activists can use this information to expose violators. For example, the environmental group Greenpeace discovered monitoring data in 1990

showing that Britain was disposing of coal ash in the North Sea. Although not explicitly forbidden by the Oslo Convention on ocean dumping, this evidence proved to be an embarrassment, and the practice was halted.

Trade sanctions can be an effective tool to compel compliance with international treaties. The Montreal Protocol, for example, bound signatory nations not to purchase CFCs or products made using them from countries that refused to ratify the treaty. Because many products employed CFCs in their manufacture, this stipulation proved to be very effective. On the other hand, trade agreements also can work against environmental protection. The World Trade Organization (WTO) was established to make international trade more fair and to encourage development. It has been used, however, to expand trade globalization and to weaken national environmental laws. In 1990, after decades of public protests and boycotts, the United States banned the import of tuna caught using methods that killed thousands of dolphins each year. Shrimp caught with nets that kill endangered sea turtles were also banned. Mexico filed a complaint with the WTO, contending that dolphin-safe tuna laws represented an illegal barrier to trade. Thailand, Malaysia, India, and Pakistan filed a similar suit over turtle-friendly shrimp laws. The WTO ordered the United States to allow the import of both tuna and shrimp from countries that allow fisheries to kill dolphins and turtles. Environmentalists point out that the WTO has never ruled against a corporate suit, in part because rules are decided by a committee composed mainly of appointed industry leaders, rather than by elected officials representing a spectrum of political and economic interests.

WHAT CAN INDIVIDUALS DO?

Whatever your skills and interests, you can participate in policy formation and contribute to understanding and protecting our common environment. If you enjoy science, there are many disciplines that contribute to environmental science. As you know

Principles of Environmental Science   http://www.mhhe.com/cunningham4e

FIGURE 15.11 Enjoying nature helps you understand what needs to be saved, and it gives you motivation to keep working.

FIGURE 15.12 Environmental education helps develop awareness and appreciation of ecological systems and how they work.

by now, biology, chemistry, geology, ecology, climatology, geography, demography, and other sciences all provide essential ideas and data to environmental science. Environmental scientists usually focus on one of these disciplines, but their work serves the others. An environmental chemist, for example, might study contaminants in a stream system, and this work might help an aquatic ecologist understand changes in a stream's food web.

You can also help seek environmental solutions if you prefer writing, art, working with children, history, politics, economics, or other areas of study. As you have read, environmental science depends on communication, education, good policies, and economics as well as on science.

Believe it or not, you can also help us understand our environment just by enjoying it (fig. 15.11). As author Edward Abbey wrote,

> It is not enough to fight for the land; it is even more important to enjoy it. While it is still there. So get out there and mess around with your friends, ramble out yonder and explore the forests, encounter the grizz, climb the mountains. Run the rivers, breathe deep of that yet sweet and lucid air, sit quietly for a while and contemplate the precious stillness, that lovely mysterious and awesome space.

When you understand how environmental systems function—from nutrient cycles and energy flows to ecosystems, climate systems, population dynamics, agriculture, and economies—you can develop well-informed opinions and help find useful answers.

Environmental education is an important tool

In 1990 Congress recognized the importance of environmental education by passing the National Environmental Education Act. The act established two broad goals: (1) to improve understanding among the general public of the natural and built environment

---

TABLE 15.4

Outcomes from Environmental Education

The natural context: An environmentally educated person understands the scientific concepts and facts that underlie environmental issues and the interrelationships that shape nature.

The social context: An environmentally educated person understands how human society is influencing the environment, as well as the economic, legal, and political mechanisms that provide avenues for addressing issues and situations.

The valuing context: An environmentally educated person explores his or her values in relation to environmental issues; from an understanding of the natural and social contexts, the person decides whether to keep or change those values.

The action context: An environmentally educated person becomes involved in activities to improve, maintain, or restore natural resources and environmental quality for all.

Source: Data from *A Greenprint for Minnesota*, Minnesota Office of Environmental Education, 1993.

---

and the relationships between humans and their environment, including global aspects of environmental problems, and (2) to encourage postsecondary students to pursue careers related to the environment. Specific objectives proposed to meet these goals include developing an awareness and appreciation of our natural and social/cultural environment, knowledge of basic ecological concepts, acquaintance with a broad range of current environmental issues, and experience in using investigative, critical-thinking, and problem-solving skills in solving environmental problems (fig. 15.12). Several states, including Arizona, Florida, Maryland, Minnesota, Pennsylvania, and Wisconsin, have successfully incorporated these goals and objectives into their curricula (table 15.4).

The Environmentalist's Bookshelf

What are some of the most influential and popular environmental books? In a survey[1] of environmental experts and leaders around the world on the best books on nature and the environment, the top ten were

A Sand County Almanac by Aldo Leopold (100)[2]
Silent Spring by Rachel Carson (81)
State of the World by Lester Brown and the Worldwatch Institute (31)
The Population Bomb by Paul Ehrlich (28)
Walden by Henry David Thoreau (28)
Wilderness and the American Mind by Roderick Nash (21)
Small Is Beautiful: Economics as If People Mattered by E. F. Schumacher (21)
Desert Solitaire: A Season in the Wilderness by Edward Abbey (20)
The Closing Circle: Nature, Man, and Technology by Barry Commoner (18)
The Limits to Growth: A Report for the Club of Rome's Project on the Predicament of Mankind by Donella H. Meadows et al. (17)

[1]Robert Merideth, 1992, G. K. Hall/Macmillan, Inc.

[2]Indicates number of votes for each book. Because the preponderance of respondents were from the United States (82 percent), American books are probably overrepresented.

Speaking in support of the National Environmental Education Act, former Environmental Protection Agency administrator William K. Reilly called for broad **environmental literacy** in which every citizen is fluent in the principles of ecology and has a "working knowledge of the basic grammar and underlying syntax of environmental wisdom." Environmental literacy, according to Reilly can help establish a stewardship ethic—a sense of duty to care for and manage wisely our natural endowment and our productive resources for the long haul. It isn't enough, he says, "for a few specialists to know what is going on while the rest of us wander about in ignorance."

You have made a great start toward learning about your environment by reading this book and taking a class in environmental science. Pursuing your own environmental literacy is a life-long process. Some of the most influential environmental books of all time examine environmental problems and suggest solutions (table 15.5). To this list we'd add some personal favorites: *The Singing Wilderness* by Sigurd F. Olson, *My First Summer in the Sierra* by John Muir, and *Remaking Society: Pathways to a Green Future* by Murray Bookchin.

Citizen science encourages everyone to participate

Many students are discovering they can make authentic contributions to scientific knowledge through active learning and undergraduate research programs. Internships in agencies or envi-

ronmental organizations are one way of doing this. Another is to get involved in organized **citizen science** projects in which ordinary people join with established scientists to answer real scientific questions. Community-based research was pioneered in the Netherlands, where several dozen research centers now study environmental issues ranging from water quality in the Rhine River, cancer rates by geographic area, and substitutes for harmful organic solvents. In each project, students and neighborhood groups team with scientists and university personnel to collect data. Their results have been incorporated into official government policies.

Similar research opportunities exist in the United States and Canada. The Audubon Christmas Bird Count, (see Exploring Science, p. 367), is a good example. Earthwatch offers a much smaller but more intense opportunity to take part in research. Every year hundreds of Earthwatch projects each field a team of a dozen or so volunteers who spend a week or two working on issues ranging from loon nesting behavior to archaeological digs. The American River Watch organizes teams of students to measure water quality. You might be able to get academic credit as well as helpful practical experience in one of these research experiences.

Environmental careers range from engineering to education

The need for both environmental educators and environmental professionals opens up many job opportunities in environmental fields. The World Wildlife Fund estimates, for example, that 750,000 new jobs will be created over the next decade in the renewable energy field alone. Scientists are needed to understand the natural world and the effects of human activity on the environment (fig. 15.13). Lawyers and other specialists are needed to develop government and industry policy, laws, and regulations

FIGURE 15.13 Many interesting, well-paid jobs are opening up in environmental fields. Here, a technician takes a well sample to test for water contamination.

Every Christmas since 1900, dedicated volunteers have counted and recorded all the birds they can find within their team's designated study site (fig. 1). This effort has become the largest, longest-running, citizen-science project in the world. For the 100th count, nearly 50,000 participants in about 1,800 teams observed 58 million birds belonging to 2,309 species. Although about 70 percent of the counts in 2000 were made in the United States or Canada, 650 teams in the Caribbean, Pacific Islands, and Central and South America also participated. Participants enter their bird counts on standardized data sheets, or submit their observations over the Internet. Compiled data can be viewed and investigated online, almost as soon as they are submitted.

Frank Chapman, the editor of *Bird-Lore* magazine and an officer in the newly formed Audubon Society, started the Christmas Bird Count in 1900. For years, hunters had gathered on Christmas Day for a competitive hunt, often killing hundreds of birds and mammals as teams tried to outshoot each other. Chapman suggested an alternative contest: to see which team could observe and identify the most birds, and the most species, in a day. The competition has grown and spread. In the 100th annual count, the winning team was in Monte Verde, Costa Rica, with an amazing 343 species tallied in a single day.

The tens of thousands of bird-watchers participating in the count gather vastly more information about the abundance and distribution of birds than biologists could gather alone. These data provide important information for scientific research on bird migrations, populations, and habitat change. Now that the entire record for a century of bird data is available on the BirdSource website (www.birdsource.org), both professional ornithologists and amateur bird-watchers can study the geographical dis-

FIGURE 1 Participating in citizen-science projects is a good way to learn about science and your local environment.

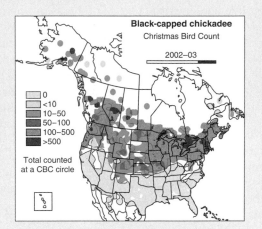

FIGURE 2 Volunteer data collection can produce a huge, valuable data set. Christmas Bird Count data, such as this map, are available online. Data from Audubon Society.

tribution of a single species over time, or they can examine how all species vary at a single site through the years. Those concerned about changing climate can look for variation in long-term distribution of species. Climatologists can analyze the effects of weather patterns such as El Niño or La Niña on where birds occur. One of the most intriguing phe-

nomena revealed by this continent-wide data collection is irruptive behavior: that is, appearance of massive numbers of a particular species in a given area in one year, and then their move to other places in subsequent years following weather patterns, food availability, and other factors.

In 2005 the 105th Christmas Bird Count collected data on nearly 70 million birds from 2,019 volunteer groups This citizen-science effort has produced a rich, geographically broad data set far larger than any that could be produced by professional scientists (fig. 2). Following the success of the Christmas Bird Count, other citizen-science projects have been initiated. Project Feeder Watch, which began in the 1970s, has more than 15,000 participants, from school-children and backyard bird-watchers to dedicated birders. The Great Backyard Bird Count of 2005 collected records on over 600 species and more than 6 million individual birds. In other areas, farmers have been enlisted to monitor pasture and stream health; volunteers monitor water quality in local streams and rivers; and nature reserves solicit volunteers to help gather ecological data. You can learn more about your local environment, and contribute to scientific research, by participating in a citizen-science project. Contact your local Audubon chapter or your state's department of natural resources to find out what you can do.

How does counting birds contribute to sustainability? Citizen-science projects are one way individuals can learn more about the scientific process, become familiar with their local environment, and become more interested in community issues. In this chapter, we'll look at other ways individuals and groups can help protect nature and move toward a sustainable society.

---

to protect the environment. Engineers are needed to develop technologies and products to clean up pollution and to prevent its production in the first place. Economists, geographers, and social scientists are needed to evaluate the costs of pollution and resource depletion and to develop solutions that are socially, culturally, politically, and economically appropriate for different parts of the world. In addition, business will be looking for a

new class of environmentally literate and responsible leaders who appreciate how products sold and services rendered affect our environment.

Trained people are essential in these professions at every level, from technical and clerical support staff to top managers. Perhaps the biggest national demand over the next few years will be for environmental educators to help train an environmentally

FIGURE 15.14 Is shopping our highest purpose?

literate populace. We urgently need many more teachers at every level who are trained in environmental education. Outdoor activities and natural sciences are important components of this mission, but environmental topics such as responsible consumerism, waste disposal, and respect for nature can and should be incorporated into reading, writing, arithmetic, and every other part of education.

How much is enough?

Our consumption of resources and disposal of wastes often have destructive impacts on the earth. Technology has made consumer goods and services cheap and readily available in the richer countries of the world. As you already know, we in the industrialized world use resources at a rate out of proportion to our percentage of the population. If everyone in the world were to attempt to live at our level of consumption, given current methods of production, the results would surely be disastrous. In this section we will look at some options for consuming less and reducing our environmental impacts. Perhaps no other issue in this book represents so clear an ethical question as the topic of responsible consumerism.

A century ago, economist and social critic, Thorstein Veblen, in his book, *The Theory of the Leisure Class,* coined the term **conspicuous consumption** to describe buying things we don't want or need just to impress others. How much more shocked he would be to see current trends. The average American now consumes twice as many goods and services as in 1950. The average house is now more than twice as big as it was 50 years ago, even though the typical family has half as many people. We need more space to hold all the stuff we buy. Shopping has become the way many people define themselves. As Marx predicted, everything has become commodified; getting and spending have eclipsed family, ethnicity, even religion as the defining matrix of our lives. But the futility and irrelevance of much American consumerism leaves a psychological void. Once we possess things, we find they don't make us young, beautiful, smart, and interesting as they promised. With so much attention on earning and spending money, we don't have time to have real friends, to cook

real food, to have creative hobbies, or to do work that makes us feel we have accomplished something with our lives. Some social critics call this drive to possess stuff **"affluenza."**

A growing number of people find themselves stuck in a vicious circle: They work frantically at a job they hate, to buy things they don't need, so they can save time to work even longer hours (fig. 15.14). Seeking a measure of balance in their lives, some opt out of the rat race and adopt simpler, less-consumptive lifestyles. As Thoreau wrote in *Walden,* "Our life is frittered away by detail . . . simplify, simplify."

The United Nations Environment Programme (UNEP) has held workshops on sustainable consumption in Paris and Tokyo. Recognizing that making people feel guilty about their lifestyles and purchasing habits isn't working, UNEP is attempting to find ways to make sustainable living something consumers will adopt willingly. The goal is economically, socially, and environmentally viable solutions that allow people to enjoy a good quality of life while consuming fewer natural resources and polluting less. A good example of this approach is a British automaker that provides a mountain bike with every car it sells, urging buyers to use the bike for short journeys. Another example cited by UNEP is European detergent makers who encourage customers to switch to low-temperature washing liquids and powders, not just to save energy but because it's good for their clothes.

Although each of our individual choices may make a small impact, collectively they can be important. The What Can You Do? box on p. 369 offers some suggestions for reducing waste and pollution.

Green consumerism has its limits

To quote Kermit the Frog, "It's not easy being green." Even with the help of endorsement programs, doing the right thing from an environmental perspective may not be obvious. Often we are faced with complicated choices. Do the social benefits of buying rain-forest nuts justify the energy expended in transporting them here, or would it be better to eat only locally grown products? In switching from Freon propellants to hydrocarbons, we spare the stratospheric ozone but increase hydrocarbon-caused smog.

What Can You Do?

Reducing Your Impact

Purchase Less

Ask yourself whether you really need more stuff.

Avoid buying things you don't need or won't use.

Use items as long as possible (and don't replace them just because a new product becomes available).

Use the library instead of purchasing books you read.

Make gifts from materials already on hand, or give nonmaterial gifts.

Reduce Excess Packaging

Carry reusable bags when shopping and refuse bags for small purchases.

Buy items in bulk or with minimal packaging; avoid single-serving foods.

Choose packaging that can be recycled or reused.

Avoid Disposable Items

Use cloth napkins, handkerchiefs, and towels.

Bring a washable cup to meetings; use washable plates and utensils rather than single-use items.

Buy pens, razors, flashlights, and cameras with replaceable parts.

Choose items built to last and have them repaired; you will save materials and energy while providing jobs in your community.

Conserve Energy

Walk, bicycle, or use public transportation.

Turn off (or avoid turning on) lights, water, heat, and air conditioning when possible.

Put up clotheslines or racks in the backyard, carport, or basement to avoid using a clothes dryer.

Carpool and combine trips to reduce car mileage.

Save Water

Water lawns and gardens only when necessary.

Use water-saving devices and fewer flushes with toilets.

Don't leave water running when washing hands, food, dishes, and teeth.

Based on material by Karen Oberhauser, Bell Museum Imprint, University of Minnesota, 1992. Used by permission.

By choosing reusable diapers over disposable ones, we decrease the amount of material going to the landfill, but we also increase water pollution, energy consumption, and pesticide use (cotton is one of the most pesticide-intensive crops grown in the United States).

When the grocery store clerk asks you, "Paper or plastic?" you probably choose paper and feel environmentally virtuous, right? Everyone knows that plastic is made by synthetic chemical processes from nonrenewable petroleum or natural gas. Paper from naturally growing trees is a better environmental choice, isn't it? Well, not necessarily. In the first place, paper making consumes water and causes much more water pollution than does plastic manufacturing. Paper mills also release air pollutants, including foul-smelling sulfides and captans as well as highly toxic dioxins.

Furthermore, the brown paper bags used in most supermarkets are made primarily from virgin paper. Recycled fibers aren't strong enough for the weight they must carry. Growing, harvesting, and transporting logs from agroforestry plantations can be as environmentally disruptive as oil production. It takes a great deal of energy to pulp wood and dry newly made paper. Paper is also heavier and bulkier to ship than plastic. Although the polyethylene used to make a plastic bag contains many calories, in the end, paper bags are generally more energy-intensive to produce and market than plastic ones.

If both paper and plastic go to a landfill in your community, the plastic bag takes up less space. It doesn't decompose in the landfill, but neither does the paper in an air-tight, water-tight landfill. If paper is recycled but plastic is not, then the paper bag may be the better choice. If you are lucky enough to have both paper and plastic recycling, the plastic bag is probably a better choice since it recycles more easily and produces less pollution in the process. The best choice of all is to bring your own reusable cloth bag.

Complicated, isn't it? We often must make decisions without complete information, but it's important to make the best choices we can. Don't assume that your neighbors are wrong if they reach conclusions different from yours. They may have valid considerations of which you are unaware. The truth is that simple black and white answers often don't exist.

Taking personal responsibility for your environmental impact can have many benefits. Recycling, buying green products, and other environmental actions not only set good examples for your friends and neighbors, they also strengthen your sense of involvement and commitment in valuable ways. There are limits, however, to how much we can do individually through our buying habits and personal actions to bring about the fundamental changes needed to save the earth. Green consumerism generally can do little about larger issues of global equity, chronic poverty, oppression, and the suffering of millions of people in the developing world. There is a danger that exclusive focus on such problems as whether to choose paper or plastic bags, or to sort recyclables for which there are no markets, will divert our attention from the greater need to change basic institutions.

HOW CAN WE WORK TOGETHER?

While a few exceptional individuals can be effective working alone to bring about change, most of us find it more productive and more satisfying to work with others.

Collective action multiplies individual power (fig. 15.15). You get encouragement and useful information from meeting regularly with others who share your interests. It's easy to get discouraged by the slow pace of change; having a support group helps maintain

FIGURE 15.15 Student volunteers plant native trees and bushes in a watershed protection project. Working together on a practical problem can be enjoyable and productive.

your enthusiasm. You should realize, however, that there is a broad spectrum of environmental and social action groups. Some will suit your particular interests, preferences, or beliefs more than others. In this section, we will look at some environmental organizations as well as options for getting involved.

Student environmental groups can have lasting effects

A number of organizations have been established to teach ecology and environmental ethics to elementary and secondary school students, as well as to get them involved in active projects to clean up their local community. Groups such as Kids Saving the Earth or Eco-Kids Corps are an important way to reach this vital audience. Family education results from these efforts as well. In a World Wildlife Fund survey, 63 percent of young people said they "lobby" their parents about recycling and buying environmentally responsible products.

Organizations for secondary and college students often are among our most active and effective groups for environmental change. The largest student environmental group in North America is the **Student Environmental Action Coalition (SEAC).** Formed in 1988 by students at the University of North Carolina at Chapel Hill, SEAC has grown rapidly to more than 30,000 members in some 500 campus environmental groups. SEAC is both an umbrella organization and grassroots network that functions as an information clearinghouse and a training center for student leaders. Member groups undertake a diverse spectrum of activities ranging from politically neutral recycling promotion to confrontational protests of government or industrial projects. National conferences bring together thousands of activists who share tactics and inspiration while also having fun. If there isn't a group on your campus, why not look into organizing one?

Another important student organizing group is the network of Public Interest Research Groups active on most campuses in the United States. While not focused exclusively on the environment, the PIRGs usually include environmental issues in their priorities for research. By becoming active, you could probably introduce environmental concerns to your local group if they are not already working on problems of importance to you. Remember that you are not alone. Others share your concerns and want to work with you to bring about change; you just have to find them. There is power in numbers. As Margaret Mead once said, "Never doubt that a small, highly committed group of individuals can change the world; indeed, it is the only thing that ever has."

Schools can be sources of information and experimentation in sustainable living. They have knowledge and expertise to figure out how to do new things, and they have students who have the energy and enthusiasm to do much of the research, and for whom that discovery will be a valuable learning experience. At many colleges and universities, students have undertaken campus audits to examine water and energy use, waste production and disposal, paper consumption, recycling, buying locally produced food, and many other examples of sustainable resource consumption. At more than 100 universities and colleges across America, graduating students have taken a pledge that reads:

> "I pledge to explore and take into account the social and environmental consequences of any job I consider and will try to improve these aspects of any organization for which I work."

Could you introduce something similar at your school?

Campuses often have building projects that can be models for sustainability research and development. Some recent examples of prize-winning sustainable design can be found at Stanford University, Oberlin College in Ohio, and the University of California at Santa Barbara. Stanford's Jasper Ridge building will provide classroom, laboratory, and office space for its biological research station. Stanford students worked with the administration to develop *Guidelines for Sustainable Buildings,* a booklet that covers everything from energy-efficient lighting to native landscaping. With 275 photovoltaic panels to catch sunlight, there should be no need to buy electricity for the building. In fact, it's expected that surplus energy will be sold back to local utility companies to help pay for building operation.

Oberlin's Environmental Studies Center, designed by architect Bill McDonough, features 370 m^2 of photovoltaic panels on its roof, a geothermal well to help heat and cool the building, large south-facing windows for passive solar gain, and a "living machine" for water treatment, including plant-filled tanks in an indoor solarium and a constructed wetland outside (fig. 15.16).

The Campus Climate Challenge, recently launched by a coalition of nonprofit groups, seeks to engage students, faculty, and staff at 500 college campuses in the United States and Canada in a long-term campaign to eliminate global warming pollution. Many campuses have invested in clean energy, set

FIGURE 15.16 The Adam Joseph Lewis Center for Environmental Studies at Oberlin College is designed to be self-sustaining even in northern Ohio's cool, cloudy climate. Large, south-facing windows let in sunlight, while 370 m² of solar panels on the roof generate electricity. A constructed wetland outside and a living machine inside purify wastewater.

FIGURE 15.17 Growth of national environmental organizations in the United States.

strict green building standards for new construction, purchased fuel-efficient vehicles, and adopted other policies to save energy and reduce their greenhouse gas emissions. Some examples include Concordia University in Austin, Texas, the first college or university in the country to purchase all of its energy from renewable sources. The 5.5 million kilowatt-hours of "green power" it uses each year will eliminate about 8 million pounds of CO_2 emissions annually, the equivalent of planting 1,000 acres of trees or taking 700 cars off the roads. Emory University in Atlanta, Georgia, has taken the lead in green building standards, with 11 buildings that are or could become LEED certified. Emory's Whitehead Biomedical Research Building was the first facility in the Southeast to be LEED certified. Like a number of other colleges, Carleton College in Northfield, Minnesota, has built its own windmill, which is expected to provide about 40 percent of the school's electrical needs. The $1.8 million wind turbine is expected to pay for itself in about ten years. The Campus Climate Challenge website at http://www.energyaction.net contains valuable resources, including strategies and case studies, an energy action packet, a campus organizing guide, and more.

At many schools, students have persuaded the administration to buy locally produced food and to provide organic, vegetarian, and fair trade options in campus cafeterias. This not only benefits your health and the environment, but can also serve as a powerful teaching tool and everyday reminder that individuals can make a difference. Could you do something similar at your school? See the Data Analysis box at the end of this chapter for other suggestions.

National organizations are influential but sometimes complacent

Among the oldest, largest, and most influential environmental groups in the United States are the National Wildlife Federation, the World Wildlife Fund, the Audubon Society, the Sierra Club, the Izaak Walton League, Friends of the Earth, Greenpeace, Ducks Unlimited, the Natural Resources Defense Council, and The Wilderness Society. Sometimes known as the "group of 10," these organizations are criticized by radical environmentalists for their tendency to compromise and cooperate with the establishment. Although many of these groups were militant—even extremist—in their formative stages, they now tend to be more staid and conservative. Members are mostly passive and know little about the inner workings of the organization, joining as much for publications or social aspects as for their stands on environmental issues. Collectively, these groups grew rapidly during the 1980s (fig. 15.17), but many of their new members had little contact with them beyond making a one-time donation.

Still, these groups are powerful and important forces in environmental protection. Their mass membership, large professional staffs, and long history give them a degree of respectability and influence not found in newer, smaller groups. The Sierra Club, for instance, with about half a million members and chapters in almost every state, has a national staff of about 400, an annual budget over $20 million, and 20 full-time professional lobbyists in Washington, D.C. These national groups have become a potent force in Congress, especially when they band together to pass specific legislation, such as the Alaska National Interest Lands Act or the Clean Air Act.

In a survey that asked congressional staff and officials of government agencies to rate the effectiveness of groups that attempt to influence federal policy on pollution control, the top five were national environmental organizations. In spite of their large budgets and important connections, the American Petroleum Institute, the Chemical Manufacturers Association, and the Edison Electric Institute ranked far behind these environmental groups in terms of influence.

Although much of the focus of the big environmental groups is in Washington, Audubon, Sierra Club, and Izaak Walton have local chapters, outings, and conservation projects. This can be a good way to get involved. Go to some meetings, volunteer, offer to help. You may have to start out stuffing envelopes or some other unglamorous job, but if you persevere, you may have a chance to do something important and fun. It's a good way to learn and meet people.

Some environmental groups, such as the Environmental Defense Fund (EDF), The Nature Conservancy (TNC), the National Resources Defense Council (NRDC), and the Wilderness Society (WS), have limited contact with ordinary members except through their publications. They depend on a professional staff to carry out the goals of the organization through litigation (EDF and NRDC), land acquisition (TNC), or lobbying (WS). Although not often in the public eye, these groups can be very effective because of their unique focus. TNC buys land of high ecological value that is threatened by development. With more than 3,200 employees and assets around $3 billion, TNC manages 7 million acres in what it describes as the world's largest private sanctuary system (fig. 15.18). Still, the Conservancy is controversial for some of its management decisions, such as gas and oil drilling in some reserves, and including executives from some questionable companies on its governing board and advisory council. The Conservancy replies that it is trying to work with these companies to bring about change rather than just criticize them.

Radical groups capture attention and broaden the agenda

A striking contrast to the mainline conservation organizations are the direct action groups, such as Earth First!, Sea Shepherd, and a few other groups that form either the "cutting edge" or the "radical fringe" of the environmental movement, depending on your outlook. Often associated with the deep ecology philosophy and bioregional ecological perspective, the strongest concerns of these militant environmentalists tend to be animal rights and protection of wild nature. Their main tactics are civil disobedience and attention-grabbing actions, such as picketing, protest marches, road blockades, and other demonstrations. Some of these actions are humorous and lighthearted, such as street theater that gets a point across in a nonthreatening way (fig. 15.19). Many of these techniques are borrowed from the civil rights movement and Mahatma Gandhi's nonviolent civil disobedience. While often more innovative than the mainstream organizations, pioneering new issues and new approaches, the tactics of these groups can be controversial.

Members of Earth First! chained themselves to a giant tree-smashing bulldozer to prevent forest clearing in Texas and blockaded roads being built into wilderness areas. In one remarkable

FIGURE 15.18 The Nature Conservancy buys land with high biodiversity or unique natural values to protect it from misuse and development.

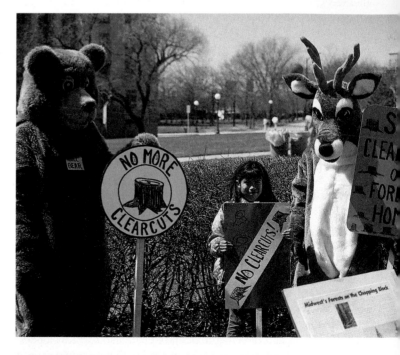

FIGURE 15.19 Street theater can be a humorous, yet effective, way to convey a point in a nonthreatening manner. Confrontational tactics get attention, but they may alienate those who might be your allies and harden your opposition.

Principles of Environmental Science

case, a 25-year-old activist named Julia "Butterfly" Hill spent more than two years sitting on a small, tarp-covered platform 55 m (180 ft) up in a redwood tree in northern California. She was protesting the clear-cutting of the Headwaters Forest by the Maxxam/Pacific Lumber Company. She came down from her perch in 1999 when her supporters raised $50,000 to buy the tree from Maxxam. While some people regard this direct action as heroic and noble, others see it as dangerous and counterproductive. What do you think? Are there circumstances in which it is appropriate or effective to use confrontation rather than cooperation and compromise to achieve your goals? Would you ever break a law to protect nature?

International nongovernmental organizations

As the opening Case Study in this chapter shows, international **nongovernmental organizations (NGOs)** can be vital in the struggle to protect areas of outstanding biological value. Without this help, local groups could never mobilize the public interest or financial support for projects, such as saving Laguna San Ignacio.

The rise in international NGOs in recent years has been phenomenal. At the Stockholm Conference in 1972, only a handful of environmental groups attended, almost all from fully developed countries. Twenty years later, at the Rio Earth Summit, more than 30,000 individuals representing several thousand environmental groups, many from developing countries, held a global Ecoforum to debate issues and form alliances for a better world.

Some NGOs are located primarily in the more highly developed countries of the north and work mainly on local issues. Others are headquartered in the north but focus their attention on the problems of developing countries in the south. Still others are truly global, with active groups in many different countries. A few are highly professional, combining private individuals with representatives of government agencies on quasi-government boards or standing committees with considerable power. Others are on the fringes of society, sometimes literally voices crying in the wilderness. Many work for political change, more specialize in gathering and disseminating information, and some undertake direct action to protect a specific resource.

Public education and consciousness-raising using protest marches, demonstrations, civil disobedience, and other participatory public actions and media events are generally important tactics for these groups. Greenpeace, for instance, carries out well-publicized confrontations with whalers, seal hunters, toxic waste dumpers, and others who threaten very specific and visible resources. Greenpeace may well be the largest environmental organization in the world, claiming some 2.5 million contributing members.

In contrast to these highly visible groups, others choose to work behind the scenes, but their impact may be equally important. Conservation International has been a leader in debt-for-nature swaps to protect areas particularly rich in biodiversity. It also has some interesting initiatives in economic development, seeking products made by local people that will provide income along with environmental protection (fig. 15.20).

FIGURE 15.20 International conservation groups often initiate economic development projects that provide local alternatives to natural resource destruction.

SUSTAINABILITY IS A GLOBAL CHALLENGE

As the developing countries of the world become more affluent, they are adopting many of the wasteful and destructive lifestyle patterns of the richer countries. Automobile production in China, for example, is increasing at about 19 percent per year, or doubling every 3.7 years. By 2030 there could be nearly as many automobiles in China than the United States. What will be the effect on air quality, world fossil fuel supplies, and global climate if that growth rate continues? Already, two-thirds of the children in Shenzhen, China's wealthiest region, suffer from lead poisoning, probably caused by use of leaded gasoline. And, as chapter 8 points out, diseases associated with affluent lifestyles—such as obesity, diabetes, heart attacks, depression, and traffic accidents—are becoming the leading causes of morbidity and mortality worldwide.

We would all benefit by helping developing countries access more efficient, less-polluting technologies. Education, democracy, and access to information are essential for sustainability (fig. 15.21). It is in our best interest to help finance protection of our common future in some equitable way. Maurice Strong, chair of the Earth Charter Council, estimates that development aid from the richer countries should be some $150 billion per year, while internal investments in environmental protection by developing countries will need to be about twice that amount. Many scholars and social activists believe that poverty is at the core of many

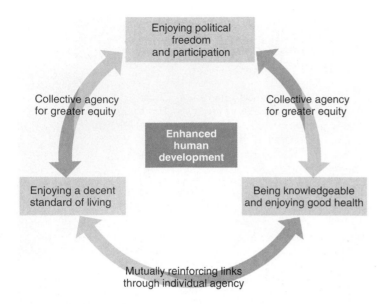

FIGURE 15.21 Human development, democracy, and education are mutually reinforcing.
Source: UN, 2002.

FIGURE 15.22 A model for integrating ecosystem health, human needs, and sustainable economic growth.
Modified from Raymond Grizzle and Christopher Barrett, personal communication.

of the world's most serious human problems: hunger, child deaths, migrations, insurrections, and environmental degradation. One way to alleviate poverty is to foster economic growth so there can be a bigger share for everyone.

Strong economic growth already is occurring in many places. The World Bank projects that if current trends continue, economic output in developing countries will rise by 4 to 5 percent per year in the next 40 years. Economies of industrialized countries are expected to grow more slowly but could still triple over that period. Altogether, the total world output could be quadruple what it is today.

That growth could provide funds to clean up environmental damage caused by earlier, wasteful technologies and misguided environmental policies. It is estimated to cost $350 billion per year to control population growth, develop renewable energy sources, stop soil erosion, protect ecosystems, and provide a decent standard of living for the world's poor. This is a great deal of money, but it is small compared to over $1 trillion per year spent on wars and military equipment.

While growth simply implies an increase in size, number, or rate of something, development, in economic terms, means a real increase in average welfare or well-being. **Sustainable development** based on the use of renewable resources in harmony with ecological systems is an attractive compromise to the extremes of no growth versus unlimited growth (fig. 15.22). Perhaps the best definition of this goal is that of the World Commission on Environment and Development, which defined sustainable development in *Our Common Future* as "meeting the needs of the present without compromising the ability of future generations to meet their own needs." Some goals of sustainable development include:

- A demographic transition to a stable world population of low birth and death rates.

- An energy transition to high efficiency in production and use, coupled with increasing reliance on renewable resources.

- A resource transition to reliance on nature's "income" without depleting its "capital."

- An economic transition to sustainable development and a broader sharing of its benefits.

- A political transition to global negotiation grounded in complementary interests between North and South, East and West.

- An ethical or spiritual transition to attitudes that do not separate us from nature or each other.

In 2000, United Nations Secretary-General Kofi Annan called for a **millennium assessment** of the consequences of ecosystem change on human well-being as well as the scientific basis for actions to enhance the conservation and sustainable use of those systems. More than 1,360 experts from around the world worked on technical reports about the conditions and trends of ecosystems, scenarios for the future, and possible responses. Each part of this study was scrutinized by governments, independent scientists, and other experts to ensure the robustness of its findings. Some of the key findings of the millennium assessment board of governors are presented in table 15.6. They serve as a good summary for this whole book. What national policies and individual actions do you think could reverse some of the threats identified in the millennium assessment and lead to a more sustainable life for everyone?

TABLE 15.6

Key Messages from the Millennium Assessment

- Everyone in the world depends on nature and ecosystem services to provide the conditions for a decent, healthy, and secure life.

- Humans have made unprecedented changes to ecosystems in recent decades to meet growing demands for food, fresh water, fiber, and energy.

- These changes have helped to improve the lives of billions, but at the same time they weakened nature's ability to deliver other key services, such as purification of air and water, protection from disasters, and the provision of medicine.

- Among the outstanding problems identified by the millennium assessment are the dire state of many of the world's fish stocks; the intense vulnerability of the 2 billion people living in the dry regions to the loss of ecosystem services, including water supply; and the growing threat to ecosystems from climate change and nutrient pollution.

- Human activities have taken the planet to the edge of a massive wave of species extinctions, further threatening our own well-being.

- The loss of services derived from ecosystems is a significant barrier to the achievement of the millennium assessment goals to reduce poverty, hunger, and disease.

- The pressures on ecosystems will increase globally in coming decades unless human attitudes and actions change.

- Measures to conserve natural resources are more likely to succeed if local communities are given ownership of them, share the benefits, and are involved in decisions.

- Even today's technology and knowledge can reduce considerably the human impact on ecosystems. They are unlikely to be deployed fully, however, until ecosystem services cease to be perceived as free and limitless, and their full value is taken into account.

- Better protection of natural assets will require coordinated efforts across all sections of governments, businesses, and international institutions. The productivity of ecosystems depends on policy choices regarding investment, trade, subsidy, taxation, and regulation, among others.

Source: Watson, Robert T., and A. H. Zakri. 2005. *Living Beyond Our Means: Natural Assets and Human Well-Being.* United Nations Publications.

REVIEW QUESTIONS

1. Why is Laguna San Ignacio important? What threatened this area?

2. What is a *policy*? How are policies formed?

3. List seven factors that can keep public policies from being rational or fair.

4. What is *NEPA*, and why is it controversial?

5. Describe three important provisions of NEPA.

6. List four important U.S. environmental laws (besides NEPA), and briefly describe what each does.

7. What's the difference between case law and civil law?

8. Compare some of the requirements of adaptive management and ecosystem management.

9. Why are international environmental conventions and treaties often ineffective? What can make them more successful?

10. Why is the World Trade Organization controversial?

11. In figure 15.10, why is part of each colored line thin while the rest is thick? What does this graph show, in general terms?

12. List two broad goals of environmental education identified by the National Environmental Education Act.

13. What is *citizen science,* and what are some of its benefits? Describe one such important project.

14. What conclusions would you draw from figure 2 on p. 367? Why are there isolated green circles in far northern Canada and Alaska? Does this mean that there are populations of birds living in these places, but not anywhere around them?

15. List five things each of us could do to help preserve our common environment.

16. Describe some things schools and students have done to promote sustainable living.

17. What are some of the benefits of joining an environmental club or group?

18. What is the environmental pledge taken by many graduating seniors?

19. Define *sustainability* assnd describe some of its principal tenets.

20. List three factors essential for sustainability or human development.

CRITICAL THINKING AND DISCUSSION

Apply the principles you have learned in this chapter to discuss these questions with other students.

1. In your opinion, how much environmental protection is too much? Think of a practical example in which some stakeholders may feel oppressed by governmental regulations. How would you justify or criticize these regulations?

2. Which is a better choice at the grocery store, paper or plastic? How would you evaluate the trade-offs between packaging choices? What evidence would you look for to make this decision in your life?

3. Do you agree with Margaret Mead that a small group of committed individuals is the only thing that can change the world?

What do you think she meant? Think of some examples of groups of individuals who have changed the world. How did they do it?

4. Suppose that you were going to make a presentation on sustainability to your school administrators. What suggestions would you make for changes on your campus? What information would you need to support these proposals?

5. Is sustainability an achievable goal? What do you think are the main barriers to attaining this objective? What have you learned in this book that gives you hope for the future?

SUGGESTED READINGS

Berry, Wendell. 2004. Compromise, hell! *Orion* 23 (6):18–25.

Blatt, Harvey. 2005. *America's Environmental Report Card: Are We Making the Grade?* MIT Press.

DeGraff, J., et al. 2005. *Affluenza: The All-Consuming Epidemic,* 2nd ed. Berrett-Koehler Publishers.

Esty, D. C., et al. 2005. *Environmental Sustainability Index: Benchmarking National Environmental Stewardship.* Yale University Press.

Gunderson, Lance H., C. S. Holling, and S. Light, eds. 1995. *Barriers and Bridges to the Renewal of Ecosystems and Institutions.* Columbia University Press.

Louv, Richard. 2005. *Last Child in the Woods: Saving Our Children from Nature-Deficit Disorder.* Algonquin Books.

McKibben, Bill. 2005. *Wandering Home.* Crown Journeys.

National Wildlife Federation. 2004. *Conservation Directory 2004: The Guide to Worldwide Environmental Organizations.* Island Press.

Schor, Juliet B. 2004. *Born to Buy: The Commercialized Child and the New Consumer Culture.* Scribner.

Taylor, M. F., et al. 2005. The effectiveness of the Endangered Species Act: A quantitative analysis. *BioScience* 55 (4):360–7.

 DATA analysis   Campus Environmental Audit

How sustainable is your school? What could you, your fellow students, the faculty, staff, and administration do to make your campus more environmentally friendly? Perhaps you and your classmates could carry out an environmental audit of your school. Some of the following items are things you could observe for yourself; other information you'd need to get from the campus administrators.

1. *Energy.* How much total energy does your campus use each year? Is any of it from renewable sources? How does your school energy use compare to that of a city with the same population? Could you switch to renewable sources? How much would that cost? How long would the payback time be for various renewable sources? Is there a campus policy about energy conservation? What would it take to launch a campaign for using resources efficiently?

2. *Buildings.* Are any campus buildings now LEED certified? Do any campus buildings now have compact fluorescent bulbs, high-efficiency fans, or other energy-saving devices? Do you have single-pane or double-pane windows? Are lights turned off when rooms aren't in use? At what temperatures (winter and summer) are classrooms, offices, and dorms maintained? Who makes this decision? Could you open a window in hot weather? Are new buildings being planned? Will they be LEED certified? If not, why?

3. *Transportation.* Does your school own any fuel-efficient vehicles (hybrids or other high-mileage models)? If you were making a presentation to an administrator to encourage him or her to purchase efficient vehicles, what arguments would you use? How many students commute to campus? Are they encouraged to carpool or use public transportation? How

might you promote efficient transportation? How much total space on your campus is devoted to parking? What's the cost per vehicle to build and maintain parking? How else might that money be spent to facilitate efficient transportation? Where does runoff from parking lots and streets go? What are the environmental impacts of this storm runoff?

4. *Water use.* What's the source of your drinking water? How much does your campus use? Where does wastewater go? How many toilets are on the campus? How much water does each use for every flush? How much would it cost to change to low-flow appliances? How much would it save in terms of water use and cost?

5. *Food.* What's the source of food served in campus dining rooms? Is any of it locally grown or organic? How much

junk food is consumed annually? What are the barriers to buying locally grown, fair-trade, organic, free-range food? Does the campus grow any of its own food? Would that be possible?

6. *Ecosystem restoration.* Are there opportunities for reforestation, stream restoration, wetland improvements, or other ecological repair projects on your campus. What percentage of the vegetation on campus is native? What might be the benefits of replacing non-native species with indigenous varieties? Have gardeners considered planting species that provide food and shelter for wildlife?

What other aspects of your campus life could you study to improve sustainability? How could you organize a group project to promote beneficial changes in your school's environmental impacts?

For Additional Help in Studying This Chapter, please visit our website at www.mhhe.com/cunningham4e. You will find practice quizzes, a chapter summary, key terms, answers to review questions, additional case studies, regional examples, and an extensive reading list, all of which will help you understand the material in this chapter.

Appendixes

Vegetation

Vegetation Regions

- Needleleaf Forest
- Broadleaf Forest
- Mixed Forest
  (Broadleaf and Needleleaf)
- Woodland and Shrub
  (Mediterranean)
- Short Grass (Steppe)
- Tall Grass (Prairie)
- River Valley and Oasis
- Highlands (Unclassified;
  Vertical Zonation)
- Desert and Desert Shrub
- Savanna Grassland and Shrub
- Wooded Savanna
- Tropical Woodland and Shrub
- Light Tropical Forest
- Tropical Rain Forest
- Heath and Moor
- Tundra and Alpine Vegetation
- Permanent Ice Cover

Scale: 1 to 125,000,000

Vegetation is the most visible consequence of the distribution of temperature and precipitation. The global distribution of vegetation types and the global distribution of climate are closely related, but not all vegetation types are the consequence of temperature and precipitation or other climatic variables. Many types of vegetation, in many areas of the world, are the consequence of human activities, particularly the grazing of domesticated livestock, burning, and forest clearance.

World Population Density

World Population Density

Numbers of persons per square mile

- Uninhabited
- Less than 2
- 2 – 25
- 26 – 50
- 51 – 150
- 151 – 300
- Over 300

Scale: 1 to 125,000,000

0   1000   2000   3000 Kilometers

0   1000   2000 Miles

No feature of human activity is more reflective of environmental conditions than where people live. In the areas of densest population, a mixture of natural and human factors have combined to allow maximum food production, maximum urbanization, and especially concentrated economic activity. Three such great concentrations appear on the map—East Asia, South Asia, and Europe—with a fourth lesser concentration in eastern North America (the "Megalopolis" region of the United States and Canada). The areas of future high density (in addition to those already existing) are likely to be in Middle and South America and Africa, where population growth rates are well above the world average. Population that is extremely dense or growing at an excessive rate when measured against a region's habitability is one of the greatest indicators of environmental deterioration.

Temperature Regions and Ocean Currents

Surface Temperature Regions

- Always cold: polar regions and high altitudes
- Cold winter and cool summer; always cool in tropical higher altitudes
- Cold winter and mild summer
- Cool winter and mild summer
- Hot summer and cold winter
- Hot summer and cool winter
- Hot summer and mild winter
- Always hot
- Always mild

Hot = above 68°F (20°C)
Mild = 50°–68°F (10°–20°C)
Cool = 32°–50°F (0°–10°C)
Cold = below 32°F (0°C)

⟶ Cool/cold current
⟶ Warm current

Scale: 1 to 125,000,000

Along with precipitation, temperature is one of the two most important environmental variables, defining the climatic conditions so essential for the distribution of human activities and the human population. Ocean currents exert a significant influence over the climate of adjacent continents and are the most important mechanism for redistributing surplus heat from the equatorial region into middle and high latitudes.

Glossary

A

acid precipitation Acidic rain, snow, or dry particles deposited from the air due to increased acids released by anthropogenic or natural resources.

acids Substances that release hydrogen atoms in water.

active solar systems Mechanical systems that use moving substances to collect and transfer solar energy.

acute effects A sudden onset of symptoms or effects of exposure to some factor.

acute poverty Insufficient income or access to resources needed to provide the basic necessities for life, such as food, shelter, sanitation, clean water, medical care, and education.

adaptation Physical changes that allow organisms to survive in a given environment.

adaptive management A management plan designed from the outset to "learn by doing" and to actively test hypotheses and adjust treatments as new information becomes available.

administrative law Executive orders, administrative rules and regulations, and enforcement decisions by administrative agencies and special administrative courts.

aerosols Minute particles or liquid droplets suspended in the air.

agency rule-making The formal process of establishing rules and standards by administrative agencies.

albedo A description of a surface's reflective properties.

allergens Substances that activate the immune system and cause an allergic response; may not be directly antigenic themselves but may make other materials antigenic.

ambient air The air immediately around us.

analytical thinking A way of systematic analysis that asks, "How can I break this problem down into its constituent parts?"

anemia Low levels of hemoglobin due to iron deficiency or lack of red blood cells.

anthropocentric Believing that humans hold a special place in nature; being centered primarily on humans and human affairs.

antigens Substances that stimulate the production of, and react with, specific antibodies.

aquifers Porous, water-bearing layers of sand, gravel, and rock below the earth's surface; reservoirs for groundwater.

arbitration A formal process of dispute resolution resulting in a legally binding decision that all parties must obey.

arithmetic growth A pattern of growth that increases at a constant amount per unit time, such as 1, 2, 3, 4 or 1, 3, 5, 7.

atmospheric deposition Sedimentation of solids, liquids, or gaseous materials from the air.

atom The smallest particle that exhibits the characteristics of an element.

atomic number The characteristic number of protons per atom of an element.

autotroph An organism that synthesizes food molecules from inorganic molecules by using an external energy source, such as light energy.

B

barrier islands Low, narrow, sandy islands that form offshore from a coastline.

bases Substances that readily bond with hydrogen ions in an aqueous solution.

Batesian mimicry Evolution by one species to resemble another species that is protected from predators by a venomous stinger, bad taste, or some other defensive adaptation.

benthic The bottom of a sea or lake.

bioaccumulation The selective absorption and concentration of molecules by cells.

biocentric preservation A philosophy that emphasizes the fundamental right of living organisms to exist and to pursue their own ends.

biocentrism The belief that all creatures have rights and values; being centered on nature rather than humans.

biochemical oxygen demand (BOD) A standard test for measuring the amount of dissolved oxygen utilized by aquatic microorganisms.

biodegradable plastics Plastics that can be decomposed by microorganisms.

biodiversity The genetic, species, and ecological diversity of the organisms in a given area.

biogeochemical cycles Movement of matter within or between ecosystems; caused by living organisms, geologic forces, or chemical reactions. The cycling of nitrogen, carbon, sulfur, oxygen, phosphorus, and water are examples.

biogeographical area A region or an ecosystem with characteristic biological, water, and land resources.

biological community The populations of plants, animals, and microorganisms living and interacting in a certain area at a given time.

biological controls Use of natural predators, pathogens, or competitors to regulate pest populations.

biomagnification Increase in concentration of certain stable chemicals (for example, heavy metals or fat-soluble pesticides) in successively higher trophic levels of a food chain or web.

biomass The accumulated biological material produced by living organisms.

biomass fuel Organic material produced by plants, animals, or microorganisms that can be burned directly as a heat source or converted into a gaseous or liquid fuel.

biomass pyramid A metaphor or diagram that explains the relationship between the amounts of biomass at different trophic levels.

biomes Broad, regional types of ecosystems characterized by distinctive climate and soil conditions and distinctive kinds of biological community adapted to those conditions.

bioremediation Use of biological organisms to remove pollution or restore environmental quality.

biosphere The zone of air, land, and water at the surface of the earth that is occupied by organisms.

biosphere reserves World heritage sites identified by the IUCN as worthy for national park or wildlife refuge status because of high biological diversity or unique ecological features.

biota All organisms in a given area.

biotic potential The maximum reproductive rate of an organism, given unlimited resources and ideal environmental conditions.

birth control Any method used to reduce births, including celibacy, delayed marriage, contraception; devices or medications that prevent implantation of fertilized zygotes and induced abortions.

blind experiments A design in which researchers don't know which subjects were given experimental treatment until after data have been gathered and analyzed.

bogs Areas of waterlogged soil that tend to be peaty; fed mainly by precipitation; low productivity; some bogs are acidic.

boreal forest A broad band of mixed coniferous and deciduous trees that stretches across northern North America (and Europe and Asia); its northernmost edge, the taiga, intergrades with the arctic tundra.

brownfields Abandoned or underused urban areas in which redevelopment is blocked by liability or financing issues related to toxic contamination.

C

cancer Invasive, out-of-control cell growth that results in malignant tumors.

capital Any form of wealth, resources, or knowledge available for use in the production of more wealth.

carbohydrate An organic compound consisting of a ring or chain of carbon atoms with hydrogen and oxygen attached; examples are sugars, starches, cellulose, and glycogen.

carbon cycle   The circulation and reutilization of carbon atoms, especially via the processes of photosynthesis and respiration.

carbon management   Projects to reduce carbon dioxide emissions from fossil fuel or to ameliorate their effects.

carbon monoxide   Colorless, odorless, nonirritating but highly toxic gas produced by incomplete combustion of fuel, incineration of biomass or solid waste, or partially anaerobic decomposition of organic material.

carbon sink   Places of carbon accumulation, such as in large forests (organic compounds) or ocean sediments (calcium carbonate).

carcinogens   Substances that cause cancer.

carnivores   Organisms that mainly prey upon animals.

carrying capacity   The maximum number of individuals of any species that can be supported by a particular ecosystem on a long-term basis.

case law   Precedents from both civil and criminal court cases.

cellular respiration   The process in which a cell breaks down sugar or other organic compounds to release energy used for cellular work; may be anaerobic or aerobic, depending on the availability of oxygen.

chain reaction   A self-sustaining reaction in which the fission of nuclei produces subatomic particles that cause the fission of other nuclei.

chaparral   A biological community characterized by thick growth of thorny, evergreen shrubs typical of a Mediterranean climate.

chemical bond   The force that holds molecules together.

chemical compounds   Molecules made up of two or more kinds of atoms held together by chemical bonds.

chemical energy   Potential energy stored in chemical bonds of molecules.

chlorofluorocarbons   Chemical compounds with a carbon skeleton and one or more attached chlorine and fluorine atoms. Commonly used as refrigerants, solvents, fire retardants, and blowing agents.

chloroplasts   Chlorophyll-containing organelles in eukaryotic organisms; sites of photosynthesis.

chronically undernourished   Receiving less than 90 percent of the minimum caloric intake needed for normal growth and development and a healthy, productive life.

chronic effects   Long-lasting results of exposure to a toxin; can be a permanent change caused by a single, acute exposure or a continuous, low-level exposure.

citizen science   Projects in which trained volunteers work with scientific researchers to answer real-world questions.

city   A differentiated community with a sufficient population and resource base to allow residents to specialize in arts, crafts, services, and professional occupations.

civil law   A body of laws regulating relations between individuals or between individuals and corporations concerning property rights, personal dignity and freedom, and personal injury.

classical economics   Modern, Western economic theories of the effects of resource scarcity, monetary policy, and competition on supply and demand of goods and services in the marketplace. This is the basis for the capitalist market system.

clear-cutting   Cutting every tree in a given area, regardless of species or size; an appropriate harvest method for some species; can be destructive if not carefully controlled.

climate   A description of the long-term pattern of weather in a particular area.

closed-canopy   A forest where tree crowns spread over 20 percent of the ground; has the potential for commercial timber harvests.

cloud forests   High mountain forests where temperatures are uniformly cool and fog or mist keeps vegetation wet all the time.

coevolution   The process in which species exert selective pressure on each other and gradually evolve new features or behaviors as a result of those pressures.

cogeneration   The simultaneous production of electricity and steam or hot water in the same plant.

cold front   A moving boundary of cooler air displacing warmer air.

coliform bacteria   Bacteria that live in the intestines (including the colon) of humans and other animals; used as a measure of the presence of feces in water or soil.

commensalism   A symbiotic relationship in which one member is benefited and the second is neither harmed nor benefited.

common law   The body of court decisions that constitutes a working definition of individual rights and responsibilities where no formal statutes define these issues.

communal resource management systems   Resources managed by a community for long-term sustainability.

community-based planning   Involving community stakeholders in pluralistic, adaptive, inclusive, proactive planning.

community ecology   The study of interactions of all populations living in the ecosystem of a given area.

competitive exclusion   A theory that no two populations of different species will occupy the same niche and compete for exactly the same resources in the same habitat for very long.

complexity   The number of species at each trophic level and the number of trophic levels in a community.

composting   The biological degradation of organic material under aerobic (oxygen-rich) conditions to produce compost, a nutrient-rich soil amendment and conditioner.

conifer   A needle-bearing tree that produces seeds in cones.

conservation of matter   In any chemical reaction, matter changes form; it is neither created nor destroyed.

conspicuous consumption   A term coined by economist and social critic Thorstein Veblen to describe buying things we don't want or need to impress others.

consumers   Organisms that obtain energy and nutrients by feeding on other organisms or their remains. See also *heterotroph*.

consumption   The fraction of withdrawn water that is lost in transmission or that is evaporated, absorbed, chemically transformed, or otherwise made unavailable for other purposes as a result of human use.

contour plowing   Plowing along hill contours; reduces erosion.

controlled studies   Comparisons made between two populations that are identical (as far as possible) in every factor except the one being studied.

control rods   Neutron-absorbing material inserted into spaces between fuel assemblies in nuclear reactors to regulate fission reaction.

convection currents   Rising or sinking air currents that stir the atmosphere and transport heat from one area to another. Convection currents also occur in water.

conventional (criteria) pollutants   The seven substances (sulfur dioxide, carbon monoxide, particulates, hydrocarbons, nitrogen oxides, photochemical oxidants, and lead) that make up the largest volume of air quality degradation; identified by the Clean Air Act as the most serious threat of all pollutants to human health and welfare.

convergent evolution   Species evolve from different origins but under similar environmental conditions to have similar traits.

coral reefs   Prominent oceanic features composed of hard, limy skeletons produced by coral animals; usually formed along edges of shallow, submerged ocean banks or along shelves in warm, shallow, tropical seas.

core   The dense, intensely hot mass of molten metal, mostly iron and nickel, thousands of kilometers in diameter at the earth's center.

Coriolis effect   The tendency for air above the earth to appear to be deflected to the right (in the Northern Hemisphere) or the left (in the South) because of the earth's rotation.

corridors   Strips of natural habitat that connect two adjacent nature preserves to allow migration of organisms from one place to another.

cost-benefit analysis (CBA)   An evaluation of large-scale public projects by comparing the costs and benefits that accrue from them.

cover crops   Plants, such as rye, alfalfa, or clover, that can be planted immediately after harvest to hold and protect the soil.

creative thinking   Original, independent thinking that asks, "How might I approach this problem in new and inventive ways?"

criminal law   A body of court decisions based on federal and state statutes concerning wrongs against persons or society.

criteria pollutants   See *conventional pollutants.*

critical factor   The single environmental factor closest to a tolerance limit for a given species at a given time.

critical thinking   An ability to evaluate information and opinions in a systematic, purposeful, efficient manner.

crude birth rate   The number of births in a year per 1000 (using the midyear population).

crude death rate   The number of deaths per thousand persons in a given year; also called crude mortality rate.

crust   The cool, lightweight, outermost layer of the earth's surface that floats on the soft, pliable underlying layers; similar to the "skin" on a bowl of warm pudding.

cultural eutrophication   An increase in biological productivity and ecosystem succession caused by human activities.

D

debt-for-nature swaps   Forgiveness of international debt in exchange for nature protection in developing countries.

deciduous   Trees and shrubs that shed their leaves at the end of the growing season.

decline spiral A catastrophic deterioration of a species, community, or whole ecosystem; accelerates as functions are disrupted or lost in a downward cascade.

decomposer Fungus or bacterium that breaks complex organic material into smaller molecules.

deductive reasoning "Top down" reasoning in which we start with a general principle and derive a testable prediction about a specific case.

deep ecology A philosophy that calls for a profound shift in our attitudes and behavior based on rejection of anthropocentric attitudes; a belief in the sacredness of nature; direct personal action to protect nature.

delta Fan-shaped sediment deposit found at the mouth of a river.

demanufacturing Disassembly of products so components can be reused or recycled.

demographic transition A pattern of falling death rates and birth rates in response to improved living conditions; typically leads to rapid then stabilizing population growth.

demography The statistical study of human populations relating to growth rate, age structure, geographic distribution, etc., and their effects on social, economic, and environmental conditions.

dependency ratio The number of nonworking members compared with working members for a given population.

desalinization (or desalination) Removal of salt from water by distillation, freezing, or ultrafiltration.

desertification Denuding and degrading a once fertile land, initiating a desert-producing cycle that feeds on itself and causes long-term changes in soil, climate, and biota of an area.

deserts Biomes characterized by low moisture levels and infrequent and unpredictable precipitation. Daily and seasonal temperatures fluctuate widely.

detritivore Organisms that consume organic litter, debris, and dung.

dew point The temperature at which condensation occurs for a given concentration of water vapor in the air.

dieback A sudden population decline; also called a population crash.

disability-adjusted life years (DALYs) A health measure that assesses the total burden of disease by combining premature deaths and loss of a healthy life that result from illness or disability.

discharge The amount of water that passes a fixed point in a given amount of time; usually expressed as liters or cubic feet of water per second.

discount rate The amount we discount or reduce the value of a future payment. When you borrow money from the bank at 10 percent annual interest, you are in effect saying that having the money now is worth 10 percent more to you than having the same amount one year from now.

disease A deleterious change in the body's condition in response to destabilizing factors, such as nutrition, chemicals, or biological agents.

dissolved oxygen (DO) content Amount of oxygen dissolved in a given volume of water at a given temperature and atmospheric pressure; usually expressed in parts per million (ppm).

divergent evolution Separation of a species into new types.

diversity The number of species present in a community (species richness), as well as the relative abundance of each species.

DNA Deoxyribonucleic acid; the long, double-helix molecule in the nucleus of cells that contains the genetic code and directs the development and functioning of all cells.

double-blind design Neither the subject (participant) nor the experimenter knows which participants are receiving the experimental or the control treatments until after data have been gathered and analyzed.

drip irrigation Uses pipe or tubing perforated with very small holes to deliver water one drop at a time directly to the soil around each plant.

E

earth charter A set of principles for sustainable development, environmental protection, and social justice developed by a council appointed by the United Nations.

earthquakes Sudden, violent movement of the earth's crust.

ecocentric (ecologically centered) A philosophy that claims moral values and rights for both organisms and ecological systems and processes.

ecofeminism A pluralistic philosophy of respect for nature based on feminist philosophies of justice and egalitarianism.

ecological development A gradual process of environmental modification by organisms.

ecological economics Application of ecological insights to economic analysis; incorporating ecological principles and priorities into economic accounting systems.

ecological niche The functional role and position of a species in its ecosystem, including what resources it uses, how and when it uses the resources, and how it interacts with other species.

ecological services Processes or materials, such as clean water, energy, climate regulation, and nutrient cycling, provided by ecosystems.

ecological succession The process by which organisms gradually occupy a site, alter its ecological conditions, and are eventually replaced by other organisms.

ecology The scientific study of relationships between organisms and their environment. It is concerned with the life histories, distribution, and behavior of individual species as well as the structure and function of natural systems at the level of populations, communities, and ecosystems.

economic development A rise in real income per person; usually associated with new technology that increases productivity or resources.

economic growth An increase in the total wealth of a nation; if population grows faster than the economy, there may be real economic growth, but the share per person may decline.

ecosystem A specific biological community and its physical environment interacting in an exchange of matter and energy.

ecosystem management An integration of ecological, economic, and social goals in a unified systems approach to resource management.

ecosystem restoration To reinstate an entire community of organisms to as near its natural condition as possible.

ecotones Boundaries between two types of ecological communities.

ecotourism A combination of adventure travel, cultural exploration, and nature appreciation in wild settings.

edge effects A change in species composition, physical conditions, or other ecological factors at the boundary between two ecosystems.

electron A negatively charged subatomic particle that orbits around the nucleus of an atom.

element A substance that cannot be broken into simpler units by chemical means.

El Niño A climatic change marked by shifting of a large warm water pool from the western Pacific Ocean toward the east. Wind direction and precipitation patterns are changed over much of the Pacific and perhaps around the world.

emergent disease A new disease or one that has been absent for at least 20 years.

emigration The movement of members from a population.

emission standards Regulations for restricting the amounts of air pollutants that can be released from specific point sources.

endangered species A species considered to be in imminent danger of extinction.

endemism A species that is restricted to a single region, country or other area.

endocrine hormone disrupters Chemicals that interfere with the function of endocrine hormones such as estrogen, testosterone, thyroxine, adrenaline, or cortisone.

energy The capacity to do work, such as moving matter over a distance.

energy recovery The incineration of solid waste to produce useful energy.

environment The circumstances or conditions that surround an organism or a group of organisms as well as the complex of social or cultural conditions that affect an individual or a community.

environmental ethics A search for moral values and ethical principles in human relations with the natural world.

environmental health The science of external factors that cause disease, including elements of the natural, social, cultural and technological worlds in which we live.

environmental impact statement (EIS) An analysis of the effects of any major program or project planned by a federal agency; required by provisions in the National Environmental Policy Act of 1970.

environmental indicators Organisms or physical factors that serve as a gauge for environmental changes. Indicator organisms, which cannot survive beyond certain environmental limits, are known as bioindicators.

environmental justice Fair access to a clean, healthy environment, regardless of class, race, income level, or other status.

environmental law Legal rules, decisions, and actions concerning environmental quality, natural resources, and ecological sustainability.

environmental literacy A basic understanding of ecological principles and the ways society affects, or responds to, environmental conditions.

environmental policy The official rules or regulations concerning the environment adopted, implemented, and enforced by some government agency.

environmental racism Decisions that unfairly expose people to polluted or degraded environments on the basis of race.

environmental science The systematic, scientific study of our environment as well as our role in it.

enzymes Molecules, usually proteins or nucleic acids, that act as catalysts in biochemical reactions.

epidemiology The study of the distribution and causes of disease and injuries in human populations.

epiphyte A plant that grows on a substrate other than the soil, such as the surface of another organism.

equilibrium community Also called a disclimax community; a community subject to periodic disruptions, usually by fire, that prevent it from reaching a climax stage.

estuaries Bays or drowned valleys where a river empties into the sea.

eutrophic Rivers and lakes rich in organic material (*eu* = well; *trophic* = nourished).

evolution A theory that explains how random changes in genetic material and competition for scarce resources cause species to change gradually.

exhaustible resources Materials present in fixed amounts in the environment, especially the earth's geologic endowment: minerals, nonmineral resources, fossil fuels.

exotic organisms Alien species introduced by human agency into biological communities where they would not naturally occur.

exponential growth Growth at a constant rate of increase per unit of time; can be expressed as a constant fraction or exponent. See also *geometric growth.*

externalizing costs Shifting expenses, monetary or otherwise, to someone other than the individuals or groups who use a resource.

extinction The irrevocable elimination of species; can be a normal process of the natural world as species outcompete or kill off others or as environmental conditions change.

extirpate To destroy totally; extinction caused by direct human action, such as hunting, trapping, etc.

F

family planning Controlling reproduction; planning the timing of birth and having only as many babies as are wanted and can be supported.

famines Acute food shortages characterized by large-scale loss of life, social disruption, and economic chaos.

fauna All of the animals present in a given region.

fecundity The physical ability to reproduce.

federal laws (statutes) Laws passed by the federal legislature and signed by the chief executive.

fens Wetlands fed mainly by groundwater.

feral A domestic animal that has taken up a wild existence.

fertility The actual number of offspring produced through sexual reproduction; usually described in terms of number of offspring of females, since paternity can be difficult to determine.

fetal alcohol syndrome A tragic set of permanent physical, mental, and behavioral birth defects that result when mothers drink alcohol during pregnancy.

fire-climax community An equilibrium community maintained by periodic fires; examples include grasslands, chaparral shrubland, and some pine forests.

first law of thermodynamics States that energy is conserved; that is, it is neither created nor destroyed under normal conditions.

First World The industrialized capitalist or market-economy countries of Western Europe, North America, Japan, Australia, and New Zealand.

flood An overflow of water onto land that normally is dry.

floodplains Low lands along riverbanks, lakes, and coastlines subjected to periodic inundation.

flora All of the plants present in a given region.

food chain A linked feeding series; in an ecosystem, the sequence of organisms through which energy and materials are transferred, in the form of food, from one trophic level to another.

food security The ability of individuals to obtain sufficient food on a day-to-day basis.

food web A complex, interlocking series of individual food chains in an ecosystem.

fossil fuels Petroleum, natural gas, and coal created by geologic forces from organic wastes and dead bodies of formerly living biological organisms.

fragmentation Disruption of habitat into small, isolated fragments.

freshwater ecosystems Ecosystems in which the fresh (nonsalty) water of streams, rivers, ponds, or lakes plays a defining role.

fuel assembly A bundle of hollow metal rods containing uranium oxide pellets; used to fuel a nuclear reactor.

fuel cells Mechanical devices that use hydrogen or hydrogen-containing fuel, such as methane, to produce an electric current. Fuel cells are clean, quiet, and highly efficient sources of electricity.

fugitive emissions Substances that enter the air without going through a smokestack, such as dust from soil erosion, strip mining, rock crushing, construction, and building demolition.

fungi Nonphotosynthetic, eukaryotic organisms with cell walls, filamentous bodies, and absorptive nutrition.

fungicide A chemical that kills fungi.

G

gamma rays Very short wavelength forms of the electromagnetic spectrum.

gap analysis A biogeographical technique of mapping biological diversity and endemic species to find gaps between protected areas that leave endangered habitats vulnerable to disruption.

gasohol A mixture of gasoline and ethanol.

gene A unit of heredity; a segment of DNA nucleus of the cell that contains information for the synthesis of a specific protein, such as an enzyme.

genetic assimilation The disappearance of a species as its genes are diluted through cross-breeding with a closely related species.

genetically modified organisms (GMOs) Organisms created by combining natural or synthetic genes using the techniques of molecular biology.

genetic engineering Laboratory manipulation of genetic material using molecular biology.

genuine progress index (GPI) An alternative to GNP or GDP for economic accounting that measures real progress in quality of life and sustainability.

geometric growth Growth that follows a geometric pattern of increase, such as 2, 4, 8, 16, etc. See also *exponential growth.*

geothermal energy Energy drawn from the internal heat of the earth, either through geysers, fumaroles, hot springs, or other natural geothermal features or through deep wells that pump heated groundwater.

global environmentalism The extension of modern environmental concerns to global issues.

grasslands Biomes dominated by grasses and associated herbaceous plants.

greenhouse effect Trapping of heat by the earth's atmosphere, which is transparent to incoming visible light waves but absorbs outgoing long-wave infrared radiation.

greenhouse gas A gas that traps heat in the atmosphere.

green plans Integrated national environmental plans for reducing pollution and resource consumption while achieving sustainable development and environmental restoration.

green political parties Political organizations based on environmental protection, participatory democracy, grassroots organization, and sustainable development.

green pricing Plans in which consumers can voluntarily pay premium prices for renewable energy.

green revolution Dramatically increased agricultural production brought about by "miracle" strains of grain; usually requires high inputs of water, plant nutrients, and pesticides.

gross domestic product (GDP) The total economic activity within national boundaries.

gross national product (GNP) The sum total of all goods and services produced in a national economy. Gross domestic product (GDP) is used to distinguish economic activity within a country from that of offshore corporations.

groundwater Water held in gravel deposits or porous rock below the earth's surface; does not include water or crystallization held by chemical bonds in rocks or moisture in upper soil layers.

gully erosion Removal of layers of soil, creating channels or ravines too large to be removed by normal tillage operations.

H

habitat The place or set of environmental conditions in which a particular organism lives.

habitat conservation plans Agreements under which property owners are allowed to harvest resources or develop land as long as habitat is conserved or replaced in ways that benefit resident endangered or threatened species in the long run. Some incidental "taking" or loss of endangered species is generally allowed in such plans.

half-life The time required for one-half of a sample to decay or change into some other form.

hazardous waste Any discarded material containing substances known to be toxic, mutagenic, carcinogenic, or teratogenic to humans or other life-forms; ignitable, corrosive, explosive, or highly reactive alone or with other materials.

health A state of physical and emotional well-being; the absence of disease or ailment.

heap-leach extraction A technique for separating gold from extremely low-grade ores. Crushed ore is piled in huge heaps and sprayed with a dilute alkaline-cyanide solution, which percolates through the pile to extract the gold, which is separated from the effluent in a processing plant. This process has a high potential for water pollution.

heat Total kinetic energy of atoms or molecules in a substance not associated with the bulk motion of the substance.

herbicide A chemical that kills plants.

herbivores Organisms that eat only plants.

hcterotroph An organism that is incapable of synthesizing its own food and, therefore, must feed upon organic compounds produced by other organisms.

high-level waste repository A place where intensely radioactive wastes can be buried and remain unexposed to groundwater and earthquakes for tens of thousands of years.

high-quality energy Intense, concentrated, and high-temperature energy that is considered high-quality because of its usefulness in carrying out work.

HIPPO Habitat destruction, Invasive species, Pollution, Population (human), and Overharvesting, the leading causes of extinction.

holistic science The study of entire, integrated systems rather than isolated parts. Often takes a descriptive or an interpretive approach.

homeostasis A dynamic, steady state in a living system maintained through opposing, compensating adjustments.

humus Sticky, brown, insoluble residue from the bodies of dead plants and animals; gives soil its structure, coating mineral particles and holding them together; serves as a major source of plant nutrients.

hurricanes Large cyclonic oceanic storms with heavy rain and winds exceeding 119 km/hr (74 mph).

hydrologic cycle The natural process by which water is purified and made fresh through evaporation and precipitation. This cycle provides all the freshwater available for biological life.

hypothesis A conditional explanation that can be verified or falsified by observation or experimentation.

I

igneous rocks Crystalline minerals solidified from molten magma from deep in the earth's interior; basalt, rhyolite, andesite, lava, and granite are examples.

inbreeding depression In a small population, an accumulation of harmful genetic traits (through random mutations and natural selection) that lowers viability and reproductive success of enough individuals to affect the whole population.

inductive reasoning "Bottom-up" reasoning in which we study specific examples and try to discover patterns and derive general explanations from collected observations.

industrial revolution Advances in science and technology that have given us power to understand and change our world.

infiltration The process of water percolation into the soil and pores and hollows of permeable rocks.

inherent value Ethical values or rights that exist as an intrinsic or essential characteristic of a particular thing or class of things simply by the fact of their existence.

inholdings Private lands within public parks, forests, or wildlife refuges.

insecticide A chemical that kills insects.

insolation Incoming solar radiation.

instrumental value The value or worth of objects that satisfy the needs and wants of moral agents. Objects that can be used as a means to some desirable end.

intangible resources Factors such as open space, beauty, serenity, wisdom, diversity, and satisfaction that cannot be grasped or contained.

Ironically, these resources can be both infinite and exhaustible.

integrated pest management (IPM) An ecologically based pest-control strategy that relies on natural mortality factors, such as natural enemies, weather, cultural control methods, and carefully applied doses of pesticides.

Intergovernmental Panel on Climate Change (IPCC) A large group of scientists from many nations and a wide variety of fields assembled by the United Nations Environment Program and World Meteorological Organization to assess the current state of knowledge about climate change.

internalizing costs Planning so that those who reap the benefits of resource use also bear all the external costs.

international treaties and conventions Agreements between nations on important issues.

interspecific competition In a community, competition for resources between members of different species.

intraspecific competition In a community, competition for resources among members of the same species.

invasive species Organisms that thrive in new territory where they are free of predators, diseases or resource limitations that may have controlled their population in their native habitat.

ionizing radiation High-energy electromagnetic radiation or energetic subatomic particles released by nuclear decay.

ionosphere The lower part of the thermosphere.

ions Electrically charged atoms that have gained or lost electrons.

irruptive growth See *Malthusian growth*.

island biogeography The study of rates of colonization and extinction of species on islands or other isolated areas based on size, shape, and distance from other inhabited regions.

isotopes Forms of a single element that differ in atomic mass due to a different number of neutrons in the nucleus.

J

J curve A growth curve that depicts exponential growth; called a J curve because of its shape.

jet streams Powerful winds or currents of air that circulate in shifting flows; similar to oceanic currents in extent and effect on climate.

joule A unit of energy. One joule is the energy expended in 1 second by a current of 1 amp flowing through a resistance of 1 ohm.

K

K-selected species Organisms whose population growth is regulated by internal (or intrinsic) as well as external factors. Large animals, such as whales and elephants, as well as top predators, generally fall in this category. They have relatively few offspring and often stabilize their population size near the carrying capacity of their environment.

keystone species A species whose impacts on its community or ecosystem are much larger and more influential than would be expected from mere abundance. This could be a top predator, a plant that shelters or feeds other organisms, or an organism that plays a critical ecological role.

kinetic energy Energy contained in moving objects, such as a rock rolling down a hill, the wind blowing through the trees, or water flowing over a dam.

kwashiorkor A widespread human protein deficiency disease resulting from a starchy diet low in protein and essential amino acids.

Kyoto Protocol An international treaty adopted in Kyoto, Japan, in 1997, in which 160 nations agreed to roll back CO_2, methane, and nitrous oxide emissions to reduce the threat of global climate change.

L

landscape ecology The study of the reciprocal effects of spatial pattern on ecological processes.

landslides Mass wasting or mass movement of rock or soil downhill. Often triggered by seismic events or heavy rainfall.

latent heat Stored energy in a form that is not sensible (detectable by ordinary senses).

LD50 A chemical dose lethal to 50 percent of a test population.

legal standing The right to take part in legal proceedings.

life expectancy The average age that a newborn infant can expect to attain in a particular time and place.

life span The longest period of life reached by a type of organism.

limiting factors Chemical or physical factors that limit the existence, growth, abundance, or distribution of an organism.

limits to growth A belief that the world has a fixed carrying capacity for humans.

lobbying Using personal contacts, public pressure, or political action to persuade legislators to vote in a particular manner.

logical thinking A rational way of thought that asks, "How can orderly, deductive reasoning help me think clearly?"

logistic growth Growth rates regulated by internal and external factors that establish an equilibrium with environmental resources. See also *S curve*.

longevity The length or duration of life; compare with survivorship.

low-quality energy Diffuse, dispersed energy at a low temperature that is difficult to gather and use for productive purposes.

LULUs Locally Unwanted Land Uses, such as toxic waste dumps, incinerators, smelters, airports, freeways, and other sources of environmental, economic, or social degradation.

M

magma Molten rock from deep in the earth's interior; called lava when it spews from volcanic vents.

malnourishment A nutritional imbalance caused by lack of specific dietary components or inability to absorb or utilize essential nutrients.

Malthusian growth A population explosion followed by a population crash; also called irruptive growth.

Man and Biosphere (MAB) program A design for nature preserves that divides protected areas into zones with different purposes. A highly protected core is surrounded by a buffer zone and peripheral regions in which multiple-use resource harvesting is permitted.

mangrove forests Diverse groups of salt-tolerant trees and other plants that grow in intertidal zones of tropical coastlines.

mantle A hot, pliable layer of rock that surrounds the earth's core and underlies the cool outer crust.

marasmus A widespread human protein deficiency disease caused by a diet low in calories and protein or imbalanced in essential amino acids.

marginal costs The cost to produce one additional unit of a good or service.

marshes Wetlands without trees; in North America, this type of land is characterized by cattails and rushes.

mass burn The incineration of unsorted solid waste.

matter Anything that takes up space and has mass.

mean Average.

mediation An informal dispute resolution process in which parties try to reach agreement through discussion and compromise; often used as an alternative to resolving disputes through lawsuits.

Mediterranean climate areas Regions defined by warm, dry summers and cool, wet winters; these areas may have distinctive and endemic species.

megacities See *megalopolis.*

megalopolis Also known as a megacity or supercity; megalopolis indicates an urban area with more than 10 million inhabitants.

megawatt (MW) Unit of electrical power equal to 1,000 kilowatts or 1 million watts.

mesosphere The atmospheric layer above the stratosphere and below the thermosphere; the middle layer; temperatures are usually very low.

metabolism All the energy and matter exchanges that occur within a living cell or organism; collectively, the life processes.

metamorphic rocks Igneous and sedimentary rocks modified by heat, pressure, and chemical reactions.

methane hydrate Small bubbles or individual molecules of methane (natural gas) trapped in a crystalline matrix of frozen water.

midoceanic ridges Mountain ranges on the ocean floor where magma wells up through cracks and creates new crust.

Milankovitch cycles Periodic variations in tilt, eccentricity, and wobble in the earth's orbit; Milutin Milankovitch suggested these are responsible for cyclic weather changes.

mineral A naturally occurring, inorganic, crystalline solid with definite chemical composition, a specific internal crystal structure, and characteristic physical properties.

minimum viable population The number of individuals needed for long-term survival of rare and endangered species.

mitigation Repairing or rehabilitating a damaged ecosystem or compensating for damage by providing a substitute or replacement area.

modern environmentalism A fusion of conservation of natural resources and preservation of nature with concerns about pollution, environmental health, and social justice.

molecules Combinations of two or more atoms.

monitored, retrievable storage Holding wastes in underground mines or secure surface facilities, such as dry casks, where they can be watched and repackaged, if necessary.

monoculture forestry Intensive planting of a single species; an efficient wood production approach, but one that encourages pests and disease infestations and conflicts with wildlife habitat or recreation uses.

monsoons Seasonal reversals of wind patterns caused by the different heating and cooling rates of the oceans and continents.

montane coniferous forests Coniferous forests of the mountains consisting of belts of different forest communities along an altitudinal gradient.

morals A set of ethical principles that guides our actions and relationships.

morbidity Illness or disease.

more-developed countries (MDC) Industrialized nations characterized by high per capita incomes, low birth and death rates, low population growth rates, and high levels of industrialization and urbanization.

mortality Death rate in a population, such as number of deaths per thousand people per year.

mulch Protective groundcover that protects the soil, saves water, and prevents weed growth; often straw, seaweed, leaves, or synthetic materials, such as heavy paper or plastic.

Müllerian (or Muellerian) mimicry Evolution of two species, both of which are unpalatable and have poisonous stingers or some other defense mechanism, to resemble each other.

multiple use Many uses that occur simultaneously; used in forest management; limited to mutually compatible uses.

mutagens Agents, such as chemicals or radiation, that damage or alter genetic material (DNA) in cells.

mutation A change, either spontaneous or by external factors, in the genetic material of a cell; mutations in the gametes (sex cells) can be inherited by future generations of organisms.

mutualism A symbiotic relationship between individuals of two different species in which both species benefit from the association.

N

National Environmental Policy Act (NEPA) The law that established the Council on Environmental Quality and that requires environmental impact statements for all federal projects with significant environmental impacts.

natural increase Crude death rate subtracted from crude birth rate.

natural resources Goods and services supplied by the environment.

natural selection The mechanism for evolutionary change in which environmental pressures cause certain genetic combinations in a population to become more abundant; genetic combinations best adapted for present environmental conditions tend to become predominant.

negative feedbacks Factors that result from a process and, in turn, reduce that same process.

neo-classical economics The branch of economics that attempts to apply the principles of modern science to economic analysis in a mathematically rigorous, noncontextual, abstract, predictive manner.

net energy yield Total useful energy produced during the lifetime of an entire energy system minus the energy used, lost, or wasted in making useful energy available.

neurotoxins Toxic substances, such as lead or mercury, that specifically poison nerve cells.

neutron A subatomic particle, found in the nucleus of the atom, that has no electromagnetic charge.

NIMBY Not-In-My-Back-Yard: the position of those opposed to LULUs.

nitrogen cycle The circulation and reutilization of nitrogen in both inorganic and organic phases.

nitrogen-fixing bacteria Bacteria that convert nitrogen from the atmosphere or soil solution into ammonia that can then be converted to plant nutrients by nitrite- and nitrate-forming bacteria.

nitrogen oxides Highly reactive gases formed when nitrogen in fuel or combustion air is heated to over 650°C (1,200°F) in the presence of oxygen or when bacteria in soil or water oxidize nitrogen-containing compounds.

noncriteria pollutants See *unconventional pollutants.*

nongovernmental organizations (NGOs) Pressure and research groups, advisory agencies, political parties, professional societies, and other groups concerned about environmental quality, resource use, and many other issues.

nonpoint sources Scattered, diffuse sources of pollutants, such as runoff from farm fields, golf courses, and construction sites.

nonrenewable resources Minerals, fossil fuels, and other materials present in essentially fixed amounts (within human time scales) in our environment.

nuclear fission The radioactive decay process in which isotopes split apart to create two smaller atoms.

nuclear fusion A process in which two smaller atomic nuclei fuse into one larger nucleus and release energy; the source of power in a hydrogen bomb.

nucleic acids Large organic molecules made of nucleotides that function in the transmission of hereditary traits, in protein synthesis, and in control of cellular activities.

nucleus The center of the atom; occupied by protons and neutrons. In cells, the organelle that contains the chromosomes (DNA).

numbers pyramid A diagram showing the relative population sizes at each trophic level in an ecosystem; usually corresponds to the biomass pyramid.

O

obese Pathologically overweight, having a body mass greater than 30 kg/m², or roughly 30 pounds above normal for an average person.

offset allowances A controversial component of air quality regulations that allows a polluter to avoid installation of control equipment on one source with an "offsetting" pollution reduction at another source.

oil shales Fine-grained sedimentary rock rich in solid organic material called kerogen. When heated, the kerogen liquefies to produce a fluid petroleum fuel.

old-growth forests Forests free from disturbance for long enough (generally 150 to 200 years) to have mature trees, physical conditions, species diversity, and other characteristics of equilibrium ecosystems.

oligotrophic Condition of rivers and lakes that have clear water and low biological productivity (*oligo* = little; *trophic* = nourished); are usually clear, cold, infertile headwater lakes and streams.

omnivores Organisms that eat both plants and animals.

open access system A commonly held resource for which there are no management rules.

open canopy A forest where tree crowns cover less than 20 percent of the ground; also called woodland.

open system A system that exchanges energy and matter with its environment.

organic compounds Complex molecules organized around skeletons of carbon atoms arranged in rings or chains; includes biomolecules, molecules synthesized by living organisms.

overharvesting Harvesting so much of a resource that it threatens its existence.

overnutrition Receiving too many calories.

overshoots The extent to which a population exceeds the carrying capacity of its environment.

oxygen sag Oxygen decline downstream from a pollution source that introduces materials with high biological oxygen demands.

ozone A highly reactive molecule containing three oxygen atoms; a dangerous pollutant in ambient air. In the stratosphere, however, ozone forms an ultraviolet absorbing shield that protects us from mutagenic radiation.

P

paradigms Overarching models of the world that shape our worldviews and guide our interpretation of how things are.

parasite An organism that lives in or on another organism, deriving nourishment at the expense of its host, usually without killing it.

parsimony A principle that says where two equally plausible explanations for a phenomenon are possible, we should choose the simpler one (also known as Ockham's razor).

particulate material Atmospheric aerosols, such as dust, ash, soot, lint, smoke, pollen, spores, algal cells, and other suspended materials; originally applied only to solid particles but now extended to droplets of liquid.

parts per billion (ppb) Number of parts of a chemical found in 1 billion parts of a particular gas, liquid, or solid mixture.

parts per million (ppm) Number of parts of a chemical found in 1 million parts of a particular gas, liquid, or solid mixture.

parts per trillion (ppt) Number of parts of a chemical found in 1 trillion parts of a particular gas, liquid, or solid mixture.

passive heat absorption The use of natural materials or absorptive structures without moving parts to gather and hold heat; the simplest and oldest use of solar energy.

pastoralists People who live by herding domestic animals.

patchiness Within a larger ecosystem, the presence of smaller areas that differ in some physical conditions and thus support somewhat different communities; promotes diversity in a system or area.

pathogens Organisms that produce disease in host organisms, disease being an alteration of one or more metabolic functions in response to the presence of the organisms.

peat Deposits of moist, acidic, semidecayed organic matter.

pelagic Zones in the vertical water column of a water body.

perennial species Plants that grow for more than two years.

permafrost A permanently frozen layer of soil that underlies the arctic tundra.

permanent retrievable storage Placing waste storage containers in a secure location where they can be inspected periodically and retrieved, if necessary, for repacking or for transfer if a better means of disposal or reuse is developed.

persistent organic pollutants (POPs) Chemical compounds that persist in the environment and retain biological activity for a long time.

pest Any organism that reduces the availability, quality, or value of a useful resource.

pesticide Any chemical that kills, controls, drives away, or modifies the behavior of a pest.

pesticide treadmill A need for constantly increasing doses or new pesticides to prevent pest resurgence.

pest resurgence Rebound of pest populations due to acquired resistance to chemicals and nonspecific destruction of natural predators and competitors by broadscale pesticides.

pH A value that indicates the acidity or alkalinity of a solution on a scale of 0 to 14, based on the proportion of H^+ ions present.

phosphorus cycle The movement of phosphorus atoms from rocks through the biosphere and hydrosphere and back to rocks.

photochemical oxidants Products of secondary atmospheric reactions. See also *smog*.

photodegradable plastics Plastics that break down when exposed to sunlight or to a specific wavelength of light.

photosynthesis The biochemical process by which green plants and some bacteria capture light energy and use it to produce chemical bonds. Carbon dioxide and water are consumed while oxygen and simple sugars are produced.

photovoltaic cell An energy-conversion device that captures solar energy and directly converts it to electrical current.

phytoplankton Microscopic, free-floating, autotrophic organisms that function as producers in aquatic ecosystems.

pioneer species In primary succession on a terrestrial site, the plants, lichens, and microbes that first colonize the site.

plankton Primarily microscopic organisms that occupy the upper water layers in both freshwater and marine ecosystems.

plasma A hot, electrically neutral gas of ions and free electrons.

poaching Hunting wildlife illegally.

point sources Specific locations of highly concentrated pollution discharge, such as factories, power plants, sewage treatment plants, underground coal mines, and oil wells.

policy A societal plan or statement of intentions intended to accomplish some social or economic goal.

policy cycle The process by which problems are identified and acted upon in the public arena.

pollution To make foul, unclean, dirty; any physical, chemical, or biological change that adversely affects the health, survival, or activities of living organisms or that alters the environment in undesirable ways.

pollution charges Fees assessed per unit of pollution based on the "polluter pays" principle.

population All members of a species that live in the same area at the same time.

population crash A sudden population decline caused by predation, waste accumulation, or resource depletion; also called a dieback.

population explosion Growth of a population at exponential rates to a size that exceeds environmental carrying capacity; usually followed by a population crash.

population momentum A potential for increased population growth as young members reach reproductive age.

positive feedbacks Factors that result from a process and, in turn, increase that same process.

potential energy Stored energy that is latent but available for use. A rock poised at the top of a hill or water stored behind a dam are examples of potential energy.

power The rate of energy delivery; measured in horsepower or watts.

precautionary principle The rule that we should leave a margin of safety for unexpected developments. This principle implies that we should strive to prevent harm to human health and the environment even if risks are not fully understood.

precedent An act or a decision that can be used as an example in dealing with subsequent similar situations.

predator An organism that feeds directly on other organisms in order to survive; live-feeders, such as herbivores and carnivores.

primary pollutants Chemicals released directly into the air in a harmful form.

primary producers Photosynthesizing organisms.

primary productivity Synthesis of organic materials (biomass) by green plants using the energy captured in photosynthesis.

primary standards Regulations of the 1970 Clean Air Act; intended to protect human health.

primary succession Ecological succession that begins in an area where no biotic community previously existed.

primary treatment A process that removes solids from sewage before it is discharged or treated further.

principle of competitive exclusion A result of natural selection whereby two similar species in a community occupy different ecological niches, thereby reducing competition for food.

probability The likelihood that a situation, a condition, or an event will occur.

producer An organism that synthesizes food molecules from inorganic compounds by using an external energy source; most producers are photosynthetic.

productivity The amount of biomass (biological material) produced in a given area during a given period of time.

prokaryotic Cells that do not have a membrane-bounded nucleus or membrane-bounded organelles.

pronatalist pressures Influences that encourage people to have children.

prospective study A study in which experimental and control groups are identified before exposure to some factor. The groups are then monitored and compared for a specific time after the exposure to determine any effects the factor may have.

proteins Chains of amino acids linked by peptide bonds.

proton A positively charged subatomic particle found in the nucleus of an atom.

proven-in-place reserves Energy sources that have been thoroughly mapped and are likely to be economically recoverable with available technology.

pull factors Conditions that draw people from the country into the city.

push factors Conditions that force people out of the country and into the city.

R

r-selected species Organisms whose population growth is regulated mainly by external factors. They tend to have rapid reproduction and high mortality of offspring. Given optimum environmental conditions, they can grow exponentially. Many "weedy" or pioneer species fit in this category.

radioactive decay A change in the nuclei of radioactive isotopes that spontaneously emit high-energy electromagnetic radiation and/or subatomic particles while gradually changing into another isotope or different element.

rainforest A forest with high humidity, constant temperature, and abundant rainfall (generally over 380 cm [150 in.] per year); can be tropical or temperate.

rain shadow Dry area on the downwind side of a mountain.

rangeland Grasslands and open woodlands suitable for livestock grazing.

rational choice Public decision making based on reason, logic, and science-based management.

reasoned judgment Thoughtful decisions based on careful, logical examination of available evidence.

recharge zones Areas where water infiltrates into an aquifer.

reclamation Chemical, biological, or physical cleanup and reconstruction of severely contaminated or degraded sites to return them to something like their original topography and vegetation.

recycling Reprocessing of discarded materials into new, useful products; not the same as reuse of materials for their original purpose, but the terms are often used interchangeably.

red tide A population explosion, or "bloom," of single-celled marine organisms called dinoflagellates. Billions of these cells can accumulate in protected bays where the toxins they contain can poison other marine life.

reduced tillage systems Farming methods that preserve soil and save energy and water through reduced cultivation; includes minimum till, conserve-till, and no-till systems.

reflective thinking A thoughtful, contemplative analysis that asks, "What does this all mean?"

reformer A device that strips hydrogen from fuels such as natural gas, methanol, ammonia, gasoline, or vegetable oil so they can be used in a fuel cell.

refuse-derived fuel Processing of solid waste to remove metal, glass, and other unburnable materials; organic residue is shredded, formed into pellets, and dried to make fuel for power plants.

regenerative farming Farming techniques and land stewardship that restore the health and productivity of the soil by rotating crops, planting ground cover, protecting the surface with crop residue, and reducing synthetic chemical inputs and mechanical compaction.

regulations Rules established by administrative agencies; regulations can be more important than statutory law in the day-to-day management of resources.

rehabilitation Rebuilding basic structure or function in an ecological system without necessarily achieving complete restoration to its original condition.

relative humidity At any given temperature, a comparison of the actual water content of the air with the amount of water that could be held at saturation.

relevé A rapid assessment of vegetation types and biodiversity in an area.

remediation Cleaning up chemical contaminants from a polluted area.

renewable resources Resources normally replaced or replenished by natural processes; resources not depleted by moderate use; examples include solar energy, biological resources such as forests and fisheries, biological organisms, and some biogeochemical cycles.

renewable water supplies Annual freshwater surface runoff plus annual infiltration into underground freshwater aquifers that are accessible for human use.

reproducibility Making an observation or obtaining a particular result consistently.

residence time The length of time a component, such as an individual water molecule, spends in a particular compartment or location before it moves on through a particular process or cycle.

resilience The ability of a community or ecosystem to recover from disturbances.

resource partitioning In a biological community, various populations sharing environmental resources through specialization, thereby reducing direct competition. See also *ecological niche*.

resources In economic terms, anything with potential use in creating wealth or giving satisfaction.

restoration To bring something back to a former condition. Ecological restoration involves active manipulation of nature to re-create conditions that existed before human disturbance.

restoration ecology Seeks to repair or reconstruct ecosystems damaged by human actions.

retrospective study A study that looks back in history at a group of people (or other organisms) who suffer from some condition to try to identify something in their past life that the whole group shares but that is not found in the histories of a control group as near as possible to those being studied but who do not suffer from the same condition.

riders Amendments attached to bills in conference committee, often completely unrelated to the bill to which they are added.

rill erosion The removing of thin layers of soil as little rivulets of running water gather and cut small channels in the soil.

risk The probability that something undesirable will happen as a consequence of exposure to a hazard.

risk assessment Evaluation of the short-term and long-term risks associated with a particular activity or hazard; usually compared with benefits in a cost-benefit analysis.

rock A solid, cohesive aggregate of one or more crystalline minerals.

rock cycle The process whereby rocks are broken down by chemical and physical forces; sediments are moved by wind, water, and gravity; sedimented and reformed into rock; and then crushed, folded, melted, and recrystallized into new forms.

rotational grazing Confining grazing animals in a small area for a short time to force them to eat weedy species as well as the more desirable grasses and forbes.

runoff The excess of precipitation over evaporation; the main source of surface water and, in broad terms, the water available for human use.

S

salinity The amount of dissolved salts (especially sodium chloride) in a given volume of water.

salinization A process in which mineral salts accumulate in the soil, killing plants; occurs when soils in dry climates are irrigated profusely.

saltwater intrusion The movement of saltwater into freshwater aquifers in coastal areas where groundwater is withdrawn faster than it is replenished.

sample To analyze a small but representative portion of a population to estimate the characteristics of the entire class.

sanitary landfills Landfills in which garbage and municipal waste are buried every day under enough soil or fill to eliminate odors, vermin, and litter.

scientific method A systematic, precise, objective study of a problem. Generally this requires observation, hypothesis development and testing, data gathering, and interpretation.

scientific theory An explanation or idea accepted by a substantial number of scientists.

S curve A curve that depicts logistic growth; called an S curve because of its shape.

secondary pollutants Chemicals modified to a hazardous form after entering the air or that are formed by chemical reactions as components of the air mix and interact.

secondary succession Succession on a site where an existing community has been disrupted.

secondary treatment Bacterial decomposition of suspended particulates and dissolved organic compounds that remain after primary sewage treatment.

second law of thermodynamics States that, with each successive energy transfer or transformation in a system, less energy is available to do work.

secure landfills Solid waste disposal sites lined and capped with an impermeable barrier to prevent leakage or leaching.

sedimentary rocks Rocks composed of accumulated, compacted mineral fragments, such as sand or clay; examples include shale, sandstone, breccia, and conglomerates.

sedimentation The deposition of organic materials or minerals by chemical, physical, or biological processes.

selection pressure Limited resources or adverse environmental conditions that tend to favor certain adaptations in a population. Over many generations, this can lead to genetic change, or evolution.

selective cutting Harvesting only mature trees of certain species and size; usually more expensive than clear-cutting but less disruptive for wildlife and often better for forest regeneration.

shade-grown coffee and cocoa Plants grown under a canopy of taller trees, which provides habitat for birds and other wildlife.

sheet erosion Peeling off thin layers of soil from the land surface; accomplished primarily by wind and water.

shelterwood harvesting Mature trees are removed from the forest in a series of two or more cuts, leaving young trees and some mature trees as a seed source for future regeneration.

sick house syndrome A cluster of allergies and other illnesses caused by sensitivity to molds, synthetic chemicals, or other harmful compounds trapped in insufficiently ventilated buildings.

significant numbers Meaningful data that can be measured accurately and reproducibly.

sinkholes A large surface crater caused by the collapse of an underground channel or cavern; often triggered by groundwater withdrawal.

sludge A semisolid mixture of organic and inorganic materials that settles out of wastewater at a sewage treatment plant.

smart growth The efficient use of land resources and existing urban infrastructure that encourages in-fill development, provides a variety of affordable housing and transportation choices, and seeks to maintain a unique sense of place by respecting local cultural and natural features.

smelting Roasting ore to release metals from mineral compounds.

smog The combination of smoke and fog in the stagnant air of London; now often applied to photochemical pollution.

social justice Equitable access to resources and the benefits derived from them; a system that recognizes inalienable rights and adheres to what is fair, honest, and moral.

soil A complex mixture of weathered rock material, partially decomposed organic molecules, and a host of living organisms.

soil horizons Horizontal layers that reveal a soil's history, characteristics, and usefulness.

southern pine forest United States coniferous forest ecosystem characterized by a warm, moist climate.

species All the organisms genetically similar enough to breed and produce live, fertile offspring in nature.

species diversity The number and relative abundance of species present in a community.

species recovery plan A plan for restoration of an endangered species through protection, habitat management, captive breeding, disease control, and other techniques that increase populations and encourage survival.

specific heat The amount of heat energy needed to change the temperature of a body. Water has a specific heat of 1, which is higher than most substances.

sprawl Unlimited, unplanned growth of urban areas that consumes open space and wastes resources.

stability In ecological terms, a dynamic equilibrium among the physical and biological factors in an ecosystem or a community; relative homeostasis.

stable runoff The fraction of water available year-round; usually more important than total runoff when determining human uses.

Standard Metropolitan Statistical Area (SMSA) An urbanized region with at least 100,000 inhabitants with strong economic and social ties to a central city of at least 50,000 people.

statistics Mathematical analysis of the collection, organization, and interpretation of numerical data.

statutory law Rules passed by a state or national legislature.

steady-state economy Characterized by low birth and death rates, use of renewable energy sources, recycling of materials, and emphasis on durability, efficiency, and stability.

stewardship A philosophy that holds that humans have a unique responsibility to manage, care for, and improve nature.

Strategic Lawsuits against Public Participation (SLAPP) Lawsuits that have no merit but are brought merely to intimidate and harass private citizens who act in the public interest.

strategic metals and minerals Materials a country cannot produce itself but that it uses for essential materials or processes.

stratosphere The zone in the atmosphere extending from the tropopause to about 50 km (30 mi) above the earth's surface; temperatures are stable or rise slightly with altitude; has very little water vapor but is rich in ozone.

stress Physical, chemical, or emotional factors that place a strain on an animal. Plants also experience physiological stress under adverse environmental conditions.

strip-cutting Harvesting trees in strips narrow enough to minimize edge effects and to allow natural regeneration of the forest.

strip-farming Planting different kinds of crops in alternating strips along land contours; when one crop is harvested, the other crop remains to protect the soil and prevent water from running straight down a hill.

strip-mining Extracting shallow mineral deposits (especially coal) by scraping off surface layers with giant earth-moving equipment; creates a huge open pit; an alternative to underground or deep open-pit mines.

subsidence Settling of the ground surface caused by the collapse of porous formations that result from withdrawal of large amounts of groundwater, oil, or other underground materials.

subsoil A layer of soil beneath the topsoil that has lower organic content and higher concentrations of fine mineral particles; often contains soluble compounds and clay particles carried down by percolating water.

sulfur cycle The chemical and physical reactions by which sulfur moves into or out of storage and through the environment.

sulfur dioxide A colorless, corrosive gas directly damaging to both plants and animals.

Superfund A fund established by Congress to pay for containment, cleanup, or remediation of abandoned toxic waste sites. The fund is financed by fees paid by toxic waste generators and by cost recovery from cleanup projects.

surface mining Some minerals are also mined from surface pits. See also *strip-mining.*

surface tension The tendency for a surface of water molecules to hold together, producing a surface that resists breaking.

survivorship The percentage of a population reaching a given age or the proportion of the maximum life span of the species reached by any individual.

sustainability Ecological, social, and economic systems that can last over the long term.

sustainable agriculture (regenerative farming) Ecologically sound, economically viable, socially just agricultural system. Stewardship, soil conservation, and integrated pest management are essential for sustainability.

sustainable development A real increase in well-being and standard of life for the average person that can be maintained over the long term without degrading the environment or compromising the ability of future generations to meet their own needs.

sustained yield Utilization of a renewable resource at a rate that does not impair or damage its ability to be fully renewed on a long-term basis.

swamps Wetlands with trees, such as the extensive swamp forests of the southern United States.

symbiosis The intimate living together of members of two species; includes mutualism, commensalism, and, in some classifications, parasitism.

synergism When an injury caused by exposure to two environmental factors together is greater than the sum of exposure to each factor individually.

T

taiga The northernmost edge of the boreal forest, including species-poor woodland and peat deposits; intergrading with the arctic tundra.

tailings Mining waste left after mechanical or chemical separation of minerals from crushed ore.

taking The unconstitutional confiscation of private property.

tar sands Geologic deposits composed of sand and shale particles coated with bitumen, a viscous mixture of long-chain hydrocarbons.

tectonic plates Huge blocks of the earth's crust that slide around slowly, pulling apart to open new ocean basins or crashing ponderously into each other to create new, larger landmasses.

temperate rainforest The cool, dense, rainy forest of the northern Pacific coast; enshrouded in fog much of the time; dominated by large conifers.

temperature A measure of the speed of motion of a typical atom or molecule in a substance.

teratogens Chemicals or other factors that specifically cause abnormalities during embryonic growth and development.

terracing Shaping the land to create level shelves of earth to hold water and soil; requires extensive hand labor or expensive machinery, but it enables farmers to farm very steep hillsides.

tertiary treatment The removal of inorganic minerals and plant nutrients after primary and secondary treatment of sewage.

thermal pollution Artificially raising or lowering of the temperature of a water body in a way that adversely affects the biota or water quality.

thermocline In water, a distinctive temperature transition zone that separates an upper layer that is mixed by the wind (the epilimnion) and a colder deep layer that is not mixed (the hypolimnion).

thermodynamics The branch of physics that deals with transfers and conversions of energy.

thermosphere The highest atmospheric zone; a region of hot, dilute gases above the mesosphere extending out to about 1,600 km (1,000 mi) from the earth's surface.

Third World Less-developed countries that are not capitalistic and industrialized (First World) or centrally planned socialist economies (Second World); not intended to be derogatory.

thorn scrub A dry, open woodland or shrubland characterized by sparse, spiny shrubs.

threatened species While still abundant in parts of its territorial range, this species has declined significantly in total numbers and may be on the verge of extinction in certain regions or localities.

tolerance limits See *limiting factors.*

topsoil The first true layer of soil; layer in which organic material is mixed with mineral particles; thickness ranges from a meter or more under virgin prairie to zero in some deserts.

total fertility rate The number of children born to an average woman in a population during her entire reproductive life.

total growth rate The net rate of population growth resulting from births, deaths, immigration, and emigration.

total maximum daily loads (TMDL) The amount of particular pollutant that a water body can receive from both point and nonpoint sources and still meet water quality standards.

Toxic Release Inventory A program created by the Superfund Amendments and Reauthorization Act of 1984 that requires manufacturing facilities and waste handling and disposal sites to report annually on releases of more than 300 toxic materials. You can find out from the EPA whether any of these sites are in your neighborhood and what toxics they release.

toxins Poisonous chemicals that react with specific cellular components to kill cells or to alter growth or development in undesirable ways; often harmful, even in dilute concentrations.

tradable permits Pollution quotas or variances that can be bought or sold.

"The Tragedy of the Commons" An inexorable process of degradation of communal resources due to selfish self-interest of "free riders" who use or destroy more than their fair share of common property. See *open access system.*

transpiration The evaporation of water from plant surfaces, especially through stomates.

trophic level Step in the movement of energy through an ecosystem; an organism's feeding status in an ecosystem.

tropical rainforests Forests near the equator in which rainfall is abundant—more than 200 cm (80 in.) per year—and temperatures are warm to hot year-round.

tropical seasonal forests Semievergreen or partly deciduous forests tending toward open woodlands and grassy savannas dotted with scattered, drought-resistant trees.

tropopause The boundary between the troposphere and the stratosphere.

troposphere The layer of air nearest to the earth's surface; both temperature and pressure usually decrease with increasing altitude.

tundra Treeless arctic or alpine biome characterized by cold, dark winters; a short growing season; and potential for frost any month of the year; vegetation includes low-growing perennial plants, mosses, and lichens.

U

unconventional pollutants Toxic or hazardous substances, such as asbestos, benzene, beryllium, mercury, polychlorinated biphenyls, and vinyl chloride, not listed in the original Clean Air Act because they were not released in large quantities; also called noncriteria pollutants.

urbanization An increasing concentration of the population in cities and a transformation of land use to an urban pattern of organization.

urban renewal Programs to revitalize old, blighted sections of inner cities.

utilitarian conservation The philosophy that resources should be used for the greatest good for the greatest number for the longest time.

V

values An estimation of the worth of things; a set of ethical beliefs and preferences that determines our sense of right and wrong.

vertical stratification The vertical distribution of specific subcommunities within a community.

visible light A portion of the electromagnetic spectrum that includes the wavelengths used for photosynthesis.

vitamins Organic molecules essential for life that we cannot make for ourselves; we must get them from our diet; they act as enzyme cofactors.

volatile organic compounds Organic chemicals that evaporate readily and exist as gases in the air.

volcanoes Vents in the earth's surface through which molten lava (magma), gases, and ash escape to create mountains.

vulnerable species Naturally rare organisms or species whose numbers have been so reduced by human activities that they are susceptible to actions that could push them into threatened or endangered status.

W

warm front A long, wedge-shaped boundary caused when a warmer advancing air mass slides over neighboring cooler air parcels.

waste stream The steady flow of varied wastes, from domestic garbage and yard wastes to industrial, commercial, and construction refuse.

water cycle The recycling and reutilization of water on earth, including atmospheric, surface, and underground phases and biological and nonbiological components.

waterlogging Water saturation of soil that fills all air spaces and causes plant roots to die from lack of oxygen; a result of overirrigation.

watershed The land surface and groundwater aquifers drained by a particular river system.

water table The top layer of the zone of saturation; undulates according to the surface topography and subsurface structure.

weather The physical conditions of the atmosphere (moisture, temperature, pressure, and wind).

weathering Changes in rocks brought about by exposure to air, water, changing temperatures, and reactive chemical agents.

wetlands Ecosystems of several types in which rooted vegetation is surrounded by standing water during part of the year. See also *swamps, marshes, bogs, fens.*

"wicked" problems Problems with no simple right or wrong answer, where there is no single, generally agreed upon definition of or solution for the particular issue.

wilderness An area of undeveloped land affected primarily by the forces of nature; an area where humans are visitors who do not remain.

wildlife refuges Areas set aside to shelter, feed, and protect wildlife; due to political and economic pressures, refuges often allow hunting, trapping, mineral exploitation, and other activities that threaten wildlife.

wind farms Large numbers of windmills concentrated in a single area; usually owned by a utility or large-scale energy producer.

Wise Use Movement A coalition of ranchers, loggers, miners, industrialists, hunters, off-road vehicle users, land developers, and others who call for unrestricted access to natural resources and public lands.

withdrawal A description of the total amount of water taken from a lake, a river, or an aquifer.

woodland A forest where tree crowns cover less than 20 percent of the ground; also called open canopy forest.

work The application of force through a distance; requires energy input.

world conservation strategy A proposal for maintaining essential ecological processes, preserving genetic diversity, and ensuring that utilization of species and ecosystems is sustainable.

World Trade Organization (WTO) An association of 135 nations organized to regulate international trade.

X

X ray Very short wavelength in the electromagnetic spectrum; can penetrate soft tissue; although it is useful in medical diagnosis, it also damages tissue and causes mutations.

Z

zero population growth (ZPG) A condition in which births and immigration in a population just balance deaths and emigration.

zone of aeration Upper soil layers that hold both air and water.

zone of leaching The layer of soil just beneath the topsoil where water percolates, removing soluble nutrients that accumulate in the subsoil; may be very different in appearance and composition from the layers above and below it.

zone of saturation Lower soil layers where all spaces are filled with water.

Credits

Chapter 8

Opener: AP photo/Avant Givon; page 179, 8.2: © William P. Cunningham; 8.3a: © Vol. 40/Corbis; 8.3b: Courtesy of Stanley Erlandsen, University of Minnesota; 8.3c: Courtesy of Donald R. Hopkins; 8.5: © Vol. 72/Corbis; p. 191: © William P. Cunningham; 8.16: © The McGraw-Hill Companies, Inc./Sharon Farmer, photographer.

Chapter 9

Opener: Courtesy of Dave Hansen, College of Agriculture Experiment Station, University of Minnesota; 9.4a: Images produced by Hal Pierce, Laboratory for Atmospheres, NASA, Goddard Space Flight Center/NOAA; 9.4b: © William P. Cunningham; 9.4c: © Marc Serota/Corbis; 9.6: © Vic Englebert/Photo Researchers; 9.12: © William P. Cunningham; 9.14: © AP/Wide World Photos; p. 214: © William P. Cunningham; 9.16: © Stonehaven CCS Canada, Inc.; 9.17: © Vol. 23/PhotoDisc/Getty Images; 9.18: Courtesy of Dr. Delbert Swanson; 9.18: © William P. Cunningham; 9.21: © China Tourism Press/Getty Images; 9.22: Image courtesy of Norman Kuring, SeaWIFS Project; 9.23: NASA; 9.25, 9.27: © William P. Cunningham; 9.28: © John D. Cunningham/Visuals Unlimited; 9.31: © William P. Cunningham.

Chapter 10

Opener: © Medioimages/PunchStock; 10.3: © Medioimages/PunchStock; 10.6: © William P. Cunningham; 10.9a: Photo by Jeff Vanuga, USDA Natural Resources Conservation Service; 10.9b: Photo by Tim McCabe, USDA Natural Resources Conservation Service; 10.9c: Photo by Lynn Betts, USDA Natural Resources Conservation Service; 10.10: Courtesy of National Renewable Energy Laboratory/NREL/PIX; 10.11a-d EROS Data Center, USGS; 10.12: Courtesy of USDA/NRCS/ Photo by Lynn Betts; 10.13: Courtesy of Tim McCabe, Soil Conservation Service, USDA; 10.14: © William P. Cunningham; 10.16: © Simon Fraser/SPL/Photo Researchers; 10.17: © Joe McDonald/Animals Animals/Earth Scenes; 10.19: © William P. Cunningham; 10.20: © Lawrence Lowry/Photo Researchers; 10.24: © Les Stone/Sygma/Corbis; 10.26: © Ben Osborne/Stone/Getty Images; 10.28: National Renewable Energy Lab.

Chapter 11

Opener: © Peter Aengst/Wilderness Society/Lighthawk; 11.7: © Vol. 16/PhotoDisc/Getty Images; 11.8: Courtesy of David McGeary; 11.10: © Bryan F. Peterson; 11.11: © AP Photo/Bob Bird; 11.13: © Joseph Nettis/Photo Researchers; 11.15: © AP Photo/Anjum Naveed; 11.16: Courtesy of Chris G. Newhall/U.S. Geological Survey; 11.18: © James Shaffer; 11.19: © Los Angeles Times Photo by Al Seib; 11.20: © Stephen Rose.

Chapter 12

Opener: © Mary Ann Cunningham; p. 284: Courtesy of Tom Finkle; 12.6: © James P. Blair/National Geographic Image Collection; 12.9: © William P. Cunningham; 12.11: © Regis Martin/The New York Times; 12.12: © Vol. 160/Corbis; 12.15: Courtesy of Office of Civilian Radioactive Waste Management, Department of Energy; 12.17: National Renewable Energy Lab; 12.18: Courtesy of Carolyn Roberts/A House of Straw; 12.20: Courtesy Capstone Turbine Corporation; 12.22: © The McGraw-Hill Companies, Inc./Doug Sherman, photographer; 12.24: National Renewable Energy Lab/NREL/PIX; 12.25: © Vol. 160/Corbis; 12.27: Courtesy of Long Island Power Authority; 12.28: Courtesy of Dr. Douglas Pratt; p. 302: William P. Cunningham; 12.29: National Renewable Energy Lab/NREL/PIX; 12.31: © The McGraw-Hill Companies, Inc./Barry Barker, photographer; 12.33: National Renewable Energy Lab; 12.35: © Ocean Power Delivery Ltd.

Chapter 13

Opener: © Ray Pfortmer/Peter Arnold, Inc.; 13.3: © Fred McConnaughey/Photo Researchers; 13.4: © Vol. 31/Photodisc/Getty Images; page 316: © Barbara Gauntt/The Clarion-Ledger, Jackson, MS; 13.7: © William P. Cunningham; 13.8: David L. Hansen, University of Minnesota Agriculture Experiment Station; 13.12: Basel Action Network (www.ban.org); 13.13: Courtesy of Urban Ore, Inc. Berkeley, CA; 13.15: © Michael Greenlar/The Image Works; 13.19: © William P. Cunningham; 13.20: © Piet Van Lier.

Chapter 14

Opener: © William P. Cunningham; p. 331: © William P. Cunningham; 14.1: © Royalty-Free/Corbis; 14.4: © William P. Cunningham; 14.5: © Louie Psihoyos/Science Faction; 14.6: © The McGraw-Hill Companies, Inc/Barry Barker, photographer; 14.7: © 2003 Regents of the University of Minnesota. All rights reserved. Used with permission of the Design Center for American Urban Landscape; 14.9, 14.10, 14.11: © William P. Cunningham; 14.12: © Roofscapes, Inc. Used by permission. All rights reserved.; 14.13: © William P. Cunningham; p. 341: © Bill Dunster; 14.17: © Vol. 6/Corbis; 14.23: © William P. Cunningham; 14.24: © Mark Luthringer.

Chapter 15

Opener: © Robert Sobol; 15.2: Courtesy of Tom Finkle; 15.4: © Vol. 160/Corbis; 15.5: © Bob Daemmrich/The Image Works; 15.6: © Vol. 25/PhotoDisc/Getty Images; 15.7, 15.11: © William P. Cunningham; 15.12: Courtesy of Dave Hansen, College of Agriculture Experiment Station, University of Minnesota; 15.13, page 367,15.15: © William P. Cunningham; 15.16: Courtesy of National Renewable Energy Laboratory/NREL/PIX; 15.18: Courtesy David L. Hansen, University of Minnesota Agricultural Extension Service; 15.19,15.20: © William P. Cunningham.

LINE ART

Chapter 3

3.28: From Verner, J. et al. *Wildlife 2000.* 1986. Reprinted by permission of the University of Wisconsin Press.

Chapter 4

4.2: Figure redrawn with permission from *Population Bulletin,* vol. 18, no. 1, 1985. Population Reference Bureau; 4.4: From Thomas Merrick, with PRB staff, "World Population in transition," *Population Bulletin,* vol. 41, no. 2 (reprinted 1991).

Chapter 5

5.23: Reprinted with permission from J. Curtis in William L. Thomas (ed.), *Man's Role in Changing the Face of the Earth.* © 1956 by The University of Chicago Press. All rights reserved.

Chapter 12

12.13: Courtesy of Northern States Power Company. Minneapolis, Minn.

Chapter 14

14.19: Reprinted from Beyond the Limits, © 1992 by Meadows, Meadows, and Randers. With permission from Chelsea Green Publishing Co., White River Junctions, Vermont; 14.20: Reprinted from Beyond the Limits, © 1992 by Meadows, Meadows, and Randers. With permission from Chelsea Green Publishing Co., White River Junctions, Vermont.

Chapter 15

15.3: Source: Data from PEW Research Center. Used by permission of Public Agenda Online; 15.8: Source: Data from United Nations Environment Programme from Global Environment Outlook-2000; 15.14: © 1990 Bruce von Alten; 15.17: Reprinted with permission of Simon & Schuster Adult Publishing Group as it appears in *Bowling Alone: The Collapse and Revival of American Community* by Robert D. Putnam. © 2000 by Robert D. Putnam.

Index

A

Atmosphere. *See also* Air pollution
  carbon dioxide in, 41
  climatic factors, 203–8
  nitrogen reentry, 43
  sulfur emissions into, 44
  water movement through, 40, 237
Atomic mass, 27
Atomic number, 27
Atomic structure, 27–28
"Atoms for Peace" speech, 291
ATP, 35
Atrazine, 192–93
Audubon Christmas Bird Count, 366, 367
Aurora australis, 203
Aurora borealis, 203
Automobiles, dependence on, 336–37, 341–42.
  See also Transportation
Average annual precipitation, 234
Average values, 8
Avian influenza, 179, 183–84
Axes for X,Y graphs, 22

B

Baby boom, 85–86
Bacillus thuringiensis (Bt), 170
Bacteria
  coliform, 247, 255
  role in nitrogen cycle, 42, 43, 167
  in soils, 161–62
Balanced diet, 157–58
Bald eagles, 120–21
Bamboo, 51
Banff National Park, 140
Bangladesh
  arsenic poisoning in, 250
  Grameen Bank, 349
  population growth, 79, 81
Bar graphs, 23, 46
Barnacles, 57
Baroclinic waves, 209
Barrier islands, 280
Barriers and Bridges, 361
Basel Convention, 315
Bases, 28
Batesian mimicry, 58
Bathypelagic zone, 103
Batteries, 299, 317
Beach erosion, 280
Becquerel, Alexandre-Edmond, 298
BedZED, 341
Beef production, 159
Beer bottles, disposable, 318
Bell curves, 8, 9
Beni Biosphere Reserve, 134
Benthic communities, 103, 107
Benzepyrene, 193
Bet Trang dumping incident, 315
Beyond the Limits, 345
Bias, 7–8
Big burp, 209
Big Cypress National Preserve, 142

Bill and Melinda Gates Foundation, 186
Bills in Congress, 359
Bingham Canyon mine, 272
Binomials, 54
Bioaccumulation, 190
Biocentric preservation, 14
Biochemical oxygen demand, 247–48
Biodegradable plastics, 323
Biodiesel, 302–3
Biodiversity. *See also* Species diversity
  benefits, 110–11
  in cities, 338
  cultural diversity and, 21
  major threats, 111–19
  protection efforts, 119–22
  role of predation in maintaining, 95–96
  types, 108–10
Biofuels production
  implications of species diversity for, 48
  options, 301–3
  renewable energy from, 167
Biogeochemical cycles, 40–45
Biogeographical areas, 142
Biological communities
  defined, 36
  dependence on species diversity,
    63–67
  dynamism, 68–70
  effects of species interactions, 55–60
  freshwater ecosystems, 106–8
  marine ecosystems, 103–6
  terrestrial biomes, 96–103
Biological oxygen demand, 247–48
Biomagnification, 190
Biomass
  defined, 36
  as energy measure, 38–39, 66–67
  as energy source, 285, 301–4
Biomes
  freshwater types, 106–8
  marine types, 103–6
  terrestrial types, 96–103
Bioremediation, 259, 327
Biosphere reserves, 146–47
Biotic potential, 60
Bird flu, 179, 183–84
BirdSource website, 367
Bird-watching data, 367
Birth control
  family planning and, 89–90
  population control based on, 74, 77
  regions of low use, 79
Birth rates (U.S.), 85
Bison
  early attempts to protect, 119
  as keystone species, 120
  near extinction, 117–18
  wolf predation on, 95–96
Bisphenol A, 192
Bituminous coal, 288
Black smokers, 33
Blind experiments, 7
Bogs, 107, 108

Bonaparte, Napoleon, 250
Bonds, chemical, 28
Books, 366
Boreal forests, 102, 128–29
Borlaug, Norman, 169
Botswana, HIV/AIDS in, 183
Bottom trawling, 112, 160
Bracken fern, 327
Brazil
  agricultural expansion in, 153–54, 159
  air pollution, 229
  deforestation, 131, 132
  ethanol production, 48, 302
  landscape ecology experiment, 144
  shade-grown coffee and cocoa, 172
  shantytowns, 335, 336
  sustainable city case study, 330, 331
Breast-feeding, 89
Breeding Bird Survey, 124
Bristle worm, 116
Britain, carbon dioxide reductions, 215
Broecker, Wallace, 207
Bronchitis, 223
Brook trout, 120
Brower, David, 14, 15
Brown, Lester, 87
Brownfields, 325–26
Brown pelicans, 117
Brown rats, 51, 52, 63
Brundtland, Gro Harlem, 15, 181
Buffers, 29
Buildings
  acid rain effects on, 225
  green design, 340–41, 350–51, 370–71
  poor housing, 335, 336
  promoting sprawl, 336–37
Bullard, Robert D., 316
Bull thistle, 50
Bureau of Land Management (U.S.),
  147, 361
Burma forest loss, 131
Bush, George W., 215
Bushmeat, 118, 127
Business development. *See* Economic
  development
Bus rapid transit, 337
By-catch, 160

C

Cacti, 118
Cadiz Township, 112
Calcium carbonate, 41
California redwoods, 102
Calment, Jeanne Louise, 82
Calories, 32, 284
Calthorpe, Peter, 339
Camel's Hump Mountain, 224–25
Cameroon
  fertility rates, 81
  forest loss, 131
  resource rights in, 21

        Principles of Environmental Science   http://www.mhhe.com/cunningham4e

Principles of Environmental Science

Hydrothermal vents, 33, 104–5
Hydroxide groups, 30
Hypolimnion, 107
Hypotheses, 6
Hypoxic zones, 248

I

Ice ages, 9, 207, 208
Ice density, 29
Iceland water supply, 238
Igneous rocks, 268
Iguazu Falls, 232
Illegal gem trade, 270–71
Incineration, 317–18, 327
Income, desire for children and, 85
Income distribution, 19–20
Income levels, life expectancy and, 82–83
Independent variables, 8, 22
Indexes, 177
India
    anemia in, 157
    arsenic poisoning in, 250
    dam projects, 241
    forest protection efforts, 134
    life expectancy gains, 82
    population control programs, 74
    population growth, 79, 80
    water pollution in, 254
Indian reservations, 315
Indicators, 50
Indicator species, 120
Indigenous peoples, 20–21
Indirect energy use, 167
Indonesia
    deforestation, 131
    ecological footprint, 78, 79
    food sources, 110
    hunger reduction, 155
    resource rights in, 21
Indoor air pollution, 220
Inductive reasoning, 5–6
Industrial energy use, 286, 295
Industrial Revolution, 76, 77
Industrial toxins, 189–90, 253. *See also*
    Toxins
Industrial wastes, 312
Industrial water use, 239
Infectious diseases. *See also* Diseases
    eradication of, 18
    flu pandemics, 179, 181
    impact of climate change on, 213
    worldwide occurrence, 181–83
Infiltration, 236
Influenza viruses, 179, 181, 182, 183–84
Injunctions, 361
Innovation from scarcity, 344–45
Inorganic nutrients, 166
Inorganic water pollutants, 248–49
Insight, 5
Integrated pest management, 168, 174
Interest groups, 356, 370–73

Intergovernmental Panel on Climate
    Change, 210
Interior Department (U.S.), 361
Internalizing costs, 348
International Conference on Population and
    Development, 88
International environmental organizations, 373
International trade, 348–49
International treaties and conventions, 122,
    363–64
International Union for the Conservation of
    Nature and Natural Resources, 142, 145
International wildlife treaties, 122
Interplanting, 174
Interpretation, 6, 9
Interspecific competition, 56
Intertidal zones, 50, 103
Intraspecific competition, 56
Inuit people, 215, 221
Invasive species, 52, 114–16
Inversions, 222–23
Invertebrates, 109
Iodine deficiency, 157
Ionic bonds, 28
Iowa Revolving Loan Fund, 300
I=PAT formula, 78
Iran, family planning in, 87, 89
Iron, 216
Irrigation
    importance to modern agriculture, 166,
        238–39, 242
    water-saving methods, 245
Irruptive behavior, 367
Island biogeography, 114
Isolation, speciation and, 54, 55
Isotopes, 27

J

Jackson, Richard, 341
Jakarta population density, 334
James Bay project, 242
Janzen, Dan, 132
Japan
    green business investment, 351
    population growth rates, 80
    recycling in, 319
Jasper Ridge building, 370
J curves, 60–61, 76
Jet streams, 207, 210
Jobs, environmental protection and, 351
Jobs in environmental sciences, 366–68
Jonah Field, 263
Joules, 32, 284
Judicial branch of government, 360–61
Junk science, 10

K

Kampuchea forest loss, 131
Kelp forests, 59–60
Kenetech, 305

Kenya, Serengeti protection efforts, 148–49
Kerala population control, 74, 87
Kerogen, 289
Kesterson Wildlife Refuge, 148, 249
Keystone species, 59–60, 66, 120
Kilocalories, 284
Kinetic energy, 31
Kingdoms, 55
King's Canyon National Park, 14
Kingston Plains, 69
Klamath River, 242
Koch Oil Company, 361
K-selected species, 62–63
Kudzu vine, 52, 115
Kuhn, Thomas, 9
Kuwait water supply, 238
Kwashiorkor, 157
Kyoto protocol, 215

L

Lackner, Klaus S., 215–16
Laguna San Ignacio, 144, 355
Lake Agassiz, 207
Lake Erie, 251, 252
Lake Manyara National Park, 149
Lake Nasser, 305
Lake Powell, 238
Lakes, 107, 237
Land conflicts, 153
Landfills, 314–15, 328
Landless Workers Movement, 153
Landsat, 37
Landscape ecology, 143–44
Landslides, 279
Land-use planning, 338–42
La Niña, 209–10
Latent heat, 204–5
Latitudes, biomes named for, 96
Lawns, 244
Laws for environmental protection, 358–61
Laws of thermodynamics, 32
Lawsuits, 6
LD50 dose, 194
Lead-acid batteries, 299
Lead as air pollutant, 220
Lead poisoning, 117, 191
Legislatures, 359–60
Legumes, 42
Leidy's comb jelly, 116
Leopold, Aldo, 13, 14
Lerner, Jamie, 331
Less-developed countries. *See* Developing
    countries
Leukemias, 110
Liability for toxic waste sites, 325–26
Liberian forest loss, 131
Liberty Link crops, 171
Lichens, 50, 58, 59
Life
    biogeochemical cycle supporting, 40–45
    earth's diversity, 3

Principles of Environmental Science

Modern environmentalism, 14. *See also* Environmentalism

Molecules, 28

Mollisols, 162

Monitored, retrievable storage, 293

Monoculture forestry, 130–31

Monsoons, 207, 208, 218

Montreal Protocol, 222, 356, 364

Moore, Stephen, 77

Morbidity, 180

More-developed countries
  desire for children in, 85–86
  population growth rates, 79, 80

Mortality
  air pollution contributions to, 220, 223
  human populations, 82, 88, 180

Mosquitoes
  as disease vectors, 182
  invasive species, 115
  pesticide resistant, 185

Motor oil, dumping of, 313–14

Mountains
  midocean ridges as, 266
  rain shadows, 100, 234
  tundra biomes, 103

Mountain Storm plant, 227

Mountaintop removal, 274, 287

Mount Mitchell, 225

Mount Pelee, 277

Mount Pinatubo, 278

Mount Rainier, 278

Mount St. Helens, 277–78

Mount Vesuvius, 277

Mozambique national parks, 144

MTBE (methyl tertiary butyl ether), 254

Mudslides, 277–78

Mudstone, 268

Muir, John, 13

Mulch, 174

Müllerian mimicry, 58

Multiple variables, graphing, 200

Mumbai growth charts, 352–53

Municipal waste, 312, 314–15. *See also* Solid waste; Wastewater treatment

Muro-ami, 2

Mussels, 33, 115–16

Mutagens, 187

Mutations, 49

Mutualism, 58

Mycorrhizae, 59

Myers, Norman, 110

Myths about health risks, 197

N

Nabarro, David, 179

Nabhan, Gary, 168

NADPH (nicotinamide adenine dinucleotide phosphate), 35

Narmada River dams, 241

National Children's Study, 191

National Environmental Education Act, 365

National Environmental Education Advancement Project, 4

National environmental organizations, 371–73

National Environmental Policy Act, 358, 359

National Flood Insurance Program, 279

National Forest Management Act, 359

National forest preserves, 13–14

National Green Pages, 348, 350

National Institute for Occupational Safety and Health, 361

National parks, 140–44, 224

National Park Service (U.S.), 14, 361

National Pollution Discharge Elimination System, 251

National Priority List, 325

National Resources Defense Council, 372

National Science Foundation, 338

Natural capital, 342, 343

Natural-care products, 350

Natural disturbances, 69

Natural experiments, 7

Natural gas
  efficiency, 286
  exploration, 273
  known reserves, 290

Natural processes, water cleanup via, 257

Natural resource accounting, 346–48

Natural resource economics, 343

Natural selection, 49–51, 54

Natural systems, 26

Nature Conservancy, 372

Nature preserves. *See* Parks and preserves

Negative feedback loops, 26

Neoclassical economics, 342–43

Net primary productivity, 36, 66

Neurotoxins, 187. *See also* Toxins

Neutrons, 27–28

Nevado del Ruíz, 277

New Madrid earthquake, 277

News media bias, 196

New source reviews, 227

Newsprint costs, 318

Newtons, 284

New urbanism, 339–42

Ngorongoro Conservation Area, 149

Nicotinamide adenine dinucleotide phosphate (NADPH), 35

Niger, age distribution in, 83

Nigeria, 79, 155

Nitrates
  formation in nitrogen cycle, 42, 43
  water pollution from, 167, 252–53, 255

Nitrites, 42

Nitrogen cycle, 41–43

Nitrogen-fixing bacteria, 42, 167

Nitrogen isotopes, 27

Nitrogen molecules, 28

Nitrogen oxides, 218–19, 226, 288

Nitrous oxide, 43, 211

Noah's flood, 9

No-fishing zones, 2, 145

Nongovernmental organizations, 373

Nonmetallic salts, 249

Nonmetal mineral resources, 269–71

Nonpoint-source emissions, 218, 245, 246, 256

Nonrenewable resources, 344

"No reasonable harm" requirement, 195

Normal distributions, 8, 9, 23

Northern spotted owl
  environmental lawsuit over, 134–35
  habitat destruction, 65, 112, 121
  as umbrella species, 120

Norton, Gale, 241

Norway rats, 51, 52

No-till systems, 174

Nouabalé-Ndoki region, 127, 133

Nuclear fission, 291

Nuclear power generation
  overview, 291–93
  percentage of current energy supply from, 285
  relative safety risks, 196–97

Nuclei, 27–28

Nucleotides, 30

Nuees ardentes, 277

Nu River dam proposal, 358

Nutrients
  human needs, 154–58
  important to plant growth, 166
  in oceans, 103, 104
  of tropical moist forests, 98–99

O

Oberlin College Environmental Studies Center, 370, 371

Obesity, 156, 181

Observational methods, 7

Observations, science's dependence on, 4, 6

Occupational Safety and Health Administration (OSHA), 361

Ocean conveyor, 202, 207

Ocean currents, 206, 207, 209–10

Ocean ecosystems, 37, 103–6

Ocean fisheries. *See* Fisheries

Oceans
  buffering effects, 212
  deep-sea organisms, 32–33, 104–5
  as ecosystems, 103–6
  as energy source, 307
  percentage of earth's water in, 235
  pollution of, 255–56
  preserves in, 145
  remote sensing, 37
  solid waste disposal in, 314

Ochre starfish, 57

Ockham's razor, 4

Off-road vehicles, 141

Ogallala Aquifer, 240

O horizon, 162, 163

Oil drilling
  exploration methods, 273
  threat to oceans, 255–56, 289
  in wildlife preserves, 142, 148

Oil resource distribution, 288–90

Population growth (human)
 brief history, 75–76
 control approaches, 74
 declines in, 18
 demographic transition, 86–88
 desire for children and, 84–86
 effects of current decisions, 90–91
 factors in, 79–84
 food production versus, 154
 in major cities, 333
 potential water conflicts and, 238
 as threat to biodiversity, 117
 varied views of, 75, 76–79
Population growth (species), 60–63, 72
Population momentum, 83
Populations, 35–36
Portland (OR) land-use planning, 339
Ports and Waterways Safety Act, 260
Positive crankcase ventilation systems, 226
Positive feedback loops, 26
Potential energy, 31
Poverty
 declines in, 348
 disease and, 186
 environmental impacts, 18–19
 hunger and, 155
Power, 284
Power lines, 299
Power towers, 297
Prairie dogs, 120
Precious metals, 270–71
Precipitation. *See also* Acid precipitation
 causes, 206
 global patterns, 234
 impact of deforestation on, 132
 infiltration, 236
 relation to biome distribution, 96–97
 seasonal, 207, 208
Precision in science, 5
Predation
 adaptations to avoid, 57–58
 impact on species relationships, 56–57
 importance to biological communities, 95–96, 111
 success rates, 38
Predator control, 118–19
Predator-mediated competition, 57
Predatory mites, 57
Predictions in science, 6
Premises, evaluating, 12
Preservation movement, 14. *See also* Environmental protection
Pressurized water reactors, 292–93
Primary consumers, 37, 38
Primary pollutants, 218
Primary productivity, 36, 37, 66–67
Primary succession, 68
Primary treatment, 257
Prior appropriation rights, 243
Probability, 6–7, 196, 197
Producers, 36, 38
Productivity, 36, 66–67
Project Feeder watch, 367

Pronatalist pressures, 84
Proofs, scientific view of, 4, 6
Prospective studies, 194
Proteins
 basic features, 30
 deficiencies of, 157
 food sources, 158–61
Protons, 27–28
Proven-in-place reserves of coal, 287
Public hearings, 359–60
Public interest groups, 356
Public Interest Research Groups, 370
Public transportation, 331, 337
Public Utilities Regulatory Policies Act, 306
Pudong, 332, 334
Pumped-hydro storage, 299
Purple loosestrife, 115
Pyrolobus fumarii, 33

R

Radical environmental groups, 372–73
Radioactive waste, 293. *See also* Hazardous waste
Radio telemetry, 113
Rain. *See* Precipitation
Rainforests. *See also* Deforestation; Tropical forests
 community complexity, 66
 Congo basin preservation, 127
 destructive trends for, 131, 132
 protection efforts, 144
 temperate, 101–2
 tropical, 98
Rain shadows, 100, 234
Rainy seasons, 207
Ranching, 139–40
Randomly distributed populations, 64
Random samples, 8
Rangelands. *See* Grasslands
Rational choice model, 357
Rats, 51, 52, 63
RBMK reactors, 292–93
Reactors, nuclear, 291–93
Recharge zones, 236–37
Reclamation projects, 242, 243, 272
Recovery plans (endangered species), 120–21, 122
Recreation, benefits of biodiversity for, 111
Recycling
 freshwater, 244–45
 metals, 275
 progressive programs for, 331
 resource renewal and, 344
 solid waste, 311, 318–20
Red deer, 140
Reduced tillage, 174
Reduction, 28
Redwood trees, 63, 102, 373
Reef populations, 2
Reflected light, land cover analysis from, 37
Reflective thinking, 11

Reforestation, 130–31, 166
Reformers, 300
Refugee camps, 156
Refuse-derived fuel, 317
Regenerative farming, 171
Regression equations, 9
Regulatory mechanisms, 26
Reich, Peter, 214
Reichhold Chemical, 326
Reilly, William K., 366
Relative toxicity, 195
Relative values, 177
Reliability of sources, 12
Remediation of water pollution, 259
Remote sensing, 37, 113, 133
Renewable energy sources
 biofuels, 48, 167
 carbon dioxide reductions via, 215
 current goals for, 18
 Denmark's use, 284
 percentage of current energy supply from, 284–85
 policies promoting, 299–300
Renewable resources, 344
Renewables portfolio standards, 299–300
Renewable water supplies, 237
Repeatability principle, 4, 5
Replacement rate, 81
Replication, 5
Reproducibility of results, 5
Reproductive rates, 62–63
Research
 citizen participation, 366, 367
 scientific principles, 4–11
 toxicology, 194–95
Residence time, 235
Resilience, 26, 66
Resource Conservation and Recovery Act, 260, 323–24, 359
Resource extraction, habitat destruction with, 112
Resources
 allocation in nature, 55–57
 assigning economic value to, 346–48
 human consumption, 78–79, 343, 368
 partitioning, 51–53, 56
 views of, 342–46
Respiration, 35
Retrievable storage, 293, 328
Retrospective studies, 194
Reuse of solid waste, 321
Revolving Loan Fund (Iowa), 300
Rhine River, 253–54
Rhyolite, 268
Ribonucleic acid, 30
Rice paddies, 172, 173, 216
Rich-poor gulf, 19–20, 79
Ricin, 195
Riffles, 108
Rill erosion, 164, 165–66
Riparian use rights, 243
Risk assessment and acceptance, 196–99
Risks, graphing, 200

Whales, 355
White House Conference on Natural
    Resources, 14
White-tailed deer populations, 119
Wilderness Act, 147–48
Wilderness area preservation, 147–48
Wilderness Society, 14, 372
Wildfires, 136–37
Wild ginseng, 118
Wildlife
    desert, 100
    diseases of, 183–85
    impact of climate change on, 212, 213
    monitoring status of, 113
    protection efforts, 142, 143–44, 148–49
    smuggling, 118
Wild species ranching, 140
Wild turkeys, 119
Willett, Walter, 157
Wilson, E. O.
    on early crop failure in Hispaniola, 59
    environmentalist efforts, 10–11
    on human threats to biodiversity, 111, 112
    island biogeography theory, 114
    on species protection, 20
Wind erosion, 165, 166
Wind farms, 305, 306

Wind power generation, 215, 284, 305–6
Winds, 205, 206–7, 208
Winged bean, 168
Withdrawal of water, 238
Wolves, 95–96
Women
    effect of gains on fertility rates, 85, 87, 88
    importance to family nutrition, 155
    indoor air pollution threat to, 220
Women's Self-Employment Project, 349
Wood burning, 129–30, 301, 303
Wood ducks, 119
Woodpecker finch, 53
Wood products, 129
Wood warblers, 53
Work, 284
World Bank, 271, 349
World Commission on Environment and
    Development, 374
World Congress on National Parks, 146
World conservation strategy, 145
World Conservation Union, 142
World Food Summit, 155
World Heritage Conference, 355
World Neighbors, 175
World Trade Organization, 349, 364
Worms, parasitic, 181–82

Y

Yamuna River, 254
Yard waste, 320
Yasuni National Park, 145
Yellowstone National Park
    creation of, 140, 141
    snowmobiles in, 141–42
    wolf reintroduction, 95–96
Yemm, Richard, 307
Yield increases, 168–69
Yosemite National Park, 14, 141
Younger Dryas, 207
Yucca Mountain waste repository, 293

Z

Zebra mussels, 115–16
Zero population growth, 81
Zimbabwe, 91, 335
Zone of aeration, 235
Zone of saturation, 235, 236